Applied Manure and Nutrient Chemistry for Sustainable Agriculture and Environment

Zhongqi He • Hailin Zhang
Editors

Applied Manure and Nutrient Chemistry for Sustainable Agriculture and Environment

Springer

Editors
Zhongqi He
Southern Regional Research Center
USDA-ARS
New Orleans, LA, USA

Hailin Zhang
Department of Plant and Soil Sciences
Oklahoma State University
Stillwater, OK, USA

ISBN 978-94-017-8806-9 ISBN 978-94-017-8807-6 (eBook)
DOI 10.1007/978-94-017-8807-6
Springer Dordrecht Heidelberg New York London

Library of Congress Control Number: 2014937316

Printed on acid-free paper

Springer is part of Springer Science+Business Media (www.springer.com)

Preface

The global agriculture sector is confronting with challenges for the sustainability of agricultural production and of the environment to accommodate population growth and living standard increase in the world. Intensive high-yielding agriculture is typically dependent on the addition of fertilizers (synthetic chemicals, animal manure, etc.). However, non-point nutrient losses from agricultural fields due to fertilization could adversely impact the environment. Increased knowledge on plant nutrient chemistry is required for improving utilization efficiency and minimizing losses from both inorganic and organic nutrient sources. For this purpose, we invited a pool of peers consisting of both insightful senior researchers and innovative junior investigators to contribute chapters that highlight recent research activities in applied nutrient chemistry geared toward sustainable agriculture and environment. This book also outlooks emerging researchable issues on alternative utilization and environmental monitoring of manure and other agricultural byproducts that may stimulate new research ideas and direction in the relevant fields.

Chapter topics of interest in this book include, but are not limited, to speciation, quantification, and interactions of various plant nutrients and relevant contributories in manure, soil, and plants. Chapter 1 overviews animal manure and waste production, the benefits of using them as nutrient sources, potential impacts of manure on environmental quality and management strategies in the US as it produces over a billion Mg of animal manure annually. The worldwide heavy use of veterinary pharmaceuticals in confined animal-feeding operations has resulted in annual discharge of 3,000–27,000 Mg of drug chemicals via livestock manure into the environment. Chapter 2 summarizes veterinary pharmaceutical uses in confined animal feeding operations, reports on presence and detection of residual veterinary medicines in manures, and reviews the environmental behaviors of pharmaceutical residues in agricultural soils. As diverse environmental problems (e.g. pathogens, greenhouse and odorous gas emissions, and phosphorus runoff) arose from animal wastes, slow pyrolysis may offer an avenue for mitigating some of these problems and reducing the waste volume prior to land application. Chapter 3 is a critical review exploring the changes in chemical speciation of nutrient elements within manure as a result of pyrolysis and other thermal conversion technologies, and recommendations are given on the critical areas where further investigation is needed on the relevant issues.

The next four chapters are with soil nitrogen and enzyme activities impacted by animal manure application. Chapter 4 provides up-to-date information on soil amino compound and carbohydrate research, and a case study of soil amino compound and carbohydrate levels impacted by organic amendments based on greenhouse manure experiment with ryegrasses. To increase the understanding of manure management in cropping systems for maximizing nitrogen use efficiency, Chap. 5 discusses the factors that can affect nitrogen mineralization and demonstrates the impact of temperature, moisture, soil wetting and drying cycles, and field spatial variability on manure nitrogen availability. Chapter 6 provides a review of the response of enzyme activities to manure applications and potential implications on soil biogeochemical cycling in agroecosystems, and also offers some perspective areas where more research may be needed and some avenues for future research. Followed Chap. 7 presents information on the most commonly studied soil phosphatases, acid and alkaline phosphomonoesterase and phosphodiesterase, and how manure application influences their activities and phosphorus cycling with a case study showing that soil application of dairy manure increases acid phosphatase activity.

Chapters 8, 9, 10, 11, 12, and 13 are dedicated to the phosphorus issue. Chapter 8 synthesizes and analyzes the basic knowledge and latest research on variety and solubility of phosphorus forms in animal manure and their effects on soil test phosphorus. Chapter 9 focuses on the major organic phosphorus form – phytate. It reviews the current knowledge of the abundance, cycling and bioavailability of phytate in soils and manure, and suggests areas where knowledge is limited, and thus where further research is needed. As a case study, Chap. 10 presents and discusses published and unpublished data on phosphorus forms and mineralization potential in Alabama cotton soils amended with poultry litter and managed as no-tilled, tilled, and mulch-tilled practices, showing poultry litter applied to soils affected many of the soil phosphorus fractions, dynamics and uptake. Chapter 11 reviews the use of iron/aluminum- and calcium/magnesium-based industrial by-products as manure amendments to reduce soluble phosphorus concentrations, and discusses the function of the chemistry of both the phosphorous sorbing materials and the receiving manure. Chapter 12 examines the effects of using bauxite residue, a by-product from the aluminum refinery industry, to modify nutrient characteristics of animal manure and manure-affected soils. Data compiled in Chap. 12 demonstrate that bauxite residues could be used as a potential amendment for reducing phosphorus and other contaminant losses in animal manures and manure-affected soils. Chapter 13 reviews fundamental basis and current state of knowledge on compound-specific isotopic effect during hydrolysis of organic phosphorus compounds. While the compound-specific isotopic study for organic phosphorus compounds is still in its infancy, Chap. 13 predicts that the future expansion of this research will develop a holistic approach to integrate transfer and transformation of organic and inorganic phosphorus and will eventually lead to sustainable agriculture and healthy ecosystem.

The last four chapters highlight impacts of animal manure and other amendments on soil and plant growth based on field experiments. Recent development in blueberry markets under organic certification has stimulated interest in production of composts specifically tailored to its edaphic requirements. Chapter 14 reports data from initial screening studies conducted in western Oregon USA to assess growth response of highbush blueberry to composts derived from diverse feedstocks and to link the response to compost chemical characteristics. An arable land in the subarctic Alaska, USA, was developed in 1978 by clearing native forest, and part of the arable land was later converted to grassland through a Conservation Reserve Program. Chapter 15 systematically presents and discusses the quantity, distribution, and features of soil water extractable organic matter as affected by the land uses to increase the understanding of soil organic matter biodegradability for new aspirations on agricultural production in the subarctic regions. The accumulation of heavy metals in biosolids amended soils and the risk of their uptake into different plant parts is a topic of great concern. Chapter 16 summarizes the accumulation of several heavy metals and nutrients in soils and in plants grown on biosolids applied soils and the use of remote sensing to monitor the metal uptake and plant stress. Research has been conducted in the southern and southeastern regions of the US to encourage the utilization of poultry litter as a row crop fertilizer away from the traditional application to pastures around chicken houses. Chapter 17 reviews results of the research on the effectiveness of poultry litter as cotton fertilizer and environmental concerns associated with its land application. Data presented in Chap. 17 demonstrate that, if effectively integrated into the cropping systems of the region, poultry litter should benefit not only cotton and other row crop farmers but also the poultry producers in the regions.

Chapter contribution was by invitation only. Each chapter that covers a specific topic was selected and decided after extensive communications between editors and chapter contributors. All chapter manuscripts were subject to the peer reviewing and revision processes. Positive comments from at least two reviewers were required to warrant the acceptance of a manuscript. We would like to thank the reviewers for their helpful comments and suggestions which certainly improved the quality of the book. These reviewers include: Nadia Carmosini, University of Wisconsin-La Crosse; Luisella Celi, Università degli Studi di Torino, Italy; Courtney Creamer, CSIRO Land and Water, Australia; Warren Dick, Ohio State University; Syam K. Dodla, Louisiana State University; Xionghan Feng, Huazhong Agricultural University, China; Thomas Forge, Agriculture and Agri-Food Canada; Mingxin Guo, Delaware State University; Fengxiang Han, Jackson State University; Donald A. Horneck, Oregon State University; Deb P. Jaisi, University of Delaware; Michael F. L'Annunziata, the Montague Group, Oceanside, CA; Philip Larese-Casanova, Northeastern University; B. Maruthi Sridhar, Texas Southern University; Daniel N. Miller, USDA-ARS; Jagadeesh Mosali, The Samuel Roberts Noble Foundation; Yvonne Oelmann, University of Tübingen, Germany; Paulo Pagliari, University of Minnesota; Po Pan, Kunming University of Science and Technology, China; John Paul, Transform Compost Systems Ltd., Canada; Chad Penn, Oklahoma State University; Thilini D. Ranatunga, Alabama A&M University; Zachary Senwo;

Alabama A&M University; Karamat Sistani, USDA-ARS; Michael Tatzber, University of Natural Resources and Life Science Vienna, Austria; Haile Tewolde, USDA-ARS; Allen Torbert, USDA-ARS; Ben J. Turner, Smithsonian Tropical Research Institute, Panama; Dexter Watts, USDA-ARS; Mingchu Zhang, University of Alaska Fairbanks; Wei Zhang, Michigan State University; and Wei Zheng, University of Illinois at Urbana-Champaign.

New Orleans, LA, USA — Zhongqi He
Stillwater, OK, USA — Hailin Zhang

Contents

Chapter 1
Animal Manure Production and Utilization in the US

Hailin Zhang and Jackie Schroder

Abstract Over a billion tons of animal manure is produced annually in the US. Animal manure is an excellent plant nutrient source and soil amendment when used properly. Manure contains plant macro- and micronutrients, supplies organic matter, improves soil quality, and maintains or increases soil pH in acid soils. However, nutrients such as phosphorus and nitrogen build up in the soil if application rates are higher than the nutrient requirements of the intended crops. Following a nutrient management plan and proven best management practices will improve manure nutrient use efficiency and reduce the impact of the land application of manure on water quality. This chapter highlights manure and animal waste production, the benefits of using them as nutrient sources, and the potential impacts of manure on environmental quality and management strategies.

1.1 Introduction

Animal production is a large segment of the economy of the United States. The United States Department of Agriculture (USDA) estimated in 2007 that there were over 2.2 billion head of livestock and poultry in the U.S. (USEPA 2013) that produced over 1.1 billion tons of wet weight manure. In another report, the United States Environmental Protection Agency (USEPA) estimated there were 1.3 million farms in 2007 and that approximately 212,000 of these farms were animal feeding operations (AFOs) (USEPA 2012). Therefore, animal manure is an abundant source of macro- and micronutrients for crop and grass production. Besides providing valuable nutrients to the soil, manure supplies organic matter to improve physical, chemical and biological properties of soils, thus improving water infiltration,

H. Zhang (✉) • J. Schroder
Department of Plant and Soil Sciences, Oklahoma State University,
368 Agriculture Hall, Stillwater, OK 74078, USA
e-mail: hailin.zhang@okstate.edu

Z. He and H. Zhang (eds.), *Applied Manure and Nutrient Chemistry for Sustainable Agriculture and Environment*, DOI 10.1007/978-94-017-8807-6_1,

enhancing retention of nutrients, reducing wind and water erosion, and promoting growth of beneficial organisms.

The majority of meat and animal products in the United States are produced by large confined animal feeding operations (CAFOs), where livestock and poultry annually generate a substantial quantity of manure (Wright et al. 1998). Many agricultural fields in the United States that have received long-term manure applications have high levels of nutrients (Chang et al. 1991). The runoff and soil erosion from those fields carry soluble and particulate nutrients to water bodies even if no additional manure is applied. Land application of manures was often based on nitrogen (N) requirements of the crops in the past. However, land application of animal manures to meet crop N needs can lead to an accumulation of phosphorus (P) in soil (Sharpley et al. 1999) because the N/P ratio of animal manures (e.g., 2:1 for broiler litter) is less than the N/P ratio of about 8:1 taken up by most crops and grasses (USDA 2001). Thus, repeated land application of manure based on plant N needs results in excessive P concentrations in soils and may saturate the soil's capacity to retain P. If not properly managed, fields that received manure can become non-point sources of sediment and nutrient losses via surface runoff, erosion, and leaching (Sharpley and Smith 1993; Sharpley 1995; Pote et al. 1996; Sharpley et al. 1996).

Thus, in the management of manure land-application, it is important to take advantage of the beneficial nutrients and organic matter while minimizing its impact on the environment. Nutrient losses to the environment can occur at the production site, during storage and after field application. Utilization of nutrients in manures in an environmentally sustainable manner is one of the most critical management issues facing the livestock industry.

Current strategies used to reduce P transport to surface water include conservation tillage, crop residue management, cover crops, buffer strips, contour tillage, runoff water impoundment, and terracing. These strategies are very effective in controlling particulate P but less effective for dissolved P in runoff water (Daniel et al. 1999; Sharpley et al. 1996). Many of those practices increase infiltration resulting in less P load in runoff. More directly, chemical treatment of manure before it is land applied will reduce the levels of soluble nutrients (Day and Funk 1998; Dao 1999, 2001; Codling et al. 2000; Dao et al. 2001; Dao and Daniel 2002). In order to control water-soluble P and metal transport to surface water, new best management practices (BMPs) must be developed, evaluated and implemented. The purpose of this chapter is to highlight manure and animal waste production, the benefits of using them as nutrient sources, and the potential impacts of manure on environmental quality and management strategies.

1.2 Manure Production and Management

Animal production is a large segment of the economy of the United States. The increased numbers of CAFOs and poultry production facilities have produced additional quantities of manure in recent years requiring proper management.

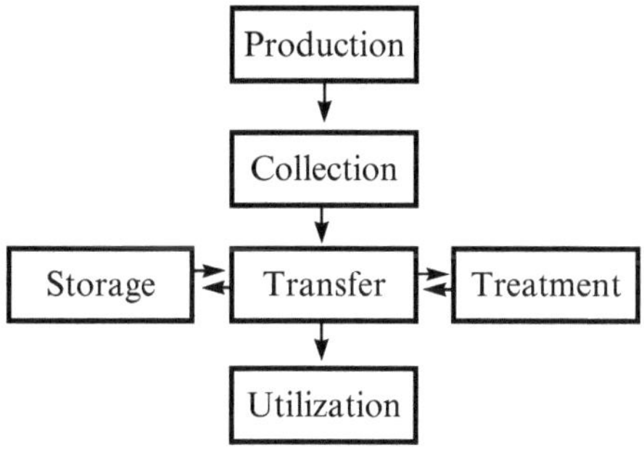

Fig. 1.1 Manure management functions

An agricultural waste management system designed for a CAFO consists of six basic functions: production, collection, storage, treatment, transfer, and utilization (Fig. 1.1). It is important to understand each of these functions since they affect the nutrient contents of the manure.

Production is the function of the amount and nature of manure generated by an AFO. The USDA Census of Agriculture used standard methods in 2007 and estimated that 2.2 billion head of beef and dairy cattle, swine, and poultry produced approximately 1.1 billion tons of wet weight manure in that year (USEPA 2013). Manure production for different categories of livestock in 2007 were: cattle (0.92 billion tons wet weight), swine (0.11 billion tons wet weight), and poultry (0.08 billion tons wet weight). Beef cattle produced more manure than any other category of livestock in 2007. The top ten states with the highest beef cattle production and associated manure generation in 2007 were (1) Texas, (2) Missouri, (3) Oklahoma, (4) Nebraska, (5) South Dakota, (6) Montana, (7) Kansas, (8) Tennessee, (9) Kentucky, and (10) Arkansas. The USEPA estimated that there were 1.3 million farms holding livestock nationwide, and that approximately 212,000 of these farms were AFOs (USEPA 2012). Besides manure, large quantities of associated animal wastes are produced at these operations, such as dead animals, wasted feed, wash water, etc. The generation of unnecessary waste should be kept to a minimum. Leaking watering facilities and spilled feed contribute to the production of waste. These problems can be reduced by careful management and maintenance of feeders, watering facilities, and associated equipment.

Collection refers to the initial capture and gathering of the waste from the point of origin or deposition to a collection point. The manure and animal waste collected could be liquid, slurry or solids depending on animal species and operating systems.

Storage is the temporary containment of the waste. The storage facility of a waste management system is the tool that gives farmers control over scheduling of transfer operation or land application. Nutrient content and forms may change during storage.

Treatment is any process designed to reduce pollution potential of the waste, including physical, biological, and chemical treatment. It includes activities that are sometimes called pretreatments, such as the separation of solids and liquids, or adding alum to poultry houses.

Transfer refers to the movement and transportation of the waste throughout the system. It includes the transfer of waste from the collection point to the storage facility, to the treatment facility, or to the utilization site. Waste may require transfer as a solid, liquid, or slurry, depending on the total solid concentration.

Utilization refers to the recycle of waste products into the environment. Agricultural wastes may be used as a source of energy, bedding, animal feed, mulch, organic matter or plant nutrients. Properly treated, they can be marketable. Most often they are land-applied as a soil amendment, therefore, the benefits and concerns of manure utilization as plant nutrients will be discussed below in detail.

1.3 Benefits of Manure Land Application

1.3.1 Manure Is a Good Source of Plant Nutrients

The actual nutrient content of manure from a particular operation will differ considerably due to the method of collection, storage and species of animal. The approximate fertilizer nutrient contents for various manures are shown in Table 1.1.

As shown in Table 1.1, the amount of N, P, K and other nutrients is significant when the manure is applied in large quantities. The nutrient value of manure may be estimated based on the prices of commercial fertilizers. However, not all the N in the manure is available to crops during the year of application because some N is in the organic form while other forms of N can be lost during application. The availability of N in the year of application may vary from 30 to 80 % compared with commercial N fertilizers depending on the type of manure and application method. Conversely, most of the P and K in manure are in the inorganic form. For all manure types, approximately 90 % of P and K in the manure are considered as available as commercial fertilizers during the first year of application.

1.3.2 Manure Improves Soil Quality

Research has shown that land application of manures can significantly impact soil chemical, physical, and biological properties, thus improving soil quality. Most of these impacts are probably due to the increase in soil organic matter (SOM) (Risse et al. 2006). Soil organic matter serves as a chelating agent and buffering material, affects the cation exchange capacity of soil, and is an important agent for soil aggregation (Eghball and Power 1999). Fraser et al. (1988) evaluated the annual application of beef feedlot manure to a grain/legume cropping system over a 5 year period and reported that manure application increased total organic C, Kjedahl N,

Table 1.1 Selected nutrient concentrations in dairy, swine, and poultry manure samples taken from random farms in the midwestern and southeastern US (Combs et al. 1998)

	Dairy		Swine		
Element	Solids	Liquid	Solids	Liquid	Poultry
	%				
N	2.3 ± 0.4	5.0 ± 1.4	1.8 ± 0.3	9.6 ± 3.2	4.1 ± 0.9
P	0.6 ± 0.2	0.8 ± 0.2	1.6 ± 0.3	2.3 ± 0.5	1.9 ± 0.2
Ca	1.6 ± 1.3	2.6 ± 1.1	2.0 ± 0.5	3.0 ± 0.7	3.7 ± 0.8
Mg	0.7 ± 0.5	0.9 ± 0.4	0.5 ± 0.1	1.1 ± 0.3	0.6 ± 0.1
Fe	0.13 ± 0.12	0.18 ± 0.14	1.5 ± 0.4	0.3 ± 0.3	0.15 ± 0.1
	mg kg^{-1}				
Zn	90 ± 74	186 ± 81	608 ± 145	1,357 ± 689	344 ± 88
Cu	27 ± 28	191 ± 286	381 ± 122	672 ± 684	481 ± 118

and potentially mineralizable N in manure-amended surface soils (0–7.5 cm) by 22–40 % as compared to the non-manured soils. Vitosh et al. (1973) showed that SOM, available P, and exchangeable K, Ca, and Mg were increased over a 9-year period with increasing rates of annual applications of feedlot manure on a loam and sandy loam soil. Annual application of cattle feedlot manure to a clay loam soil over 8 years significantly increased SOM and total N, and lowered the C/N ratio in the top 30 cm of soil (Sommerfeldt et al. 1988). Numerous researchers have shown long-term application of poultry litter increased soil organic carbon (Sharpley and Smith 1993; Mitchell and Tu 2006; Adeli et al. 2007).

The Magruder Plots established in 1892 are the oldest continuous soil fertility wheat experiment west of the Mississippi River (Magruder Plots 2008). Beef feedlot manure was applied every 4 years at a rate to supply 134 kg N ha^{-1} from 1899 to 1969 and at rate of 269 kg N ha^{-1} from 1969 to present (Davis et al. 2003). Continued application of manure not only supplied plant nutrients but also slowed down the depletion of soil organic matter content as demonstrated by this long-term experiment in Stillwater, Oklahoma (Fig. 1.2).

1.3.3 Manure Maintains Soil pH

Manures, especially poultry litter and feedlot manure, may raise or maintain pH in acidic and near neutral soils via a liming effect because they contain some $CaCO_3$, which originates in the animal diet (Eghball 1999; Moore and Edwards 2005). Eghball (1999) evaluated the effect of composted and uncomposted feedlot manure and ammonium nitrate applied annually to corn over a 4-year period. The study found that feedlot manure and composted manure raised soil pH but ammonium nitrate significantly decreased soil pH. In another study, poultry litter applied annually to tall fescue for 7 years increased soil pH. Additionally, numerous other researchers have shown that addition of animal manures to acid soils increased pH (Hue 1992; Cooper and Warman 1997; Wong et al. 1998; Whalen et al. 2000; Tang et al. 2007).

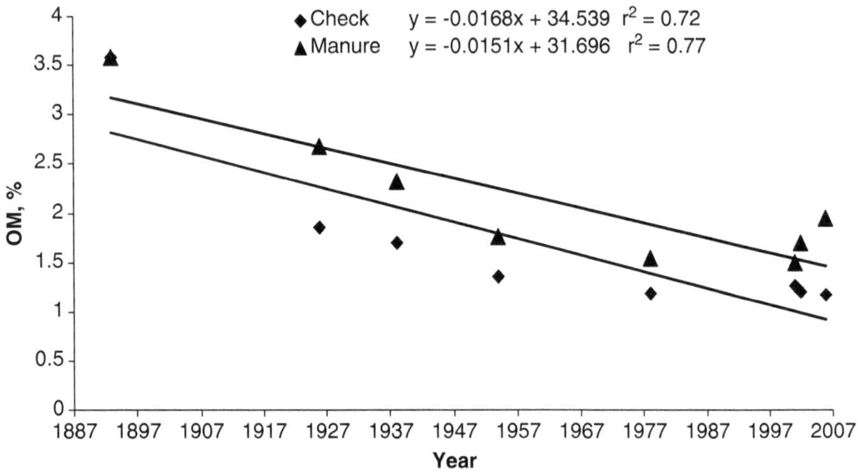

Fig. 1.2 Difference of organic matter reduction in manured and check treatments from the Magruder Plots, Stillwater, OK (Girma et al. 2007). Manure application slowed down organic matter depletion due to cultivation

Table 1.2 Effects of manure and chemical fertilizer application on soil pH of the Magruder Plots

Treatments	Soil pH
Manure	6.32
Check	5.83
P	5.66
NP	5.21
NPK	5.26
NPKL	5.51

The Magruder Plots in Stillwater have received inputs of beef feedlot manure every 4 years for many decades. The soil pH of the top 6 in. of the manured plot is greater than any other treatments as illustrated in Table 1.2. Manure maintained soil pH in the ideal range for most field crops. However, plots that received other treatments required lime to correct the low pH for optimum crop production.

Tang et al. (2007) found both poultry litter and feedlot manure increased soil pH, reduced exchangeable Al (Fig. 1.3), and increased wheat biomass. Wheat biomass was positively correlated with soil pH (r = 0.76), but was negatively correlated with exchangeable Al (r = −0.87). A path analysis showed significant direct effects ($p < 0.01$) between wheat growth and OC added and between wheat growth and P_2O_5 added. This suggests that animal manures have the potential to reduce Al toxicity in acidic soils as evidenced by greenhouse and field studies.

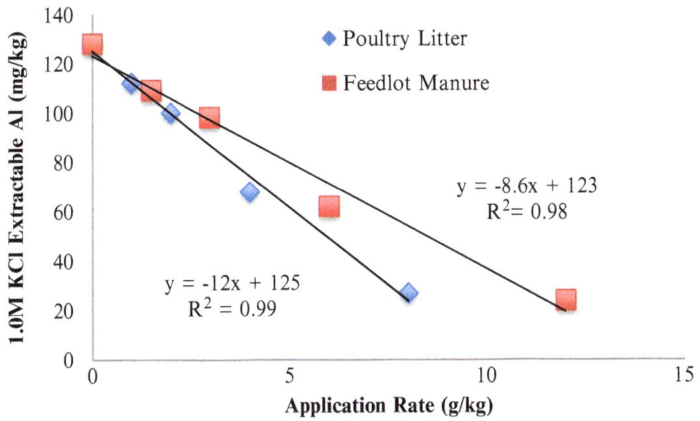

Fig. 1.3 Poultry litter application to an acid soil significantly reduced exchangeable Al concentration (Tang et al. 2007)

1.3.4 Manure Improves Soil Physical Properties

Studies have shown that water stable aggregates (WSA) increase infiltration (Roberts and Clanton 2000), porosity (Kirchmann and Gerzabek 1999), and water holding capacity (Mosaddeghi et al. 2000). Therefore, water stable aggregates greatly affect soil physical properties (Haynes and Naidu 1998). Tiarks et al. (1974) reported the application of cattle feedlot manure increased the geometric mean diameter of water-stable aggregates from 80 to 800 μm. Mikha and Rice (2004) demonstrated that manure application significantly increased soil aggregation and aggregate-associated C and N. Whalen et al. (2003) found the application of composted cattle manure increased the amount of WSA within 1 year of application and the mean weight diameter of aggregates increased with increasing compost application rates. Other studies have shown manure application reduced soil compaction, and increased friability (Schjønning et al. 1994; Mosaddeghi et al. 2000).

Mueller et al. (1984) found manure application reduced runoff and soil losses with different tillage systems. Giddens and Barnett (1980) used rainfall simulation to study the effect of application of poultry litter on runoff water quality and soil loss and reported runoff and soil loss were both decreased by litter application. Conversely, studies by Sauer et al. (1999) and Gilley and Eghball (1998) showed that application of cattle manure did not decrease runoff. Gilley and Risse (2000) conducted an extensive review of natural runoff plot data including more than 70 plot-years worth of data from seven locations under a variety of tillage and cropping conditions. They concluded that plots annually treated with manure reduced runoff from 2 to 62 % and soil loss from 15 to 65 % compared to untreated plots. Furthermore, the reductions occurred at all locations and the measured runoff and soil loss values were reduced substantially as manure application rates increased.

Fig. 1.4 Winter wheat response to poultry litter applied in an acidic soil. The amount of poultry litter increased from 0 to 18 tons/ha from left to right (Tang et al. 2007)

1.3.5 Manure Application Increases Crop Yields

Numerous studies have shown land application of manures will result in crop yields that are either equivalent or superior to those achieved with commercial fertilizers (Xie and MacKenzie 1986; Motavalli et al. 1989; Zhang et al. 1998; Badaruddin et al. 1999; Lithourgidisa et al. 2007; Butler et al. 2008). Higher crop yields with manure as compared to commercial fertilizers have been credited to manure-supplied nutrients or to improved soil conditions not provided by commercial fertilizers (CAST 1996). For example, Badaruddin et al. (1999) evaluated studies conducted in Sudan, Bangladesh, and Mexico and found farmyard manure (10 t ha^{-1}) gave the highest yield response (14 %) and approximately equivalent levels of NPK gave the lowest (5.5 %), suggesting organic fertilizer contributed to other growth factors in addition to nutrients. Several researchers have shown that the addition of farmyard manure increased wheat grain yield by improving soil water holding capacity and chemical conditions (Gill and Meelu 1982; Sattar and Gaur 1989). Tang et al. (2007) demonstrated the dramatic response of winter wheat biomass to poultry litter in a strongly acidic soil with a pH of 4.5 (Fig. 1.4).

1.4 Concerns with Land Application of Manure

1.4.1 Phosphorus Buildup in Soils

Application of manure based on crop N requirements or over-application often results in a buildup of soil test P (STP) and/or other nutrients beyond sufficient levels for optimal crop yields. This is because the amount of manure-P is considerably greater

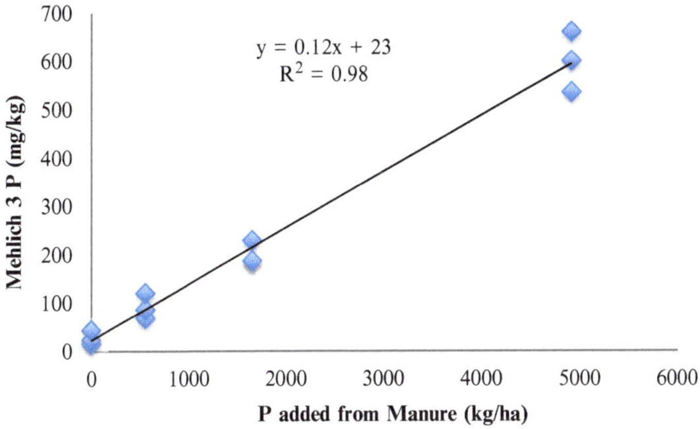

Fig. 1.5 The relationship between Mehlich 3 P and P added from manure for the Richfield soil. ***$p < 0.001$

than the amount of P removed with harvested crops when the application rate is based on crop N needs. For example, the N:P ratio of most poultry litter and feedlot manure is close to 2:1, but most crops require an N to P ratio of 8:1. Therefore, while N and some P are used by the crops, most of the excess P stays in the soil. A long-term beef feedlot manure application study conducted at the Experiment Station in Guymon, Oklahoma showed a strong relationship between soil test P and P added from manure (Fig. 1.5). The site received annual applications of manure at equivalent N rates of 0, 56, 168, and 504 kg ha^{-1}. The slope of 0.12 indicates that approximately 8 kg ha^{-1} of manure-borne P would be required to increase Mehlich 3 P by 1.0 mg kg^{-1} under normal corn production practices.

The buildup and potential loss of P is probably soil specific, because the adsorption capacity for P is different for different soils as shown in Fig. 1.6. Soil texture and organic matter contents as well as aluminum and iron oxides are important factors determining the adsorption capacity of an individual soil. Zhang et al. (2005) used multiple regression techniques and path analysis to determine the soil properties most directly related to P sorption in 28 Oklahoma benchmark soils, and found that aluminum and iron oxides were the most important soil properties for the direct estimation of P sorption. The potential for P loss will probably be higher if the soil has reached its adsorption maximum. Therefore, it is imperative to prevent soil P from building up. Once P is built up in the soil, remediation techniques and efficiency are limited.

1.4.2 Elevated Concentrations of Metals

Trace minerals such as As, Se, Cr, Cu, and Zn are sometimes added to feeds to prevent diseases, improve weight gains and feed conversion, and increase egg production for poultry (Miller et al. 1991; Tufft and Nockels 1991; Schroder

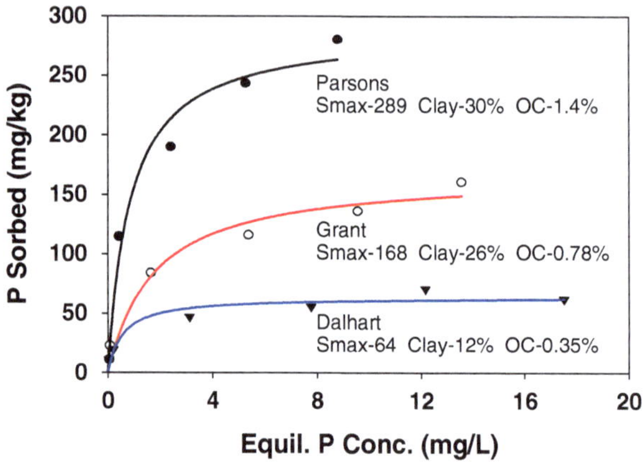

Fig. 1.6 Phosphorus adsorption isotherms of three Oklahoma Benchmark soils with differing soil clay and organic matter contents

et al. 2011). Because most of the metals ingested by livestock are excreted, the concentration of metals in manures is dependent on the concentrations of these metals in the animal's diet (Krishnamachari 1987; Miller et al. 1991). Thus, elevated concentrations of trace minerals were found in some manured soils (Li et al. 1997). The primary danger associated with manure-borne metals is that they do not degrade (Bolan et al. 2004).

Repeated applications of manure may enrich metal levels in soil to exceed crop requirements and possibly lead to phytotoxicity (Bolan et al. 2004). Several researchers have reported metal toxicity to ruminants grazing on pastures which had received manure applications (Bremner 1981; Lamand 1981; Poole 1981; Eck and Stewart 1995). Elevated concentrations of As, Cu, and Zn have been observed in soils that have received long-term application of manures (Kingery et al. 1994; Schroder et al. 2011; van der Watt et al. 1994). In another study, Christen (2001) found a strong relationship between water-extractable As in soils and the amount of poultry litter applied. Researchers have reported high concentrations of metals in runoff from soils that had received manure applications (Edwards et al. 1997; Moore et al. 1998). Thus, a potential exists for manure-treated soils to serve as non-point sources of metal pollution through leaching, runoff or erosion. However, metal additions to feeds have been reduced or eliminated in recent years due to environmental concerns or the discovery of replacement feed additives.

Most studies indicate manure Cd, Cu, and Zn exist primarily in the organically complexed form (Bolan et al. 2004). Several different chemical extractions including mineral acids, salt solutions, buffer solutions, and chelating agents have been used to predict bioavailability of metals in manure treated soils (Sutton et al. 1984; Payne et al. 1988; van der Watt et al. 1994). Chelating agents (e.g., EDTA and DTPA) are more effective in removing soluble metal–organic complexes that are potentially bioavailable and have often been found to be more reliable in predicting plant availability (Sims and Johnson 1991). Several studies have found that application of manures increased DTPA-extractable metals (Wallingford et al. 1975;

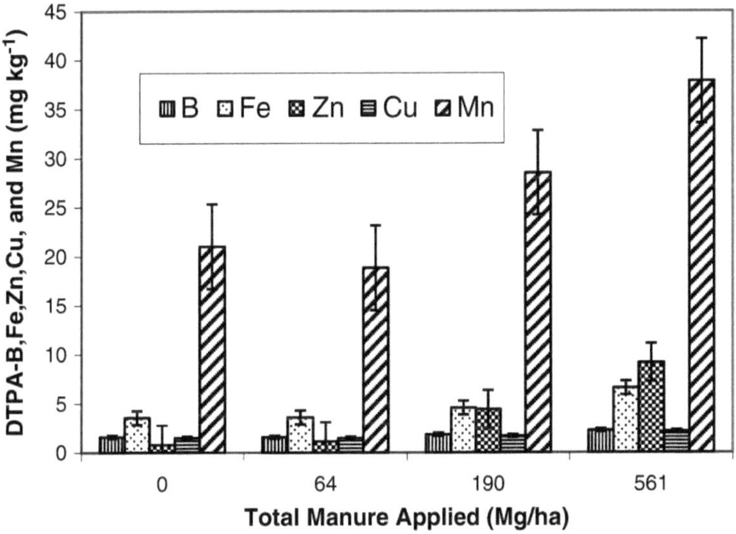

Fig. 1.7 The relationships between DTPA-extractable micronutrients and amount of total manure applied for the Richfield soil (Richards et al. 2011)

Payne et al. 1988; Anderson et al. 1991; Zhu et al. 1991; Narwal and Singh 1998; Arnesen and Singh 1998). The results of long-term research in Guymon, Oklahoma (Richards et al. 2011) agree well with these studies. Long-term application of beef feedlot manure increased concentrations of DTPA-extractable micronutrients (Fig. 1.7), which are beneficial to most crops.

1.5 Sustainable Manure Utilization

Utilization of nutrients and organic matter in manures in an environmentally sustainable manner is one of the most critical management issues facing the U.S. livestock industry. The key to avoiding environmental problems associated with manure application is to apply manure based on crop nutrient requirements, by developing a practical nutrient management plan and implementing available best management practices on the farm.

1.5.1 Nutrient Management Plan Development

The USEPA and the Department of Agriculture (USDA) announced a joint strategy to implement comprehensive nutrient management plans (CNMPs) on AFOs in 1998. A CNMP is a conservation farm plan that is specific to AFOs. The CNMP incorporates practices to utilize animal manure as beneficial resources and documents the management and strategies adopted by the AFO to address natural resource concerns related to soil erosion, animal manure, and disposal of organic by-products. The CNMP normally contains six different elements: (1) manure and wastewater handling and storage, (2) land treatment practices, (3) nutrient management plan (NMP), (4) record keeping, (5) feed management, and (6) other waste utilization options. The most important component of CNMP is to develop and follow a NMP when manure is applied. Recently, the USDA Natural Resources Conservation Service (NRCS) revised its 590 Nutrient Management Standard, so that each state is required to use a phosphorus risk assessment index (P Index) (USDA-NRCS 2011). Currently, 48 U.S. states have adopted a P Index as a site assessment tool to identify critical source areas and to target practices to reduce P loss. The P index ranks fields according to their vulnerability to potential P loss (Sharpley et al. 2003).

Many factors influence the loss of P from watersheds and its influence on water quality. In addition to the source factors and transport factors, many states have modified the P index to improve the assessment of site vulnerability to P loss by including the use of soil properties to modify soil test P calculations, estimates of availability/solubility of P applied, flooding frequency, BMPs, and ranking of the sensitivity of receiving water bodies.

Overall, the P index is site specific, ranks a site's vulnerability to P loss, identifies critical areas where there is a significant risk of P loss, and targets low-risk areas for manure application to build soil productivity. The proper use of the P index along with other farming practices allows farmers to utilize manures and fertilizers in an environmentally and agronomically sound manner.

1.5.2 Best Management Practice Implementation

There are a number of suggested management practices to improve nutrient use efficiency and to minimize the impact of manure application on the environment. A list of best management practices (BMPs) and brief descriptions related to P and manure management can be found on the website of the Southern Extension and Research Activity Group 17 (2013): Minimize Nutrient Losses from Agriculture (http://www.sera17.ext.vt.edu/SERA_17_Publications.htm). A selected few of these BMPs will be discussed below.

1.5.2.1 Use of Plants to Remove Excess Nutrients from Soils

Grasses are known to remove nutrients including P and K from soils to various degrees. One important management option for removing nutrients from soils is to use a bioaccumulator crop which removes the maximum amounts of nutrients from soil. Growing a high dry matter yielding forage crop is one method of managing

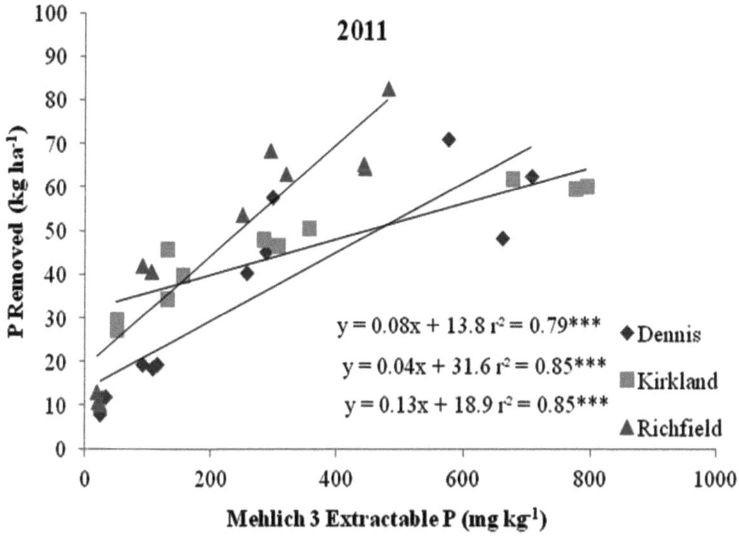

Fig. 1.8 The relationship between P removed by crabgrass grown in Dennis, Kirkland, and Richfield soil series and four levels of increasing Mehlich 3 P in 2011. ***Significant at the 0.001 alpha level (Barrett 2012)

nutrient-loaded sites (phytoremediation). The amount of nutrients taken up by the crop increases as dry matter increases, thus upon harvest more nutrients can be removed from the field. Bermudagrass (*Cynadon dactylon* L.) is an example of a forage crop with high yield characteristics which may be utilized in a forage system designed for nutrient removal. An alternative strategy is to use forages that have high nutrient uptake for specific nutrients even though dry matter yield may be less than some other forages.

The amount of nutrients removed from the field is a function of the concentration of nutrients in the plant and the plant biomass removed from the field. A greenhouse study conducted at Oklahoma State University revealed that crabgrass can be a good forage and P remover since it has a high yield potential, good forage quality and high P content (Barrett 2012). In a 2-year study, Barrett (2012) evaluated the ability of Red River crabgrass to remove excess soil P from nutrient-loaded soils using four levels of P that ranged from zero to 1,135 mg P kg^{-1} in three different soil series, the Dennis, Kirkland, and Richfield soils. During the first year of the study, the crabgrass was able to reduce water soluble P (WSP) across the four soil P levels by 48 % (29–69 %) in the Dennis soil, by 59 % (32–62 %) in the Kirkland soil, and by 51 % (48–68 %) in the Richfield soil. Additionally, the growth of crabgrass reduced Mehlich-3 P (M3P) by 28 % (13–50 %) in the Dennis soil, by 28 % (11–39 %) in the Kirkland soil, and by 30 % (14–53 %) in the Richfield.

In Barrett's (2012) study, crabgrass removed an average of 49.1 mg P kg^{-1} soil over the 2-year period and the P removed was positively correlated with M3P (Fig. 1.8).

1.5.2.2 Using Amendments to Reduce Dissolved and Particulate Nutrients

Most of the P in runoff from pastures is in the soluble form (Edwards and Daniel 1993; DeLaune et al. 2004). In a laboratory study, Moore and Miller (1994) evaluated the capability of different Al, Ca, and Fe amendments to reduce P solubility in poultry litter. Their study found the treatments formed insoluble metal phosphate minerals and that soluble P levels in the poultry litter were reduced from >2,000 to <1 mg P kg^{-1} litter with the addition of alum (aluminum sulfate), quick lime, slaked lime, ferrous chloride, ferric chloride, ferrous sulfate, and ferric sulfate under favorable pH conditions. In a field study, Shreve et al. (1995) reported that P runoff from tall fescue plots fertilized with poultry litter treated with alum and ferrous sulfate was reduced by 87 and 77 %, respectively. Dao (1999) reported that the addition of 10 % alum amendment reduced soluble P in stockpiled manure by 85 % and reduced soluble P in composted manure by 93 %. Field studies by Moore et al. (1999, 2000) found that P in runoff from pastures fertilized with alum-treated litter was 75 % less as compared to normal litter. Sims and Luka-McCafferty (2002) conducted a study in the Delmarva (Delaware–Maryland–Virginia) Peninsula evaluating the effect of alum on litter properties and elemental composition, and on the solubility of several elements in litter that are of particular concern for water quality (Al, As, Cu, P, and Zn). Their study confirmed earlier work by Moore and Miller (1994) by finding that alum treatment decreased litter pH and the water solubility of P. Additionally, their study reported decreased water solubility of As, Cu, and Zn. Similarly, a study by Moore et al. (1998) found application of alum to poultry litter reduced concentrations of As, Cu, Fe, and Zn in runoff water as compared to untreated poultry litter. In a long-term paired watershed study, Moore and Edwards (2007) found that cumulative P loads in runoff from normal litter were 340 % greater than that from alum-treated litter over the 10-year period (15.0 vs. 4.45 kg P ha^{-1}).

Another amendment that has been reported in the literature for the reduction of P in runoff water is the application of drinking water treatment residuals (WTRs). Water treatment residuals are rich sources of amorphous Al or Fe oxides and have a high sorption capacity for P. Water treatment residuals are generated by the coagulation/flocculation using Al salts, Fe salts, or mixed polymers to suspend particles and speed sedimentation to purify source water for municipal drinking water. Because WTRs predominately contain sediment and organic matter removed from the source water, they have soil-like properties. However, compared to natural soils, WTRs contain large amounts of amorphous Al or Fe and thus have a high capacity for P sorption. Several studies have been conducted over the last few years to evaluate land application of WTRs to reduce P loss from agricultural land. These studies may be categorized based on the method of application: surface application to intercept and remove P in runoff, incorporation into soil to reduce soil test P, and co-blending of WTR with organic waste such as manure to reduce P solubility (Dao 1999; Codling et al. 2000; Dao et al. 2001; Dao and Daniel 2002; Dayton and Basta 2005).

Coal combustion by-products such as fly ash rapidly accumulate and may cause disposal issues unless useful ways are found to utilize such by-products. Fly ash has relatively high levels of calcium, iron, and aluminum oxides and has been suggested as a phosphorus immobilization amendment in animal manure. Studies by Dao (1999, 2001) indicated co-blending fly ash with beef and dairy manure was effective in reducing dissolved-reactive P levels in beef and dairy cattle manure prior to land application. Additionally, applying fly ash-treated stockpiled or composted manure significantly reduced water-soluble P and Mehlich-3 P in manure-amended soils. In another study, Stout et al. (1998) found three different types of coal combustion by-products reduced Mehlich-3 and water extractable P by 45 % and 72 %, respectively. Generally, the behavior and effects of P or metals in animal manure are similar to those from biosolids, so the chemical amendments that work on manured soil should also work on soils that have received biosolids (Bolan et al. 2004).

Bauxite residue, a by-product from the aluminum refinery industry, has been used to modify nutrient characteristics of animal manure and manure-affected soils. More details of using bauxite to minimize nutrient loss from animal manure and manure amended soils are provided in Chap. 12 of this book. Some of these industrial by-products have been placed in filter ditches or retention cells to remove P before it enters water bodies. See Chap. 11 for more details.

1.5.2.3 Establishing Conservation Buffers and Filter Strips

Conservation buffers are small areas or strips of land in permanent vegetation, which are designed to slow water runoff, and to reduce nutrient and soil losses. Riparian buffers, filter strips, grassed waterways, shelterbelts, windbreaks, living snow fences, contour grass strips, cross-wind trap strips, shallow water areas for wildlife, field borders, alley cropping, wind barriers, and vegetative barriers are examples of conservation buffers. The major benefits of buffers include the removal of pesticides, nutrients, pathogens, and sediments. Other benefits include reduction of wind erosion, slowed water movement, reduction of down-stream flooding, stabilization of streambanks, establishment of vegetation, improvement of air quality. Conservation buffers are strategically placed along the edge of fields in a watershed to effectively minimize the movement of sediments, nutrients, and pesticides.

Several studies have been conducted to evaluate the effectiveness of using a buffer strip to reduce nutrients in runoff. Mayer et al. (2006) evaluated 14 comprehensive and regional reviews of riparian buffer literature containing N data from approximately 60 different studies. Their study found that N removal effectiveness varied widely among studies. In their study, the experimental data were fitted with a non-linear regression model to make predictions on removal effectiveness and buffer width. The non-linear regression model indicated buffer strips were effective at removing large amounts of N (i.e., approximately 74 % from runoff water). The model showed 50, 75, and 90 % N removal efficiencies would occur in buffers 3 m,

28 m, and 122 m wide, respectively. Although wider buffers are more effective in remove nutrients, more land is taken out of intended production. Their study indicated that while greater consistency of N removal occurred with increasing buffer width, other factors such as flow pattern and vegetation type affected N removal. For example, they found that forest buffers were more effective than grass buffers in removing N.

Additionally, numerous studies have investigated the effectiveness of removing P in runoff using buffer strips. Daniels and Gilliam (1996) evaluated the use of a 6 m vegetated filter strip (VFS) in natural rainfall conditions and found reductions of approximately 60 % for total P and 50 % of the soluble P load in runoff. In another study, Patty et al. (1997) used VSFs of 6, 12, and 18 m, and reported the average P removal increased with buffer width and was 40, 52, and 87 % for the 6, 12, and 18 m width, respectively. Abu-Zrieg et al. (2003) used VFSs and evaluated P removal in artificial runoff from cropland. The VSFs were 2, 5, 10, and 15 m in width and their study found average P trapping efficiencies ranged from 31 to 89 %. Their study concluded filter width was the predominant factor affecting P trapping and the primary mechanisms involved in P removal were sediment deposition, infiltration, and plant removal.

The effectiveness of buffers is affected by buffer width, slope, vegetative species, soil texture, and flow velocity. In a review of over 80 BMP experiments involving buffers, Liu et al. (2008) reported the factors most affecting the efficacy of buffers on sediment trapping were buffer width and slope. The use of buffers combined with crop residue management and nutrient management will allow farmers to maintain environmental sustainability. Riparian zones are the interfaces between land and streams; as such they are the last areas for uptake of nutrients prior to steam entry. Riparian zones must be maintained with a continuous cover crop so that adequate uptake of nutrients and chemicals is achieved to protect water quality.

1.6 Future Work

Animal manure has been proven to be a valuable nutrient source and soil amendment. The key for a sustainable manure land application is to develop and follow nutrient management plans including phosphorus risk assessment tool. The existing P indexes vary among different states. More standardized and quantitative tools need to be developed and implemented. Manure applied to the soil surface without incorporation, such as on pastures and no-till fields, is subject more to nutrient losses. Therefore, improved manure applicators are needed to increase nutrient use efficiency and decrease the impact of manure use on environmental quality. The effectiveness of various BMPs to minimize nutrient transport needs to be further studied and the effective ones should be promoted to farmers.

References

Abu-Zrieg M, Rudra RP, Whiteley HR, Lalonde MN, Kaushik NK (2003) Phosphorus removal in vegetated filter strips. J Environ Qual 32:613–619

Adeli A, Sustani KR, Rowe DE, Tewolde H (2007) Effects of broiler litter applied to no-till and tillage cotton on selected soil properties. Soil Sci Soc Am J 71:974–983

Anderson MA, McKenna JR, Martens DC, Donohue SJ, Kornegay ET, Lindemann MD (1991) Long-term effects of copper rich swine manure application on continuous corn production. Commun Soil Sci Plant Anal 22:993–1002

Arnesen AKM, Singh BR (1998) Plant uptake and DTPA-extractability of Cd, Cu, Ni and Zn in a Norwegian alum shale soil as affected by previous addition of dairy and pig manures and peat. Can J Soil Sci 78:531–539

Badaruddin M, Reynolds MP, Ageeb OAA (1999) Wheat management in warm environments: effect of organic and inorganic fertilizers, irrigation frequency, and mulching. Agron J 91:975–983

Barrett MJ (2012) Phytoremediation of soil phosphorous with crabgrass. Thesis, Oklahoma State University

Bolan NS, Adriano DC, Mahimairaja S (2004) Distribution and bioavailability of trace elements in livestock and poultry manure by-products. Crit Rev Environ Sci Technol 34:291–338

Bremner I (1981) Effects of the disposal of copper-rich slurry on the health of grazing animals. In: Hermite PL, Dehandtschutter J, Reidel D (eds) Copper in animal wastes and sewage sludge. Reidel, Springer Netherlands, Dordrecht, pp 245–260

Butler TJ, Han KJ, Muir JP, Weindorf DC, Lastly L (2008) Dairy manure compost effects on corn silage production and soil properties. Agron J 100:1541–1545

CAST (1996) Integrated animal waste management. Council for Agricultural Science and Technology, Ames

Chang C, Sommerfeldt TG, Entz T (1991) Soil chemistry after eleven annual applications of cattle feedlot manure. J Environ Qual 20:475–480

Christen K (2001) Chickens, manure, and arsenic. Environ Sci Technol 35:184A–185A

Codling EE, Chaney RL, Mulchi CL (2000) Use of aluminum- and iron-rich residues to immobilize phosphorus in poultry litter and litter-amended soils. J Environ Qual 29:1924–1931

Combs SM, Peters JB, Zhang LS (1998) Manure: trace element content and impact on soil management. New Horizons Soil Science. Number 7. Univ Wisconsin, Madison, WI, 10 p. http://www.uwex.edu/ces/forage/wfc/proceedings2001/micronutrient_status_of_manure.htm. Accessed 14 Aug 2013

Cooper JM, Warman PR (1997) Effects of three fertility amendments on soil dehydrogenase activity, organic C and pH. Can J Soil Sci 77:281–283

Daniel TC, Hausten G, Moore PA, Nichols D (1999) Effect of land applied alum residuals and HCA on runoff phosphorus concentrations from high-P soils. In: WEF/AWWA joint residuals and biosolids management conference abstracts, January 27–30, 1999, Charlotte, NC

Daniels RB, Gilliam JW (1996) Sediment and chemical load reduction by grass and riparian filters. Soil Sci Soc Am J 60:246–251

Dao TH (1999) Co-amendments to modify phosphorus extractability and N:P ratio in feedlot manure and composted manure. J Environ Qual 28:1114–1121

Dao TH (2001) Chemical treatments for particulate aggregation and P reduction in dairy manure. In: SERA-IEG 17, State College, PA, 9–11 July 2001

Dao TH, Daniel TC (2002) Chemical separation of particulate and dissolved phosphorus and release from dairy manure. J Environ Qual 31:1388–1398

Dao TH, Sikora LJ, Hamasaki A, Chaney RL (2001) Manure P extractability as affected by aluminum- and iron by-products and aerobic composting. J Environ Qual 30:1693–1698

Davis RL, Patton JJ, Teal RK, Tang Y, Humphreys MT, Mosali J, Girma K, Lawles JW, Moges SM, Malapati A, Si J, Zhang H, Deng S, Johnson GV, Mullen RW, Raun WR (2003) Nitrogen

balance in the Magruder Plots following 109 years in continuous winter wheat. J Plant Nutr 26:1561–1580
Day DL, Funk T (1998) Processing manure: physical, chemical and biological treatment. In: Hatfield JL, Stewart BA (eds) Animal waste utilization: effective use of manure as a soil resource. Ann Arbor Press, Chelsea, pp 243–282
Dayton EA, Basta NT (2005) Use of drinking water treatment residuals as a potential best management practice to reduce phosphorus risk index scores. J Environ Qual 34:2112–2117
DeLaune PB, Moore PA Jr, Carman DK, Sharpley AN, Haggard BE, Daniel TC (2004) Development of a phosphorus index for pastures fertilized with poultry litter-factors affecting phosphorus runoff. J Environ Qual 33:2183–2191
Eck HV, Stewart BA (1995) Manure. In: Rechcigl JE (ed) Soil amendments and environmental quality. Lewis Publishers, Boca Raton, pp 169–198
Edwards DR, Daniel TC (1993) Effects of poultry litter application rate and rainfall intensity on quality of runoff from fescuegrass plots. J Environ Qual 22:361–365
Edwards DR, Moore PA Jr, Daniel TC, Srivastava P, Nichols DJ (1997) Vegetative strip removal of metals in runoff from poultry litter-amended fescuegrass plots. Trans ASAE 40:121–127
Eghball B (1999) Liming effects of beef cattle feedlot manure or compost. Commun Soil Sci Plant Anal 30:2563–2570
Eghball B, Power JF (1999) Phosphorus- and nitrogen-based manure and compost applications: corn production and soil phosphorus. Soil Sci Soc Am J 63:895–901
Fraser DG, Doran JW, Sahs WW, Lesoing GW (1988) Soil microbial populations and activities under conventional and organic management. J Environ Qual 17:585–590
Giddens J, Barnett AP (1980) Soil loss and microbiological quality of runoff from land treated with poultry litter. J Environ Qual 9:518–520
Gill HS, Meelu OP (1982) Studies on the substitution of inorganic fertilizers with organic manure and their effect on soil fertility in rice-wheat rotation. Fertil Res 3:303–314
Gilley JE, Eghball B (1998) Runoff and erosion following field application of beef cattle manure and compost. Trans ASAE 41:1289–1294
Gilley JE, Risse LM (2000) Runoff and soil loss as affected by the application of manure. Trans ASAE 43:1583–1588
Girma K, Holtz SL, Arnall DB, Tubana BS, Raun WR (2007) The Magruder Plots: untangling the puzzle. Agron J 99:1191–1198
Haynes RJ, Naidu R (1998) Influence of lime, fertilizer and manure applications on soil organic matter content and soil physical conditions: a review. Nutr Cycl Agroecosyst 51:121–137
Hue NV (1992) Correcting soil acidity of a highly weathered ultisol with chicken manure and sewage sludge. Commun Soil Sci Plant Anal 23:241–264
Kingery WL, Wood CW, Delaney DP, Williams JC, Mullins GL (1994) Impact of long-term land application of broiler litter on environmentally related soil properties. J Environ Qual 23:139–147
Kirchmann H, Gerzabek MH (1999) Relationship between soil organic matter and micropores in a long-term experiment at Ultuna, Sweden. J Plant Nutr Soil Sci 162:493–498
Krishnamachari K (1987) Fluorine. In: Mertz W (ed) Trace elements in human and animal health, vol 1, 5th edn. Academic, San Diego, pp 265–416
Lamand M (1981) Copper toxicity in sheep. In: Hermite PL, Dehandtschutter J, Reidel D (eds) Copper in animal wastes and sewage sludge. Reidel, Springer Netherlands, Dordrecht, pp 261–272
Li M, Hue NV, Hussain SKG (1997) Changes of metal forms by organic amendments to Hawaii soils. Commun Soil Sci Plant Anal 28:381–394
Lithourgidisa AS, Matsib T, Barbayiannisb N, Dordasc CA (2007) Effect of liquid cattle manure on corn yield, composition, and soil properties. Agron J 99:1041–1047
Liu X, Zhang X, Zhang M (2008) Major factors influencing the efficacy of vegetated buffers on sediment trapping: a review and analysis. J Environ Qual 33:1667–1674

Magruder Plots (2008) Long-term application of N, P, K, lime and manure, 1892–2005. http://nue.okstate.edu/Long_Term_Experiments/Magruder_Plots_Yield_Summary.htm. Accessed 14 Aug 2013

Mayer PM, Reynolds SK, McCutchen MD, Canfield TJ (2006) Riparian buffer width, vegetative cover, and nitrogen removal effectiveness: a review of current science and regulations. EPA/600/R-05/118. United States Environmental Protection Agency, Cincinnati

Mikha MM, Rice CW (2004) Tillage and manure effects on soil and aggregate-associated carbon and nitrogen. Soil Sci Soc Am J 68:809–816

Miller RE, Lei X, Ullrey DE (1991) Trace elements in animal nutrition. In: Mortvedt JJ (ed) Micronutrients in agriculture, 2nd edn. Soil Science Society of America, Madison, pp 593–662

Mitchell CC, Tu S (2006) Nutrient accumulation and movement from poultry litter. Soil Sci Soc Am J 70:2146–2153

Moore PA Jr, Edwards DR (2005) Long-term effects of poultry litter, alum-treated litter, and ammonium nitrate on phosphorus availability in soils. J Environ Qual 36:163–174

Moore PA Jr, Edwards DR (2007) Long-term effects of poultry litter, alum-treated litter, and ammonium nitrate on phosphorus availability in soils. J Environ Qual 36:163–174

Moore PA Jr, Miller DM (1994) Decreasing phosphorus solubility using aluminum, calcium, and iron amendments. J Environ Qual 23:325–330

Moore PA Jr, Daniel TC, Gilmore JT, Shreve BR, Edwards DR, Wood BH (1998) Decreasing metal runoff from poultry litter with aluminum sulfate. J Environ Qual 27:92–99

Moore PA Jr, Daniel TC, Edwards DR (1999) Reducing phosphorus runoff and improving poultry production with alum. Poult Sci 78:692–698

Moore PA Jr, Daniel TC, Edwards DR (2000) Reducing phosphorus runoff and inhibiting ammonia loss from poultry manure with aluminum sulfate. J Environ Qual 29:37–49

Mosaddeghi MR, Hajabbasi MA, Hemmat A, Afyuni M (2000) Soil compactibility as affected by soil moisture content and farmyard manure in central Iran. Soil Tillage Res 55:87–97

Motavalli PP, Kelling KA, Converse JC (1989) First-year nutrient availability from injected dairy manure. J Environ Qual 18:180–185

Mueller DH, Wendt RC, Daniel TC (1984) Soil and water losses as affected by tillage and manure application. Soil Sci Soc Am J 48:896–900

Narwal RP, Singh BR (1998) Effect of organic materials on partitioning extractability and plant uptake of metals in an alum shale soil. Water Air Soil Pollut 103:405–410

Patty L, Rheal BR, Gril JJ (1997) The use of grass buffer strips to remove pesticides, nitrate, and soluble phosphorus compounds from runoff water. Pestic Sci 49:243–251

Payne GG, Martens DC, Kornegay ET, Lindemann MD (1988) Availability and form of copper in three soils following eight annual applications of copper enriched swine manure. J Environ Qual 17:740–746

Poole DBR (1981) Implications of applying copper rich pig slurry to grassland-effects on the health of grazing sheep. In: Hermite PL, Dehandtschutter J, Reidel D (eds) Copper in animal wastes and sewage sludge. Reidel, Springer Netherlands, Dordrecht, pp 273–282

Pote DH, Daniel TC, Sharpley AN, Moore PA Jr, Edwards DR, Nichols DJ (1996) Relating extractable soil P to P losses in runoff. Soil Sci Soc Am J 60:855–859

Richards J, Zhang H, Schroder JL, Hattey JA, Raun WR, Payton ME (2011) Micronutrient availability as affected by the long-term application of phosphorus fertilizer and organic amendments. Soil Sci Soc Am J 75:927–939

Risse LM, Cabrera ML, Franzluebbers AJ, Gaskin JW, Gilley JE, Killorn R, Radcliffe DE, Tollner WE, Zhang H (2006) Land application of manure for beneficial reuse. In: Rice JM et al (eds) Animal agriculture and the environment: National Center for Manure and Animal Waste Management white papers. American Society of Agricultural and Biological Engineers, St. Joseph, pp 283–316

Roberts RJ, Clanton CJ (2000) Surface seal hydraulic conductivity as affected by livestock manure application. Trans ASAE 43:603–613

Sattar MA, Gaur AC (1989) Effect of VA-mycorrhiza and phosphate dissolving microorganism on the yield and phosphorus uptake of wheat (*Triticum vulgare*) in Bangladesh. Bangladesh J Agric Res 14:233–239

Sauer TJ, Daniel TC, Moore PA Jr, Coffey KP, Nichols DJ, West CP (1999) Poultry litter and grazing animal effects on runoff water quality. J Environ Qual 28:860–865

Schjønning P, Christensen BT, Carstensen B (1994) Chemical and physical properties of a sandy loam receiving animal manure, mineral fertilizer, or no fertilizer for 90 years. Eur J Soil Sci 45:245–268

Schroder JL, Zhang H, Richards JL, He Z (2011) Sources and contents of heavy metals and other trace elements in animal manure. In: He Z (ed) Chemistry of animal manure. Nova Science Publishers, New York, pp 385–414

Sharpley AN (1995) Dependence of runoff phosphorus on extractable soil phosphorus. J Environ Qual 24:920–926

Sharpley AN, Smith SJ (1993) Nitrogen and phosphorus fate from long-term poultry litter applications to Oklahoma soils. Soil Sci Soc Am J 57:131–1137

Sharpley AN, Daniel TC, Sims JT, Pote DH (1996) Determining environmentally sound soil phosphorus levels. J Soil Water Conserv 51:160–166

Sharpley AN, Daniel TC, Sims JT, Lemunyon, Stevens JR, Parry R (1999) Agricultural phosphorus and eutrophication. USDA Miscellaneous Publications ARS-149. U.S. Government Printing Office, Washington, DC

Sharpley AN, Weld JL, Beegle KPJA, Gburek WJ, Moore PA Jr, Mullins G (2003) Development of phosphorus indices for nutrient management planning strategies in the United States. J Soil Water Conserv 58:137–152

Shreve BR, Moore PA Jr, Daniel TC, Edwards DR, Miller DM (1995) Reduction of phosphorus in runoff from field-applied poultry litter using chemical amendments. J Environ Qual 24:106–111

Sims JT, Johnson GV (1991) Micronutrient soil test. In: Mortvedt JJ (ed) Micronutrients in agriculture, 2nd edn. Soil Science Society of America, Madison, pp 427–476

Sims JT, Luka-McCafferty NJ (2002) On-farm evaluation of aluminum sulfate (alum) as a poultry litter amendment: effects on litter properties. J Environ Qual 31:2066–2073

Sommerfeldt TG, Chang C, Entz T (1988) Long-term annual manure applications increase soil organic matter and nitrogen, and decrease carbon to nitrogen ratio. Soil Sci Soc Am J 52:1668–1672

Southern Extension and Research Activity Group 17 (2013) Minimize nutrient losses from agriculture. http://www.sera17.ext.vt.edu/SERA_17_Publications.htm. Accessed 14 Aug 2013

Stout WL, Sharpley AN, Pionke HB (1998) Reducing soil phosphorus solubility with coal combustion by-products. J Environ Qual 27:111–118

Sutton AL, Nelson DW, Mayrose VB, Nye JC, Kelly DT (1984) Effect of copper levels in swine manure on corn and soil. J Environ Qual 13:198–203

Tang Y, Zhang H, Schroder JL, Payton ME, Zhou D (2007) Animal manure reduces aluminum toxicity in an acid soil. Soil Sci Soc Am J 71:1699–1707

Tiarks AE, Mazurak AP, Chesnin L (1974) Physical and chemical properties of soil associated with heavy applications of manure from cattle feedlots. Soil Sci Soc Am J 38:826–830

Tufft LS, Nockels CF (1991) The effects of stress, Escherichia coli, dietary ethylene diamine tetraacetic acid, and their interaction on tissue trace elements in chicks. Poult Sci 70:2439–2449

U.S. Department of Agriculture (2001) Comprehensive nutrient management planning technical guidance. USDA-NRCS, Washington, DC

U.S. Department of Agriculture-Natural Resources Conservation Service (2011) Natural Resources Conservation Service. Conservation Practice Standard. Nutrient Management Code 590. U.S. Department of Agriculture, Washington, DC

USEPA (2012) NPDES permit writers' manual for CAFOs. EPA 833-F-12-001. Office of Water, Office of Wastewater Management, Washington, DC. http://www.epa.gov/npdes/pubs/cafo_permitmanual_chapter1.pdf. Accessed 14 Aug 2013

USEPA (2013) Literature review of contaminants in livestock and poultry manure and implications for water quality. EPA 820-R-13–002. Office of Water, Washington, DC

van der Watt HVH, Sumner ME, Cabrera ML (1994) Bioavailability of copper, manganese and zinc in poultry manure. J Environ Qual 23:43–49

Vitosh ML, Davis JF, Knezek BD (1973) Long-term effects of manure, fertilizer, and plow depth on chemical properties of soils and nutrient movement in a monoculture corn system. J Environ Qual 2:296–299

Wallingford GW, Murphy LS, Powers WL, Manges HL (1975) Effects of beef feedlot manure and lagoon water on iron, zinc, manganese and copper content in corn and in DTPA soil extracts. Soil Sci Soc Am J 39:482–487

Whalen JK, Chang C, Clayton GW, Carefoot JP (2000) Cattle manure amendments can increase the pH of acid soils. Soil Sci Soc Am J 64:962–966

Whalen JK, Hub Q, Liua A (2003) Compost applications increase water-stable aggregates in conventional and no-tillage systems. Soil Sci Soc Am J 67:1842–1847

Wong MTF, Nortcliff S, Swift RS (1998) Method for determining the acid ameliorating capacity of plant residue compost, urban waste compost, farmyard manure, and peat applied to tropical soils. Commun Soil Sci Plant Anal 29:2927–2937

Wright RJ, Kemper WD, Millner PD, Power JF, Korcak RF (1998) Agricultural uses of municipal, animal, and industrial byproducts. Conservation Research Report 44. U.S. Department of Agriculture, ARS, Beltsville

Xie RJ, MacKenzie AF (1986) Urea and manure effects on soil nitrogen and corn dry matter yields. Soil Sci Soc Am J 50:1504–1509

Zhang H, Smeal D, Tomko J (1998) Nitrogen fertilization value of feedlot manure for irrigated corn production. J Plant Nutr 21:287–296

Zhang H, Schroder JL, Fuhrman JK, Basta NT, Storm DE, Payton ME (2005) Path analysis and multiple regression of phosphorus sorption capacity. Soil Sci Soc Am J 69:96–106

Zhu YM, Berry DF, Martens DC (1991) Copper availability in two soils amended with 11 annual applications of copper-enriched hog manure. Commun Soil Sci Plant Anal 22:769–783

Chapter 2
Residual Veterinary Pharmaceuticals in Animal Manures and Their Environmental Behaviors in Soils

Weiping Song and Mingxin Guo

Abstract The worldwide heavy use of veterinary pharmaceuticals in confined animal-feeding operations has resulted in annual discharge of 3,000–27,000 tons of drug chemicals via livestock manure into the environment. More than 50 major antibiotics have been detected in poultry, swine, cattle, and horse manures at 0.01–765 mg kg^{-1} dry manure mass. In animal manures, most veterinary pharmaceuticals degrade rapidly via biochemical reactions, demonstrating a half-life time 2–30 days. In soils, veterinary pharmaceuticals interact with soil minerals, organic matter, and organisms and are subject to sorption, photohydrolysis, oxidation, and biodegradation. The soil distribution coefficient (K_d) values of animal pharmaceuticals range from 0.3 to 6,300 L kg^{-1}, varying with the chemical species and soil properties. The persistence of veterinary pharmaceuticals in soils is influenced by soil type, organic matter content, pH, moisture content, and temperature. Though certain antibiotics such as roxithromycin, sarafloxacin, and virginiamycin are persistent, the vast majority of veterinary pharmaceuticals are degradable (half-life <30 days) in soils. The sorption, rapid degradation, and physical attenuation limit residual pharmaceuticals in the top 30-cm soil of agricultural land at generally less than 1 μg kg^{-1}, posing little impacts on soil microorganisms, fauna, and plants. Nevertheless, veterinary pharmaceuticals could migrate from manured fields to water bodies via surface runoff and leaching. In North American drainage ditches and streams, up to 290 ng L^{-1} of animal antibiotics had been detected, although the concentrations were far below the no-observed-effect concentration levels of veterinary pharmaceuticals to aquatic organisms. Antibiotic-resistant bacteria have been identified in animal manures and livestock-handling workers, indicating the risk of antibiotic-resistant genes spread in association with

W. Song
Department of Chemistry, Delaware State University, Dover, DE 19901, USA

M. Guo (✉)
Department of Agriculture and Natural Resources, Delaware State University, Dover, DE 19901, USA
e-mail: mguo@desu.edu

Z. He and H. Zhang (eds.), *Applied Manure and Nutrient Chemistry for Sustainable Agriculture and Environment*, DOI 10.1007/978-94-017-8807-6_2,

veterinary pharmaceutical overuse and manure disposal. Future research should focus on developing standard composting protocols to eliminate residual veterinary pharmaceuticals and antibiotic-resistant pathogens from animal manures and on cultivating animal-feeding methods alternative to drug administration.

2.1 Introduction

Veterinary pharmaceuticals are chemical drugs administered to domestic animals to treat diseases, prevent infections, increase weight gain, or improve feed efficiency. Common veterinary pharmaceuticals include antibiotics, antiparasitics, anti-inflammatory medicines, anesthetics, pain relievers, and specialized products used to manage animal reproductive or metabolic conditions. These medications are prepared in a variety of forms such as pills, liquids, injections, or powders and can be applied to animals via feed or drinking water, by injection or skin insertion, or simply through drenching (OTA 1979).

The U.S. confined animal feeding operations rely heavily on veterinary pharmaceuticals to maintain healthy, productive livestock. Veterinary antibiotics are regular feed supplements of poultry, swine, cattle, equine, and aquaculture (Henderson and Coats 2010). Considering chemical structures, most veterinary pharmaceuticals are amphiphilic or amphoteric, ionizable organic compounds consisting of a nonpolar core and multiple polar functional groups (Thiele-Bruhn 2003). After imposed to livestock, these pharmaceuticals are typically absorbed through animals' digestive and circulatory systems and discharged in waste from the excretory system. The pharmaceuticals are generally metabolized and deactivated for biological functions after the animal body passage. A significant portion (10–90 %) of the applied quantities, however, may remain intact as parent compounds and deposit in animal tissues and excrement (Kumar et al. 2005a). Certain metabolites are also biologically active (Halling-Sørensen et al. 1998). Residues of veterinary pharmaceuticals and their active metabolites in animal tissues and excreta have been exclusively detected (Kumar et al. 2005a; Furtula et al. 2010). Responding to the residual pharmaceuticals, microorganisms such as *Enterococcus* spp., *Staphylococcus* spp., and *E. coli* in animal manures may develop antibiotic resistance (Hayes et al. 2004; Furtula et al. 2010). Through land application of animal waste as an organic fertilizer, these residual veterinary pharmaceuticals and antibiotic-resistant microorganisms enter into soil and water and may influence aquatic ecosystem and accumulate in food crops (Solomon et al. 2010; Carlsson et al. 2013). To assess the potential risks posed by veterinary pharmaceuticals from land application of animal waste, the occurrence of animal drug compounds in animal manures and their fate and transport in agricultural ecosystems need to be addressed. This chapter is to summarize veterinary pharmaceutical uses in confined animal feeding operations, reports on presence and detection of residual veterinary medicines in manures, and review the environmental behaviors of pharmaceutical residues in agricultural soils.

2.2 Uses of Veterinary Pharmaceuticals in Animal Production

More than 400 active chemical ingredients have been manufactured into nearly 2,000 veterinary pharmaceutical products to treat various species of animals including pigs, cattle, horses, sheep, goats, birds, fish, deer, cats, and dogs (FDA 2012). These chemicals are conventionally placed into five groups: anthelmintics (dewormers), tranquilizers, antibiotics, hormones, and agonists. According to their functions, they can be further categorized as therapeutic medicines (to treat animals for preventing diseases, combating infections, or alleviating pain or injury. Examples include coccidiostatics, trimetoprim, and sulfamethizol) and growth promoters (to help with animal feed digestion and growth efficiency. Examples are tylosin, monensin, and virginiamycin) (Garrido Frenich et al. 2010).

More than 70 % of the consumed veterinary pharmaceuticals are antibiotics – chemicals that can inhibit the growth of other microorganisms even at extremely low concentrations (Halling-Sørensen et al. 1998). There are over 150 antibiotics in use today, of which more than 90 % are natural products of bacteria and fungi (molds) and semisynthetic modifications of natural compounds, and a few such as sulfonamides are completely synthetic (von Nussbaum et al. 2006). The first commercially manufactured antibiotic was penicillin, a chemical compound derived from *Penicillium* fungi. Antibiotics were initially and are continuously used for therapeutically treating human and animal diseases and infections. In 1949, the U.S. officially approved the use of antibiotics as a feed additive in the rearing of domestic animals for human consumption, so did the United Kingdom in 1953 (Witte 2000). Today, supplementing animal feed with antibiotics has been practiced in nearly all livestock and aquaculture operations in most countries. Antibiotics added in feed serve predominantly as growth promoters. It is believed that the antibiotics inhibit subclinical pathogenic bacterial infections, increase uptake and utilization of nutrients through the intestinal wall, and suppress the activity and population of bacteria in the intestines and thus, preserve the energy in feed that would be lost due to microbial fermentation, promoting animal growth through nutrient and energy availability enhancement (Gaskins et al. 2002). Registered animal antibiotics for use as growth promoter/feed efficiency in Australia, Denmark, European Union (EU), and Canada are well summarized in Sarmah et al. (2006). The antibiotics approved for use in U.S. food-producing animals are given in Table 2.1. Relative usage of these chemicals is illustrated in Fig. 2.1.

Addition of antibiotics to animal feed is recommended at dose ranging from 3 to 220 mg kg^{-1}, depending on the species and growth stage of the animal and the type of antibiotics (McEwen and Fedorka-Cray 2002). Multiple antibiotics are often supplemented in combination. Some antibiotics are added for a specific growth stage of animals but some could be fed continuously up to the point of slaughter (Kumar et al. 2005a). Furtula et al. (2009) reported that chicken feeds in British Columbia, Canada contained multiple antibiotics at concentrations varying with bird growth phases, typically 22 mg kg^{-1} virginiamycin, 99 mg kg^{-1} monensin, 120 mg kg^{-1} salinomycin, 80 mg kg^{-1} narasin, 80 mg kg^{-1} nicarbazin, 165 mg kg^{-1}

Table 2.1 Antimicrobials drugs approved for use in food-producing animals in the U.S.

Antimicrobial class	Basic chemical structure	Individual drugs
Aminocoumarins	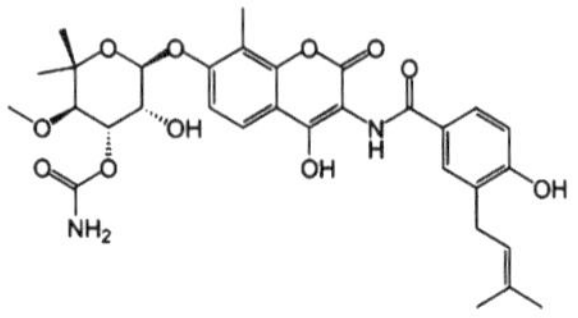	Novobiocin
Aminoglycosides	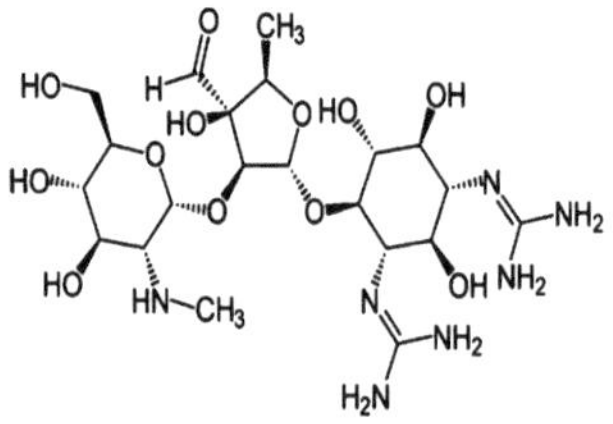	Apramycin[a] Dihydrostreptomycin Efrotomycin Gentamicin Hygromycin B Neomycin[a] Spectinomycin Streptomycin
Amphenicols	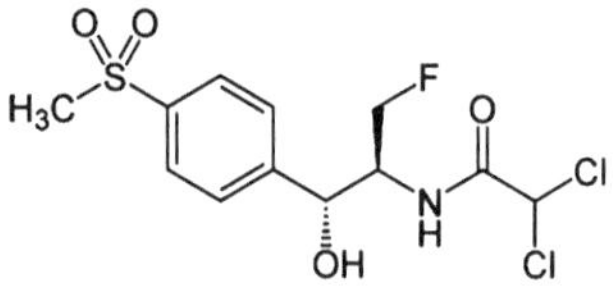	Florfenicol
Cephalosporins	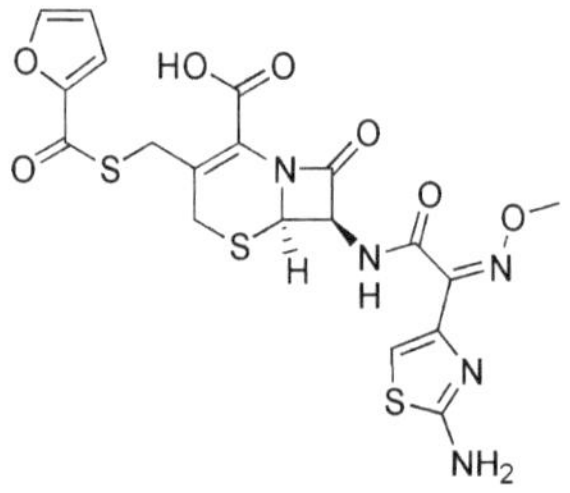	Ceftiofur Cephapirin
Diaminopyrimidines	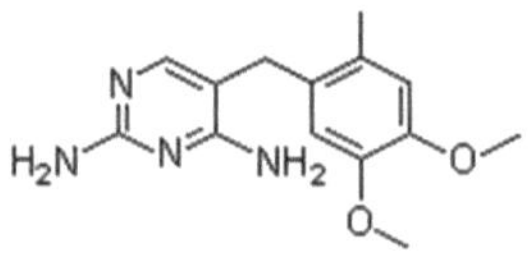	Ormetoprim
Fluoroquinolones	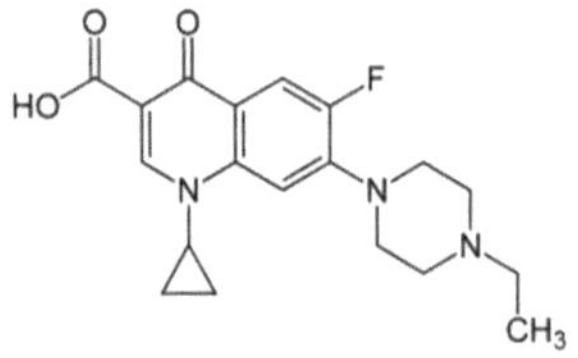	Danofloxacin Enrofloxacin
Glycolipids	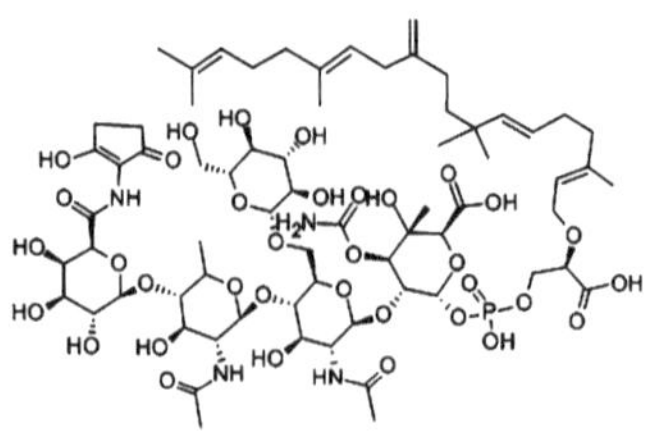	Bambermycin[a]

(continued)

Table 2.1 (continued)

Antimicrobial class	Basic chemical structure	Individual drugs
Ionophores	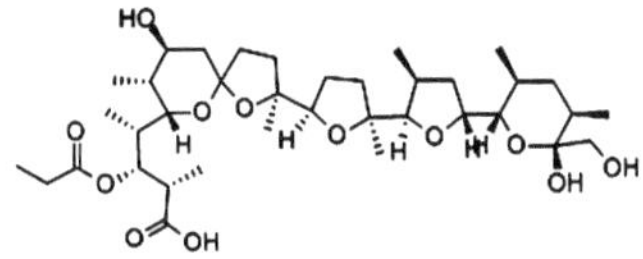	Laidlomycin Lasalocid Monensin Narasin Salinomycin Semduramicin
Lincosamides	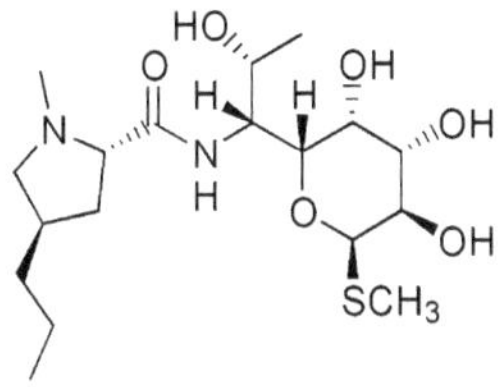	Lincomycin[a] Pirlimycin
Macrolides	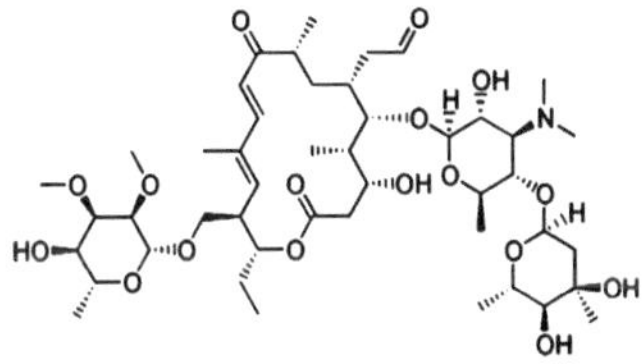	Carbomycin Erythromycin Oleandomycin Tilmicosin Tulathromycin Tylosin[a]
Penicillins	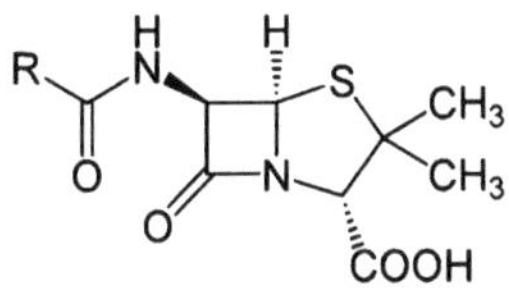	Amoxicillin Ampicillin Cloxacillin Hetacillin Penicillin[a]
Pleuromutilins	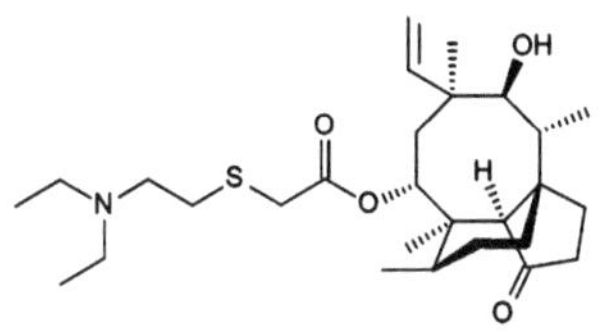	Tiamulin[a]
Polypeptides	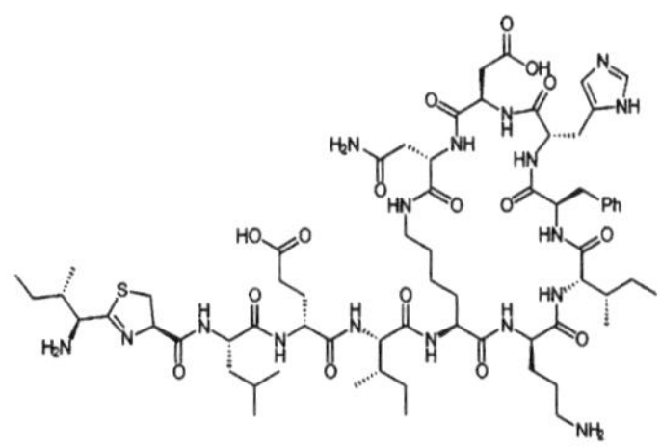	Bacitracin[a] Polymixin B
Quinoxalines	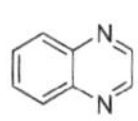	Carbadox[a]

(continued)

Table 2.1 (continued)

Antimicrobial class	Basic chemical structure	Individual drugs
Streptogramins	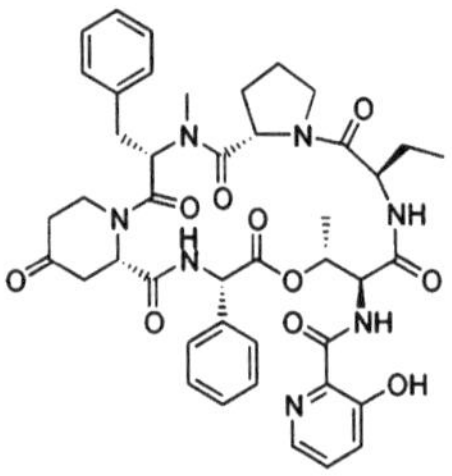	Virginiamycin[a]
Sulfas	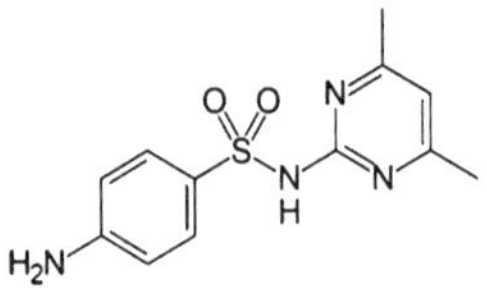	Sulfachlorpyridazine Sulfadimethoxine Sulfamerazine Sulfamethazine[a] Sulfaquinoxaline Sulfathiazole[a]
Tetracyclines	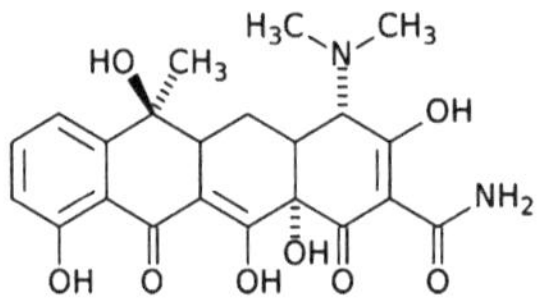	Chlortetracycline[a] Oxytetracycline[a] Tetracycline

Source: FDA (2010)

[a]Approved for use as **swine** feed supplements for therapeutic purposes (carbadox, sulfamethazine, and sulfathiazole) and for promoting pig growth (others)

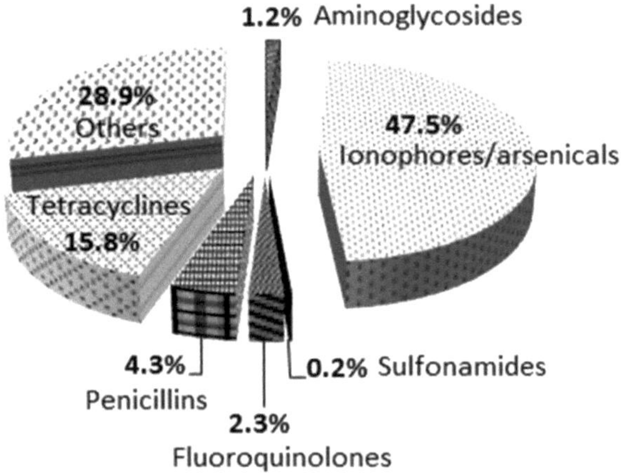

Fig. 2.1 Relative percentages of the used antibiotics for animal production in the US in 1999 (Source: Sarmah et al. 2006)

bacitracin, and 22 mg kg^{-1} penicillin in empirical combinations. Currently the U.S. Food and Drug Administration (FDA) approves 14 antibiotics for use in swine feed (Table 2.1), of which 11 are recommended as growth promoters at 2–150 mg kg^{-1} of feed (Holt 2008). Due to the development of bacterial antibiotic resistance, however, animal feeds often contain antibiotics at contents higher than the recommended levels. A survey revealed that 25 % of the 3,000 tested swine feeds in the U.S. contained antibiotics at higher-than-the-recommended concentrations (Dewey et al. 1997). The widespread use of antibiotics at increasing rates may facilitate the evolution of bacteria toward antibiotics-resistant strains and consequently, induce new, untreatable livestock diseases (Kumar et al. 2005a). Antibiotics inhibit or destroy sensitive bacteria, providing an environment for those resistant variants to flourish and become dominant. The antibiotic resistance can be further transferred via plasmids to other bacteria.

Globally it is unclear what veterinary pharmaceuticals and in what quantities are being used, as data on the annual production and consumption of animal medicines are not readily available in many countries. The U.S. uses 13,067 tons of veterinary antibiotics in domestic animal agriculture and exports 1,632 tons to other countries annually (FDA 2010). In China, more than 6,000 tons of veterinary antibiotics are consumed annually (Zhao et al. 2010); the most common antibiotics are tetracyclines, sulfonamides, tylosin, and fluoroquinolones (Li et al. 2013). In the United Kingdom, 897 tons of antibiotics were applied to animal production in 2000 (Thiele-Bruhn and Aust 2004). The annual EU consumption of veterinary antibiotics was approximately 5,000 tons by 2005 (Kumar et al. 2005a). Since 2006, the use of antibiotics as a feed supplement of food-producing animals has been banned in EU countries (Europa 2005).

2.3 Residual Veterinary Pharmaceuticals in Animal Manures

The use of veterinary pharmaceuticals is vitally important in confined food animal production. Nevertheless, animals do not utilize all the applied pharmaceuticals. Depending on the chemical and the animal species, 10–90 % of the feed-supplemented pharmaceuticals are excreted in animal urine and feces as intact parent compounds or bioactive metabolites (Kumar et al. 2005a). Through controlled-feeding trials with broilers eating antibiotics-supplemented feeds for 36 days, Kumar et al. (2004) found that 3–60 % of the antibiotics penicillin, salinomycin, bacitracin, chlortetracycline, virginiamycin I, virginiamycin II, monensin, and narasin added separately or in combination at 2–110 mg kg^{-1} in feed were excreted in bird manure. With the advances of analytical techniques, antibiotics such as tetracyclines, tylosin, monensin, sulfadimidine, and sulfathiazole have been detected in swine slurry, cattle manure, poultry litter, and fish farm sediment from different countries at a wide concentration range from trace to 200 mg kg^{-1} or mg L^{-1} (Kumar et al. 2005a). Literature reported concentrations of residual veterinary pharmaceuticals in manure wastes from confined food-producing animals are summarized in Table 2.2.

Detection of residual veterinary pharmaceuticals in manures is typically achieved by extracting animal waste with nonpolar and polar solvent extractants, purifying

Table 2.2 Reported concentrations of residual veterinary pharmaceuticals in animal manures

Manure type	Pharmaceuticals	Concentration	Country	References
Swine manure	Sulfonamides	0.01–29 mg kg^{-1}	China	Pan et al. (2011)
	Tetracyclines	0.03–765 mg kg^{-1}		
	Macrolide	0.05–0.11 mg kg^{-1}		
Swine manure	Tetracyclines	0.3–56.8 mg kg^{-1}	China	Li et al. (2013)
	Sulfas	0.1–4.8 mg kg^{-1}		
	Tylosin	0.2–1.9 mg kg^{-1}		
Manure from mother pigs with farrows	Sulfamethazine	3.3–8.7 mg kg^{-1}	Switzerland	Haller et al. (2002)
	Sulfathiazole	0–12.4 mg kg^{-1}		
	Trimethoprim	Traces		
Manure from fattening pigs	Sulfamethazine	0.13–0.23 mg kg^{-1}	Switzerland	Haller et al. (2002)
	Sulfathiazole	0.10–0.17 mg kg^{-1}		
Swine slurry	Tetracyclines	0.1–46 mg kg^{-1}	Australia	Carballo et al. (2007)
	Sulfadimidine	0.1–20 mg kg^{-1}		
Swine slurry	Tetracycline	14–41 mg kg^{-1}	Germany	Hamscher et al. (2005)
	Sulfamethazine	0–7.2 mg kg^{-1}		
	Sulfadiazine	3.5–11.3 mg kg^{-1}		
Swine slurry	Tetracycline	0.04–0.70 mg L^{-1}	Denmark	Sengeløv et al. (2003)
Swine lagoon slurry	Chlortetracycline	0.068–1.0 mg L^{-1}	USA	Campagnolo et al. (2002)
	Tetracycline	0.025–0.41 mg L^{-1}		
	Sulfamethazine	0.1–0.4 mg L^{-1}		
	Lincomycin	0.07–0.24 mg L^{-1}		
Swine manure lot liquid	Chlortetracycline	3.5–5.2 mg L^{-1}	USA	Kumar et al. (2004)
	Tylosin	3.3–7.9 mg L^{-1}		
Swine manure lot liquid	Tetracycline	3.2–4.0 mg L^{-1}	Germany	Hamscher et al. (2002)
	Chlortetracycline	0.09–0.10 mg L^{-1}		
Poultry litter	Salinomycin	0.32–4.4 mg kg^{-1}	Canada	Furtula et al. (2010)
	Bacitracin	0.01–1.76 mg kg^{-1}		
	Narasin	2.2–33.0 mg kg^{-1}		
	Nicarbacin	5.4–22.4 mg kg^{-1}		

Poultry manure[a]	Tetracyclines	0.5–13.4 mg kg^{-1}	China	Li et al. (2013)
	Sulfas	0.1–7.1 mg kg^{-1}		
	Tylosin	0.2–0.4 mg kg^{-1}		
Poultry manure[a]	Tetracyclines	0.05–0.5 mg kg^{-1}	Turkey	Karci and Balcioglu (2009)
	Sulfas	3–37 mg kg^{-1}		
	Enrofloxacin	0.01–0.08 mg kg^{-1}		
Poultry manure[a]	Chlortetracycline	23 mg kg^{-1}	Canada	Warman and Thomas (1981)
Broiler manure[a]	Tetracyclines	0.1–1.7 mg kg^{-1}	Australia	Carballo et al. (2007)
	Sulfadiazine	3.1–51 mg kg^{-1}		
	Enrofloxacin	0.2–2.8 mg kg^{-1}		
Turkey manure[a]	Sulfadiazine	3.1–91 mg kg^{-1}	Australia	Carballo et al. (2007)
	Enrofloxacin	0.2–8.3 mg kg^{-1}		
Dairy cow manure[a]	Tetracyclines	0.2–10.4 mg kg^{-1}	China	Li et al. (2013)
	Sulfas	0.1–1.0 mg kg^{-1}		
	Tylosin	0.2–0.3 mg kg^{-1}		
Beef cattle manure[a]	Chlortetracycline	5.3 mg kg^{-1}	USA	Patten et al. (1980)
	Oxytetracycline	11.3 mg kg^{-1}		
Cattle manure[a]	Monensin	1–5 mg kg^{-1}	Canada	Donoho (1984)
Fresh cattle manure[a]	Oxytetracycline	872 mg kg^{-1}	Italy	De Liguoro et al. (2003)
	Tylosin	116 mg kg^{-1}		
Newly removed cattle bedding	Oxytetracycline	367 mg kg^{-1}	Italy	De Liguoro et al. (2003)
	Tylosin	32.8 mg kg^{-1}		
Mixed swine and cattle slurries	Sulfas	20 mg kg^{-1}	Switzerland	Haller et al. (2002)
	Sulfamethazine	0.13–8.7 mg kg^{-1}		
Aged cattle manure and bedding mixture	Tetracyclines	0.05–0.4 mg kg^{-1}	Turkey	Karci and Balcioglu (2009)
	Sulfas	0.1–8 mg kg^{-1}		

[a]Manure is equivalent to feces

and concentrating the extracts using solid phase extraction (SPE) columns, and determining the processed extracts for concentrations of target chemicals using liquid chromatography-mass spectrometry (LC-MS) or high performance liquid chromatography (HPLC) techniques, with chemical standards for calibration. Due to low extraction efficiency and poor recovery of pharmaceuticals from background-complex waste media, accurate measurement of residual veterinary drugs in animal waste is always challenging. Extractants consisting of weakly acidic buffers and organic solvents (e.g., 1:1 methanol: pH 4.0 EDTA-McIlvaine buffer) are recommended to recover residual antibiotics from lyophilized manure materials (Thiele-Bruhn 2003). For certain veterinary chemicals, other formulated extractants (e.g., deionized water, a methanol-water mixture, or pure methanol) may be more efficient (Michelini et al. 2012). Methods such as repeated extraction, accelerated solvent extraction (ASE), ultrasonic liquid extraction, and Soxhlet extraction are commonly used to improve the extraction efficiency. Experiments with artificial antibiotic spiking disclosed 64–107 % efficiency of extracting pharmaceutical from manures (Furtula et al. 2009; Li et al. 2013). As instantaneous chemical spiking does not incorporate the "aging effect" of antibiotic-manure matrix interactions (i.e., chemicals being more tightly sorbed in the media and over time becoming more difficult to extract), the measured recovery rates may have over-estimated the actual extraction efficiency from genuine samples.

Most veterinary pharmaceuticals are biodegradable and therefore, storage or composting of animal manure helps eliminate the residual chemicals (Dolliver et al. 2008). De Liguoro et al. (2003) reported that the concentrations of oxytetracycline and tylosin in cattle manure decreased from 366.8 to 2.1 mg kg^{-1} and from 32.8 to <0.1 mg kg^{-1}, respectively, after heaping the waste outdoor for 135 days. Degradation of veterinary pharmaceuticals in animal manures is chiefly a result of microbial activity. As a result, manure sterilization significantly inhibited degradation of the residual veterinary chemicals (Wang et al. 2006). Degradation of veterinary pharmaceuticals in animal manures by the inherent microorganisms poses challenges for accurately measuring the residual level of animal drugs in excreta. Meanwhile it suggests an effective approach for reducing bioactive chemical contamination simply by stockpiling or composting animal waste for adequate time (e.g., 2–3 months) prior to land disposal.

2.4 Environmental Behaviors of Veterinary Pharmaceuticals in Soils

Veterinary pharmaceuticals are introduced into the environment through discharge, handling, storage, and disposal of animal waste. Antibiotics in excreta from grazing livestock and in water of aquaculture ponds reach the environment by direct discharge. Collected manures (in storage sheds, slurry tanks, and lagoons) and processed products (e.g., compost, sewage sludge) are commonly applied via

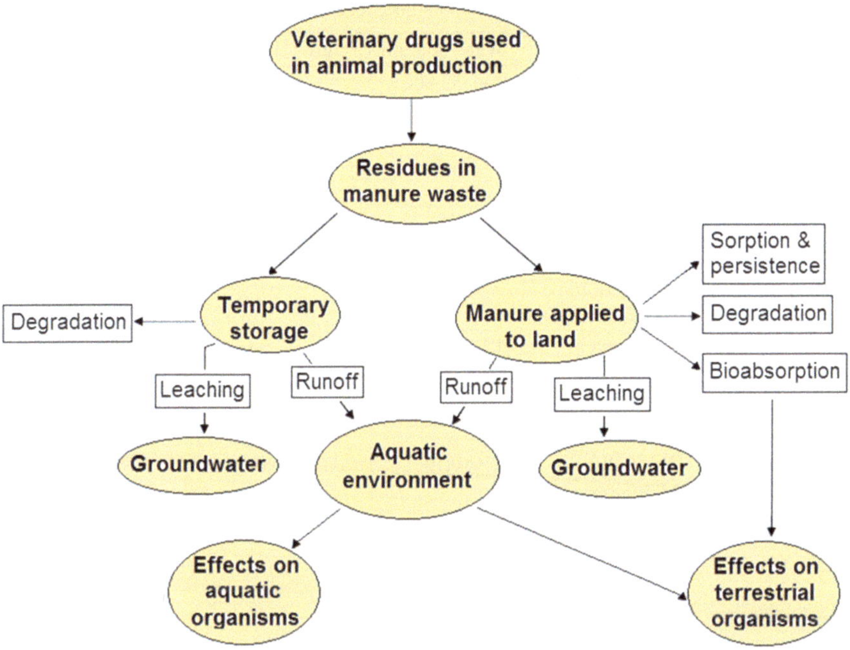

Fig. 2.2 Expected fate, transport, and exposure pathways for veterinary pharmaceuticals in the environment

spreading, injection, or irrigation to cropland as an organic fertilizer. In the U.S., the food animal industry generates annually 8.5 million dry tons of swine manure, 16.2 million dry tons of poultry litter, and 107.4 million dry tons of cattle waste, of which the vast majority is disposed of through land application (Kumar et al. 2005a). If assume 50 % (in the range of 10–90 %) of the 13,067 tons of veterinary antibiotics consumed annually (FDA 2010) are excreted in animal waste, there were more than 6,500 tons of drug chemicals entering the U.S. soil and water environments.

Anticipated transport and exposure pathways of animal waste-derived veterinary pharmaceuticals are outlined in Fig. 2.2. In soil, veterinary pharmaceuticals may be adsorbed by soil solids, degraded by soil microorganisms, taken up by plants, transport to surface water via runoff and to groundwater via leaching, and generate adverse impacts on terrestrial and aquatic ecosystems.

2.4.1 Sorption of Veterinary Pharmaceuticals in Soils

In soils, veterinary pharmaceuticals interact with clay minerals and organic matter, resulting in sorption, binding, and fixation of the chemicals in the soil matrix. The strength of the interaction is dependent on the chemical species and the soil

Table 2.3 Water solubility and soil distribution coefficient of veterinary pharmaceuticals

Drug class	Antibiotics	Water solubility (mg L^{-1})	K_d (L kg^{-1})	K_{OC} (L kg^{-1})
Tetracyclines	Tetracycline	230–52,000	420–1,030	27,800–93,300
	Chlortetracycline			
	Oxytetracycline			
Sulfonamides	Sulfanilamide	7.5–00	0.6–4.9	60–200
	Sulfadiazine			
	Sulfadimidine			
	Sulfadimethoxine			
	Sulfapyridine			
	Sulfamethoxazole			
Aminoglycosides	Efrotomycin	10–500	8–290	580–11,000
	Kanamycin			
	Neomycin			
	Streptomycin			
β-lactams	Ampicillin	22–10,100	NA	NA
	Meropenem			
	Penicillin G			
	Ceftiofur			
	Cefotiam			
Macrolides	Erythromycin	0.45–15	8.3–128	770–7,990
	Oleandomycin			
	Tylosin			
Fluorquinolones	Ciprofloxacin	3.2–17,790	260–6,310	16,500–770,000
	Enrofloxacin			
	Flumequin			
	Sarafloxacin			
	Oxolinic acid			
Imidazoles	Fenbendazole	6.3–407	0.54–0.67	38–56
	Metronidazole			
	Oxfendazole			
Polypeptides	Avermectin	Insoluble – miscible	18–134	4,760–6,600
	Bacitracin			
	Ivermectin			
	Virginiamycin			
Polyethers	Lasalocid	<0.003	1–210	61–15,700
	Monensin			
	Salinomycin			
Glycopeptides	Vancomycin	>1,000	0.3–0.7	NA
Quinoxalines	Olaquindox	1,000,000	0.69–1.67	46–116

Source: Yeager and Halley (1990), Tolls (2001), Thiele-Bruhn (2003), Sassman and Lee (2007)

property and is influenced by temperature, moisture, and the soil solution chemistry (Kumar et al. 2005a). A parameter "distribution (partition) coefficient" K_d (the ratio of the quantity of an adsorbate sorbed per unit mass of sorbent solid to the amount of the adsorbate remaining in solution at equilibrium; normally predicted from the linear portion of the adsorption isotherms) is commonly used to measure the sorption of a solute to soil. It is believed that sorption of organic contaminants in soil is mainly via interactions with soil organic matter (SOM) and thus, the adsorption coefficient K_{OC} (K_d normalized by soil organic carbon (OC) content f_{OC}: $K_{OC} = K_d/f_{OC}$) also serves as a measure of sorption (Kishi et al. 1990). Veterinary chemicals with increased aromaticity and electropolarity demonstrate higher K_d and K_{OC} values and tend to be strongly bound to soils (Thiele-Bruhn et al. 2004). The K_d and K_{OC} values of selected veterinary pharmaceuticals are given in Table 2.3.

Sorption of veterinary pharmaceuticals to soil constituents occurs naturally and ubiquitously, especially for those with high K_d or K_{OC} values (Table 2.3). The sorption is generally rapid: in antibiotics-spiked (400–12,000 mg kg^{-1}) soil slurry systems under agitation, more than 95 % of the chlortetracycline adsorption to a sandy loam and a clay soil occurred within 10 min and 95 % of the tylosin adsorption occurred within 3 h (Allaire et al. 2006). Soil active sorption sites include SOM and the broken edges, basal planes, and interlayer space of clay minerals (Nowara et al. 1997). For compounds with significant hydrophobic moieties (i.e., compounds with high octanol/water partition coefficients), sorption to soil is mainly via hydrophobic partitioning into SOM. Gruber et al. (1990) found that sorption of avermectin B1a to three different textured soils increased as the SOM increased and the K_d showed an evidently positive relationship with f_{OC}, suggesting predominance of hydrophobic partitioning in this particular antibiotic sorption. By studying migration of carbamazepine, naproxen and diclofenac in field soil profiles with artificial irrigation, Chefetz et al. (2008) concluded that sorption of the pharmaceuticals to soil was governed by SOM in its quantity and physiochemical nature. For hydrophilic, ionizable antibiotics, however, hydrophobic partitioning may not be dominant. Instead, physiochemical interactions with soil minerals become more important. A review by Tolls (2001) noticed that the K_d values of animal antibiotics varied greatly with soil types, but translation to K_{OC} by normalizing the values with soil f_{OC} did not significantly decrease the variation, suggesting clay minerals were also important in binding veterinary pharmaceuticals. The binding mechanisms include van der Walls interaction, electrostatic attraction, cation bridging, and anion exchange. Studies showed that sorption of tetracycline, oxytetracycline, oxolinic acid, and enrofloxacin to pure clay minerals and sediments increased with increasing the sorbent surface area (Sithole and Guy 1987; Nowara et al. 1997), indicating a surface-related process of antibiotics to soil. Many pharmaceuticals contain functional groups such as amines, carboxyls, and hydroxyls (Table 2.1). Protonation or deprotonation of these groups in pH-specific media engenders positive or negative charges. Positively-charged antibiotics bind to soil particles through electrostatic attraction/cation exchange (Gao and Pedersen 2005; Wang et al. 2012). Anionic pharmaceutical molecules form complexes with cations that are adsorbed on negatively charged soil constituents; the cation bridging enables the pharmaceuticals to be retained in soils (Tolls 2001). In alkaline solutions, some antibiotics can even form complexes with clay minerals through anion exchange whereby the carboxylic groups of the chemicals directly replace the hydroxyl groups on mineral surfaces. This type of sorption is so strong that a phosphate solution is required to extract the sorbed antibiotics (Sassman and Lee 2007).

Furthermore, sorption of antibiotics to soil minerals is affected by the media pH, ionic strength, and types of exchangeable cations (Pils and Laird 2007; Wang et al. 2012). The media pH influences drug-soil interactions by altering the charges of pharmaceuticals and the cation exchange capacity (CEC) of soils. For example, at pH 5.0 oxytetracycline has zero charges and interacts with organic matter mainly via hydrophobic partitioning; at lower and higher pH, the chemical becomes positively and negatively charged, respectively, and was sorbed to soil minerals mainly via cation exchange and cation bridging, respectively (Kulshrestha et al. 2004). Sassman and Lee (2005, 2007) noticed that the CEC-normalized sorption of tetracyclines and the f_{OC}-normalized sorption of monensin and lasalocid to soils decreased with increasing soil pH in the range of 4.2–7.5. Wang et al. (2008) observed that the presence of Cu^{2+} enhanced sorption of tetracycline to montmorillonite in a wide pH range. Similar to clay minerals, SOM contains various functional groups and is typically negatively charged. Polar pharmaceutical compounds are also sorbed to soil through interactions with SOM via van der Waals force, electric attraction, cation bridging, and anion exchange (MacKay and Canterbury 2005; Gu et al. 2007; Sibley and Pedersen 2008). Agricultural soils are typically low in OC content (i.e., $f_{OC} < 3\ \%$) and therefore, these types of interaction may not be as contributing as with soil minerals.

Considering that most veterinary pharmaceuticals consist of both hydrophobic and hydrophilic moieties (Table 2.1), it can be concluded that sorption of veterinary pharmaceuticals in soils is a result of interactions of the chemicals with soil clay minerals and SOM chiefly through hydrophobic partitioning, electric attraction, and cation bridging. The interactions are determined by the physiochemical nature of the pharmaceuticals and the soils (e.g., the quantity and the type of soil clay and SOM) and are impacted by the soil solution chemistry. As such, Jones et al. (2005) identified soil texture, CEC, and iron oxide content as the most important factors that determined the K_d values of oxytetracycline in 26 OC $< 4\ \%$ soils. Sorption of sulfamethazines in five different mineral soils was influenced by soil OC content, soil surface area, and soil solution pH (Lertpaitoonpan et al. 2009).

2.4.2 Transport of Veterinary Pharmaceuticals in Soil

Once in soil through land application of animal waste, residual veterinary pharmaceuticals can be transported from the manure-applied fields to groundwater through percolation (leaching) and to surface waters through runoff. The chemicals weakly bound to soil materials (small K_d, Table 2.3) are likely to migrate out of the field in runoff water or be leached down in the soil profile by percolation water, whereas those strongly sorbed by soil solids (high K_d, Table 2.3) can move to other locations together with associated soil particles eroded by runoff water (Davis et al. 2006; Kim et al. 2010; Chen et al. 2011). The mobility of antibiotic chemicals in soils is controlled by their sorptivity and life time and influenced by soil solution pH and ionic strength (Rabølle and Spliid 2000; Blackwell et al. 2007; Chen et al. 2011).

Furthermore, sorptive veterinary pharmaceuticals can form associations with colloids and dissolved organic matter (DOM) and migrate readily in the soil profile through preferential flow channels (Zitnick et al. 2011; Ding et al. 2013; Zou and Zheng 2013).

To reduce runoff losses of veterinary chemicals, immediate soil incorporation of land applied animal waste becomes important (Boxall 2008). Runoff losses of antibiotics (e.g., sulfonamides) could be one to two orders of magnitude higher from grassland than from cultivated land receiving surface application of manure slurry (Kreuzig et al. 2005). Moreover, surface applied manure can significantly increase the amount of runoff water from the treated field, likely due to the surface sealing effect of manure particulates (Burkhardt et al. 2005). Surface runoff of veterinary pharmaceuticals from animal waste has spread the chemicals to the general water environment. Lissemore et al. (2006) surveyed seven tributaries of an agricultural watershed in Southern Ontario, Canada for occurrence of veterinary pharmaceuticals in surface water and detected 14 antibiotics in 125 stream samples, with prevalence of lincomycin, monensin, carbamazepine, and sulfamethazine at median concentrations by 44 ng L^{-1}. In Lansing (Michigan, USA) where animal manures from confined livestock feeding operations were extensively applied to cropland, antibiotics were detected in waters from 11 farm drainage tile channels and surface ditches, with amprolium by 288 ng L^{-1} and monensin by 189 ng L^{-1} at approximately 50 % detection frequency of 109 samples (Song et al. 2010).

2.4.3 Transformation of Veterinary Pharmaceuticals in Soils

In agricultural soils, residual veterinary pharmaceuticals introduced by land application of animal manure are subject to abiotic and biotic transformation and degradation. The degradability and degradation pathways vary significantly with veterinary chemicals and the transformation rate is influenced by a number of environmental factors including soil type, soil conditions (temperature, moisture, and oxygen status), manure type, soil-manure ratio, pH, and light (Boxall 2008; Lin and Gan 2011). Reported studies on degradation of veterinary pharmaceuticals in soils are summarized in Table 2.4.

Under general conditions, most animal antibiotics are degradable in soil, especially in the presence of manure waste, with a half-life time <30 days. Certain antibiotics such as roxithromycin, sarafloxacin, and virginiamycin, however, can be persistent and remain largely unchanged in soil over 120 days (Table 2.4). Degradation of veterinary pharmaceuticals in agricultural soils is a comprehensive result of microbial decomposition, organic transformation, oxidation, photolysis, and hydrolysis. Chee-Sanford et al. (2009) reviewed the possible degradation pathways of manure-introduced antibiotics in soil. As water is always present in animal waste and natural soils, hydrolysis may be an important mechanism for animal pharmaceuticals to dissipate in the environment. It is known that the antibiotics β-lactams, macrolides, and sulfonamides are susceptible to hydrolysis

Table 2.4 Reported studies on degradation of veterinary pharmaceuticals in soils

Pharmaceuticals	Methods/degradation conditions	Degraded, (%)	Half life $t_{1/2}$ (day)	References
Ceftiofur sodium	Three agricultural soils (70 % moisture capacity) were fortified with ceftiofur sodium at 10 mg C/50 g soil and incubated at 22 °C for 7 weeks	49–79	22–49	Gilbertson et al. (1990)
Bambermycin Barcitracin Chlortetracycline Erythromycin Tylosin	Spiked individual antibiotics to chicken feces-amended soils at 5.6 mg kg^{-1} and incubated the mixtures for 30 days at 30 °C, 20 °C, and 4 °C, respectively	100, 100, 10 71, 67, 77 56, 12, 0 100, 75, 3 100, 100, 60[a]		Gavalchin and Katz (1994)
Sarafloxacin	Fortified three agricultural soils (50–70 % field capacity with sarafloxacin at 3.4 mg kg^{-1}) and incubated the soils at 22 °C in the dark for 80 days	0.5–0.6 %	Persistent	Marengo et al. (1997)
Virginiamycin	Incubated [14C] virginiamycin at 0.5 mg C/50 g soil with six agricultural soils (pH 5.4–8.2, clay 5–33 %, 50–70 % of field capacity) that were amended with glucose at 9.5 mg C/50 g soil at room temperature for 64 days with daily air flushing	12–40	87–173	Weerasinghe and Towner (1997)
Sulfadimethoxine	A 20 % moisture silt loam was amended with sulfadimethoxine-fortified steer manure and incubated at 25 °C for 70 days	89	3–11	Wang et al. (2006)
Erythromycin Oleandomycin Roxithromycin Salinomycin Tiamulin Tylosin	Spiked a 12 % moisture sandy loam with combined six antibiotics each at 2 mg kg^{-1} and incubated the soil at 20 °C in the dark for 120 days	98 95 25 100 99 100	20 27 Persistent 5 16 8	Schlüsener and Bester (2006)
Sulfamethazine Sulfachloropyridine	Applied sulfonamides to two moist soils at by 100 mg kg^{-1} and incubated the soils at 25 °C in the dark for 40 days	70–83 61–66	19 21	Accinelli et al. (2007)

(continued)

Table 2.4 (continued)

Pharmaceuticals	Methods/degradation conditions	Degraded, (%)	Half life $t_{1/2}$ (day)	References
Oxytetracycline Sulfachloropyridine	200 L of liquid swine manure were fortified with 7.08 g oxytetracycline and 5.24 g sulfachloropyridine and surface applied to a 120-m^2 sandy loam field plot. 127 days	83 98	21–23 3–4	Blackwell et al. (2007)
Tylosin	Incubated 50 mg kg^{-1} tylosin-spiked sandy loam (field capacity) at 20 °C in the dark for 30 days	93	7–8	Hu and Coats (2007)
Monensin Lasalocid	Spiked two contrasting soils (field capacity) with separately monensin and lasalocid at 2 mg kg^{-1} and incubated the soils at 23 °C in the dark for 30 days	100 99	1.2–1.9 1.4–3.6	Sassman and Lee (2007)
Diclofenac[b] Ibuprofen[b] Naproxen[b] Sulfamethoxazole[b]	Spiked a sandy and a loamy moist soil with the compounds each at 40 μg kg^{-1} and incubated the soils at 21 °C in dark under aerobic conditions for 84 days	92–100 96–100 59–97 90–95	4.8–29.6 10.4–15.2 17.4–84.8 9.0–11.4	Lin and Gan (2011)

[a]Degraded at 30, 20, and 4 °C, respectively
[b]Human medicines. Not used for animal treatments

(Huang et al. 2001). Under mild acidic or basic conditions, penicillin G can be readily transformed to penicillinoic acid through hydrolysis (Huang et al. 2001):

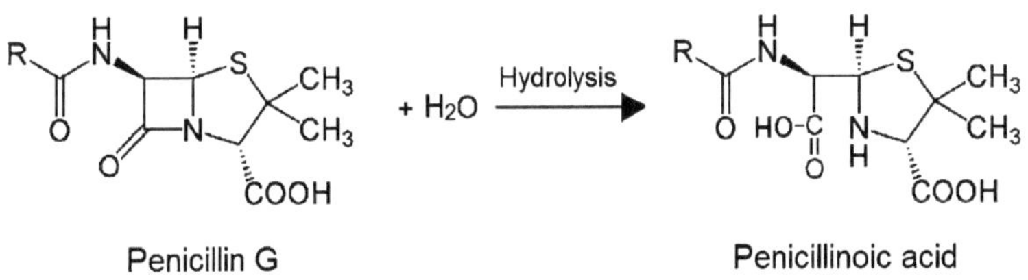

If exposed to daylight, antibiotics may undergo photolysis at the soil-atmosphere interface. Quinolones and tetracyclines are particularly sensitive to photo irradiation

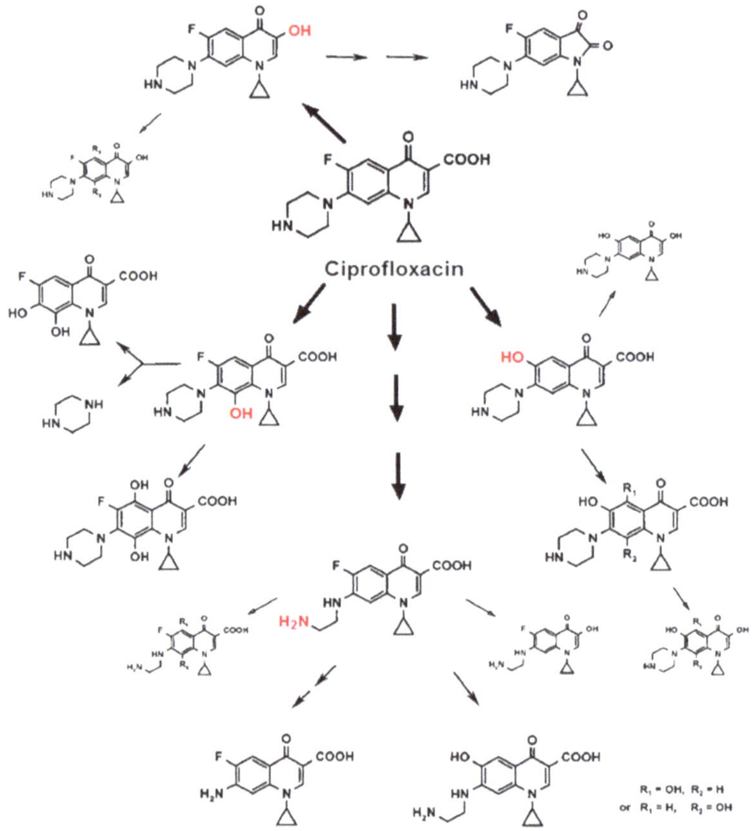

Fig. 2.3 Proposed pathways for the microbial degradation of ciprofloxacin by the fungus *G. striatum* in cultural solutions. Primary hydroxylation at one of the several alternative sites initiates four principal degradation routes. Metabolites identified by HPLC-MS are included at reduced size (Source: Wetzstein et al. 1999)

(Doi and Stoskopf 2000). The photolysis reaction of tetracycline in water under light can be described as (Jiao et al. 2008):

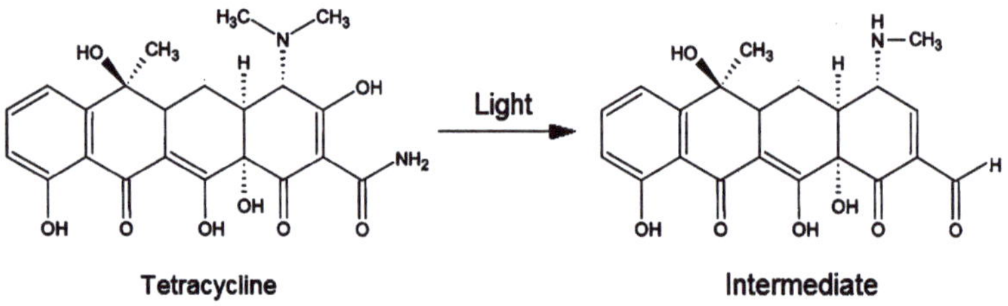

Compared to other reactions, however, photodegradation of antibiotics may be minor under field conditions due to limited light exposure (Beausse 2004). Instead, biodegradation is the major pathway for antibiotics transformation in soil. Many veterinary chemicals are susceptible to enzymatic degradation reactions such as oxidative decarboxylation and hydroxylation (Al-Ahmad et al. 1999). Hydroxylation of the antibiotic ciprofloxacin initiated by the fungus *G. striatum* in a cultural solution lead to further biodegradation of the chemical to a number of metabolites (Fig. 2.3). Microbial addition by inoculating surface water with activated sewage sludge, recycled beef farm water, or lake sediments resulted in significantly accelerated degradation of the inherent veterinary antibiotics (Gartiser et al. 2007; Li et al. 2011), whereas microbial depletion by sterilizing agricultural soils greatly inhibited degradation of various antibiotic chemicals (Accinelli et al. 2007; Lin and Gan 2011).

In addition to biodegradation, chemical processes other than hydrolysis and photolysis are also important for antibiotic transformation in soil. Soil organic matter possesses a variety of functional groups such as amines, carboxyls, carbonyls, ethers, hydroxyls, nitriles, nitrosos, sulfides, and sulfonyls (Swift 1996) and is active in reacting with veterinary chemicals. Hu and Coats (2007) found that in autoclaved soils (OC 1.6 %), tylosin degraded rapidly and demonstrated a half-life of 8 days. Hydrolysis or photolysis could not explain the rapid degradation, as in irradiated water tylosin showed a half-life of 200 days. The same authors also noticed that tylosin degraded in nonsterilized soils at a rate nearly equal to that in the sterilized soils. This does not indicate that biodegradation of tylosin was insignificant. Likely, the spiked antibiotic at 50 mg kg^{-1} soil had inhibited the microbial activity, creating an equivalently sterile environment. The inhibitory effect of antibiotics on the resident microorganisms has been noticed (Gartiser et al. 2007).

Temperature influences degradation of veterinary pharmaceuticals in soils. Gavalchin and Katz (1994) spiked chlortetracycline and erythromycin to chicken feces-amended soils at 5.6 mg kg^{-1} and incubated the soils at different temperatures. After 30 days of 30, 20, and 4 °C incubation, 56 %, 12 %, and 0 % of the spiked chlortetracycline dissipated, respectively and for erythromycin, the levels were 100 %, 75 %, and 3 %, respectively. Li et al. (2011) observed that ceftiofur hydrolyzed to desfuroylceftiofur in deionized water, with a half-life time of 289 days at 15 °C. The half-life time was shortened to 96, 21, and 5 days, respectively, as the hydrolysis temperature increased to 25, 35, and 45 °C. Degradation of veterinary pharmaceuticals is also influenced by soil oxygen availability. Dissipation of diclofenac, ibuprofen, naproxen, sulfamethoxazole, and trimethoprim from two mineral soils under anaerobic conditions was substantially slower than under aerobic conditions (Lin and Gan 2011). Sorption to soil minerals and SOM preserves veterinary antibiotics and enhances their persistence in soils (Zitnick et al. 2011). Smith et al. (1992) reported that microbial utilization of [14C] quinoline from solution was 30 times more rapid than from the bound on surfaces of suspended clay particles. A higher soil moisture content allows more chemicals in the solution phase, enhancing the accessibility to microorganisms. Wang et al. (2006) reported that the half-life time of sulfadimethoxine in a silt loam decreased from 10.4 days to 6.9 days and further to 4.9 days as the soil moisture content was elevated from 15 % to 20 % and additionally to 25 %, respectively. Degradation of veterinary pharmaceuticals is further influenced by soil pH. For example, the half-life time of virginiamycin in different agricultural soils ranged from 87 to 173 days and was negatively correlated to soil pH (Weerasinghe and Towner 1997). In neutral or alkaline solutions, tylosin hydrolyzed to tylosin aldol; at pH <4, however, the main hydrolysis product became desmycosin (Paesen et al. 1995). As organic matter and microorganisms promote antibiotic degradation, amendment of soil with more animal manure typically accelerate the dissipation of veterinary pharmaceuticals (Wang et al. 2006). Clearly, many environmental factors influence the stability and persistence of animal pharmaceuticals in soils and subsequently, affect their transport and potential ecological impacts in agricultural systems.

2.4.4 Uptake and Accumulation of Veterinary Pharmaceuticals in Crop Plants

In soil, residual veterinary pharmaceuticals can be absorbed by plants and accumulated in soil fauna. The uptake and accumulation, however, may vary with organism species and pharmaceutical compounds. Bioaccumulation of antibiotics by plants has been confirmed using *in-vitro* laboratory research. The plants millet (*Panicum miliaceum*), maize (*Zea mays*), and pea (*Pisum sativum*) grown in a culture medium containing 300 mg L^{-1} sulfadimethoxine for 8–18 days showed concentrations of the antibiotic ranging from 178 to 2,070 mg kg^{-1} in their roots and 12.5–110 mg kg^{-1} in their stalk/leaves (Migliore et al. 1995). In a greenhouse trial, barley (*Hordeum distichum L.*) grown in soils spiked with sulfadimethoxine at 109 mg kg^{-1} for 45 days accumulated the antibiotic 79.0 mg kg^{-1} in its roots and 18.2 mg kg^{-1} in its leaves (Migliore et al. 1996). Corn, cabbage, and green onion grown in swine manure-amended soil pots containing chlortetracycline and tylosin each at 0.58–1.58 mg kg^{-1} soil for 3–6 weeks accumulated chlortetracycline in plant tissues at 2–17 μg kg^{-1} fresh weight but not tylosin (Kumar et al. 2005b). Bioaccumulation of sulfamethazine by corn, lettuce, and potato from swine slurry-fertilized, 1.25–2.50 mg kg^{-1} sulfamethazine-fortified potting soils ranged from 0.1–1.2 mg kg^{-1} dry weight in above-ground plant tissues (Dolliver et al. 2007). Willow and maize grown in greenhouse potting soils spiked with 10 mg kg^{-1} sulfadiazine for 40 days showed presence of the chemical in the roots at 333 and 26.5 mg kg^{-1} dry weight, respectively, but not in the above-ground tissues (Michelini et al. 2012). Irrigation of soybean in greenhouse pots with water containing the pharmaceuticals carbamazepine, diphenhydramine, and fluoxetine each at 10 μg L^{-1} for 60–110 days resulted in accumulation of carbamazepine but not diphenhydramine and fluoxetine in plant roots and leaves at 1.9–3.4 μg kg^{-1} dry weight (Wu et al. 2010). As the concentrations of residual pharmaceuticals in agricultural soils receiving animal manures would be far lower compared with the levels tested in laboratory and greenhouse research, bioaccumulation of animal drugs in food crops, if there any, should be rather insignificant. So far it is not clear that bioaccumulation of veterinary chemicals in field crops poses health hazard to consumers of tainted plants.

2.4.5 *Concentrations of Veterinary Pharmaceuticals in Manured Agricultural Soils*

A few studies reported on low concentrations of veterinary pharmaceuticals detected in soils of cropland repeatedly receiving animal manure applications. The top 30-cm soils collected from a crop field immediately after cattle manure application at 96 ton ha^{-1} demonstrated concentrations of oxytetracycline 6–7 μg kg^{-1} and tylosin below the detection limit (De Liguoro et al. 2003). Hamscher et al. (2002, 2005) determined residual antibiotics in agricultural fields with sandy soils that were fertilized annually with livestock manure slurry at 30–50 m^3 ha^{-1} for many years in Northern Germany and detected 43–199 μg kg^{-1} tetracycline, 3.7–7.3 μg kg^{-1} chlortetracycline, but not any oxytetracycline, sulfamethazine, or tylosin in the top 30 cm soils. No antibiotics were present in deeper soils and only sulfamethazine was detected at 0.05–0.24 μg L^{-1} in groundwater 140 cm below the surface. In Lansing (Michigan, USA) MI where concentrated livestock feeding operations exist, 0.03–0.26 μg kg^{-1} amprolium and 0.004–0.50 μg kg^{-1} monensin were detected in the top 10-cm sandy loam soils of the agricultural land (Song et al. 2010). Due to attenuation and degradation, veterinary pharmaceuticals generally have rather low concentrations (e.g., <10 μg kg^{-1}) and are mostly undetectable in soils of agricultural land receiving animal manure applications. So far contamination of food crops by plant absorption of the residual chemicals in manured agricultural soils has not been reported.

2.5 Ecological Impacts of Residual Veterinary Pharmaceuticals from Manures

Pharmaceutical antibiotics affect microorganisms even at low concentrations. The effects and effective doses vary with exposure time, microorganism species, and drug chemicals. The half maximal effective concentration (EC_{50}, the effective concentration that causes toxicity to 50 % of the test population) of chlortetracycline is 2.2 mg L^{-1} for sewage sludge bacteria and that of tylosin is 54.7 mg L^{-1} (Halling-Sørensen 2001). The EC_{50} of monensin for soil respiration is 176 mg kg^{-1} (Thiele-Bruhn 2003). Nevertheless, species of soil fauna such as earthworms, springtails, and enchytraeid worms are not influenced by antibiotics even at concentrations >100 mg kg^{-1} (Bauger et al. 2000). The presence of chlotetracycline and oxytetracycline at 160 mg kg^{-1} in loamy soils stimulated the nutrient uptake and promoted the growth of radish, wheat, and corn, but had little influence on bean growth (Batchelder 1982). In a loamy soil containing 109 mg kg^{-1} sulfadimethoxine, the growth of barley was slightly inhibited (Migliore et al. 1996). Sulfadimethoxine at 300 mg L^{-1} in a culture solution also restricted the growth of roots, hypocotyls, and leaves of pea, corn, and millet (Migliore et al. 1995). The antibiotics oxytetracycline and chlortetracycline at 5.3 and 11.3 mg kg^{-1}, respectively, in beef cattle feces applied to a sandy loam at 18.8 dry ton ha^{-1} enhanced soil respiration but did not influence the growth of corn seedlings (Patten et al. 1980). The fact is that land application of animal waste rarely results in residual antibiotics exceeding 0.2 mg kg^{-1} in agricultural soils. These results suggest that at environmentally relevant concentrations, residual veterinary pharmaceuticals in manure-fertilized soils will not adversely affect soil microorganisms, fauna, and plants.

Residual veterinary pharmaceuticals may be transported to surface waters from manure-applied cropland. In surface ditches and drainage tile channels surrounding the cropland that repeatedly receives animal manure, an array of antibiotics have been detected in the water. Streams in agricultural watersheds also show the trace presence of veterinary pharmaceuticals. Fourteen antibiotics have been detected at concentrations by 44 ng L^{-1} in seven tributaries of an agricultural watershed in Southern Ontario, Canada (Lissemore et al. 2006). A reconnaissance of 139 streams in the U.S. main continent detected 22 veterinary antibiotics in water at 0.014–0.10 μg L^{-1} (Kolpin et al. 2002). Laboratory studies indicate that aquatic plants are the non-target organisms most sensitive to antibiotic chemicals. Of the 25 tested antibiotics of different chemical classes, lomefloxacin, sulfamethoxine, and chlortetracycline are most phytotoxic to duckweed (*Lemna gibba*), with EC_{25} values of 38, 37, and 114 mg L^{-1}, respectively (Brain et al. 2004). At low concentrations (e.g., 0.3–3 mg L^{-1}) in water, antibiotics may enhance the growth of aquatic macrophytes (Solomon et al. 2010). In aquatic microcosm trials, effects of multiple antibiotics at concentrations below 0.22 μM were not observed on zooplankton, phytoplankton, and fishes (Richards et al. 2004; Wilson et al. 2004). The no-observed-effect concentrations of 15 tested veterinary pharmaceuticals in water to zebrafish (*Danio rerio*) embryos were around 0.02 mg L^{-1} (Carlsson et al. 2013). This value of the parasiticide ivermectin to zooplankton was measured at 0.1 μg L^{-1} (Boonstra et al. 2011). These no-observed-effect concentrations, however, were 40–100 times higher than maximum environmentally-relevant concentrations. Adverse exposure to veterinary pharmaceuticals in natural waters, if there is any, would be arid and semi-arid headwater streams adjacent to concentrated, confined livestock feeding operations (Brooks et al. 2006).

There is great concern on development and spread of antimicrobial resistance through heavy use and dispersion of veterinary pharmaceuticals in the environment. Common genes may mutate in the presence of antibiotics and resistance genes can transfer among diverse microorganisms. In animal intestinal guts live up to 10^{14} commensal bacteria of several hundred species (Andremont 2003). Antibiotic-resistant commensal bacteria may be selected every time when an antibiotic drug is administered. The bacteria are then excreted in animal feces and reach the environment through manure storage, handling, and disposal. In stored swine manure, 4–32 % of the bacteria, 71 % of the *Enterococcus fecalis* and 97 % of the *E. coli* were found resistant to at least one of the following antibiotics: tylosin, tetracycline, ampicillin, furatrizine, chloramphenicol, kanamycin, streptomycin, or sulfonamides (Haack and Andrews 2000; Cotta et al. 2003). These microorganisms survive the transition from manure storage lagoon into soil (Boes et al. 2005) and can be transported to surface water and groundwater via runoff and leaching processes (Unc and Goss 2003). Bacteria resistant to kanamycin and neomycin have been found in a coastal plain stream in South Carolina (Left et al. 1993). A strain of superbug methicillin-resistant bacteria *Staphylococcus aureus* associated with antibiotic-fed animals was detected in the noses of livestock-handling workers in North Carolina and Iowa (Harrison et al. 2013; Rinsky et al. 2013). Transport of antibiotic-resistant bacteria and transfer of antibiotic-resistant genes in the environment were well reviewed by Chee-Sanford et al. (2009). Through conjugation (transfer of DNA between a donor and a recipient cell), transduction (bacteriophage-mediated transfer of DNA between bacterial cells), and transformation (transporting exogenous DNA into the cell cytoplasm and integrating the DNA into the recipient genome), antibiotic-resistant genes can be spread between different microbial genera and species (Chee-Sanford et al. 2009). Once antibiotic-resistant pathogens become predominant, the original drugs turn to be ineffective in treating the related infections. Disastrous losses of health and economy may occur if alternative treatment methods are not available. Banning the use of antibiotics in animal feed should root up the risk. To the bottom line, anaerobic digestion or thermophilic composting of animal waste can dramatically reduce the population of antibiotic-resistant microorganisms in manures (Sobsey et al. 2001; Cote et al. 2006).

2.6 Conclusion

The extensive use of veterinary pharmaceuticals, especially antibiotics as feed supplements in domestic animal production has resulted in significant discharge of the chemicals into the environment. In poultry, swine, and cattle manures, the residual contents of more than 50 major antibiotics were detected at 0.01–765 mg kg^{-1} dry manure mass. On average, 50 % (10–90 %) of the administrated veterinary pharmaceuticals were excreted as parent chemicals or bioactive metabolites in animal faeces and therefore, it was estimated that worldwide 3,000–27,000 tons of animal drugs entered into the natural ecosystem every year through manure disposal. In animal manures, pharmaceutical residues undergo microbial degradation and chemical transformation such as photolysis, hydrolysis, and decarboxylation. The half-lives of veterinary pharmaceuticals in animal manures ranged from less than 2 days to longer than 30 days. Composting or simply heaping for 60–90 days prior to manure land application is an effective way to eliminate the pharmaceutical residues. In soils, veterinary pharmaceuticals interact with soil minerals, organic matter, aqueous components, and organisms and are subject to sorption, photohydrolysis, oxidation, and biodegradation. Soil clay minerals and organic matter are both active agents for binding pharmaceuticals through a variety of mechanisms including van der Walls interaction, hydrophobic partitioning, electrostatic attraction, anion exchange, and cation bridging. However, the soil distribution coefficient (K_d) values of animal pharmaceuticals vary dramatically with the chemical species, from 0.3 to 6,300 L kg^{-1}. The persistence of veterinary pharmaceuticals in soils was influenced by a number of environmental factors such as soil type, organic matter content, pH, moisture content, and temperature. Most pharmaceuticals are degradable in soils, demonstrating a half-life <30 days; whereas certain antibiotics such as roxithromycin, sarafloxacin, and virginiamycin were persistent and might remain unchanged in soils over months. In soils repeatedly receiving livestock manures for years, up to 200 μg kg^{-1} of animal antibiotics were detected in the top 30-cm cultivation layer. Veterinary pharmaceuticals at such concentrations in soil would not impose discernible impacts on soil microorganisms, fauna, and plants, though plants and soil fauna could uptake and accumulate veterinary pharmaceuticals. So far contamination of food crops by residual pharmaceuticals in manured soils has not been reported. Even so, soil residual veterinary pharmaceuticals could diffuse from manured fields via surface runoff and deep percolation. Up to 290 ng L^{-1} of animal antibiotics were detected in drainage water from manure-fertilized cropland. In US and Canada, 22 animal antibiotics were detected in rivers and streams at up to 100 ng L^{-1}, though a concentration lower than the no-observed-effect concentrations of veterinary pharmaceuticals to aquatic organisms. The standing concern associated with heavy use of animal pharmaceuticals is development of antimicrobial resistance and spread antibiotic-resistant bacteria through manure disposal. To eliminate the risk, reducing and eventually banning the use of antibiotics in animal feed should be enacted.

2.7 Outlook on Practical Applications

Eliminating residual veterinary pharmaceuticals and antibiotic-resistant pathogens from livestock manure prior to land disposal of the waste is essential. Development of standard composting protocols and other effective methods for sterilizing animal manure should be a focus of future research. Composting is practical for on-farm operations and will also be able to eradicate animal-borne hormones. Prevention of pathogens from spreading from manure storage, handling, and composting facilities via runoff should be reinforced. Liquid from manure lagoons should be disinfected before environmental discharge. Since the long-term ecological impacts of veterinary pharmaceuticals at low concentrations in soil, water, and plants are largely unknown and are difficult to monitor, use of veterinary pharmaceuticals as feed additives should be gradually phased out. Research should advocate and assist in "sustainable, green animal-feed operations" by identifying alternative methods for enhancing animal health and growth.

References

Accinelli C, Koskinen WC, Becker JM, Sadowsky MJ (2007) Environmental fate of two sulfonamide antimicrobial agents in soil. J Agric Food Chem 55:2677–2682

Al-Ahmad A, Daschner FD, Kümmerer K (1999) Biodegradability of cefotiam, ciprofloxacin, meropenem, penicillin G, and sulfamethoxazole and inhibition of waste water bacteria. Arch Environ Contam Toxicol 37:158–163

Allaire SE, Del Castillo J, Juneau V (2006) Sorption kinetics of chlortetracycline and tylosin on sandy loam and heavy clay soils. J Environ Qual 35:969–972

Andremont A (2003) Commensal flora may play key role in spreading antibiotic resistance. ASM News 69:601–607

Batchelder AR (1982) Chlorotetracycline and oxytetracycline effects on plant growth and development in soil systems. J Environ Qual 11:675–678

Bauger AJ, Jensen J, Krogh PH (2000) Effects of the antibiotics oxytetracycline and tylosin on soil fauna. Chemosphere 40:751–757

Beausse J (2004) Selected drugs in solid matrices: a review of environmental determination, occurrence and properties of principal substances. Trends Anal Chem 23:753–761

Blackwell PA, Kay P, Boxall ABA (2007) The dissipation and transport of veterinary antibiotics in a sandy loam soil. Chemosphere 67:292–299

Boes J, Alban L, Bagger J, Mongelmose V, Baggsen DL, Olsen JE (2005) Survival of *Escherichia coli* and *Salmonella typhimurium* in slurry applied to clay soil on a Danish swine farm. Prev Vet Med 69:213–228

Boonstra H, Reichman EP, van den Brink PJ (2011) Effects of the veterinary pharmaceutical ivermectin in indoor aquatic microcosms. Arch Environ Contam Toxicol 60:77–89

Boxall BAB (2008) Fate of veterinary medicines applied to soil. In: Kümmerer K (ed) Pharmaceuticals in the environment: sources, fate, effects and risks. Springer, Berlin/Heidelberg, pp 103–119

Brain RA, Johnson DJ, Richards SM, Sanderson H, Sibley PK, Solomon KR (2004) Effects of 25 pharmaceutical compounds to Lemna gibba using a 7-day static renewal test. Environ Toxicol Chem 23:371–382

Brooks BW, Riley TM, Taylor RD (2006) Water quality of effluent dominated stream ecosystems: ecotoxicological, hydrological, and management considerations. Hydrobiologia 556:365–379

Burkhardt M, Stamm C, Waul C, Singer H, Müller S (2005) Surface runoff and transport of sulfonamide antibiotics and tracers on manured grassland. J Environ Qual 34:1363–1371

Campagnolo ER, Hohnson KR, Karpari A et al (2002) Antimicrobial residues in animal water and water resources proximal to large-scale swine and poultry feeding operations. Sci Total Environ 299:89–95

Carballo EM, Barreiro CG, Scharf S, Gans O (2007) Environmental monitoring study of selected veterinary antibiotics in animal manure and soils in Austria. Environ Pollut 148:570–579

Carlsson G, Patring J, Kreuger J, Norrgren L, Oskarsson A (2013) Toxicity of 15 veterinary pharmaceuticals in zebrafish (*Danio rerio*) embryos. Aquat Toxicol 126:30–41

Chee-Sanford JC, Mackie RI, Koike S et al (2009) Fate and transport of antibiotic residues and antibiotic resistance genes following land application of manure waste. J Environ Qual 38:1086–1108

Chefetz B, Mualem T, Ben-Ari J (2008) Sorption and mobility of pharmaceutical compounds in soil irrigated with reclaimed wastewater. Chemosphere 73:1335–1343

Chen H, Gao B, Li H, Ma LQ (2011) Effects of pH and ionic strength on sulfamethoxazole and ciprofloxacin transport in saturated porous media. J Contam Hydrol 126:29–36

Cote C, Masse DI, Quessy S (2006) Reduction of indicator and pathogenic microorganisms by psychrophilic anaerobic digestion in swine slurries. Bioresour Technol 97:686–691

Cotta MA, Whitehead TR, Zeltwanger RL (2003) Isolation, characterization and comparison of bacteria from swine feces and manure storage pits. Environ Microbiol 5:737–745

Davis JG, Truman CC, Kim SC, Ascough JC, Carlson K (2006) Antibiotic transport via runoff and soil loss. J Environ Qual 35:2250–2260

De Liguoro M, Cibin V, Capolongo F et al (2003) Use of oxytetracycline and tylosin in intensive calf farming: evaluation of transfer to manure and soil. Chemosphere 52:203–212

Dewey CE, Cox BD, Straw BE, Budh EJ, Hurd HS (1997) Association between off-label feed additives and farm size, veterinary consultant use, and animal age. Prev Vet Med 31:133–146

Ding Y, Teppen BJ, Boyd SA, Li H (2013) Measurement of associations of pharmaceuticals with dissolved humic substances using solid phase extraction. Chemosphere 91:314–319

Doi AM, Stoskopf MK (2000) The kinetics of oxytetracycline degradation in deionized water under varying temperature, pH, light, substrate, and organic matter. J Aquat Anim Health 12:246–253

Dolliver H, Kumar K, Gupta S (2007) Sulfamethazine uptake by plants from manure-amended soil. J Environ Qual 36:1224–1230

Dolliver H, Gupta S, Noll S (2008) Antibiotic degradation during manure composting. J Environ Qual 37:1245–1253

Donoho AL (1984) Biochemical studies on the fate of monensin in animals and in the environment. J Anim Sci 58:1528–1539

Europa (2005) Ban on antibiotics as growth promoters in animal feed enters into effect. Europa press release, December 22, 2005. http://europa.eu/rapid/press-release_IP-05-1687_en.htm. Accessed 10 Dec 2012

FDA (2010) 2009 summary report on antimicrobials sold or distributed for use in food-producing animals. Food and Drug Administration, Rockville. www.fda.gov/downloads/ForIndustry/UserFees/AnimalDrugUserFeeActADUFA/UCM231851.pdf. Accessed 5 Mar 2013

FDA (2012) The Green Book – FDA approved products. Food and Drug Administration, Rockville. http://www.accessdata.fda.gov/scripts/animaldrugsatfda

Furtula V, Huang L, Chambers PA (2009) Determination of veterinary pharmaceuticals in poultry litter and soil by methanol extraction and liquid chromatography-tandem mass spectrometry. J Environ Sci Health B 44:717–723

Furtula V, Farrell EG, Diarrassouba F et al (2010) Veterinary pharmaceuticals and antibiotic resistance of *Escherichia coli* isolates in poultry litter from commercial farms and controlled feeding trials. Poult Sci 89:180–189

Gao J, Pedersen JA (2005) Adsorption of sulfonamide antimicrobial agents to clay minerals. Environ Sci Technol 39:9509–9516

Garrido Frenich A, Plaza-Bolanos P, Aguilera-Luiz MM, Martinez-Vidal JL (2010) Veterinary drugs and growth-promoting agent analyses. Nova Science, Hauppauge

Gartiser S, Urich E, Alexy R, Kümmerer K (2007) Ultimate biodegradation and elimination of antibiotics in inherent tests. Chemosphere 67:604–613

Gaskins HR, Collier CT, Anderson DB (2002) Antibiotics as growth promotants: mode of action. Anim Biotechnol 13:29–42

Gavalchin J, Katz SE (1994) The persistence of fecal-borne antibiotics in soil. J AOAC Int 77:481–485

Gilbertson TJ, Hornish RE, Jaglan PS et al (1990) Environmental fate of ceftiofur sodium, a cephalosporin antibiotic. Role of animal excreta in its decomposition. J Agric Food Chem 38:890–894

Gruber VF, Halley BA, Hwang SC, Ku CC (1990) Mobility of avermectin B1a in soil. J Agric Food Chem 38:886–890

Gu C, Karthikeyan KG, Sibley SD, Pedersen JA (2007) Complexation of the antibiotic tetracycline with humic acid. Chemosphere 66:1494–1501

Haack BJ, Andrews RE (2000) Isolation of Tn916-like conjugal elements from swine lot effluent. Can J Microbiol 46:542–549

Haller MY, Müller SR, McArdell CS, Alder AC, Suter MJ (2002) Quantification of veterinary antibiotics (sulfonamides and trimethoprim) in animal manure by liquid chromatography–mass spectrometry. J Chromatogr A 952:111–120

Halling-Sørensen B (2001) Inhibition of aerobic growth and nitrification of bacteria in sewage sludge by antibacterial agents. Arch Environ Contam Toxicol 40:451–460

Halling-Sørensen B, Nors Nielsen S, Lanzky PF et al (1998) Occurrence, fate and effects of pharmaceutical substances in the environment – a review. Chemosphere 36:357–393

Hamscher G, Sczesny S, Höper H, Nau H (2002) Determination of persistent tetracycline residues in soil fertilized with liquid manure by high-performance liquid chromatography with electrospray ionization tandem mass spectrometry. Anal Chem 74:1509–1518

Hamscher G, Pawelzick HT, Höper H, Nau H (2005) Different behavior of tetracyclines and sulfonamides in sandy soils after repeated fertilization with liquid manure. Environ Toxicol Chem 24:861–868

Harrison EM, Paterson GK, Holden MTG et al (2013) Whole genome sequencing identifies zoonotic transmission of MRSA isolates with the novel *mecA* homologue *mecC*. EMBO Mol Med 5:509–515

Hayes JR, English LL, Carr LE, Wagner DD, Joseph SW (2004) Multiple-antibiotic resistance of *Enterococcus* spp. isolated from commercial poultry production environments. Appl Environ Microbiol 70:6005–6011

Henderson KD, Coats JR (2010) Veterinary pharmaceuticals in the environment: an introduction. In: Henderson KD, Coats JR (eds) Veterinary pharmaceuticals in the environment, vol 1018, ACS symposium series. American Chemical Society, Washington, DC, pp 3–7

Holt JP (2008) Growth performance and the development of antibiotic resistant bacteria in swine fed growth-promoting antimicrobials. PhD thesis, North Carolina State University, Raleigh

Hu D, Coats JR (2007) Aerobic degradation and photolysis of tylosin in water and soil. Environ Toxicol Chem 26:884–889

Huang CH, Renew JE, Smeby KL et al (2001) Assessment of potential antibiotic contaminants in water and preliminary occurrence analysis. In: Proceedings of the 2nd international conference on pharmaceuticals and endocrine disrupting chemicals in water. Groundwater Association, Westerville, pp 46–57

Jiao S, Zheng S, Yin D, Wang L, Chen L (2008) Aqueous photolysis of tetracycline and toxicity of photolytic products to luminescent bacteria. Chemosphere 73:377–382

Jones AD, Bruland GL, Agrawal SG, Vasudevan D (2005) Factors influencing the sorption of oxytetracycline to soils. Environ Toxicol Chem 24:761–770

Karci A, Balcioglu IA (2009) Investigation of the tetracycline, sulfonamide, and fluoroquinolone antimicrobial compounds in animal manure and agricultural soils in Turkey. Sci Total Environ 407:4652–4664

Kim SC, Davis JG, Truman CC, Ascough Ii JC, Carlson K (2010) Simulated rainfall study for transport of veterinary antibiotics – mass balance analysis. J Hazard Mater 175:836–843

Kishi H, Kogure N, Hashimoto Y (1990) Contribution of soil constituents in adsorption coefficient of aromatic compounds, halogenated alicyclic and aromatic compounds to soil. Chemosphere 21:867–876

Kolpin DW, Furlong ET, Meyer MT et al (2002) Pharmaceuticals, hormones and other organic wastewater contaminants in US streams, 1999–2000: a national reconnaissance. Environ Sci Technol 36:1202–1211

Kreuzig R, Holtge S, Brunotte J et al (2005) Test-plot studies on runoff of sulfonamides from manured soils after sprinkler irrigation. Environ Toxicol Chem 24:777–781

Kulshrestha P, Giese RF, Aga DS (2004) Investigating the molecular interactions of oxytetracycline in clay and organic matter: insights on factors affecting its mobility in soil. Environ Sci Technol 38:4097–4105

Kumar K, Thompson A, Singh AK, Chander Y, Gupta SC (2004) Enzyme-linked immunosorbent assay for ultratrace determination of antibiotics in aqueous samples. J Environ Qual 33:250–256

Kumar K, Gupta SC, Baidoo SK, Chander Y, Rosen CJ (2005a) Antibiotic uptake by plants from soil fertilized with animal manure. J Environ Qual 34:2082–2085

Kumar K, Gupta SC, Chander Y, Singh AK (2005b) Antibiotic use in agriculture and its impact on the terrestrial environment. Adv Agron 87:1–54

Left LG, Dana JR, McArthur JV, Shimkets LJ (1993) Detection of Tn5-1ike sequences in Kanamycin-resistant stream bacteria and environmental DNA. Appl Environ Microbiol 59:417–421

Lertpaitoonpan W, Ong SK, Moorman TB (2009) Effect of organic carbon and pH on soil sorption of sulfamethazine. Chemosphere 76:558–564

Li XL, Zheng W, Machesky ML, Yates SR, Katterhenry M (2011) Degradation kinetics and mechanism of antibiotic ceftiofur in recycled water derived from a beef farm. J Agric Food Chem 59:10176–10181

Li Y, Zhang X, Li W, Lu X, Liu B, Wang J (2013) The residues and environmental risks of multiple veterinary antibiotics in animal faeces. Environ Monit Assess 185:2211–2220

Lin K, Gan J (2011) Sorption and degradation of wastewater-associated non-steroidal anti-inflammatory drugs and antibiotics in soils. Chemosphere 83:240–246

Lissemore L, Hao C, Yang P et al (2006) An exposure assessment for selected pharmaceuticals within a watershed in southern Ontario. Chemosphere 64:717–729

MacKay AA, Canterbury B (2005) Oxytetracycline sorption to organic matter by metal-bridging. J Environ Qual 34:1964–1971

Marengo JR, Kok RA, O'Brien K et al (1997) Aerobic biodegradation of (14C)-Sarafloxacin hydrochloride in soil. Environ Toxicol Chem 16:462–471

McEwen SA, Fedorka-Cray PJ (2002) Antimicrobial use and resistance in animals. Clin Infect Dis 34:S93–S106

Michelini L, Reichel R, Werner W et al (2012) Sulfadiazine uptake and effects on *Salix fragilis L.* and *Zea mays L.* plants. Water Air Soil Pollut 223:5243–5257

Migliore L, Brambilla G, Cozzolina S, Gaudio L (1995) Effect on plants of sulphadimethoxine used in intensive farming (*Panicum miliaceum*, *Pisumm sativum* and *Zea mays*). Agric Ecosyst Environ 52:103–110

Migliore L, Brambilla G, Casoria P et al (1996) Effect of sulphadimethoxine contamination on barley (*Hordeum distichum* L., Poaceae, Liliopsida). Agric Ecosyst Environ 60:121–128

Nowara A, Burhenne J, Spiteller M (1997) Binding of fluoroquinolone carboxylic acid derivatives to clay minerals. J Agric Food Chem 45:1459–1463

OTA (1979) Drugs in livestock feed. NTIS order #PB-298450. Office of Technology Assessment, Congress of the United States, Washington, DC

Paesen J, Cypers W, Busson R et al (1995) Isolation of decomposition products of tylosin using liquid chromatography. J Chromatogr A 699:99–106

Pan X, Qiang Z, Ben W, Chen M (2011) Residual veterinary antibiotics in swine manure from concentrated animal feeding operations in Shandong Province, China. Chemosphere 84:695–700

Patten DK, Wolf DC, Kunkle WE, Douglass LW (1980) Effects of antibiotics in beef cattle feces on nitrogen and carbon mineralization in soil and plant growth and composition. J Environ Qual 9:167–172

Pils JRV, Laird DA (2007) Sorption of tetracycline and chlortetracycline on K- and Ca-saturated soil clays, humic substances, and clay–humic complexes. Environ Sci Technol 41:1928–1933

Rabølle M, Spliid NH (2000) Sorption and mobility of metronidazole, olaquindox, oxytetracycline, and tylosin in soil. Chemosphere 40:715–722

Richards SM, Wilson CJ, Johnson DJ et al (2004) Effects of pharmaceutical mixtures in aquatic microcosms. Environ Toxicol Chem 23:1035–1042

Rinsky JL, Naimpalli M, Wing S et al (2013) Livestock-associated methicillin and multidrug resistant *Staphylococcus aureus* is present among industrial, not antibiotic-free livestock operation workers in North Carolina. PLoS One 8(7):e67641. doi:10.1371/journal.pone.0067641

Sarmah AK, Meyer MT, Boxall ABA (2006) A global perspective on the use, sales, exposure pathways, occurrence, fate and effects of veterinary antibiotics (VAs) in the environment. Chemosphere 65:725–759

Sassman SA, Lee LS (2005) Sorption of three tetracyclines by several soils: assessing the role of pH and cation exchange. Environ Sci Technol 39:7452–7459

Sassman SA, Lee LS (2007) Sorption and degradation in soils of veterinary ionophore antibiotics: monensin and lasalocid. Environ Toxicol Chem 26:1614–1621

Schlüsener MP, Bester K (2006) Persistence of antibiotics such as macrolides, tiamulin and salinomycin in soil. Environ Pollut 143:565–571
Sengeløv G, Halling-Sørensen AB, Baloda SB et al (2003) Bacterial antibiotic resistance levels in Danish farmland as a result of treatment with pig manure slurry. Environ Int 28:587–595
Sibley SD, Pedersen JA (2008) Interaction of the macrolide antimicrobial clarithromycin with dissolved humic acid. Environ Sci Technol 42:422–428
Sithole BB, Guy RD (1987) Models for tetracycline in aquatic environments: I. Interaction with Bentonite clay systems. Water Air Soil Pollut 32:303–314
Smith SC, Ainsworth CC, Traina SJ, Hicks RJ (1992) Effect of sorption on the biodegradation of quinoline. Soil Sci Soc Am J 56:737–746
Sobsey MD, Khatib LA, Hill MR et al (2001) Pathogens in animal wastes and the impacts of waste management practices on their survival, transport and fate. White Paper Summaries. www.cals.ncsu.edu/waste_mgt/natlcenter/whitepaperstimmaries/pathogens.pdf. Accessed 29 Dec 2012
Solomon KR, Hillis DG, Lissemore L, Sibley PK (2010) Risks of agricultural pharmaceuticals in surface water systems and soils. In: Henderson K et al (ed) Veterinary pharmaceuticals in the environment, ACS symposium series. American Chemical Society, Washington, DC, pp 191–204
Song W, Ding Y, Chiou CT, Li H (2010) Selected veterinary pharmaceuticals in agricultural water and soil from land application of animal manure. J Environ Qual 39:1211–1217
Swift RS (1996) Organic matter characterization. In: Sparks DL (ed) Methods of soil analysis, Part 3. Chemical methods. Soil Science Society of America, Madison, pp 1011–1069
Thiele-Bruhn S (2003) Pharmaceutical antibiotic compounds in soils: a review. J Plant Nutr Soil Sci 166:145–167
Thiele-Bruhn S, Aust MO (2004) Effects of pig slurry on the sorption of sulfonamide antibiotics in soil. Arch Environ Contam Toxicol 47:31–39
Thiele-Bruhn S, Seibicke T, Schulten HR, Leinweber P (2004) Sorption of sulfonamide pharmaceutical antibiotics on whole soils and particle-size fractions. J Environ Qual 33:1331–1342
Tolls J (2001) Sorption of veterinary pharmaceuticals in soils: a review. Environ Sci Technol 35:3397–3406
Unc A, Goss MJ (2003) Movement of fecal bacteria through the vadose zone. Water Air Soil Pollut 149:327–337
von Nussbaum F, Brands M, Hinzen B et al (2006) Medicinal chemistry of antibacterial natural products – exodus or revival? Angew Chem Int Ed 45:5072–5129
Wang Q, Guo M, Yates SR (2006) Degradation kinetics of manure-derived sulfadimethoxine in amended soil. J Agric Food Chem 54:157–163
Wang YJ, Jia DA, Sun RJ, Zhu HW, Zhou DM (2008) Adsorption and cosorption of tetracycline and copper(II) on montmorillonite as affected by solution pH. Environ Sci Technol 42:3254–3259
Wang C, Teppen BJ, Boyd SA, Li H (2012) Sorption of lincomycin at low concentrations from water by soils. Soil Sci Soc Am J 76:1222–1228
Warman PR, Thomas RL (1981) Chlortetracycline in soil amended with poultry manure. Can J Soil Sci 61:161–163
Weerasinghe CA, Towner D (1997) Aerobic biodegradation of virginiamycin in soil. Environ Toxicol Chem 16:1873–1876
Wetzstein HG, Stadler M, Tichy HV et al (1999) Degradation of ciprofloxacin by basidiomycetes and identification of metabolites generated by the brown rot fungus *Gloeophyllum striatum*. Appl Environ Microbiol 65:1556–1563
Wilson CJ, Brain RA, Sanderson H et al (2004) Structural and functional responses of plankton to a mixture of four tetracyclines in aquatic microcosms. Environ Sci Technol 38:6430–6439
Witte W (2000) Selective pressure by antibiotic use in livestock. Int J Antimicrob Agents 16:S19–S24

Wu CX, Spongberg AL, Witter JD et al (2010) Uptake of pharmaceutical and personal care products by soybean plants from soils applied with biosolids and irrigated with contaminated water. Environ Sci Technol 44:6157–6161

Yeager RL, Halley BA (1990) Sorption/desorption of [14C]efrotomycin with soils. J Agric Food Chem 38:883–886

Zhao L, Dong YH, Wang H (2010) Residues of veterinary antibiotics in manures from feedlot livestock in eight provinces of China. Sci Total Environ 408:1069–1075

Zitnick KK, Shappell NW, Hakk H et al (2011) Effects of liquid swine manure on dissipation of 17beta-estradiol in soil. J Hazard Mater 186:1111–1117

Zou Y, Zheng W (2013) Modelling manure colloid-facilitated transport of the weakly hydrophobic antibiotic florfenicol in saturated soil columns. Environ Sci Technol 47:5185–5192

Chapter 3
Changes in Nutrient Content and Availability During the Slow Pyrolysis of Animal Wastes

Minori Uchimiya

Abstract Although a large number of reports are available on the total and soluble phosphorus, potassium, and nitrogen (and other nutrient elements) content of manure biochars, information is lacking to understand the changes in chemical speciation of different elements during pyrolysis. Manure is intrinsically heterogeneous, and is composed of feces, urine, bedding materials, waste feed (and their degradation products formed during storage) and in some cases soils. Molecular level understandings in thermochemical transformation pathways are necessary to predict the utility of manure biochars as a sterile, renewable, organic fertilizer for different soil types. This critical review will utilize micro/spectroscopic characterization of manure biochars from slow pyrolysis and gasification as well as manure ash to understand the phosphorus speciation. Inorganic (ash) component is enriched with phosphorus and potassium, and can comprise over 50 wt% (on a dry weight basis) of manure biochar. "Ash" is a complex mixture of amorphous, semi-crystalline, and crystalline inorganic phases as well as organic (char) components. These ash components may be originally present in the manure feedstock or thermochemically produced (e.g., apatite). Organic carbon fraction of biochar likely stabilizes ash by various organo-mineral interactions, e.g., by serving as a template for the formation of nano-crystals during pyrolysis. Depending on the thermochemical conversion procedure, stable minerals like hydroxyapatite and whitlockite may form. Much like phosphate rock, acidulation and other pretreatments such as blending with more soluble fertilizer and compost may be necessary to fully utilize less soluble P of manure biochars under alkaline conditions.

M. Uchimiya (✉)
ARS Southern Regional Research Center, United States Department of Agriculture (USDA), New Orleans, LA 70124, USA
e-mail: sophie.uchimiya@ars.usda.gov

Z. He and H. Zhang (eds.), *Applied Manure and Nutrient Chemistry for Sustainable Agriculture and Environment*, DOI 10.1007/978-94-017-8807-6_3,

3.1 Introduction

In the United States, manure is traditionally land-applied. Recent trend towards fewer and larger, concentrated animal feeding operations has resulted in an excessive land-applied manure (Cantrell et al. 2007). Diverse environmental problems arose from pathogens, greenhouse and odorous gas emission, and phosphorus runoff (e.g., Centner and Newton 2008). Slow pyrolysis offers an avenue for mitigating some of these problems and reducing the waste volume prior to land application. Slow pyrolysis of manure produces the solid product called biochar having high phosphorus, potassium, and nitrogen (PKN) content (Cao and Harris 2010; Jha et al. 2010; Singh et al. 2010). Manure-derived biochar can serve as a sterile fertilizer that may release nutrients in a more controlled manner, compared to untreated and composted manures. In addition, stable carbon components of biochar offer a carbon sequestration potential. A large number of reviews and reports have described the agronomic and environmental benefits of manure biochar soil amendment (Lehmann et al. 2003; Chan et al. 2008; Uchimiya et al. 2012).

Biomass is a complex heterogeneous structure composed of cellulose, lignin, and other organic (primarily C, H, N, S, O) as well as inorganic components that can exceed 40 % of dry weight (Vassilev et al. 2010). Inorganic components (SiO_2, Al_2O_3, Fe_2O_3, TiO_2, CaO, K_2O, P_2O_5, MgO, SO_3, Na_2O) may originally exist in biomass or are produced during thermal treatments (Song et al. 2011). The publically accessible North Carolina State University website (NCSU 1994) lists manure characterization methods and results for diverse manure varieties. Compared to plant biomass, manure contains much higher moisture as well as ash composed of alkali metals (30–40 % dry weight) (Vassilev et al. 2010). At an elevated temperature (even below 900 °C), alkali metals such as K react to form alkaline silicates and sulfates (Wang et al. 2008). Alkali silicates and sulfates have melting points below 700 °C and volatilize Na and K (Wang et al. 2008). Compared to lignocellulosic feedstock, animal waste (either manure or litter) is inherently heterogeneous and is composed of feces, urine, bedding materials, waste feed, fermentation products (Miller and Varel 2002; Miller and Varel 2003), and in some cases soil. Dairy and swine manure streams often contain less than 10 % total solid and are typically discharged into open-air lagoons (Cantrell et al. 2008). Drier manures such as poultry litter and cattle feedlot waste contain a mixture of feces, bedding, waste/undigested feed, and urine (Cantrell et al. 2007). Composting is commonly employed for drier (60–80 % total solid) manures (Cantrell et al. 2007). Chemical analyses of poultry litter showed a large fraction of protein and ash, and smaller fraction of cellulose and hemicellulose (Singh et al. 2008). Bedding materials (e.g., wood shavings, sawdust, straw, peat or paper pieces) are used to maintain the floor dry (Wheeler and Smith Zajaczkowski 2002).

Phosphorus is commonly used as a flame retardant to increase the char yields for textiles and wood (Nowakowski et al. 2008). Phosphorus is used to increase the char yield, and catalyzes biomass pyrolysis by allowing chemical reactions to occur at lower pyrolysis temperatures (Nowakowski et al. 2008). Unfortunately, most studies focused on the identification of gas-phase pyrolysis products using GC-MS (Nowakowski et al. 2008), and limited information is available to understand the chemical speciation of P (either naturally present in manure feedstock, or added to lignocellulosic feedstock) in the solid char product. A large number of ultimate analysis and bioassay results have emerged to understand the total/ bioavailable nutrient content of manure biochars (Chan et al. 2008; Cantrell et al. 2012; Wang et al. 2012). However, there is a knowledge gap to understand how high-temperature treatments change the chemical speciation of nutrients. Such fundamental knowledge is necessary to predict the solubility and reactivity of P and other nutrient elements in biochar-amended agricultural soils.

This chapter is a critical review exploring the changes in chemical speciation of nutrient elements within manure as a result of pyrolysis and other thermal conversion technologies. A particular emphasis will be given to phosphorus. Because of limited literature reports on the P speciation of slow pyrolysis biochars, this chapter will partly rely on spectroscopic analyses of gasified and incinerated manures. Implication for manure biochars will be discussed, and recommendations will be given on the critical areas where further investigation is particularly needed.

3.2 Nutrient Composition of Manure Biochars

3.2.1 Total and Bioavailable Nutrient Contents in Manure Biochars and Related Thermal Conversion Products

A large number of reports are available on the total and bioavailable nutrient contents of manure biochars (Chan et al. 2008; Gaskin et al. 2008; Lima et al. 2009; Novak et al. 2009; Cantrell et al. 2012; Enders and Lehmann 2012; Tsai et al. 2012; Uchimiya et al. 2012; Wang et al. 2012; Dai et al. 2013; Xu et al. 2013; Zhao et al. 2013). It must be noted that, depending on the digestion method, nutrient recovery from manure biochars can vary by orders of magnitude (Fig. 3.1) (Enders and Lehmann 2012). Volatilization loss during dry-ashing is particularly problematic for P, K, S, As, Se, and Hg (Enders and Lehmann 2012). After comparing various wet-digestion and dry-ashing methods, Enders and Lehmann (2012) recommended a modified dry-ashing method for reproducibly recovering P, K, S, Ca, Mg, and Zn from biochars to provide the ultimate analysis results. In order to (1) minimize volatilization loss of certain elements and (2) ensure oxidation of recalcitrant C, the modified dry-ashing procedure employed a wet-digestion after the dry-ashing step (Enders and Lehmann 2012). A closed-vessel microwave digestion is another way to minimize the volatilization loss of certain elements (Enders and Lehmann 2012), and has been widely utilized to characterize manure biochars.

As an illustrative example for the pyrolysis temperature-dependent nutrient content of manure biochars, Table 3.1 presents results for dairy, feedlot, poultry, turkey, and swine manure feedstocks (Cantrell et al. 2012). Values in Table 3.1 were

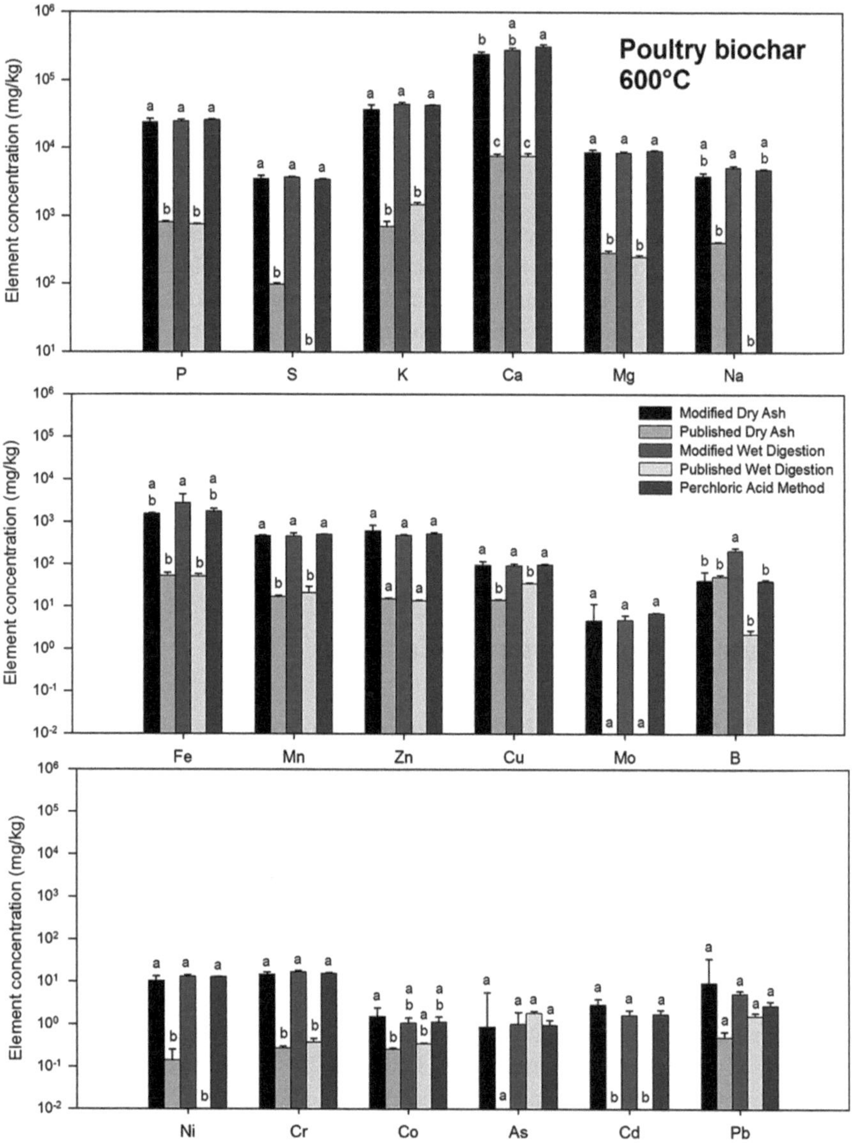

Fig. 3.1 Total elemental contents of poultry biochar produced at 600 °C obtained by different digestion methods (Enders and Lehmann 2012)

obtained by wet acid digestion using concentrated HNO_3 and 30 % H_2O_2, followed by ICP-AES quantification (Cantrell et al. 2012). In agreement with other literature sources (Wang et al. 2012), total P content of biochar (in g kg^{-1} on a dry weight basis) increased as a function of pyrolysis temperature (Table 3.1). Soluble P, in contrast, consistently decreased as a function of pyrolysis temperature (Table 3.1). Of

Table 3.1 Mean mineral analyses of manure and biochars produced at 350 and 700 °C

Feedstock	P	Soluble P	Al	Ca	Fe	K	Mg	Na
Dairy (MD)	5.61 (0.31)	1.47 (0.15)	1.61 (0.21)	16.0 (0.3)	2.29 (0.17)	6.70 (1.4)	6.94 (0.36)	2.51 (1.32)
350	10.0 (0.3)	0.38 (0.05)	2.30 (0.04)	26.7 (1.3)	3.64 (0.16)	14.3 (2.2)	12.2 (0.73)	5.62 (0.85)
700	16.9 (0.5)	0.10 (0.0)	4.92 (0.22)	44.8 (1.1)	6.48 (0.19)	23.1 (0.8)	20.6 (0.4)	8.79 (0.44)
Paved-feedlot (FL)	7.07 (1.9)	3.07 (0.15)	1.31 (0.33)	14.0 (0.3)	1.55 (0.32)	20.2 (0.2)	4.55 (0.3)	2.80 (0.11)
350	11.4 (0.2)	0.46 (0.02)	1.86 (0.08)	22.7 (0.4)	2.26 (0.20)	32.0 (0.1)	7.66 (0.14)	4.88 (0.15)
700	17.6 (0.3)	0.18 (0.2)	3.95 (0.09)	35.0 (0.9)	3.45 (0.19)	49.1 (0.6)	12.2 (0.2)	7.60 (0.14)
Poultry litter (PL)	13.9 (0.6)	2.90 (0.20)	0.400 (0.02)	18.0 (0.7)	0.68 (0.03)	30.5 (2.2)	6.40 (0.61)	9.20 (0.41)
350	20.8 (0.5)	0.43 (0.06)	0.399 (0.01)	26.6 (0.1)	1.32 (0.36)	48.5 (4.0)	9.46 (0.47)	14.8 (0.92)
700	31.2 (0.0)	0.90 (0.14)	0.988 (0.02)	40.2 (0.2)	1.89 (0.13)	74.0 (3.2)	14.5 (0.0)	22.2 (0.51)
Swine solids (SW)	24.7 (0.8)	11.0 (1.3)	0.786 (0.04)	23.9 (1.1)	3.15 (0.42)	10.9 (0.3)	15.0 (0.5)	3.62 (0.15)
350	38.9 (0.4)	0.39 (0.08)	1.17 (0.02)	39.1 (0.5)	4.84 (0.04)	17.8 (0.2)	24.4 (0.4)	5.98 (0.15)
700	59.0 (2.7)	0.06 (0.01)	2.14 (0.02)	61.5 (3.2)	7.48 (0.45)	25.7 (2.5)	36.9 (2.7)	9.35 (0.90)
Turkey litter (TL)	16.1 (1.1)	9.11 (0.48)	1.17 (0.05)	24.1 (3.3)	1.47 (0.24)	25.0 (0.2)	5.27 (0.33)	3.80 (0.16)
350	26.2 (1.1)	1.06 (0.07)	2.14 (0.52)	40.4 (1.1)	2.78 (0.22)	40.1 (4.4)	8.50 (0.39)	6.60 (0.11)
700	36.6 (3.8)	1.75 (1.16)	3.84 (0.33)	56.1 (6.8)	3.65 (0.50)	55.9 (6.8)	12.4 (1.2)	9.24 (0.65)

Table adapted from Cantrell et al. (2012)

All values are on a dry weight basis in g $kg^{-1}{}_{db}$ (standard deviation in *parentheses*) except for Mg in mg $kg^{-1}{}_{db}$. For P, soluble concentration is provided, in addition to the total content particular interest is swine solid biochar that contained lower soluble P especially when pyrolyzed at 700 °C, despite the highest total P content among all manure varieties at both 350 and 700 °C (Table 3.1). Cantrell et al. (2012) attributed the uniquely low P solubility of swine biochars to P-stabilization by the polymer flocculent during pyrolysis. Amongst unpyrolyzed manure feedstocks, swine solids had the highest soluble P (Table 3.1), further indicating the importance of polyacrylamide-phosphate interactions during pyrolysis. Interestingly, polyacrylamide hydrogels have been utilized as growth media to crystallize calcium phosphate (Iijima and Moriwaki 1998; Yokoi et al. 2010a, b).

Animal waste is among the highest ash containing biomass enriched in CaO (57 wt% by high temperature ash analysis for chicken litter), P_2O_5 (15 %), and K_2O (12 %) (Vassilev et al. 2010). Ash in manure contains higher P_2O_5 and lower SiO_2, compared to the plant biomass (Vassilev et al. 2010). Pyrolysis results in the enrichment of some elements in the biochar product, and volatilization loss of the same or other elements. Pyrolysis of mallee tree at 350–750 °C resulted in an enrichment of alkali and alkaline earth metals (AAEM; Na, K, Mg, Ca) and P in biochar, as opposed to a substantial loss of S, N, and Cl (Wu et al. 2011). For wastewater sludge containing 34 % ash, greater pyrolysis temperature (300–700 °C) increased Ca, Fe, Mg, S, Cu, and Zn contents (Hossain et al. 2011). In contrast, Cr, Ni, and Pb contents were increased up to 500 °C, and then decreased by a further temperature increase (500–700 °C), and the greatest loss was observed for Cr (Hossain et al. 2011). Pyrolysis of coal resulted in the volatilization of Na and Cl separately (most likely as carboxylate complexes), and monovalent species (Na) was volatilized more easily than divalent species (Mg, Ca) (Quyn et al. 2002).

In addition to the changes in the total amount of each element, pyrolysis of biomass significantly impacts the chemical speciation and bioavailability of plant nutrients such as P and K (Bridle and Pritchard 2004; Silber et al. 2010). For example, pyrolysis at 450 °C decreased the water soluble P fraction of sewage sludge from >60 to <20 % (Bridle and Pritchard 2004). Calcium-bound P phases (HCl-extractable) increased to >30 % in the char fraction, whereas Fe, Al-bound fractions (NaOH extractable) did not change significantly (Bridle and Pritchard 2004). Fast pyrolysis of cornstraw increased exchangeable and acid-Extractable K, at the expense of water soluble K fractions (Silber et al. 2010). Nearly an order of magnitude greater bioavailable (Colwell; pH 8.5 sodium bicarbonate (Coventry et al. 2001)) P content was observed in poultry litter biochar produced at 450 °C than the steam activated 550 °C biochar (Chan et al. 2008). In contrast, Mehlich I (50 mM HCl + 12.5 mM H_2SO_4) P did not significantly differ for poultry litter biochars produced at 400 and 500 °C (Gaskin et al. 2008). However, for both biochars, steam activation decreased the Mehlich I P concentrations (Gaskin et al. 2008). In a separate report, Mehlich I P was higher for poultry litter biochar formed at 700 °C than 350 °C (Novak et al. 2009). Thermal treatment, e.g., forest fire, converts organic phosphorus to more bioavailable orthophosphate (Cade-Menun et al. 2000; Knicker 2007). Thermal treatment of orthophosphate in turn results in a less soluble polyphosphate species (Cade-Menun et al. 2000). These mineralization, or ashing, processes lead to an increase in available P immediately after a fire event; however this increase is often short-termed and leads to the long-term P loss and reduced forest productivity (Cade-Menun et al. 2000). Acidulation (Steffens 1994) and other pretreatments such as blending with more soluble fertilizer and compost (Nishanth and Biswas 2008) may be necessary to fully utilize less soluble P of manure biochars under alkaline conditions, much like phosphate rock. The following section will utilized micro/spectroscopic characterization reports to understand the chemical speciation of P and other nutrient elements in manure biochars produced by slow pyrolysis. Particular emphasis will be given to x-ray diffraction (XRD), x-ray absorption near edge structure (XANES), and high resolution transmission electron microscope (HETEM) techniques.

3.2.2 XRD Determination of Phosphorus Speciation in Manure Biochars

Mined phosphate rock is the major chemical phosphorus fertilizer worldwide (van Straaten 2002). It is widely accepted that world's reserve of rock phosphate will become depleted in the next several decades (Gilbert 2009). Phosphate rock is composed primarily of apatite, and is commercially graded by the phosphorus pentoxide (P_2O_5) content (van Straaten 2002). Hydroxyapatite ($Ca_{10}(PO_4)_6(OH)_2$; OH may be replaced by F or Cl) is a calcium phosphate mineral. Greater carbonate substitution and lower crystallinity make hydroxyapatite more soluble and bioavailable (van Straaten 2002). Figure 7.20 of Stumm and Morgan (1996) provided the speciation diagram for soluble phosphate, based on the stability constant for various solubility limiting phases in Table 7.5 of the book (Stumm and Morgan 1996). Readers are also referred to Chapter 5.3 of Sposito (1989) for fundamental chemistry controlling phosphate solubility in calcareous soils (Sposito 1989). Phosphate solubility strongly depends on pH and E_h (Sposito 1989; Stumm and Morgan 1996), and under reducing conditions, secondary minerals like vivianite ($Fe_3(PO_4)_2{\cdot}8H_2O$) may form (Peretyazhko and Sposito 2005). XRD is widely employed to characterize P mineralogy in phosphate fertilizers.

Wang et al. (2012) produced biochars from 50:50 (by dry weight) mixture of cattle manure and wood chips at 250, 350, 450, 550 °C by slow pyrolysis with a very short residence time: kiln was allowed to cool to room temperature when the peak temperature was reached (Wang et al. 2012). The XRD spectra of biochars (Fig. 3.2) did not show crystalline P, except for minor peaks attributable to struvite ($MgNH_4PO_4{\cdot}6H_2O$) (Wang et al. 2012). Struvite recovered from biosolids has shown bioavailability equivalent to single superphosphate (Plaza et al. 2007). Standard bioavailable P extraction methods for New Zealand (2 % formic acid), EU (2 % citric acid), and US and Australia (1 M neutral ammonium citrate) showed an increase in absolute P concentration as a function of pyrolysis temperature, concurrently with the increase in total P content of biochar (Wang et al. 2012).

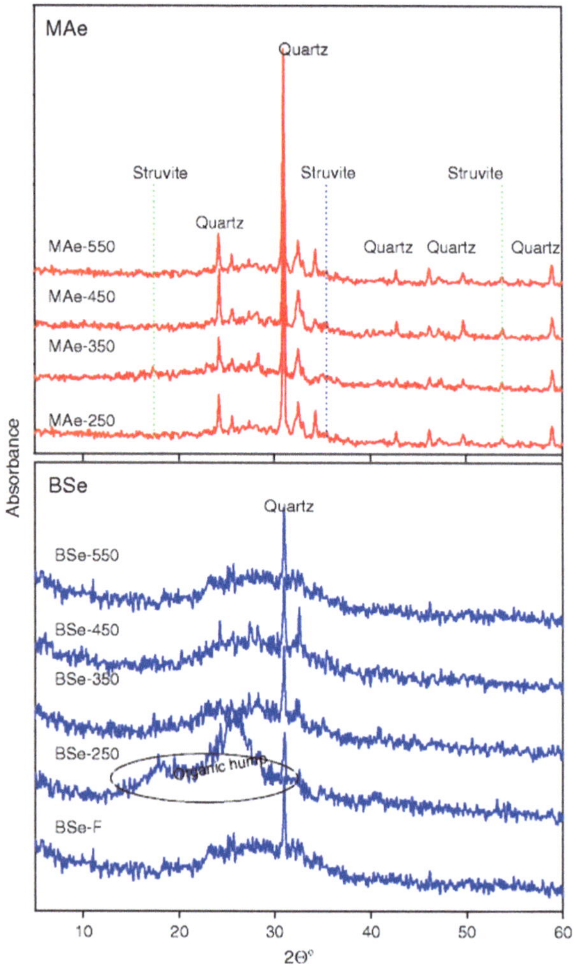

Fig. 3.2 XRD spectra of 250–550 °C biochars and biosolid feedstock (BSe-F). Possible struvite peaks in manure-wood (MAe) biochars were lined out by *dotted lines*; the region in the ellipse in BSe was attributed to "organic hump" (Wang et al. 2012)

Zhao et al. (2013) produced slow pyrolysis biochars at 500 °C (4 h residence time) from 20 different biomass of manure, wood, crop waste, food waste, aquatic plants, and municipal waste origins (Zhao et al. 2013). In contrast to Cantrell et al. (2012) (Table 3.1), swine biochar contained by far the highest water soluble P concentration (16 h shaking at 1:2,000 (w/v) of biochar to water) (Zhao et al. 2013). Unlike poultry and cow manures, swine solids are separated from liquid waste using polyacrylamide as a flocculent (Cantrell et al. 2012). In Zhao et al. (2013), some nutrients (N, K) were higher in non-manure biochars. Water Extractable N was an order of magnitude higher in food waste (bone dregs) and sludge biochars than cow, pig, or chicken manure biochars (Zhao et al. 2013). Water Extractable K was orders of magnitude higher in grass and aquatic plant biochars than manure biochars (Zhao et al. 2013). Limited solubility of P and Ca was attributed to (1) whitlockite ($(Ca, Mg)_3(PO_4)_2$) in manure and food waste biochars and (2) calcite ($CaCO_3$) and calcium oxalate

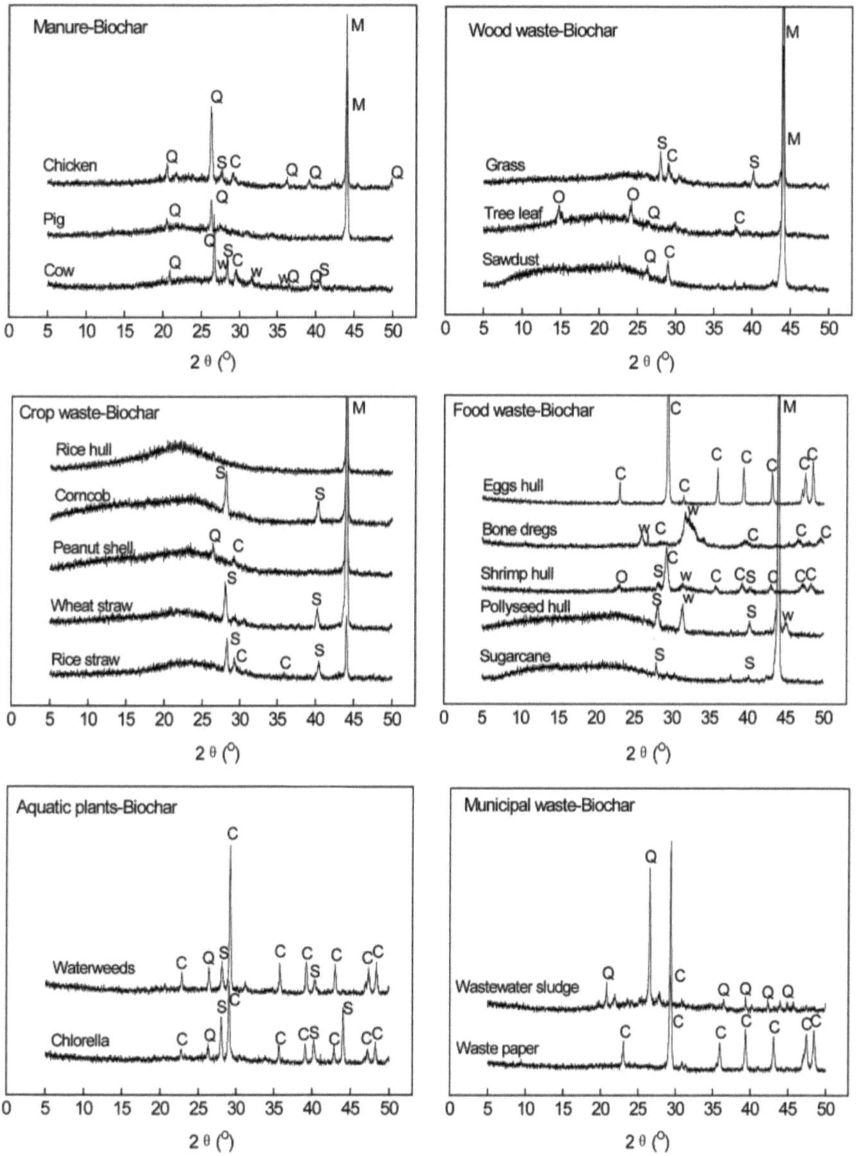

Fig. 3.3 X-ray diffraction patterns of biochars derived from different waste biomasses. *C* calcite, *M* X-ray mount, *O* calcium oxalate, *Q* quartz, *S* sylvite, *W* whitlockite (Zhao et al. 2013)

($Ca(COO)_2$) in most biochars from XRD analyses (Fig. 3.3). Sufficiently high pyrolysis temperature (≥500 °C) is however necessary to form whitlockite from dairy manure (Cao and Harris 2010). High solubility of K was in agreement with the XRD signal for sylvite (KCl) (Fig. 3.3).

XRD study of manure ash from incineration (Herrera et al. 2010; Komiyama et al. 2013) and gasification (Ksawery et al. 2010) provided further insights into the thermal formation of phosphate minerals. In particular, provided comprehensive reviews on ash produced from combustion of manure and other biomass (Vassilev et al. 2013a, b). XRD analyses on gasification by-product (ash) of anaerobically digested pig manure showed hydroxyapatite of varying Ca (20–49 %) and P (8–18 %) contents as the main P phases (Ksawery et al. 2010). Minor presence of quartz, K-feldspar ($KAlSi_3O_8$), and ilmenite was observed (Ksawery et al. 2010). Removal of calcite from ash using pH 5 acetate increased P extractability in water by ten times (Ksawery et al. 2010). Ksawery et al. (2010) reported that apatite in ash had the mean crystal dimension of 40–50 nm. Although microscopic image of this apatite nanocrystal is not available, nano-sized precipitates are expected to be highly reactive, and are ubiquitous in nature (Hochella et al. 2008). Formation of MgO nanocrystals within biochar has been observed for a plant biomass pre-treated with $MgCl_2$ and then pyrolyzed at 600 °C for 1 h (Fig. 3.4) (Zhang et al. 2012). Nano-sized MgO (Fig. 3.4) was deemed responsible for the ability of biochar to sorb phosphate as well as nitrate (Zhang et al. 2012).

Herrera et al. (2010) observed hydroxyapatite in XRD spectra of ashed dairy manure of cows fed with high amount of soluble $CaCl_2$ (Herrera et al. 2010). More soluble P phase, Mg-substituted whitlockite, was observed in ashed manure for cows fed with less soluble $CaCO_3$ (Herrera et al. 2010). Neither hydroxyapatite nor whitlockite was detected in XRD spectra of dried manure (prior to ashing to remove organic matter) (Herrera et al. 2010). Their observation emphasizes the importance of organic matter removal for thermal formation of hydroxyapatite from manure. Above-described literature reviews on XRD analyses indicated the presence of apatite in ashed or gasified manures, but not in manure biochars containing as much as >40 % C by dry weight. It must be noted that compared to manure ash, limited XRD reports are available for manure biochars to confirm the existence of hydroxyapatite and other thermally formed phosphorus minerals.

X-ray absorption near-edge structure (XANES) spectroscopy analyses of poultry manure showed the dominance of soluble CaP species and free or weakly bound phosphate, and the absence of crystalline P minerals (Sato et al. 2005). In a historically (>25 years) manure amended soil, more stable CaP species such as β-tricalcium calcium phosphate were observed; however, crystalline CaP was still not observed (Sato et al. 2005). In a separate report, hydroxyapatite was observed in XANES analysis of turkey manure having high Ca:P ratio (>2) (Toor et al. 2005).

Hydroxyapatite can form microbially (Omelon and Grynpas 2008) and chemically (Sposito 1989) under biogeochemical settings, or thermally. Calcination (heating in air) and other high temperature processing are commonly employed to produce crystalline apatite, especially for the synthesis of biomaterials (Tõnsuaadu et al. 2012). Thermal stability of apatite is controlled by the Ca/P ratio, degree of substitution by carbonate and other ions, and heating atmosphere (Tõnsuaadu et al. 2012). Figure 3.5 summarizes increasing thermal stability from carbonate-substituted apatite, hydroxyapatite, to fluorapatite. As shown in Fig. 3.5, carbonate

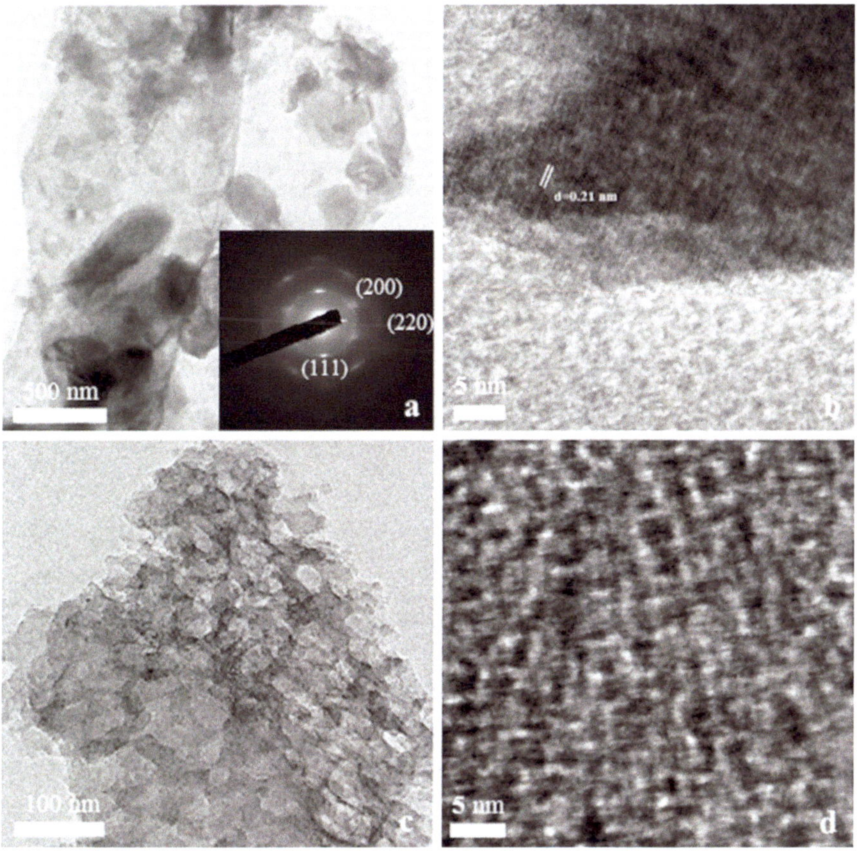

Fig. 3.4 TEM image of MgO-sugar beet tailing biochar nanocomposites (**a**), HR-TEM image of nanosized MgO particles in biochar matrix (**b**), TEM image of the bulk biochar matrix (**c**), and HR-TEM image of the nano-MgO surface (**d**). The inset in (**a**) shows the selected area electron diffraction (*SAED*) pattern (Zhang et al. 2012)

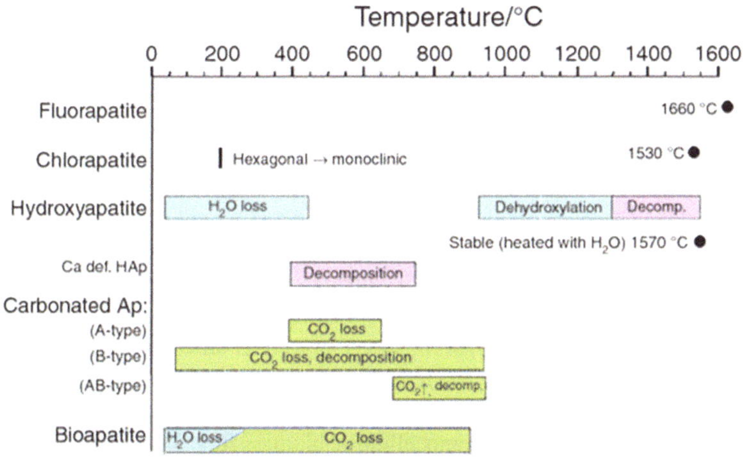

Fig. 3.5 Thermal events for apatites (FAp, ClAp, HAp, CO_3Ap, BioAp) showing a phase change (ClAp), loss of structural groups, decomposition or melting (indicated by a *black circle*) (Tõnsuaadu et al. 2012)

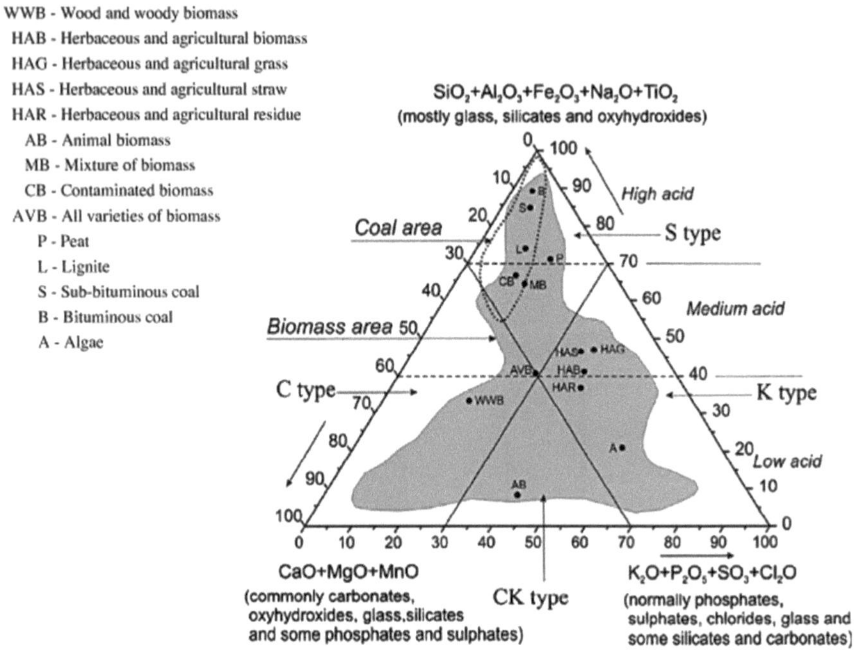

Fig. 3.6 Position areas of 86 biomass varieties and 38 solid fossil fuels in the chemical classification system of biomass ash (Vassilev et al. 2013a)

evolves as CO_2 at 600–1,000 °C to form crystalline apatite minerals. Hydroxyapatite is also the main component of bone char, a pyrolysis product of fish/animal bones (Fuller et al. 2003; Sternitzke et al. 2012). Bone char is a commercially available P fertilizer today. The ability of bone char to sorb Cd and other heavy metals adds benefit to this renewable P fertilizer (Siebers and Leinweber 2013). Similarly, manure biochars are able to bind heavy metals and release soluble phosphorus at the same time (Uchimiya et al. 2012). In addition, liming effects of manure biochar can reduce aluminum phytotoxicity (Qian et al. 2013). Gilbert (2009) emphasized the need for such alternative, renewable P sources to ultimately replace rapidly depleting rock phosphate reserves worldwide (Gilbert 2009).

Manure is enriched with carbonates, phosphates, and other inorganic components that are highly water soluble, reactive, and decompose at lower temperature (Vassilev et al. 2013a). These inorganic (ash) constituents of manure cause agglomeration and fouling during thermochemical conversion processes (Vassilev et al. 2013a). The original mineral components (e.g., quartz) as well as thermal transformation products (e.g., apatite) form manure ash, a complex mixture of amorphous, semi-crystalline, and crystalline inorganic phases as well as un-combusted organic components (char) (Vassilev et al. 2013a). Based on the unique thermal transformation behavior of inorganic components, Vassilev et al. (2013a, b) categorized manure as a “CK type” biomass, apart from lignocellulosic biomass (Fig. 3.6). Table 3 of Vassilev et al. (2013a, b) lists over 20 different phosphate phases found in biomass ashes.

Literature reviews provided above indicated the following controlling factors of phosphorus (and additional nutrient elements) chemical speciation in manure biochars:

1. Removal of organic matter by combustion is likely necessary to thermally form stable CaP minerals such as the crystalline hydroxyapatite phases.
2. Ash (inorganic) component of manure biochar will contain residual organic (char) fraction that will remain even when combusted. Phosphorus and other nutrient elements may exist as organo-mineral complexes that have distinctive release behaviors from purely inorganic minerals. Recalcitrant organic C-mineral complexes may partly explain orders of magnitude discrepancy (depending on the digestion method) in reported nutrient contents of manure biochars (Enders and Lehmann 2012).
3. Pyrolysis of manure will likely produce reactive nanometer-sized precipitates on the biochar surface. These nano-sized crystals can contain P and other nutrient elements that may be released into aqueous phase in a more controlled fashion. Such nanomaterials are reactive sorbents for inorganic (e.g., heavy metals) and organic (e.g., agrochemicals) solutes, and will add further complexity to chemical interactions within rhizosphere that will ultimately influence the crop growths.
4. Speciation of biochar ash (inorganic constituents) is controlled by the speciation of feedstock ash as well as the pyrolysis conditions.

3.3 Conclusion and Recommendation for Future Research

Phosphorus-rich manure is today a widely applied organic fertilizer. Improvements are needed to reduce problems specific to the manure fertilizer. Pyrolysis will thermally decompose anthropogenic pollutants (e.g., veterinary pharmaceuticals) and pathogens, reduce waste volume and odor, and will stabilize carbon. Phosphorus is a known catalyst for biomass pyrolysis, and alters thermochemical conversion pathways. Depending on the pyrolysis temperature, P in manure feedstock may (i) bind with biochar C to form organic P, (ii) crystallize to form inorganic P minerals, or (iii) volatilize and then condense upon cooling to remain on biochar surface as labile P species. Characterization of different P forms in manure biochar is necessary for their utilization as a controlled-release fertilizer.

The following areas need to be explored to fill knowledge gap in evaluating the quality of manure biochars as a renewable, sterile, controlled-release fertilizer:

1. HRTEM imaging to understand the occurrence and fate (solubility, sorption capacity) of nano-sized precipitates within biochar matrices. Spectroscopic characterization (e.g., solid-state NMR and XANES) is necessary to understand P speciation in manure biochars.
2. Long-term field study to evaluate the utility of manure biochars as a replacement for conventional organic and chemical P fertilizers. Manure is typically less

effective than the chemical P fertilizer in the first application season; however, similar crop yield has been observed in the long term (Smith et al. 1998). Similar longer-term effectiveness of biochar P must be explored.

3. Composting is often used to saturate biochars with nutrients, and to biodegrade undesirable volatile matter components prior to soil amendment. Influence of composting and other post-pyrolysis processing (e.g., sizing) must be taken into consideration when biochar is used as a PKN fertilizer.

References

Bridle TR, Pritchard D (2004) Energy and nutrient recovery from sewage sludge via pyrolysis. Water Sci Technol 50:169–175

Cade-Menun BJ, Berch SM, Preston CM, Lavkulich LM (2000) Phosphorus forms and related soil chemistry of Podzolic soils on northern Vancouver Island. II. The effects of clear-cutting and burning. Can J For Res 30:1726–1741

Cantrell K, Ro K, Mahajan D, Anjom M, Hunt PG (2007) Role of thermochemical conversion in livestock waste-to-energy treatments: obstacles and opportunities. Ind Eng Chem Res 46:8918–8927

Cantrell KB, Ducey T, Ro KS, Hunt PG (2008) Livestock waste-to-bioenergy generation opportunities. Bioresour Technol 99:7941–7953

Cantrell KB, Hunt PG, Uchimiya M, Novak JM, Ro KS (2012) Impact of pyrolysis temperature and manure source on physicochemical characteristics of biochar. Bioresour Technol 107:419–428

Cao XD, Harris W (2010) Properties of dairy-manure-derived biochar pertinent to its potential use in remediation. Bioresour Technol 101:5222–5228

Centner TJ, Newton GL (2008) Meeting environmental requirements for the land application of manure. J Anim Sci 86:3228–3234

Chan KY, Van Zwieten L, Meszaros I, Downie A, Joseph S (2008) Using poultry litter biochars as soil amendments. Aust J Soil Res 46:437–444

Coventry JL, Halliwell DJ, Nash DM (2001) The orthophosphate content of bicarbonate soil extracts. Aust J Soil Res 39:415–421

Dai Z, Meng J, Muhammad N, Liu X, Wang H, He Y, Brookes PC, Xu J (2013) The potential feasibility for soil improvement, based on the properties of biochars pyrolyzed from different feedstocks. J Soil Sediment 13:989–1000

Enders A, Lehmann J (2012) Comparison of wet-digestion and dry-ashing methods for total elemental analysis of biochar. Commun Soil Sci Plant Anal 43:1042–1052

Fuller CC, Bargar JR, Davis JA (2003) Molecular-scale characterization of uranium sorption by bone apatite materials for a permeable reactive barrier demonstration. Environ Sci Technol 37:4642–4649

Gaskin JW, Steiner C, Harris K, Das KC, Bibens B (2008) Effect of low-temperature pyrolysis conditions on biochar for agricultural use. Trans ASABE 51:2061–2069

Gilbert N (2009) The disappearing nutrient. Nature 461:716–718

Herrera D, Harris WG, Nair VD, Josan M, Staples CR (2010) Effect of dietary modifications of calcium and magnesium on reducing solubility of phosphorus in feces from lactating dairy cows. J Dairy Sci 93:2598–2611

Hochella MF Jr, Lower SK, Maurice PA, Penn RL, Sahai N, Sparks DL, Twining BS (2008) Nanominerals, mineral nanoparticles, and earth systems. Science 319:1631–1635

Hossain MK, Strezov V, Chan KY, Ziolkowski A, Nelson PF (2011) Influence of pyrolysis temperature on production and nutrient properties of wastewater sludge biochar. J Environ Manag 92:223–228

Iijima M, Moriwaki Y (1998) Lengthwise and oriented growth of octacalcium phosphate crystal in polyacrylamide gel in a model system of tooth enamel apatite formation. J Cryst Growth 194:125–132

Jha P, Biswas AK, Lakaria BL, Rao AS (2010) Biochar in agriculture – prospects and related implications. Curr Sci 99:1218–1225

Knicker H (2007) How does fire affect the nature and stability of soil organic nitrogen and carbon? A review. Biogeochemistry 85:91–118

Komiyama T, Kobayashi A, Yahagi M (2013) The chemical characteristics of ashes from cattle, swine and poultry manure. J Mater Cycles Waste Manag 15:106–110

Ksawery K, John GR, Gorm PT, Baiq Emielda Y (2010) The composition and dissolution in citric extractants of ash from the thermal gasification of pig manure. Chem Eng J 163:1–9

Lehmann J, da Silva JP, Steiner C, Nehls T, Zech W, Glaser B (2003) Nutrient availability and leaching in an archaeological anthrosol and a ferralsol of the central amazon basin: fertilizer, manure and charcoal amendments. Plant Soil 249:343–357

Lima IM, Boateng AA, Klasson KT (2009) Pyrolysis of broiler manure: char and product gas characterization. Ind Eng Chem Res 48:1292–1297

Miller DN, Varel VH (2002) An in vitro study of manure composition on the biochemical origins, composition, and accumulation of odorous compounds in cattle feedlots. J Anim Sci 80:2214–2222

Miller DN, Varel VH (2003) Swine manure composition affects the biochemical origins, composition, and accumulation of odorous compounds. J Anim Sci 81:2131–2138

Nishanth D, Biswas DR (2008) Kinetics of phosphorus and potassium release from rock phosphate and waste mica enriched compost and their effect on yield and nutrient uptake by wheat (Triticum aestivum). Bioresour Technol 99:3342–3353

North Carolina Department of Agriculture and Consumer Services (1994) http://www.bae.ncsu.edu. Biological and Agricultural Engineering Department, Agronomic Division, North Carolina State University

Novak JM, Lima I, Xing B, Gaskin JW, Steiner C, Das KC, Watts DW, Busscher WJ, Schomberg H (2009) Characterization of designer biochar produced at different temperatures and their effects on a loamy sand. Ann Environ Sci 3:195–206

Nowakowski DJ, Woodbridge CR, Jones JM (2008) Phosphorus catalysis in the pyrolysis behaviour of biomass. J Anal Appl Pyrolysis 83:197–204

Omelon SJ, Grynpas MD (2008) Relationships between polyphosphate chemistry, biochemistry and apatite biomineralization. Chem Rev 108:4694–4715

Peretyazhko T, Sposito G (2005) Iron(III) reduction and phosphorous solubilization in humid tropical forest soils. Geochim Cosmochim Acta 69:3643–3652

Plaza C, Sanz R, Clemente C, Fernández JM, González R, Polo A, Colmenarejo MF (2007) Greenhouse evaluation of struvite and sludges from municipal wastewater treatment works as phosphorus sources for plants. J Agric Food Chem 55:8206–8212

Qian L, Chen B, Hu D (2013) Effective alleviation of aluminum phytotoxicity by manure-derived biochar. Environ Sci Technol 47:2737–2745

Quyn DM, Wu H, Li CZ (2002) Volatilisation and catalytic effects of alkali and alkaline earth metallic species during the pyrolysis and gasification of Victorian brown coal. Part I. Volatilisation of Na and Cl from a set of NaCl-loaded samples. Fuel 81:143–149

Sato S, Solomon D, Hyland C, Ketterings QM, Lehmann J (2005) Phosphorus speciation in manure and manure-amended soils using XANES spectroscopy. Environ Sci Technol 39:7485–7491

Siebers N, Leinweber P (2013) Bone char: a clean and renewable phosphorus fertilizer with cadmium immobilization capability. J Environ Qual 42:405–411

Silber A, Levkovitch I, Graber ER (2010) pH-dependent mineral release and surface properties of cornstraw biochar: agronomic implications. Environ Sci Technol 44:9318–9323

Singh K, Risse M, Worley J, Das KC, Thompson S (2008) Effect of fractionation on fuel properties of poultry litter. Appl Eng Agric 24:383–388

Singh B, Singh BP, Cowie AL (2010) Characterisation and evaluation of biochars for their application as a soil amendment. Aust J Soil Res 48:516–525

Smith KA, Chalmers AG, Chambers BJ, Christie P (1998) Organic manure phosphorus accumulation, mobility and management. Soil Use Manag 14:154–159
Song G, Shen L, Xiao J (2011) Estimating specific chemical energy of biomass from basic analysis data. Ind Eng Chem Res 50:9758–9766
Sposito G (1989) The chemistry of soils. Oxford University Press, New York
Steffens D (1994) Phosphorus release kinetics and extractable phosphorus after long-term fertilization. Soil Sci Soc Am J 58:1702–1708
Sternitzke V, Kaegi R, Audinot JN, Lewin E, Hering JG, Johnson CA (2012) Uptake of fluoride from aqueous solution on nano-sized hydroxyapatite: examination of a fluoridated surface layer. Environ Sci Technol 46:802–809
Stumm W, Morgan JJ (1996) Aquatic chemistry. Wiley-Interscience, New York
Tõnsuaadu K, Gross KA, Pluduma L, Veiderma M (2012) A review on the thermal stability of calcium apatites. J Therm Anal Calorim 110:647–659
Toor GS, Peak JD, Sims JT (2005) Phosphorus speciation in broiler litter and turkey manure produced from modified diets. J Environ Qual 34:687–697
Tsai WT, Liu SC, Chen HR, Chang YM, Tsai YL (2012) Textural and chemical properties of swine-manure-derived biochar pertinent to its potential use as a soil amendment. Chemosphere 89:198–203
Uchimiya M, Bannon DI, Wartelle LH, Lima IM, Klasson KT (2012) Lead retention by broiler litter biochars in small arms range soil: impact of pyrolysis temperature. J Agric Food Chem 60:5035–5044
van Straaten P (2002) Rocks for crops: agrominerals of sub-Saharan Africa. International Centre for Research in Agroforestry, Nairobi
Vassilev SV, Baxter D, Andersen LK, Vassileva CG (2010) An overview of the chemical composition of biomass. Fuel 89:913–933
Vassilev SV, Baxter D, Andersen LK, Vassileva CG (2013a) An overview of the composition and application of biomass ash. Part 1. Phase-mineral and chemical composition and classification. Fuel 105:40–76
Vassilev SV, Baxter D, Andersen LK, Vassileva CG (2013b) An overview of the composition and application of biomass ash.: Part 2. Potential utilisation, technological and ecological advantages and challenges. Fuel 105:19–39
Wang L, Weller CL, Jones DD, Hanna MA (2008) Contemporary issues in thermal gasification of biomass and its application to electricity and fuel production. Biomass Bioenergy 32:573–581
Wang T, Camps-Arbestain M, Hedley M, Bishop P (2012) Predicting phosphorus bioavailability from high-ash biochars. Plant Soil 357:173–187
Wheeler E, Smith Zajaczkowski J (2002) Horse stable manure management. College of Agricultural Sciences. G-97. Penn State Cooperative Extension, University Park, PA, USA
Wu H, Yip K, Kong Z, Li CZ, Liu D, Yu Y, Gao X (2011) Removal and recycling of inherent inorganic nutrient species in mallee biomass and derived biochars by water leaching. Ind Eng Chem Res 50:12143–12151
Xu X, Cao X, Zhao L (2013) Comparison of rice husk- and dairy manure-derived biochars for simultaneously removing heavy metals from aqueous solutions: role of mineral components in biochars. Chemosphere 92:955–961
Yokoi T, Kawashita M, Kikuta K, Ohtsuki C (2010a) Biomimetic mineralization of calcium phosphate crystals in polyacrylamide hydrogel: effect of concentrations of calcium and phosphate ions on crystalline phases and morphology. Mater Sci Eng 30:154–159
Yokoi T, Kawashita M, Kikuta K, Ohtsuki C (2010b) Crystallization of calcium phosphate in polyacrylamide hydrogels containing phosphate ions. J Cryst Growth 312:2376–2382
Zhang M, Gao B, Yao Y, Xue Y, Inyang M (2012) Synthesis of porous MgO-biochar nanocomposites for removal of phosphate and nitrate from aqueous solutions. Chem Eng J 210:26–32
Zhao L, Cao X, Wang Q, Yang F, Xu S (2013) Mineral constituents profile of biochar derived from diversified waste biomasses: implications for agricultural applications. J Environ Qual 42:545–552

Chapter 4
Soil Amino Compound and Carbohydrate Contents Influenced by Organic Amendments

Zhongqi He, Daniel C. Olk, and Heidi M. Waldrip

Abstract Amino compounds (i.e. amino acids and sugars), and carbohydrates are labile organic components and contribute to the improvement of soil fertility and quality. Animal manure and other organic soil amendments are rich in both amino compounds and carbohydrates, hence organic soil amendments might affect soil processes through these labile components. This chapter first reviews the advances in research on soil concentrations of amino compounds and carbohydrates as impacted by animal manure and other organic amendments. The published papers are mainly on the amino compounds and carbohydrate changes in long-term field trials and laboratory or greenhouse incubations, tracking the fate of amendment-derived amino compounds and carbohydrates in soils. To investigate the amino compound and carbohydrate changes in the short-term (16 weeks), we present a greenhouse pot study (4–16 weeks) which demonstrated that poultry manure increased extractable amino compounds when applied to soil at a high rate, and depressed them at a low application rate. The increase at the high rate was further amplified in rhizospheric soils. Whereas microbial activities promoted carbon (C) and nitrogen (N) mineralization, the presence of ryegrass resulted in lower concentrations of amino compounds and carbohydrates. Finally, we recommend that more types of organic amendments be evaluated for their impacts on soil amino compound and carbohydrate levels, and the relationship in changes between the two types of compounds. Such cumulative knowledge would provide a basis for establishing the specific contributions of amino compounds and carbohydrates to soil N and C dynamics over the course of agricultural seasons and beyond.

Z. He (✉)
Southern Regional Research Center, USDA-ARS, New Orleans, LA 70124, USA
e-mail: zhongqi.he@ars.usda.gov

D.C. Olk
National Laboratory for Agriculture and the Environment, USDA-ARS, Ames, IA 50011, USA

H.M. Waldrip
Conservation and Production Research Laboratory, USDA-ARS, Bushland, TX 79012, USA

Z. He and H. Zhang (eds.), *Applied Manure and Nutrient Chemistry for Sustainable Agriculture and Environment*, DOI 10.1007/978-94-017-8807-6_4,

4.1 Introduction

It is reported that up to 50 % of organic carbon (C) and 80 % of organic nitrogen (N) within soil can exist as amino acids (AA), amino sugars, and carbohydrates (Creamer et al. 2012). Amino acids are the basic structural units of proteins and peptides and serve as substrates for soil enzymes involved in N turnover in soil, are energy sources for soil microorganisms, and are important sources of N for plants. Senwo and Tabatabai (1998) investigated the AA composition of soil organic matter (OM) extracted from ten surface soils and surface soils collected from three long-term cropping systems (continuous corn, corn-soybean-corn-soybean, and corn-oats-meadow) in Iowa, USA. They reported that the total AA extracted from the ten soils was positively correlated with organic C and clay content, but not with total N, pH, or sand content. Expressed as percentages of organic C and N in soils, the amounts extracted ranged from 11 to 32 %, and 32 to 50 %, respectively. The type of crop rotation did not significantly affect the total AA content of soils that received the same N treatment, but it did affect the total AA content of soils from the unamended plots. Werdin-Pfisterer et al. (2009) examined the composition, concentration, and seasonal patterns of soil AAs across a primary successional sequence in interior Alaska that encompassed a natural gradient of plant productivity and soil physicochemical characteristics. They found that the AA pool was dominated by glutamic acid (Glu), glutamine (Gln), aspartic acid (Asp), asparagine (Asn), alanine (Ala), and histidine (His). These six AAs accounted for approximately 80 % of the total AA pool. Their work demonstrates that AAs are important constituents of the biogeochemically diverse soil N pool in the boreal forest of interior Alaska.

Carbohydrates represent a very active organic component of soil and are a readily available energy source for microorganisms (Lima et al. 2009). Soil carbohydrate content influences soil physical characteristics, cation exchange and metal complexing reactions, anion retention, synthesis of humic substances, and biological activity in general. Derrien et al. (2006) measured the dynamics of C in neutral carbohydrates by natural ^{13}C labeling of soils collected from an experimental wheat-maize field at Versailles, France over a 9-year period. They reported that fractions of particulate OM >50 μm were dominated by glucose and xylose, the two main structural carbohydrates of plants. The mean age of sugar-C in these fractions was slightly younger than that of total organic C or the C in plant residue after hydrolysis. In this fraction, xylose-C was younger than glucose-C. The fine fraction (0–50 μm) accounted for 75 % of the soil sugars, which were primarily microbial in origin and consisted of glucose, arabinose, galactose, xylose, fucose and mannose. The mean age of carbohydrate C in the fine fraction was between 60 and 100 years, and was similar to the age of total organic C in this fraction.

Animal manure plays an important role in soil fertility and plant nutrient management. This is especially true where the effective use of nutrient sources is crucial to maintaining productivity, such as in organic and low-input agriculture (He 2011, 2012). Amino acids and amino sugar compounds [hereafter referred to as amino compound(s), AC or ACs] are the major ingredients of organic N in animal manure (He and Mao 2011; He and Olk 2011). This chapter briefly reviews the advances in research on the influence of animal manure and other soil amendments on soil concentrations of ACs and carbohydrates, and then presents a greenhouse study in which ryegrass was grown in a soil with poultry manure applied at the agronomically recommended rate (43 Mg manure ha^{-1}), at an excessive rate (215 Mg manure ha^{-1}), and in unamended soil. Concentrations of ACs and carbohydrates were measured in both the rhizosphere and non-rhizosphere soils in these pots and in similarly amended pots that did not contain growing plants.

4.2 Effects of Organic Amendments on Soil Amino Compounds

At Rothamsted (U.K.), Beavis and Mott (1996) compared AA composition of manured and unmanured soils under continuous wheat. The authors measured 15 well-resolved and reliably identifiable AAs [i.e. lysine (Lys), His, arginine (Arg), Asp, threonine (Thr), serine (Ser), Glu, glycine (Gly), Ala, valine (Val), methionine (Met), isoleucine (Ile), leucine (Leu), tyrosine (Tyr), and phenylalanine (Phe)]. They found that the content of all the measured AAs, with the exception of Ile, Met and Tyr, were higher in farmyard manure treated soil than in untreated soil. Their results showed a measurable distinction in AC distribution pattern between soil treated with farmyard manure and untreated soil. Furthermore, the authors claimed that a treatment effect on soil AC distribution could be measured in the relative proportions (i.e. % of individual AA in total molar AA concentration) after removal of the absolute concentration trends, thereby revealing that differences in soil AA distribution does not simply reflect changes in OM concentrations. Discriminant analysis of the treatment groups, after the effect of absolute AA concentration had been removed, suggested that it may be possible to identify an AA "fingerprint" for classifying soils to reflect land use. Scheller and Raupp (2005) evaluated soil AC content of topsoils in a long-term (15-year) experiment with mineral fertilizer (MIN), composted cattle manure (CM), and composted cattle manure with application of biodynamic preparations (CMBD). The authors measured a total of 18 hydrolyzable AC compounds, not including cysteine (Cys) and tryptophan (Trp). Their data showed that about 42.9–53.7 % of total N in these soils was bound in the 18 measured AC compounds. Apart from the less-abundant His and Met, the difference in AA contents between the mineral treatment and the two manure treatments were statistically significant ($P = 0.05$). Except for Arg and proline (Pro), there was also significant difference between MIN and CM, and between CM and CMBD in AA composition. However, increasing application rates of either mineral fertilizer or composted manure had no clear influences on the 18 AA content levels. These authors (Scheller and Raupp 2005) concluded that treatment differences were due to altered AA metabolism in the soil and were not solely the result of manure AA contribution.

Simpson et al. (1999) reported AA δ^{15} N values of manure and topsoils (15-cm depth) collected from manured and unmanured grassland from an experimental farm in Northumberland (U.K.). The AAs Val, Ala, Leu and Ile were isotopically lighter in unmanured grassland than in manured grassland, with differences between 1.8 and 7.3 ‰. These data demonstrated that a distinction could be made between soil OM derived from manure or from vegetative cover based on δ^{15} N values of the hydrophobic soil AA compounds.

Using δ^{15} N measurement, Bol et al. (2008) examined the long-term dynamics of AA compounds in a bare fallow soil in an experiment that began in 1928 in Versailles, France. The soil was treated with ammonium sulphate, ammonium nitrate, sodium nitrate, or manure (a horse manure and wheat straw mix) as amendments. Bol et al. (2008) measured total C and N, δ^{15} N, and distribution of 18 AA compounds in topsoil (0–25 cm). Their data showed that, compared to the other fertilization regimes, soils that had received long-term manure applications were more enriched in Leu. The variation of Ala δ^{15} N values among treatments was related to source input, with the order of control < N fertilizers < manure. Similar to the previous observation by Simpson et al. (1999), the Ala δ^{15} N value of manured soil was about 2 ‰ higher than unmanured soils. Manure application also tended to increase His δ^{15} N values compared to the other four treatments. The authors attributed this increase to the presence of relatively undecomposed, but enriched in δ^{15} N, plant-derived material in the manure.

A recently improved method of acid hydrolysis and anion chromatography–pulsed amperometry developed by Martens and Loeffelmann (2003) can detect and quantify 18 proteinous AAs, one unusual AA [ornithine (Orn)], and two amino sugars (glucosamine and galactosamine) in manure, soil, and plant residues (Fig. 4.1). Using this method, Olk et al. (2008) observed changes in soil ACs during a 28-day laboratory incubation of nine dairy manure-amended soils from six U.S. states. Immediately after manure application, concentrations for 18 out of 19 ACs were found to numerically increase in soils that had received manure. After 28 days, the concentrations in soil decreased for 16 of the 19 ACs by an average of 11 %, suggesting either modest mineralization of ACs or an increase in chemical stabilization by soil that limited their extractibility. He et al. (2014) determined the content of 21 ACs in 23 poultry litters (PL), 15 soils (three depths) from pastureland that had received PL for 0, 5, 10, 15 and 20 years, and five samples of grass that was grown in the experimental pastureland. Their data (He et al. 2014) indicated that repeated PL application not only contributed to AC-N, but also accelerated AC-N transformations in soil. Grass shoots harvested from the pasture that received PL for 5 years contained the most AC-N (23.7 mg g^{-1} of dry matter). Some studies have proposed that plants can take up intact AAs directly from soil (Rothstein 2009; Scott and Rothstein 2011); however, in this study (He et al. 2014) it was not conclusive from the plant AC-N data that grass took up intact soil AAs.

Using the same method of Martens and Loeffelmann (2003), Creamer et al. (2013) examined the responses of the soil ACs and soil-bound protein-N acquiring enzymes (arylamidase and β-N-acetylglucosaminidase) along a grassland-to-leguminous woodland successional chronosequence in order to

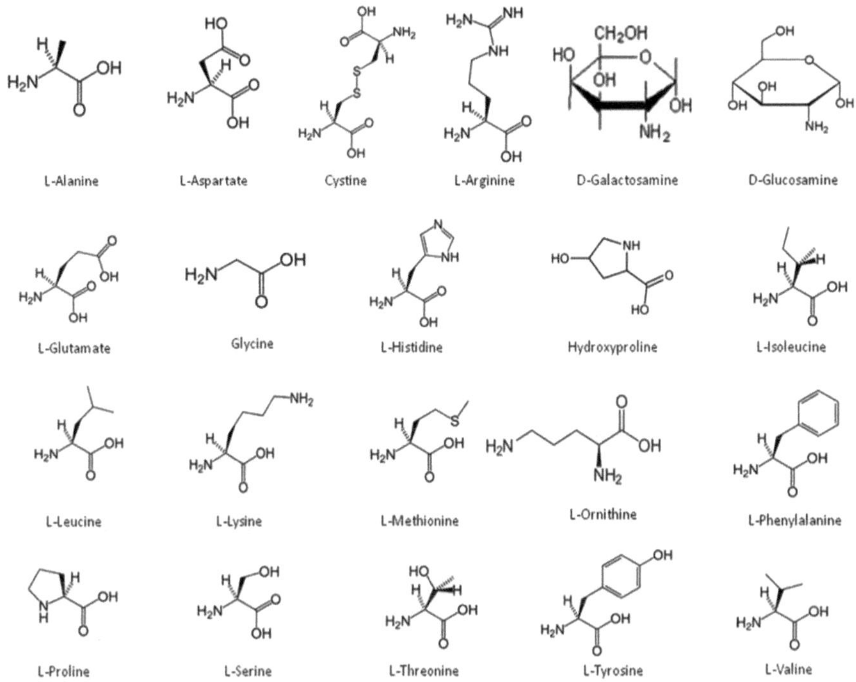

Fig. 4.1 Chemical structures of 21 amino compounds separated and quantified by methanesulfonic acid (*MSA*) extraction and anion chromatography

determine changes to soil N chemistry and extractability. They found that the proportion of total N held within ACs was lower in the woodland soil (47 %) than in grassland soils (62 %). The relative activities of both N-acquiring enzymes were also higher in woodland soils. Therefore, Creamer et al. (2013) concluded that, although encroachment of N-fixing trees and shrubs into grasslands and savannas increases soil C and N, this additional organic N appears to be less extractable by chemical hydrolysis and thus potentially in more refractory forms.

4.3 Effects of Organic Amendments on Soil Carbohydrates

Animal manure is rich in OM, and a significant portion of manure OM is in the form of carbohydrates (He 2011; He and Ohno 2012). Lima et al. (2009) investigated the effects of long-term application of different organic fertilizers (sewage sludge, farmyard manure, compost) on the structure of soil OM. Among the three organic amendments, they observed that farmyard manure increased the concentrations of lignin and lignin-like products in soil OM, whereas compost contributed to an increase of protein and protein-like compounds in addition to carbohydrates. However, according to their results, concentrations of the monosaccharides they measured (rhamnose, xylose, glucose, mannose, arabinose, fucose and galactose) were not significantly influenced by the different organic fertilizers. Mbah et al. (2007) measured soil OM and carbohydrate contents in a Dystric Leptosol under organic fertilizer management in the Abakaliki area of southeastern Nigeria. The materials used for soil amendment were poultry droppings (PD), cow dung (CD), sewage sludge (SS) and swine waste (SW) and were applied at 0, 10 and 20 Mg ha^{-1} for 2 years. The authors observed that soil OM was higher in plots that received organic amendments than in unamended plots by 122–179 % and 166–226 % in the first and second cropping seasons, respectively. The organic amendments increased soil carbohydrate content, but the response differed between amendment types and rate of application. On average, concentrations of dilute acid-soluble carbohydrates in soils were higher than cold water-soluble carbohydrates by 354 % (SS), 489 % (SW), 692 % (CD) and 676 % (PD) in the first planting season. The increase in dilute acid-soluble carbohydrate relative to the hot water soluble carbohydrate in the first season was 9 % (SS), 20 % (SW), 19 % (CD) and 14 % (PD). In the second cropping season increases of 193 % (cold water soluble), 86 % (hot water soluble) and 93 % (dilute acid soluble) were observed relative to the control. However, the authors pointed out that the soil carbohydrate pool was not related to soil aggregation, whereas total soil OM was correlated to aggregate stability.

Dungait et al. (2009) investigated the fate of cow manure-derived carbohydrates in a temperate grassland soil using compound-specific stable isotope analysis. In this field-scale experiment, manure from cows fed a corn silage diet (bulk $\delta^{13}C$ value of -12.6 ‰) was applied to a temperate grassland soil (bulk $\delta^{13}C$ value of -30.3 ‰) in April. In this study, the surface soil directly beneath cow pats was sampled at seven dates over a year (day 7, 14, 28, 56, 112, 224, and 372) and results indicated that manure-derived carbohydrates comprised nearly 60 % of the total C contribution from the manure to the soil at maximum manure C incorporation (56 days) and also after 14 days. For the remaining dates, glucose, xylose, arabinose, galactose and mannose constituted 20–30 % of the manure-derived soil C. In addition, the movement of manure-derived carbohydrates within the soil was inequivalent between the 0–1 and 1–5 cm horizons, as manure-derived xylose and arabinose increased in the 0–1 cm depth at day 224, but was higher in the 1–5 cm depth at day 112.

For carbohydrate analysis, Martens and Loeffelmann (2002) reported use of a modified solublization and autoclave digestion method coupled with ion chromatography. Their optimum solublization conditions were extraction with concentrated H_2SO_4 (18 M) for 15–30 min, followed by autoclave hydrolysis in 1–1.5 M H_2SO_4. With this improved method, these researchers were able to recover 82–97 % of purified cellulose–glucose and a major portion of plant-derived glucose that had not been accounted for when using previously published methods. Application of this methodology to soils resulted in additional release of glucose equivalents if the hemicellulose fraction was first extracted with 1.0 M H_2SO_4 (30 min autoclave digestion). Adopting this improvement, Creamer et al. (2012) analyzed monosaccharides derived from cellulose and hemicellulose in long-term (1-year) laboratory incubated soil samples. This work was done as part of an examination of the changes in soil OM chemistry in response to woody plant encroachment into grass-dominated ecosystems in southern Texas (U.S.). Soils were taken from three landscape scenarios (grassland, younger woody cluster, older woody cluster). Following incubation, these researchers found that concentrations of extractable carbohydrates were about 20 % lower in grassland and younger cluster soils than in soils where woody plants were older and more established. Of the three scenarios, soil from younger woody clusters exhibited the greatest drop in carbohydrate content, shifting from 230 mg carbohydrate-C g^{-1} soil organic carbon (SOC) before incubation (the highest among the three landscape scenarios) to 187 mg carbohydrate-C g^{-1} SOC after incubation (the lowest post-incubation values). Specifically, glucose was the only carbohydrate compound significantly lost during incubation of all samples; however, with the exception of fucose, all other carbohydrates also tended to decrease. Galactose, xylose, and arabinose were lost significantly from the soils from at least one of the landscape scenarios. Galactose was lost from the younger woody cluster and grassland soils, xylose from both younger and older woody cluster soils, and arabinose was lost from younger woody clusters soils. Fucose was the only carbohydrate that increased and, although it increased for all landscape scenarios, the increase was significant only for grassland soils. Based on their observations, the authors concluded that during the 1-year incubation the total carbohydrate C decreased somewhat faster than bulk soil C, with preferential loss of plant-derived carbohydrates and/or production of microbial carbohydrates most strongly expressed in grassland and younger woodland soils.

4.4 Case Study: Soil Concentrations of Amino Compounds and Carbohydrates Following Poultry Manure Application

The purpose of the work was to explore the scientific questions of (1) whether manure application changes the types and amounts of labile compounds in soil, represented by ACs and carbohydrates, and (2) if the chemical nature of manure-bound ACs affect their accumulation and short-term mineralization. This case study is only a preliminary analysis of these data. A full presentation and discussion of the relevant data will be published in a peer-reviewed journal later.

4.4.1 Experimental Procedure

This experiment was conducted in a greenhouse using natural lighting conditions. Background information on this study are found in Waldrip et al. (2011). Briefly, a Bangor silt loam was amended with layer hen manure (PM) at 0, 43 (the agronomically recommended rate, Low-PM), or 215 Mg ha^{-1} (High-PM). Aliquots of 1.5 kg (soil/sand

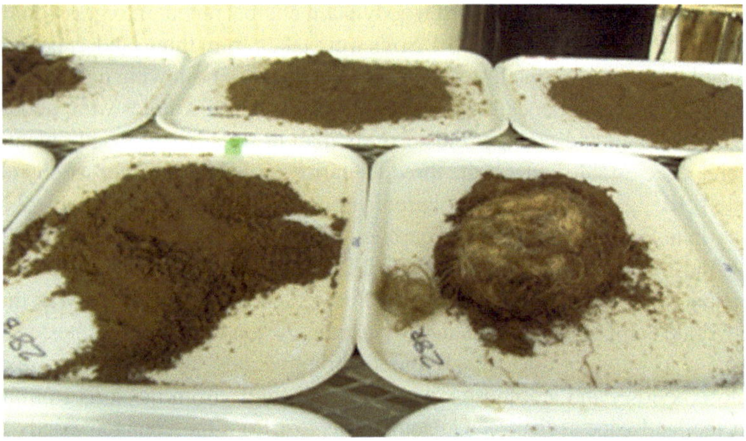

Fig. 4.2 Collection of non-rhizospheric (*left front* and *middle* and *right* on back row) and rhizospheric (*right front*) soils in the study by Waldrip et al. (2011)

plus appropriate manure and chemical amendments) were placed in 6-in. diameter pots, gravimetrically moistened to 60 % water-filled pore space (wfps) with distilled water, and packed to a bulk density of approximately 1.2 Mg m^{-3}. After pre-incubation, 12 pre-germinated ryegrass (*Lolium perenne* L. cv. 'Fiesta 3') seedlings were planted at a depth of ½ inch. The study also included pots that did not contain ryegrass (unplanted) in order to account for changes in soil chemistry and AC content due to the presence of growing plants. Contents of the pots were destructively sampled at 4, 8, and 16 weeks after experiment initiation. There were four replicate pots per treatment at each sampling time. At each sampling, soil and plant material were gently tipped out of the pots and plant material was manually removed from the bulk soil. Rhizosphere soil was considered to be material attached to roots, and was collected with a small brush and by thumb and forefinger manipulation of root and associated soil aggregates (Fig. 4.2). Both bulk and rhizosphere soils were sieved (2 mm) and stored at 4 °C in plastic bags until analysis.

4.4.2 Changes of Amino Compound Contents in Soils Amended with Poultry Manure

The content of ACs were measured using the methods of Martens and Loeffelmann (2003) and (Olk 2008). The distribution of the measured 21 ACs in the three types of soil (non-rhizospheric, rhizospheric, unplanted) after 4 weeks of incubation following the lower rate of PM application are shown in Fig. 4.3. The distribution of the 21 ACs were typical for soils, although the relative abundance of the ACs was not always in the same order as those reported in previous studies (He et al. 2014; Olk et al. 2008). In Fig. 4.3, we also present the assumed AC contents in manured soils based on data

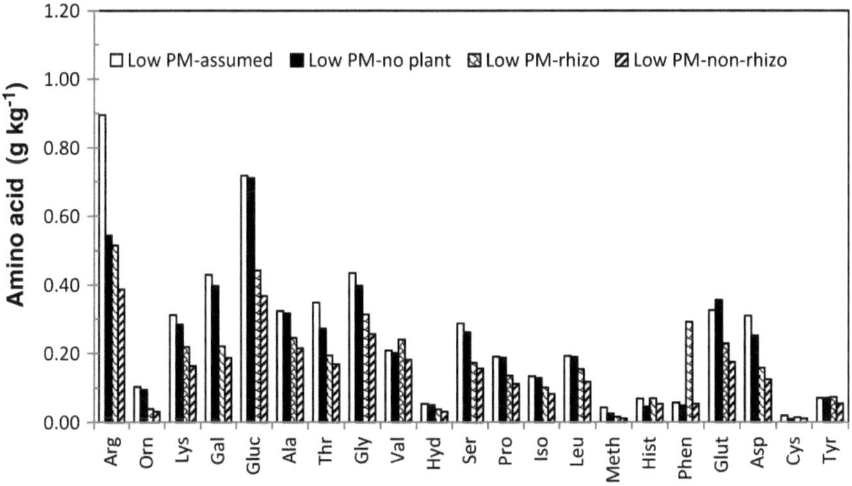

Fig. 4.3 Amino compound (AC) contents in soils that received poultry manure (PM) at the agronomically recommended rate (Low PM) 4 weeks after experiment initiation and ryegrass seedling transplant., Samples include soil with no ryegrass planted (no plant), rhizospheric (rhizo), and non-rhizospheric (non-rhizo) soils. The Low PM-assumed data were calculated based on the AC contents of the non-planted control soil and the AC contents of soil with applied manure

from the unplanted control soil (no manure amendment) and the input from poultry manure. The measured levels of most of the 21 ACs in unplanted manured soils were at the same levels as the theoretically calculated (i.e. assumed) contents. The exceptions were Arg, Thr, and aspartic acid, which were 20–40 % lower than their calculated contents. In both rhizospheric and non-rhizospheric soils, the content of about 15 ACs was lower than in unplanted soils as well as what was theoretically calculated, indicating that growing ryegrass stimulated mineralization of these ACs. There was also a general trend for ACs to be more abundant in rhizospheric soils than in non-rhizospheric soils. This phenomenon might be due to the direct secretion of AC compounds by plant roots. This hypothesis may be also applicable to explaining the unusually higher Phe and Val contents in rhizospheric soils than the other two soil types (non-rhizospheric, unplanted).

These trends with the lower application rate of manure (Low PM) were not reproduced at the higher application rate (High PM) (Fig. 4.4). In the high manured soil, the contents of all AC compounds were lower in soil lacking ryegrass than theoretically calculated values. In fact, the contents of 20 of the 21 ACs were even lower than their corresponding contents in the low manured soil. For example, the greatest observed concentrations were 0.46, 0.60, and 0.34 g AC kg^{-1} soil for Arg, glucosamine, and Gly in the high manured soils, lower than their corresponding contents of 0.54, 0.71, and 0.40 g AC kg^{-1} soil, respectively, in the low manured soils (Fig. 4.3). These data indicated that high manure application rates resulted in more efficient mineralization of native and manure ACs. This is consistent with a decrease in soil C noted by Angers et al. (2010) after 20 years of liquid swine

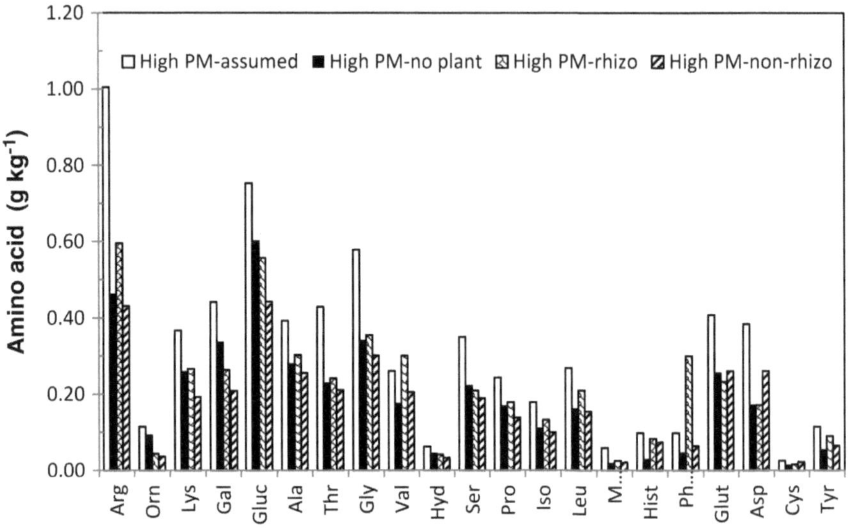

Fig. 4.4 Amino compound (AC) contents in soils that received poultry manure (PM) at a higher rate than agronomically recommended (High PM) 4 weeks after experiment initiation and ryegrass seedling transplant Samples include soil with no ryegrass planted (no plant), rhizospheric (rhizo), and non-rhizospheric (non-rhizo) soils. The High PM-assumed data were calculated based on the AC contents of the non-planted control soil and the AC contents of soil with applied manure

manure application to soil. Whereas the mechanism is unclear, these findings in the current study may be due to the stimulation of OM decomposition following application of manure at a higher rate, likely due to an increase in soluble C, N, and other substrates necessary for microbial growth. At the high rate of manure application, the contents of most ACs in rhizospheric and non-rhizospheric soils were equal to or greater than those in the unplanted soil, implying that no additional AC mineralization occurred in ryegrass-planted soils than in the absence of growing plants. Thus, AC mineralization was not driven by the presence of growing plants in soils that received high rates of manure. Further analyses of the AC contents in soils at 8 and 16 weeks after planting should provide more insight on how the manure application impacted the soil AC profiles.

4.4.3 Changes of Carbohydrate Contents in Soils Amended with Poultry Manure

Seven carbohydrates (fucose, arabinose, rhamnose, galactose, glucose, xylose, and mannose) were measured in the soils from the PM/ryegrass study of Waldrip et al. (2011) using an improved method (Martens and Loeffelmann 2002; Olk 2008). Figure 4.5 presents the total carbohydrate contents in unplanted and planted soils at 4 and 16 weeks after manure incorporation and ryegrass planting. In both

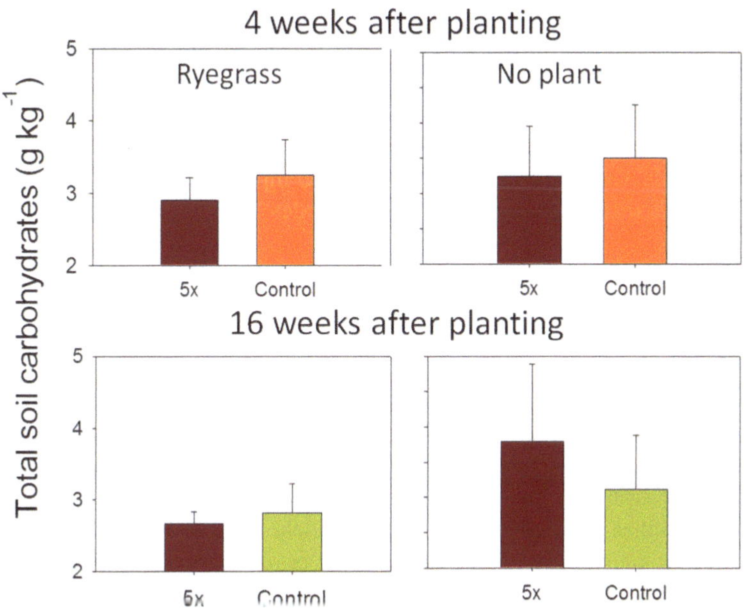

Fig. 4.5 Total soil carbohydrate contents in high-manured (5×) and non-manured (control) soils from the poultry manure/ryegrass study by Waldrip et al. (2011) at 4 and 16 weeks after planting of ryegrass. Label "ryegrass" and "no plant" represent soil samples with or without ryegrass planted

unplanted and planted soil, carbohydrate content was lower in unamended controls than in soils that received PM at week 4. Growing ryegrass further reduced the total carbohydrate contents of both manured and non-manured soils. These changes in total carbohydrate content with manure application and plant growth were quite similar to the corresponding trends observed in AC contents (Sect. 4.4.2, this chapter). Thus, the same mechanism of plant-stimulated mineralization due to increased availability of microbial substrates may well have played a major role in the decomposition of labile soil organic matter (SOM) compounds early in the incubation. The week 16 data showed that total carbohydrate content in the unplanted unmanured control soil was lower than at week 4, which reflected active mineralization of C compounds by the soil biota. The total carbohydrate content of manured soils increased over time, indicating a contribution by the microbial community to the soil carbohydrate pool. In the soils with ryegrass grown, the week 16 data showed that total carbohydrate content decreased over time, indicating utilization of these carbohydrates during plant growth.

Oades (1984) reported that the ratio of galactose + mannose/arabinose + xylose (g + m/a + x) is a useful parameter for distinguishing microbial from plant residues, where the carbohydrate ratio is low (<0.5) for plant polysaccharides and high (>2.0) for microbial polysaccharides. In the current study, PM application had no overall effect on the calculated carbohydrate ratio. However, when data of all manured soils and all sampling times (4, 8, and 16 weeks) were combined, the ratio was 1.17 for soils without ryegrass, 1.03 for non-rhizospheric soils, and 0.88 for rhizospheric soils. The decrease of the carbohydrate ratio in soils that contained ryegrass could have resulted from input of plant-derived sugars and/or a loss of microbial sugars. Considering the fact that the total carbohydrate level was lower in planted non-rhizospheric soils than unplanted soils (Fig. 4.5), we attribute the decreasing value of the ratio to faster decomposition of microbially derived carbohydrates than of plant-derived carbohydrates.

4.5 Conclusion and Future Work

Effective use of nutrient sources is crucial to maintaining agricultural productivity. Manure can play an important role in soil fertility management. Manures are rich in ACs and carbohydrates, which are both major organic components in the soils of terrestrial ecosystems. As reviewed in this chapter, only a limited number of published studies have conducted long-term field trials or laboratory incubations to study how amendment of soil with manure or other organic compounds impact AC and carbohydrate contents. The research on short-term impacts of manure application is even rarer. Our case study of a greenhouse pot study (16 weeks in duration) demonstrated that poultry manure increased soil extractable ACs when added at a high application rate and depressed them when added at a low application rate. The increase in extractable ACs at the high rate was further amplified in the soils taken from the rhizosphere. The presence of ryegrass decreased concentrations of ACs and carbohydrates, more so through decrease in microbial compounds.

Although both ACs and carbohydrates in manure and other organic amendments could contribute to the improvement of soil fertility and soil physical properties, reports on characterization and quantification of both of these compounds in the same study both such as the greenhouse case study, are rare. Clearly, more studies are required to develop a more complete picture of changes in soil ACs and carbohydrates, and the relationships between the two compounds, during laboratory or greenhouse incubations and field studies with different types of organic amendments. Such cumulative knowledge would provide a basis for establishing their specific contributions to soil N and C dynamics over the time course of agricultural seasons and beyond. This type of information will lead toward a more process-level understanding of the relationships between soil fertility, quality, and metabolic functioning following manure application, and could be incorporated into more-complex process-level models that simulate soil nutrient transformations, emissions of gases of environmental concern, and plant growth, such as the Dentrification-Decomposition (DNDC) model (Li et al. 1994). This information might also help resolve the sometimes contradictory effects of manure application on total soil C levels (Angers et al. 2010). The mineralization/immobilization of N when organic sources are added to soil is represented in many simulation models as the outcome of decomposition of the added material and synthesis of SOM, such as the

Agricultural Production Systems Simulation Model (APSIM) (Probert et al. 2005). One of the key parameters regulating decomposition in these models is the C:N ratio of the added manure. Probert et al. (2005) reported that C:N ratios of carbohydrate fractions (e.g. soluble and fiber) are better indicators of manure quality than is the C:N ratio of whole manure. Similarly, Daud et al. (2010) reported that the overall manure C:N ratio was not a good indicator of manure quality and proposed that model simulations of manure mineralization could be improved by analyses of non-fibrous carbohydrate components of manure, such as the sugar and starch fractions. Broader and more in-depth research similar to the greenhouse case study presented in this chapter should be encouraged to provide further insight for modeling efforts and to improve the general understanding of effects of manure on soil quality and fertility.

References

Angers DA, Chantigny MH, MacDonald JD, Rochette P, Côté D (2010) Differential retention of carbon, nitrogen and phosphorus in grassland soil profiles with long-term manure application. Nutr Cycl Agroecosyst 86:225–229

Beavis J, Mott CJB (1996) Effects of land use on the amino acid composition of soils. 1. Manured and unmanured soils from the Broadbalk continuous wheat experiment, Rothamsted, England. Geoderma 72:259–270

Bol R, Ostle NJ, Petzke KJ, Chenu C, Balesdent J (2008) Amino acid in long-term bare fallow soils: influence of annual N fertilizer and manure applications. Eur J Soil Sci 59:617–629

Creamer CA, Filley TR, Olk DC, Plante A, Peltre C, Top SM, Boutton TW (2012) Degree of woody encroachment into grasslands controls soil carbohydrate and amino compound changes during long term laboratory incubation. Org Geochem 52:23–31

Creamer CA, Filley TR, Olk DC, Stott DE, Dooling V, Boutton TW (2013) Changes to soil organic N dynamics with leguminous woody plant encroachment into grasslands. Biogeochemistry 113:307–321

Daud MF, Parsons D, Malau-Aduli A, Lane P (2010) Improved modeling of manure mineralisation through new methods for characterising the carbohydrate pools. In: Food security from sustainable agriculture. http://www.regional.org.au/au/asa/2010/crop-production/nutrients/7181_daud.htm?print=2011

Derrien D, Marol C, Balabane M, Balesdent J (2006) The turnover of carbohydrate carbon in a cultivated soil estimated by ^{13}C natural abundances. Eur J Soil Sci 57:547–557

Dungait JAJ, Bol R, Bull ID, Evershed RP (2009) Tracking the fate of dung-derived carbohydrates in a temperate grassland soil using compound-specific stable isotope analysis. Org Geochem 40:1210–1218

He Z (ed) (2011) Environmental chemistry of animal manure. Nova, New York

He Z (ed) (2012) Applied research of animal manure: challenges and opportunities beyond the adverse environmental concerns. Nova, New York

He Z, Mao J (2011) Functional groups identified by solid state ^{13}C NMR spectroscopy. In: He Z (ed) Environmental chemistry of animal manure. Nova, New York, pp 41–59

He Z, Ohno T (2012) Fourier transform infrared and fluorescence spectral features of organic matter in conventional and organic dairy manure. J Environ Qual 41:911–919

He Z, Olk DC (2011) Manure amino compounds and their bioavailability. In: He Z (ed) Environmental chemistry of animal manure. Nova, New York, pp 179–199

He Z, Senwo ZN, Zou H, Tazisong IA, Martens DA (2014) Amino compounds in poultry litter, litter-amended pasture soils and grass shoots. Pedosphere 24:178–185
Li C, Frockling S, Harriss RC (1994) Modeling carbon biogeochemistry in agricultural soils. Glob Biogeochem Cycles 8:237–254
Lima DL, Santos SM, Scherer HW, Schneider RJ, Duarte AC, Santos EB, Esteves VI (2009) Effects of organic and inorganic amendments on soil organic matter properties. Geoderma 150:38–45
Martens DA, Loeffelmann KL (2002) Improved accounting of carbohydrate carbon from plants and soils. Soil Biol Biochem 34:1393–1399
Martens DA, Loeffelmann KL (2003) Soil amino acid composition quantified by acid hydrolysis and anion chromatography-pulsed amperometry. J Agric Food Chem 51:6521–6529
Mbah CN, Anikwe MAN, Onweremadu EU, Mbagwu JSC (2007) Soil organic matter and carbohydrate contents of dystric leptosol under organic waste management and their role in structural stability of soil aggregates. Int J Soil Sci 2:268–277
Oades JM (1984) Soil organic matter and structural stability: mechanisms and implications for management. Plant Soil 76:319–337
Olk DC (2008) Improved analytical techniques for carbohydrates, amino compounds, and phenols: tools for understanding soil processes. Soil Sci Soc Am J 72:1672–1682
Olk DC, Fortuna A, Honeycutt CW (2008) Using anion chromatography-pulsed amperometry to measure amino compounds in dairy manure-amended soils. Soil Sci Soc Am J 72:1711–1720
Probert ME, Delve R, Kimani S, Dimes J (2005) Modelling nitrogen mineralization from manures: representing quality aspects by varying C:N ratio of sub-pools. Soil Biol Biochem 37:279–287
Rothstein DE (2009) Soil amino-acid availability across a temperate-forest fertility gradient. Biogeochemistry 92:201–215
Scheller E, Raupp J (2005) Amino acid and soil organic matter content of topsoil in a long term trial with farmyard manure and mineral fertilizers. Biol Agric Hortic 22:379–397
Scott EE, Rothstein DE (2011) Amino acid uptake by temperate tree species characteristic of low- and high-fertility habitats. Oecologia 167:547–557
Senwo ZN, Tabatabai MA (1998) Amino acid composition of soil organic matter. Biol Fertil Soils 26:235–242
Simpson IA, Bol R, Bull IA, Evershed RP, Petzke KJ (1999) Compound specific stable isotope signals in anthropogenic soils as indicators of early land management. Rapid Commun Mass Spectrom 13:1315–1319
Waldrip HM, He Z, Erich MS (2011) Effects of poultry manure amendment on soil phosphorus uptake by ryegrass, soil phosphorus fractions, and phosphatase activity. Biol Fertil Soils 47:407–418
Werdin-Pfisterer NR, Kielland K, Boone RD (2009) Soil amino acid composition across a boreal forest successional sequence. Soil Biol Biochem 41:1210–1220

Chapter 5
Nitrogen Mineralization in Soils Amended with Manure as Affected by Environmental Conditions

Dexter B. Watts and H. Allen Torbert

Abstract Nitrogen (N) is the most deficient nutrient in most agricultural production systems; therefore, the economic sustainability of most crops is dependent on adequate supply of nitrogen. Consideration for N availability must be taken into account when incorporating animal manure into a cropping system's management practice. Since N mineralization is the process by which manure N is transformed into readily available forms for plant uptake, understanding how environmental conditions affect the N mineralization process is essential for making manure N availability prescriptions. This chapter discusses the factors that can affect N mineralization and demonstrate the impact of temperature, moisture, soil wetting and drying cycles, and field spatial variability on manure N availability. Our objective is to increase the understanding of manure management in cropping systems in order to maximize N use efficiency.

5.1 Introduction

Nitrogen (N) plays key roles in the synthesis of amino acids, proteins, ribonucleic acid (RNA), and deoxyribonucleic acid (DNA) needed to promote and sustain plant growth and is, therefore, the single most important nutrient needed for agricultural production. However, N availability is often limited in managed soil ecosystems, or existing in forms that are unavailable for plant uptake. Within the soil-plant root zone, N continuously cycles among various organic (unavailable) and inorganic (available) forms. Thus, careful management of this cycle is crucial to sufficiently supply plant-available N for sustainable agricultural production.

Nitrogen mineralization has long been recognized as the central process of the soil N cycle and the single most important component controlling N availability to

---

D.B. Watts (✉) • H.A. Torbert
National Soil Dynamics Laboratory, USDA-ARS, Auburn, AL 36832, USA
e-mail: dexter.watts@ars.usda.gov

Z. He and H. Zhang (eds.), *Applied Manure and Nutrient Chemistry for Sustainable Agriculture and Environment*, DOI 10.1007/978-94-017-8807-6_5,

plants. Understanding the dynamics of soil N mineralization is essential when refining prescriptions for organic source usage for crop and forage production. To determine N availability recommendations for different production systems, mineralization studies should be conducted on manure amended agricultural soils for different regions, climates, and soil textures to develop fertilizer application rates (inorganic or organic), for crop uptake and use efficiency.

Animal manure is the most common organic amendment applied for crop production. Manure amendments, as well as legumes, have been used to increase soil fertility since the beginning of agricultural production. In developed countries during the 1950s, reliance on manure and legumes as the primary N source for plant production decreased in the wake of commercially produced fertilizers. However, recent increases in commercial N fertilizer prices and high demand for organic production (Olson 2001; Jawson and Bull 2002; Wood et al. 2002) have generated renewed interest in animal manure as a nutrient source. There are also strong indications that manure generation from confined animal feeding operations (CAFOs) will continue to play important roles in agricultural production in coming years.

Overall, US animal production has more than doubled since 1960, with poultry production experiencing the greatest gains (USDA-Economic Research Service 2009). The demographics of animal production have changed from small family farm animal production units to large scale integrated CAFOs. These large-scale animal operations generate large amounts of manure that can be used as nutrient sources for crop production. Presently in the US, there are approximately 315 million acres of cropland in use, and manure is applied to 15.8 million acres, equivalent to 5 % (USDA-Economic Research Service 2012). Corn acreage (approximately one-quarter of US cropland) accounts for more than half of the land receiving manure application (USDA-Economic Research Service 2012). It is estimated that 70 % of manure is applied to major crops such as corn, soybeans, wheat, cotton, oats, barley, peanuts, and sorghum, while the remainder is applied to hay and grasses (USDA-Economic Research Service 2012). Generally, the manure used as a nutrient source for crop production is representative of animal production within the region. For instance, large-scale US animal production occurs in regions of the Southeast (poultry and hogs), the High Plains (fed cattle, dairy, and hogs), and the West (dairy), where manure production far exceeds the farm production nutrient needs.

Similarly to undisturbed soil ecosystems, manure-N mineralization is impacted by biotic and abiotic factors. Animal size and species, housing and rearing management, feed ration, manure storage, climate (Eghball et al. 2002), and whether chemical amendments were also added (Watts et al. 2012) all affect manure composition and potentially impact manure N mineralization rates. Manure has a long residence time in soil, providing N to crops for several years after application; since manure is an organic material, N availability is a function of the decomposition rate. For instance, poultry manure has an organic N availability rate of 55 %, broiler litter 55 %, swine manure 40 %, cattle feedlot manure 30 %, composted cattle feedlot manure 18 %, and dairy manure 21 % during the first year after application, with the remaining N becoming available in succeeding years; organic N availability in year two has been reported as 2 %, 5 %, 2 %, 15 %, 8 %, and 14 %, respectively (Eghball et al. 2002). Thus, crop management must take into account the residual impact of manure N mineralization in succeeding years.

Soil organisms are responsible for transforming N into plant-available forms, thereby controlling N mineralization. Microfauna and mircoflora are responsible for the initial breakdown of soil organic N through decomposition. Soil bacteria are responsible for immobilization and mineralization of soil N, the processes that controls the supply of NH_4^+ and NO_3^- for plant uptake. Mineralization occurs in a two-step process where microorganisms first transform organic N compounds into NH_4^+ and then into NO_3^-. More specifically, heterotrophic microorganisms convert organic N in the organic material to NH_4–N through a process called ammonification, while autotrophic microorganisms transform NH_4–N to NO_3–N in a two-step process called nitrification [(1) NH_4^+ to NO_2^-; (2) NO_2^- to NO_3^-]. The carbon (C) fraction of the organic matter is used as the energy source for soil microbes, thereby controlling the rate of mineralization. Once N is in the inorganic form, it becomes susceptible to losses through leaching, ammonia volatilization, or returned to the atmosphere in the forms of N_2O, NO, or N_2 as a result of denitrification process (Tisdale et al. 1985). Thus, careful consideration is needed when managing agricultural systems to increase N retention and improve N-use efficiency for crop production. In this chapter, we will discuss some of the factors affecting N mineralization in manures, and the management implications in agriculture systems.

5.2 Factors that Influence N Mineralization

Mineralization of soil organic N is directly related to microbial activity and biomass inputs, which are affected by biotic and abiotic factors. Among the environmental factors influencing N mineralization are temperature, soil moisture (Katterer et al. 1998; Watts et al. 2010a; Torbert et al. 1992), soil wetting and drying cycles (Kruse et al. 2004; Watts et al. 2007), soil texture (Torbert and Wood 1992), soil organic matter and nutrient concentration (Watts et al. 2010b), residue composition (Cabrera et al. 1993), and the interacting effects of soil C cycling (Griffin 2008). These factors must be considered when estimating N mineralization of manures and manure management for N availability to maximize N-use efficiency.

Factors specific to the soil conditions can also impact N mineralization rates of manure. For example, soil texture affects N mineralization since a soil texture that is more conducive to the formation of stable aggregates will increase the retention of soil organic C and N (Beare et al. 1994). In soils with high amounts of aggregates, the clay-sized particles are bound around organic material, thereby protecting the soil organic matter from decay (Beare et al. 1994; Jastow 1996). When soil aggregates are destroyed, the soil organic matter is exposed to microbial attack.

Soil bulk density can also have a major impact on N mineralization. In a laboratory incubation study, Linn and Doran (1984) reported slightly less respiration with soil of different levels of soil compaction across a range of water filled pore space (WFPS) compared with loose, porous soils. Torbert and Wood (1992) observed that respiration (measure by amount of CO_2 evolved from soil) from sandy Coastal Plain soils decreased 65 % when bulk density was increased from 1.4 to 1.8 Mg m^{-3} at 60 % water-filled pore space (WFPS), indicating that bulk density could impact microbial activity and therefore N mineralization.

The interacting effect of soil organic matter C and N is probably the most important factor affecting N mineralization. For example, Franzluebbers et al. (1994) found that N mineralization is related to the amount of organic matter in soil. Therefore, agronomic practices that build up soil organic matter by adding crop residues back to the soil greatly impact the potential for N mineralization (Wood and Edwards 1992; Torbert et al. 1999; Watts et al. 2010b). Mineralization and immobilization occurs simultaneously within soil, continuously cycling N from organic to inorganic forms and vice versa. The rate of N mineralization (transformation of organic N to inorganic forms) to immobilization (transformation of inorganic N to organic forms) is a function of the C and N fractions in the material being decomposed. Soil microbes use C primarily as an energy source, while N is needed for protein synthesis. Thus, immobilization could occur if organic matter is added to soil with a wide C:N ratio limiting the supply of N. For example, undisturbed topsoil is at equilibrium with its environment when the soil C:N is 12:1. When fresh organic matter is added to soil, the dividing line C:N between immobilization and N release is 20:1; however, when organic substances with C:N greater than 30:1 are added to soil, immobilization occurs during initial decomposition. In other words, when organic material additions with C:N between 20:1 and 30:1 are incorporated into soil, there is a balance between immobilization and mineralization, maintaining soil solution N is at a constant level. Conversely, when organic materials with ratio of less than 20:1 are present, N is released into soil solution (Tisdale et al. 1985).

The factors influencing the quality of the organic matter and its recalcitrant nature can also influence the rate of N mineralization. Research has indicated that tillage intensity can impact C and N cycling and the resulting levels of N mineralization (Torbert et al. 1997, 1998; Follett and Schimel 1989; Watts et al. 2010b). For example, Torbert et al. (1997) reported that increased tillage intensity decreased N mineralization during incubation of soils in a tillage intensity system study. In this study, the no-till system had significantly higher N mineralization than the more intensively tilled systems (Table 5.1).

The ratio of C mineralization to N mineralization of incubated soil is considered to be an index of recalcitrant C levels in the soil, with an increase in the ratio indicating an increase in the recalcitrant C level (Nadelhoffer et al. 1991). In the study reported by Torbert et al. (1998), the C to N mineralization ratio decreased as tillage intensity decreased (Table 5.1). This indicated that the accumulated organic C in the conservation tillage systems may be of less recalcitrant forms. Therefore, the more intensively tilled soil had a lower potential for C mineralization and a

Table 5.1 Tillage intensity impact on N mineralization

Tillage intensity	C mineralization (mg kg^{-1})	C turnover	N mineralization	C:N mineralization ratio (g g^{-1})
Conventional	806 a[a]	18.0 a	8.89 a	156 a
Strip-tillage	833 a	17.9 a	11.60 b	80 ab
Grazed	837 a	18.5 a	9.97 ab	149 ab
Fallow	845 a	10.9 b	13.61 c	70 b
CRP-pine	776 a	25.1 c	8.44 a	99 ab

Data from Torbert et al. (1998)
[a]Means followed by the same letter in each column do not differ significantly (p = 0.10 level)

lower potential for N mineralization compared to less intensively tilled systems. The influences of some of these factors on N mineralization of animal manure used for fertilizer in agricultural systems are further discussed in this chapter.

5.2.1 Temperature, Moisture, Wetting and Drying Cycles

Nitrogen mineralization in agricultural systems has been shown to accelerate with increasing soil temperature (Cassman and Munns 1980; Eghball 2000). Vigil and Kissel (1995) reported that decomposition was 3.7 times faster at 25 °C compared to 15 °C and 13 times faster at 15 °C compared to 5 °C. Temperature can also affect N mineralization differently depending on change in latitude. Campbell et al. (1984) reported that northern US soils were affected more by incubation temperatures than were Southern soils. Campbell et al. (1984) concluded that this was probably due to the decomposition rate, as a function of temperature, which is related to the amount of readily decomposable organic matter in soil.

Nitrogen mineralization tends to increase when soil moisture approaches field capacity and declines with decreasing moisture (Cassman and Munns 1980). According to Linn and Doran (1984), soil microbial activity is optimum when soil WFPS is near 60 %, resulting in maximum organic matter decomposition. Anaerobic conditions tend to occur at a WFPS greater than 60 %, thereby slowing the decomposition process. Knoepp and Swank (2002) found a significant interaction between soil moisture content and soil temperature in N mineralization.

Surface soils, in most ecosystems, experience periods of drying followed by relatively rapid rewetting cycles. It has been shown that N mineralization rates are generally increased for a few days following rewetting of a dry soil (Birch 1958; Bloem et al. 1992; Cui and Caldwell 1997; Franzluebbers et al. 2000). Soil undergoes complex physical, chemical, and biological changes under the impact of drying and re-wetting, including changes to soil structure (aggregation), soil organic matter (SOM) and microflora (Soulides and Allison 1961; Sorensen 1974; Utomo and Dexter 1982).

Wetting and drying cycles have an influence on microbial activity, thereby affecting decomposition of SOM (Soulides and Allison 1961; Bloem et al. 1992;

Magid et al. 1999). Drying followed by rapid rewetting cycles generally causes an increase in organic substrate available for microbial attack (Soulides and Allison 1961; Sorensen 1974). These substrates are partially derived from the death of a portion of the soil organisms upon drying (Lund and Goksoyr 1980; Bottner 1985) by causing microbes to undergo osmotic shock, which can induce microbial cell lysis (Bottner 1985; Van Gestel et al. 1992) or lead to the release of intracellular solutes (Halverson et al. 2000). The labile substrates that become available (i.e., C and N compounds in the form of organic material) are rapidly mineralized by the remaining soil microbes, which cause a pulse in mineralization of N and C (Birch 1959; Keift et al. 1987). Wetting and drying cycles have also been shown to cause soil aggregates to break apart, exposing physically protected organic matter to become available for further degradation (Adu and Oades 1978; Lundquist et al. 1999). The organic matter that was previously unavailable can be rapidly mineralized by the microbial community (Appel 1998).

Presently, it is estimated that manure N use efficiency is about 30–50 % in most agricultural soils, leaving 50–70 % in soil subject to leaching or runoff (Delgado 2002). However, a better understanding of soil under different textural, moisture, and temperature conditions from different climatic regions is needed in order to develop management practices that optimize from manure derived N. Information is needed on the relative rate of N mineralized in soils amended with manure as affected by different climatic conditions. The importance of these factors was demonstrated in the research reported by Watts et al. (2007), who took a closer look at the impact of manure application on N mineralization in soils subjected to different temperatures and wetting and drying cycles under laboratory incubation conditions.

In research reported by Watts et al. (2007), an incubation study compared soils which were closely located in the same field but varied in chemical and physical characteristics to a soil from a different region of the country. The soils which were proximate to each other were collected from a precision agriculture experiment established at Auburn University's E.V. Smith Experiment Station located in Macon County, Alabama and compared to soil under long-term pasture management located in Champaign County, Illinois. Bulk samples (0–15 cm) were collected from field plots and transported back to the laboratory for analysis. The two soil series evaluated from Macon County, AL were Bama and Goldsboro. The Bama series (fine-loamy, siliceous, subactive, thermic Typic Paleudults) consisted of very deep, well-drained soils formed in loamy sediments. The Goldsboro series (fine-loamy, siliceous, subactive, thermic Aquic Paleudults) consisted of very deep, moderately well drained soils formed in loamy sediments. The soil from Champaign County, IL was a Catlin soil series (fine-silty, mixed, mesic Typic Arguidolls) consisting of moderately well drained, moderately permeable soil on till plains and moraines. These soils formed in loess and in the underlying loam glacial till. Evaluating the soil characteristics (Table 5.2), it can be observed that the Catlin soil greatly varied from the other two soils, with the Catlin soil having a higher CEC, total C and total N.

An aerobic incubation experiment was performed similar to the procedure described by Honeycutt et al. (2005). Treatments were replicated three times with

Table 5.2 Properties of the soils used for the incubation study

		BD	CEC	Total C	Total N	Sand	Silt	Clay
Soil series	pH	g cm^{-3}	cmol kg^{-1}	g kg^{-1}		%		
Bama	6.7	1.46	5.8	7.53	0.72	66.3	21.2	12.5
Goldsboro	6.5	1.58	6.3	7.08	0.78	33.7	48.8	17.5
Catlin	7.2	1.20	26.8	40.66	3.14	18.7	66.3	15.0

Data from Watts et al. (2007)

250 g of soil (oven-dried weight basis) amended with the appropriate amount of composted dairy manure to give 350 organic N kg^{-1} soil. Samples were thoroughly mixed and placed into 2 L Mason jars. Deionized water was added to bring the soil moisture to approximately 60 % water-filled pore space (WFPS). The controls were treated with deionized water only. Samples were incubated at three different temperatures (11, 18, and 25 °C) and exposed to two wetting/drying regimes. Half were maintained at constant moisture (60 % WFPS) and the other half were cycled from 60 to 30 % WFPS. Samples were aerated daily (2 h), at which time moisture content was adjusted gravimetrically as required. Cycled soils were allowed to reach 30 % WFPS to complete a drying cycle. After undergoing a drying cycle, soils were re-wetted by adding the appropriate amount of water to bring the soil back to 60 % WFPS. On the day that cycled soils were rewetted to 60 % WFPS, a sub-sample was taken from the cycled soil and its corresponding soil at constant moisture.

The total amount of inorganic N which was mineralized in the Catlin soil (Table 5.3) was 175 and 215 % higher than the Goldsboro and Bama soil, respectively. Even though the Goldsboro soil had higher clay content and was considered to be a better soil than the Catlin, N mineralization was probably influenced by CEC, total N and C, and organic matter, resulting from the climate where the soil was originated. Soil under cool wet climates tend to build up organic matter (e.g. Catlin), while soil from hot humid climates (Goldsboro and Bama) tend to promote organic matter decomposition and high rainfall events tend to cause leaching. Since N mineralization is a process that is mediated by microbial activity, the microbial populations of soil from the Northern climatic environments were probably more adapted to relatively low soil temperature, therefore, causing a higher mineralization in the Catlin soil at 11 °C. As soil temperature increased the Catlin soil experienced faster decomposition due to the residual organic matter.

The amount of inorganic N mineralized was significantly affected by temperature, with the highest two temperatures mineralizing the most in this experiment (Table 5.3). This suggests that the increase in N pool was related to temperature, indicating an influence of thermal conditions on the degree of microbial exploitation of organic N sources. The Catlin soil showed the greatest difference between temperature regimens (Table 5.3). This is similar to the finding of Dalias et al. (2002) who found that the optimum temperature for nitrification showed a good correlation with the geographic origin of the soil. This was also in accordance with Campbell et al. (1984) who stated that the decomposition rate (as a function of

Table 5.3 Interaction between soil type and temperature on N mineralization over time

		Total N in form of NO_3^- and NH_4^+				
	Temp	Day 0	Cycle 1	Cycle 2	Cycle 3	Cycle 4
Soil	°C	(mg kg^{-1})				
Catlin	11	3.40 a[a]	68.44 a	68.55 a	77.32 a	79.01 a
	18	3.60 a	85.49 a	91.64 b	99.28 b	106.87 b
	25	2.48 a	86.67 a	102.79 b	110.27 b	120.61 b
	11	1.02 a	24.32 a	27.38 a	30.05 a	35.82 a
Goldsboro	18	1.04 a	30.14 a	29.95 a	48.60 b	43.01 b
	25	1.39 a	29.54 a	27.46 a	35.11 ab	38.41 ab
	11	1.09 a	20.50 a	20.83 a	23.79 a	29.01 a
Bama	18	1.36 a	31.09 a	21.18 a	30.07 a	31.15 ab
	25	1.28 a	29.20 a	27.42 a	37.80 a	44.17 b

Data from Watts et al. (2007)
[a]Means followed by the same letter do not differ significantly ($p = 0.10$ level)

temperature) was related to the amount of readily decomposable organic matter in the soil. Not all organic material decomposes at the same rate. For example, water-soluble carbohydrates, amino acids, and amino-sugars have a short residence time in a soil. With an active microbial population under typical summer field conditions, these components are usually consumed within a week or two. Organic matter that is humified or derived from more resistant materials (lignin associated) has longer residence time under the same field conditions and may last years. In northern climates, where summers are cooler and shorter, less decomposition of organic plant litter and native soil organic matter will occur in a given year, resulting in a buildup of more easily decomposed material over time. In southern climates, summer is longer and warmer, allowing for more complete decomposition of most of the litter material. Over a long period in Southern soils, only the most resistant organic materials are left behind, whereas in Northern soils, both easily decomposable and resistant materials can accumulate (Campbell et al. 1984).

5.2.2 *Field Study Spatial Variability*

To develop improved manure management practices, scientists must understand the impact of soil conditions and their variability across a field on N mineralization from manure. Most large-scale agricultural fields have varying degrees of soil types resulting from different soil forming factors and landscape positions. Nitrogen mineralization processes in one area of a field may differ from another due to inherent soil-forming properties. It has been reported that soil texture influences how rapidly N mineralization occurs (Hubbard et al. 2008; Sistani et al. 2008). Similarly, Thomsen and Olesen (2000) observed that increased sand content generally led to increased N mineralization as a response to increased aeration.

Table 5.4 Physical properties of the soil studied by Watts et al. (2010a)

Soil	BD	Sand	Silt	Clay
	g cm^{-3}	%		
Bama	1.68	66.25	21.25	12.50
Lynchburg	1.64	46.25	41.25	12.50
Goldsboro	1.61	33.75	48.75	17.50

A study by Watts et al. (2010a) demonstrated how N mineralization of manure can be impacted by spatial variability across an agricultural field and how the impacts could differ depending on the time of year. In this study, soil samples were collected from an ongoing precision agriculture experiment located at Auburn University's E.V. Smith Experiment Station located in Macon County, Alabama (Terra et al. 2006). The three soil series evaluated in the study were Bama, Goldsboro, and Lynchburg (Table 5.4). These three soils were found in close proximity to one another in the same field, yet differ in texture. The soils were located on three different landscape positions consisting of the summit (Bama), drainageway (Lynchburg), and sideslope (Goldsboro). The Bama series consisted of very deep, well-drained soils that formed in loamy sediments. Slopes ranged from 0 to 5 %. These soils were fine-loamy, siliceous, subactive, thermic Typic Paleudults. The Goldsboro series consisted of very deep, moderately well drained soils that formed in loamy sediments. Slopes ranged from 0 to 2 %. These soils were fine-loamy, siliceous, subactive, thermic Aquic Paleudults. The Lynchburg series consisted of very deep, somewhat poorly drained soils that formed in loamy sediments. Slopes ranged from 0 to 2 %. These soils were Aquic Paleudults and Paleaquults.

This study used the *in situ* resin core method to observe N mineralization rates. The use of the *in situ* core method can be costly and labor intensive, but valuable in understanding N mineralized under natural field conditions. The *in situ* resin core method has been used to observe N mineralization rates in forest ecosystems (Distefano and Gholz 1986; Binkley et al. 1992). This method was also used to study mineralization rates in dryland agroecosystems (Kolberg et al. 1997) and Eghball (2000) performed an N mineralization study using this method on beef cattle manure and beef cattle compost during the growing season.

The field *in situ* mineralization method entailed placing polyvinyl chloride plastic cylinders in the surface 20 cm of the soil profile. These *in situ* soil core (microplot cylinders) incubation chambers were 6.25 cm in diameter and 20.32 cm in length. Intact cores were collected by driving the polyvinyl chloride cylinder into the top 20 cm of the soil profile using a hydraulic core sampler. The top 4 cm of soil in the microplot cylinders was removed and an appropriate amount of composted dairy manure was added and thoroughly mixed to give 350 kg N ha^{-1} (Fig. 5.1). Anion and cation exchange resin was placed in the bottom of each cylinder to capture leachate. The soil cores were inserted back in the fallow ground in the same place from where they were originally taken. The microplot cylinders were sampled periodically through a 70-day period in both the winter and the summer season.

The results indicated that N mineralization of the manure was not only greatly affected by soil types but these impacts of soil types differed by season (Table 5.5).

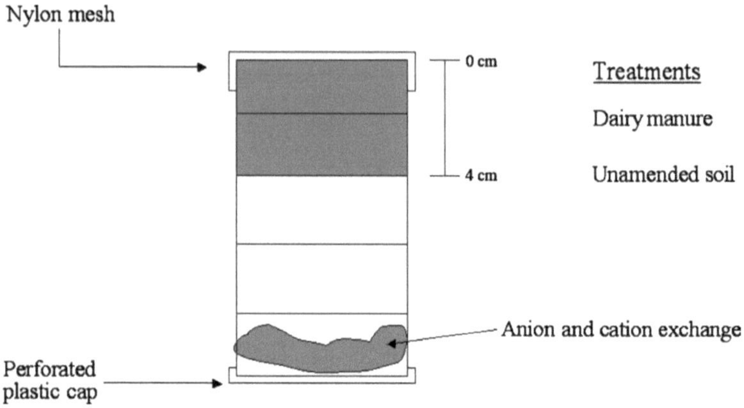

Fig. 5.1 Design of the microplot cylinder used in the winter 2004 and summer 2005 *in situ* mineralization study (Figure from Watts et al. 2010a)

Table 5.5 Amount of mineralized N (NO_3^- + NH_4^+) retained in soil and captured in the resin after manure application

	Days after dairy compost addition							
	0	7	14	21	28	49	70	Mean
	(mg kg^{-1})							
Winter 2004								
Bama	12.6	13.6	13.6	13.8	13.5	13.8	14.5	13.5 a[a]
Lynchburg	11.8	15.7	17.7	19.2	19.4	19.8	22.7	17.8 b
Goldsboro	13.1	14.2	12.8	13.3	13.0	14.2	14.6	13.5 a
Summer 2005								
Bama	13.1	20.0	18.6	39.3	51.8	73.8	81.5	39.6 b
Lynchburg	11.2	21.4	19.1	47.0	49.2	74.0	79.9	40.3 b
Goldsboro	12.7	24.9	36.3	46.5	63.0	80.9	89.7	46.4 a

Data from Watts et al. (2010a)
[a]Means followed by the same letter do not differ significantly (0.10 level)

Averaged across sampling days for the winter season, the Lynchburg soil, a loam soil (located in a depression), had the highest total N mineralization. The Lynchburg soil had higher CEC, total C, and total N compared with the other soils. As a result of the soil fertility, microbial activity, which mediates the mineralization process, most likely contributed to a higher N availability. In addition, this soil, which is located in a depression area, has higher infiltration due to the more even distribution of sand, silt and clay than the other soils. Increased soil aeration most likely also contributed to higher mineralization. It is important to note that most of the nitrogen measured for Lynchburg soil was observed in the anion exchange resin. This suggests that sandy soils located in depression areas in agronomic fields are potentially susceptible to leaching during winter. This influence of soil texture was documented by Delgado et al. (1999) and Delgado (2001) who observed more N leaching in sandy soils (Follett and Delgado 2002).

On the other hand, the Goldsboro soil, a loam soil located on a slight sideslope, mineralized more N than the other two soils during the summer months (Table 5.5). This was likely caused by the ability of this soil to maintain a higher soil moisture content than the other soils. The percentage of sand, silt, and clay in the Goldsboro soil was more evenly distributed with a slightly higher silt and clay content than the other soils. This suggests that the greater silt and clay content of this loam soil was integral in maintaining higher moisture content than the other soils, corresponding to a higher N mineralization capacity in this soil.

Although other researchers have observed higher N mineralization rates occurring in manure-amended soil with the highest percentage of sand (Sistani et al. 2008; Hubbard et al. 2008; Thomsen and Olesen 2000), moisture was the overriding factor affecting mineralization in this study. For instance, the Bama soil (summit) had the highest percentage of sand, but the Goldsboro soil had the highest mineralization capacity. Given that soil moisture is extremely important in the Southeastern US during the summer months, the higher moisture content maintained in the Goldsboro soil was more conducive for microbial activity which mediates N mineralization, providing the soil with more plant-available N. This study demonstrates that not only can spatial variability play an important role in N mineralization of manure; seasonal impacts should also be considered when developing manure management recommendations.

Inorganic N concentrations observed in the soil compartment of the *in situ* cores for all soil types during the winter months (Fig. 5.2) linearly decreased over time. This decrease in inorganic N with time suggests that N mineralization was minimal, and that the inorganic N fraction from the composted dairy manure was being leached due to the dynamic properties of N in the soil. However, (Fig. 5.2) the amount of inorganic N observed in the soil compartment during the summer months experienced fluctuating concentrations from Day 0 to Day 70. This fluctuating inorganic N concentration observed during the summer months might be attributable to mineralization and the accumulation of plant-available N followed by periods of leaching caused by rainfall events. Under normal conditions, inorganic N continues to accumulate in the soil until it is taken up by a plant or lost to leaching.

Concentrations of inorganic N extracted from the ion exchange resins increased over time during the summer and winter months for all soils (Fig. 5.3). The inorganic N observed in the resins during the summer months was significantly higher than that collected during the winter months, suggesting that as mineralization increases, so does the potential for N loss. Because mineralization was minimal (maximum mineralization was <2 % of the added dairy compost N) during the winter months, inorganic N collected in the resins was mainly attributable to leaching of NO_3^- contained in the composted dairy manure. In general, the water content of a soil is normally higher during winter months while plant

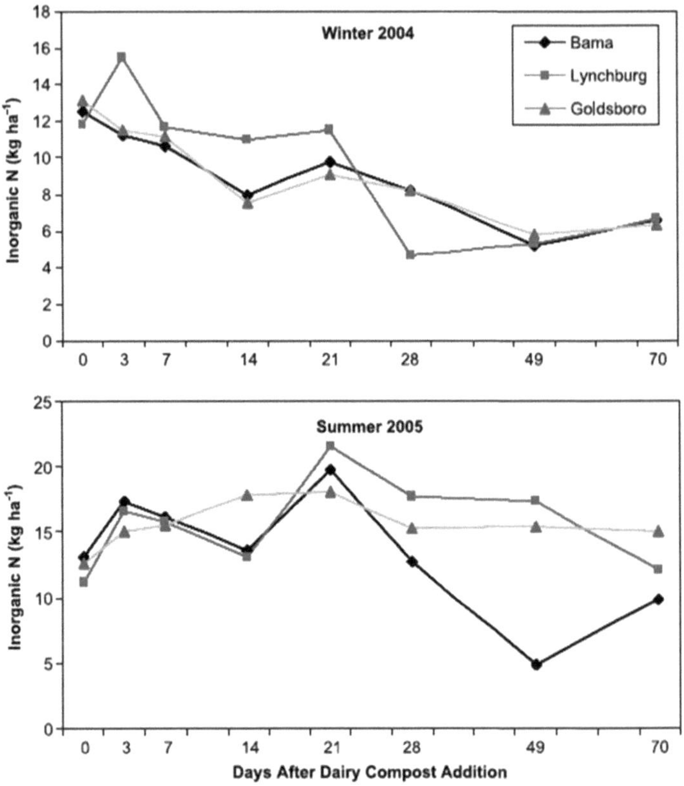

Fig. 5.2 Seasonal N mineralization (winter 2004 and summer 2005) of organic N in the soil compartment of the micro-plot cylinder for the three different soil types amended with composted dairy manure at eight sampling times during the season (Data from Watts et al. 2010a)

growth is minimal, suggesting that the NO_3^- retained by or added to the soil has a greater chance of being lost. Inorganic N concentrations observed in the resins during summer months were mainly attributable to the mineralization of the added composted dairy manure.

Although N mineralization and leaching occurred among all soils during the winter months, the soil type located in a low-lying area (Lynchburg), significantly mineralized and lost more of the added N from dairy compost (80–90 % more) compared with the other soils (Fig. 5.3). During the summer, the loam soil (Goldsboro) with the greatest water-holding capacity mineralized the most N, mineralizing significantly more (9–10 %) than the other soils. These results show that soil variability, temperature, and landscape need to be considered when applying manure to agricultural fields. This study effectively shows the importance of considering the spatial variability of an agricultural field when managing N release from manure.

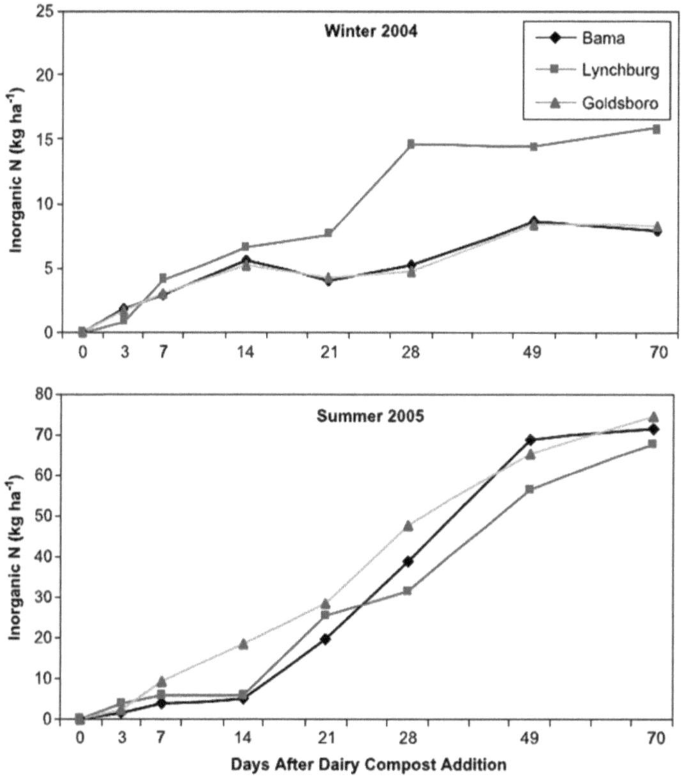

Fig. 5.3 Seasonal concentrations (winter 2004 and summer 2005) of inorganic N collected by the ion exchange resin for the three different soil types amended with composted dairy manure at eight sampling times during the season (Data from Watts et al. 2010a)

5.3 Future Research Needs on N Mineralization

Increased N inputs have dramatically changed the rate of N cycling in most agricultural systems. For instance, worldwide use of industrially produced inorganic N and manure from concentrated animal feeding operations have created new pathways for N redistribution and cycling in soil; however, soil N retention capacity has not drastically increased. Thus, more research is needed to develop management practices that will control N cycling, mineralization, and storage in soil. The importance of N retention in soil and the impact of N mineralized from organic matter will continue to increase in coming years. This will be partially driven by the need to feed the growing population as well as increased demand for organic production. Thus, national and international exchange of scientific research will be needed to better predict N availability and losses from ecosystems receiving organic sources of nutrients to improve our understanding of N mineralization and to refine soil N availability models.

Historically, most N mineralization studies have been conducted in agricultural systems used for silviculture, forage, and row crop production; however, there has been limited research using organic sources as the sole N supply in specialty industries such as in horticulture (nursery, greenhouse), and turf production. Further research is also needed on the impact that belowground plant residues (roots) have on N supply to succeeding crops and the effect of mixing legumes with cash crops (e.g., inter seeding clover with wheat or corn). Research on improvements of N fixing capacity of legume as well as the use of mixed crop residues with different C: N ratios is lacking. The development of plant growth regulators containing microbial inoculates that fix and supply N to plants is also needed.

References

Adu JK, Oades JM (1978) Physical factors influencing decomposition of organic materials in soil aggregates. Soil Biol Biochem 10:109–115

Appel T (1998) Non-biomass soil organic N: the substrate for N mineralization flushes following soil drying-rewetting and for organic N rendered CaCl2-extractable upon soil drying. Soil Biol Biochem 30:1445–1456

Beare MH, Hendrix PF, Coleman DC (1994) Water-stable aggregates and organic matter fractions in conventional and notillage soils. Soil Sci Soc Am J 58:777–786

Binkley D, Bell R, Sollins P (1992) Comparison of methods for estimating soil nitrogen transformations in adjacent conifer and alder-conifer forests. Can J For Res 22:858–863

Birch HF (1958) The effect of soil drying on humus decomposition and nitrogen availability. Plant Soil 10:9–31

Birch HF (1959) Further observations on humus decomposition and nitrification. Plant Soil 9:262–286

Bloem J, de Ruiter PC, Koopman GJ, Libbink GJ, Brussaard L (1992) Microbial numbers and activity in dried and rewetted arable soil under integrated and conventional management. Soil Biol Biochem 24:655–665

Bottner P (1985) Response of microbial biomass to alternate moist and dry conditions in a soil incubated with ^{14}C and ^{15}N labeled plant material. Soil Biol Biochem 17:329–337

Cabrera ML, Chiang SC, Merka WC, Thompson SA, Poncorbo OC (1993) Nitrogen transformations in surface-applied poultry litter: effect of litter physical characteristics. Soil Sci Soc Am J 57:1519–1525

Campbell CA, Jame YW, Winkleman GE (1984) Mineralization rate constants and their use for estimating mineralization in some Canadian prairie soils. Can J Soil Sci 64:333–343

Cassman KG, Munns DN (1980) Nitrogen mineralization as affected by soil moisture, temperature, and depth. Soil Sci Soc Am J 44:1233–1237

Cui M, Caldwell MM (1997) A large ephemeral release of nitrogen upon wetting of dry soil and corresponding root responses in the field. Plant Soil 191:291–299

Dalias P, Anderson JM, Bottner P, Couteaux M (2002) Temperature responses of net nitrogen mineralization and nitrification in conifer forest soils incubated under standard laboratory conditions. Soil Biol Biochem 34:691–701

Delgado JA (2001) Use of simulations for evaluations of best management practices on irrigated cropping systems. In: Shaffer MJ, Ma L, Hansen S (eds) Modeling carbon and nitrogen dynamics for soil management. Lewis Publishers, Boca Raton, pp 355–381

Delgado JA (2002) Quantifying the loss mechanisms of nitrogen. J Soil Water Conserv 57:389–398

Delgado JA, Sparks RT, Follet RF, Sharkoff JL, Riggenbach RR (1999) Use of winter cover crops to conserve water and water quality in the San Luis Valley of south central Colorado. In: Lal R (ed) Soil quality and soil erosion. CRC Press, Boca Raton, pp 125–142

Distefano JF, Gholz JL (1986) A proposed use of ion exchange resin to measure nitrogen mineralization and nitrification in intact soil cores. Commun Soil Sci Plant Anal 17:989–998

Eghball B (2000) Nitrogen mineralization from field-applied beef cattle feedlot manure and compost. Soil Sci Soc Am J 64:2024–2030

Eghball B, Weinhold BJ, Gulley JE, Eigenberg RA (2002) Mineralization of manure nutrient. J Soil Water Conserv 57:470–473

Follett RF, Delgado JA (2002) Nitrogen fate and transport in agricultural systems. J Soil Water Conserv 57:402–408

Follett RF, Schimel DS (1989) Effect of tillage practices on microbial biomass dynamics. Soil Sci Soc Am J 53:1091–1096

Franzluebbers K, Weaver RW, Juo ASR, Franzluebbers AJ (1994) Carbon and nitrogen mineralization form cowpea plants part decomposing in moist and in repeatedly dried and wetted soil. Soil Biol Biochem 26:1379–1387

Franzluebbers AJ, Haney R, Honeycutt CW, Schomberg H, Hons F (2000) Flush of carbon dioxide following rewetting of dried soil relates to active organic pools. Soil Sci Soc Am J 64:613–623

Griffin TS (2008) Nitrogen availability. In: Schepers JS, Raun WR (eds) Nitrogen in agricultural soils, vol 49, Agronomy monograph. ASA, CSSA, SSSA, Madison, pp 616–646

Halverson LJ, Jones TM, Firestone MK (2000) Release of intercellular solutes by four soil bacteria exposed to dilution stress. Soil Sci Soc Am J 64:1630–1637

Honeycutt CW, Griffin TS, Weinhold BJ, Eghball B, Albrecht SL, Powell JM, Woodbury BL, Sistani KR, Hubbard RK, Torbert HA, Eigenberg RA, Wright RJ, Jawson MD (2005) Protocols form nationally coordinated laboratory and field research on manure nitrogen mineralization. Commun Soil Sci Plant Anal 36:2807–2822

Hubbard RK, Bosch DD, Marshall LK, Strickland TC, Rowland D, Griffin TS, Honeycutt CW, Albrecht SL, Sistani KR, Torbert HA, Woodbury BJ, Powell JM (2008) Nitrogen mineralization from broiler litter applied to southeastern coastal plain soils. J Soil Water Conserv 63:182–192

Jastow JD (1996) Soil aggregation formation and the accrual of particulate and mineral-associated organic matter. Soil Biol Biochem 28:656–676

Jawson MD, Bull CT (2002) USDA research into organic farming. Am J Altern Agric 17:201–202

Katterer T, Reichstein M, Anren O, Lomander A (1998) Temperature dependence of organic matter decomposition: a critical review using literature data analyzed with different models. Biol Fertil Soils 27:258–262

Keift TL, Soroker E, Firestone MK (1987) Microbial biomass response to a rapid increase in water potential when dry soil is wetted. Soil Biol Biochem 19:119–126

Knoepp JD, Swank WT (2002) Using soil temperature and moisture to predict forest soil nitrogen mineralization. Biol Fertil Soils 36:177–182

Kolberg RL, Rouppet B, Westfall DG, Peterson GA (1997) Evaluation of an in situ net soil nitrogen mineralization method in dryland agroecosystems. Soil Sci Soc Am J 61:504–508

Kruse J, Kissel DE, Cabrera ML (2004) Effects of drying and rewetting on carbon and nitrogen mineralization in soils and incorporated residues. Nutr Cycl Agroecosyst 69:247–256

Linn DM, Doran JW (1984) Effect of water-filled pore space on carbon dioxide and nitrous oxide production in tilled and notilled soils. Soil Sci Soc Am J 48:1267–1272

Lund V, Goksoyr J (1980) Effects of water fluctuations on microbial mass and activity in soil. Microb Ecol 6:115–123

Lundquist EJ, Jackson LE, Scow KM (1999) Wet dry cycles affect DOC in two California agricultural soils. Soil Biol Biochem 31:1031–1038

Magid J, Kjaergaard C, Gorissen A, Kuikman PJ (1999) Drying and rewetting of a loamy sand soil did not increase the turnover of native organic matter, but retarded the decomposition of added 14-C labeled plant material. Soil Biol Biochem 31:595–602

Nadelhoffer KI, Giblin AE, Shaver GR, Laundre LA (1991) Effects of temperature and substrate quality on element mineralization in six arctic soils. Ecology 72:242–253

Olson J (2001) Metering manure: new manure application technology conserves resource and protects yield. Farm Ind News 34:72–73

Sistani KR, Adeli A, McGowen SL, Tewolde H, Brink GE (2008) Laboratory and field evaluation of broiler litter nitrogen mineralization. Bioresour Technol 99:2603Y261

Sorensen LH (1974) Rate of decomposition of organic matter in soil as influenced by repeated air drying-rewetting and repeated additions of organic material. Soil Biol Biochem 6:31–37

Soulides DA, Allison FE (1961) Effect of drying and freezing soils on carbon dioxide production, available mineral nutrients, aggregation, and bacterial population. Soil Sci 91:291–298

Terra JA, Shaw JN, Reeves DW, Raper RL, van Santen E, Schwab EB, Mask PL (2006) Soil management and landscape variability affects field-scale cotton productivity. Soil Sci Soc Am J 70:94–107

Thomsen IK, Olesen JE (2000) C and N mineralization of composted and anaerobically stored ruminant manure in differently textured soils. J Agric Sci 135:151–159

Tisdale SL, Nelson WL, Beaton JD (1985) Soil fertility and fertilizers, 4th edn. Macmillan Publishing Company, New York

Torbert HA, Wood C (1992) Effects of soil compaction and water-filled pore space on soil microbial activity and N losses. Commun Soil Sci Plant Anal 23:1321–1331

Torbert HA, Hoeft RG, Vanden Heuvel RM, Mulvaney RL (1992) Effect of moisture regime on recovery and utilization of fertilizer N applied to corn. Commun Soil Sci Plant Anal 23:1409–1426

Torbert HA, Potter KN, Morrison JE Jr (1997) Tillage intensity and fertility level effects on nitrogen and carbon cycling in a Vertisol. Commun Soil Sci Plant Anal 28:699–710

Torbert HA, Potter KN, Morrison JE Jr (1998) Tillage intensity and crop residue effects on nitrogen and carbon cycling in a Vertisol. Commun Soil Sci Plant Anal 29:717–727

Torbert HA, Prior SA, Reeves DW (1999) Land management effects on nitrogen and carbon cycling in an ultisol. Commun Soil Sci Plant Anal 30:1345–1359

USDA-Economic Research Service (2009) Manure use for fertilizer and for energy: report to congress. Administrative publication no. AP-037. U.S. Government Print Office, Washington, DC. Available at: http://www.ers.usda.gov/Publications/AP/AP037/. Verified 18 June 2012

USDA-Economic Research Service (2012) Manure use for fertilizer and for energy: report to congress. Administrative publication no. (AP-037) U.S. Gov. Print. Office, Washington D.C. Available at: http://www.ers.usda.gov/Publications/AP/AP037/. Verified 18, June 2012.

Utomo WH, Dexter AR (1982) Changes in soil aggregate water stability induced by wetting and drying cycles in non-saturated soil. J Soil Sci 33:141–163

Van Gestel M, Ladd JN, Amato M (1992) Microbial biomass response to seasonal change and imposed drying regimes at increasing depths of undisturbed topsoil profiles. Soil Biol Biochem 24:102–111

Vigil MF, Kissel DE (1995) Rate of nitrogen mineralized from incorporated crop residues as influenced by temperature. Soil Sci Soc Am J 59:1636–1644

Watts DB, Torbert HA, Prior SA (2007) Mineralization of N in soils amended with dairy manure as affected by wetting/drying cycles. Commun Soil Sci Plant Anal 38:2103–2116

Watts DB, Torbert HA, Prior SA (2010a) Soil property and landscape position effects on seasonal nitrogen mineralization of composted dairy manure. Soil Sci 175:27–35

Watts DB, Torbert HA, Prior SA, Huluka G (2010b) Long-term tillage and poultry litter impacts soil carbon and nitrogen mineralization and fertility. Soil Sci Soc Am J 74:1239–1247

Watts DB, Smith KE, Torbert HA (2012) Impact of poultry litter cake, cleanout, and bedding chemical amendments on soil C and N mineralization. Int J Agron 2012(2):1–8, Article ID 204629

Wood CW, Edwards JH (1992) Agroecosystem management effects on soil carbon and nitrogen. Agric Ecosyst Environ 39:123–138

Wood M, Chavez L, Comis D, Arnold J (2002) Organic grows on America. Agric Res 50:4–9

Chapter 6
Soil Enzyme Activities as Affected by Manure Types, Application Rates, and Management Practices

Veronica Acosta-Martínez and Heidi M. Waldrip

Abstract Manure application can restore soil ecosystem services related to nutrient cycling and soil organic matter (SOM) dynamics through biochemical transformations mediated by soil enzymes. Soil enzymes are crucial in soil metabolic functioning, as they drive the decomposition of organic residues, humification processes, transformations leading to the release of plant available nutrients, stabilization of soil structure, and degradation of xenobiotic (foreign or strange) compounds. However, despite the fact that there is an exhaustive amount of literature available on the effects of manure on soil enzyme activities, there is no comprehensive overview of recent research findings that compares different management scenarios, manure types, and potentially new manure products or management. The purpose of this chapter is to provide a review of the response of enzyme activities to manure applications and their potential implications on soil biogeochemical cycling in agroecosystems. Additionally, this chapter intends to provide some perspective on specific areas where more information is warranted and pinpoint avenues for future research.

6.1 Introduction

Most (80–90 %) soil processes that are involved in the decomposition and transformation of nutrients from organic compounds occur through biochemical reactions mediated by enzymes. As summarized in Fig. 6.1, soil enzymes can be either intracellular (enzymes inside microbial cells) or extracellular (enzymes that have been released into soil solution and are attached to soil surfaces) (Kiss et al. 1975;

V. Acosta-Martínez (✉)
USDA-ARS, Cropping Systems Research Laboratory, Lubbock, TX 79415, USA
e-mail: veronica.acosta-martinez@ars.usda.gov

H.M. Waldrip
USDA-ARS, Conservation and Production Research Laboratory, Bushland, TX 79012, USA
e-mail: heidi.waldrip@ars.usda.gov

Z. He and H. Zhang (eds.), *Applied Manure and Nutrient Chemistry for Sustainable Agriculture and Environment*, DOI 10.1007/978-94-017-8807-6_6,

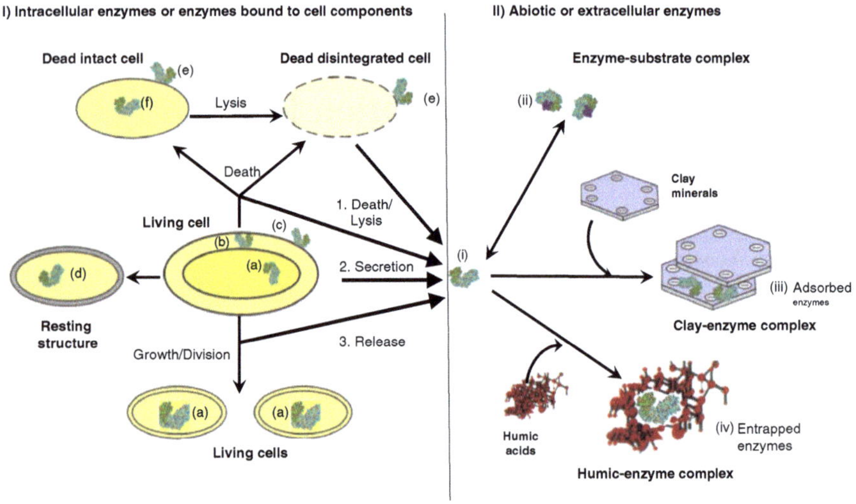

Fig. 6.1 Soil enzymes are derived from two major sources: (**I**) Intracellular enzymes or enzymes bound to cell components which include (*a*) enzymes functioning within the cytoplasm of proliferating microbial, animal and plant cells, (*b*) enzymes restricted to the periplasmatic space of proliferating Gram-negative bacteria, (*c*) enzymes attached to the outer surface of viable cells with active sites extending into the soil environment, (*d*) enzymes within non-proliferating cells such as fungal spores, protozoa cysts, plant seeds and bacterial endospores, (*e*) enzymes attached to whole dead cells and cell debris, and (*f*) enzymes located inside intact dead cells; (**II**) Abiotic or extracellular enzymes, including (*i*) enzymes leaking from intact cells or released from dead or lysed cells that originated from the cell membrane or within the cell and which may survive for a short period in the soil solution, (*ii*) enzymes existing temporally as soluble or insoluble enzyme-substrate complexes, (*iii*) enzymes absorbed to the external or internal (i.e., within the lattices of 2:1 layer silicates) surfaces of clay minerals, or (*iv*) enzymes complexed with humic colloids via absorption, entrapment, or co-polymerization during humification. (This conceptual visual of the location of enzymes in soil was first developed by Klose 2003, and it is modified from Acosta-Martínez and Klose 2008)

Burns 1978; Ladd and Butler 1975; Nannipieri et al. 2002). The total enzyme activity of a soil (intracellular and extracellular) drives ecosystem services related to metabolic functioning, including: (1) decomposition of organic residues, (2) humification processes that stabilize soil organic matter (SOM), (3) the release of plant available nutrients, (4) the transformations of carbon (C), nitrogen (N), phosphorus (P), and sulfur (S) compounds via various processes, (5) stabilization of soil structure, and (6) degradation of xenobiotic (foreign or strange) compounds.

The application of manure to soil can restore ecosystem services related to nutrient cycling and SOM dynamics through the activities of enzymes involved in carbon (C), nitrogen (N), phosphorus (P), and sulfur (S) cycling. However, the response of soil enzyme activities to anthropogenic disturbances, including manure application, is dependent upon inherent soil properties (e.g. pH, SOM and nutrient content, texture) and environmental factors (e.g. precipitation, temperature). The stabilization and protection of enzymes in soils occurs primarily via association with humic substances and clay complexes (Fig. 6.1), which can increase their resistance to changes in environmental and climatic conditions that affect soil properties. In general, an enzyme-mediated reaction can increase about twofold for every 10 °C increase in temperature between 10 and 50 °C; however, very high temperatures can reduce soil enzyme activities due to inactivation (denaturation) at temperatures higher than 60–70 °C (Tabatabai 1994). In addition to temperature effects, each enzyme has a specific pH value at which the reaction rate is optimal, and at each side of this pH optimum the rate is lower. For example, phosphomonoesterases are classified as acid and alkaline phosphomonoesterases because there are two iso-enzymes that contribute to the total activity with optimal activities in soils under acid and alkaline pH ranges, respectively. In short, the total enzyme activity of soil is very complex and depends on enzyme kinetics and stoichiometry of the different enzyme pools, as well as specific soil properties. Generally, enzyme activities are higher in clay and loam soils than in sandy soils following the clay and organic C contents (Acosta-Martínez and Klose 2008). Similarly, enzyme activities tend to decrease with increasing soil depth due to decreasing amounts of organic C and N along the soil profile (Senwo et al. 2007). Growing plants can also influence soil enzymes, and enzyme activities are often higher in the rhizosphere than in bulk soil (Waldrip et al. 2011).

Enzyme activities have been used as an early indicator of changes in soil quality, as they are more sensitive to changes in land use or management practices than other soil properties (Gregorich et al. 2006; Acosta-Martínez et al. 2007). One of the most common enzyme groups evaluated in soil are the dehydrogenases, which are oxido-reductases that are involved in the oxidation of multiple organic molecules with metabolic cofactors such as NAD^+ or $NADP^+$ as acceptor molecules (Dixon and Webb 1979). The study of dehydrogenase activity can provide information on overall organic matter oxidation through decomposition processes. Dehydrogenase activity is strictly intracellular, and thus reflects the total oxidative activities of the entire soil microbial community. Another commonly investigated group of enzymes are the hydrolases, which catalyze the hydrolysis of various chemical bonds (e.g., ester, glycosidic, ether and peptide bonds) by reaction with water. In soils, hydrolases play a key role in the cycling of C (e.g., β-glucosidase, α-galactosidase), C and N (e.g., β-glucosaminidase), N (e.g., urease, asparaginase, aspartase), P (e.g., alkaline phosphatase, acid phosphatase, phosphodiesterase), and S (e.g., arylsulfatase). While it has been noted that analysis of a number of enzymes involved in different reactions can provide an improved characterization of overall soil biogeochemical cycling, Dick (1994) emphasized the importance of carefully selecting enzymes that may reflect the influence of fertilization or other management practices on soil quality.

The application of manure and other organic amendments to soil has been suggested to exert a more important influence in maintaining soil microbial activity and diversity than other management practices, including conservation tillage (Dick et al. 1988). Most studies agree that significant increases in the activities of soil enzymes are observed in soils amended with animal manures (Khan 1970; Verstraete and Voets 1977; Dick et al. 1988; Martens et al. 1992) and green manures/crop residues (Verstraete and Voets 1977; Dick et al. 1988; Martens et al. 1992), as compared to unamended soils (Parham et al. 2002; Larkin et al. 2005;

Pérez-Piqueres et al. 2006). Some studies have reported a more pronounced increase in enzyme activities under manure applications than inorganic fertilizer (Bolton et al. 1985), while others have reported that repeated applications of inorganic fertilizers can suppress production of certain soil enzymes that are involved in cycling of a given nutrient (e.g. urease and amidase activities) compared to soils that receive manure additions (Dick et al. 1988; Dick 1992).

The response of soil enzyme activities may be short-lived with organic amendment applications, and many studies have reported that (Perucci 1990; Perucci and Giusquiani 1990; Martens et al. 1992). The effects of amendments on soil enzyme activities are difficult to predict because enzymes are substrate specific and the particular response of each enzyme to organic amendment may vary depending on the amendment type (e.g. swine lagoon slurry, solid beef cattle feedyard manure, poultry litter, sewage sludge, crop residues, etc.) and chemical composition (e.g. nutrient and OM content, pH, presence of inhibitory metals). Thus, manure types, application rates, timing and duration of application, application techniques (e.g. surface application of solid or liquid manure, fertigation, injection, incorporation via tillage), and management practices (e.g. tillage, no-till) may affect differently the response of enzyme activities, and this may vary for different soils and climatic and management conditions.

Generally, studies utilize a diverse group of enzymes involved in the mineralization of various N, P, C, and S compounds in order to evaluate the effects of manure on overall nutrient cycling and the different biochemical reactions that occur in soil. However, despite the fact that there is an exhaustive amount of literature on the effects of manure on soil enzyme activities, at present there is no comprehensive overview of recent research findings that compare different management scenarios on soil enzyme activities. The purpose of this chapter is to provide a review of the response of enzyme activities to manure applications and their implications on biogeochemical cycling in agroecosystems. In this chapter, we will provide an overview of the general response trends of various enzyme activities from studies that utilized different types of manure, rates of application, and application techniques (e.g. surface application or incorporation with tillage). Additionally, this chapter provides perspectives and suggestions for avenues of future research.

6.2 Mechanisms Involved in Changing Soil Enzyme Activities due to Manure Applications

There are two possible mechanisms for the effects of manure on enzyme activities of soils (Fig. 6.2). The first mechanism is via manure-induced changes to soil physicochemical properties, including bulk density, moisture content, pH, and soil temperature (Fig. 6.2a). This mechanism also includes the addition of nutrients

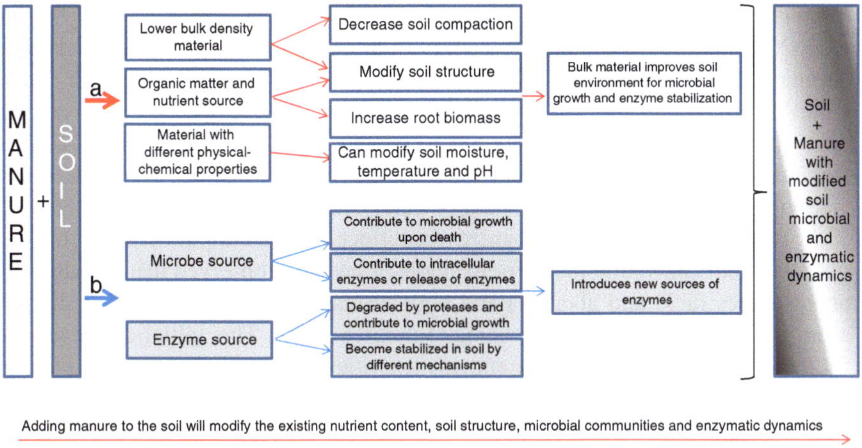

Fig. 6.2 Two possible mechanisms for the effects of manure on enzyme activities of soils: (**a**) via manure-induced changes to soil physicochemical properties, including bulk density, moisture content, pH, and soil temperature; and (**b**) via microbial and enzyme load from the manure, which can contribute to microbial growth upon substrate release following death of manure-derived microbes, directly contribute microbes with intracellular enzymes and/or manure-derived extracellular enzymes

derived from manure, which can improve the soil conditions for microbial growth and increase enzyme stabilization (i.e., increase root biomass, decrease soil compaction). The second mechanism is related to the microbial and enzyme load from the manure, which can contribute to microbial growth upon substrate release following death of manure-derived microbes, directly contribute microbes with intracellular enzymes, and/or contribute manure-derived extracellular enzymes (Fig. 6.2b).

Although organic amendments often contain enzymes, studies have emphasized that the increase observed in enzyme activity in soils amended with organic residues is most likely due to stimulation of microbial activity by changes in soil properties (Fig. 6.2a), rather than the direct addition of amendment-derived enzymes or microbes (Fig. 6.2b) (Martens et al. 1992; Dick 1994). For example, Saison et al. (2006) explained that the effect of a compost amendment on soil enzyme activities was mainly due to the physicochemical characteristics of the compost matrix rather than compost-borne microorganisms. In addition, these researchers (Saison et al. 2006) saw no resilience of microbial characteristics during the study (6–12 months) after amendment with a high rate of compost. However, it has been noted that enzyme activities in manure can be comparable or higher than those found in soils. A recent study reported levels of dehydrogenase activity during initial stages of pig manure compositing are comparable to levels found in soils [160–250 mg triphenyl formazan (TPF) kg^{-1} soil 24 h^{-1}] (Tiquia 2005). In addition, phosphatase activities in poultry manure were more than two orders of magnitude higher than commonly found in soils (Chap. 7, Sect. 7.4). Although studies consider that organic amendment-borne microorganisms may not have long-term survival and therefore may have little effect on soil processes (Saison et al. 2006), the fact that a diverse pool of microbes are found in manure and other organic amendments (Durso et al. 2011) increases the possibility of some direct contribution of manure to soil enzyme pools. Thus, it is possible that among different organic amendments, the microbial and enzyme loads of manure can be significant enough to exert certain influence on the total enzyme activities, especially in low organic matter soils and/or sandy soils.

It is also important to recognize the effects of manure processing on the microbial and enzymatic load. For example, it has been observed that activities of phosphatases (acid and alkaline) and β-glucosidase in manure decreased sharply after just 3 days of active composting (Vuorinen 2000). This initial decrease was followed by an increase in these enzyme activities during the early curing phases, and then finally by an overall decrease in all activities. The study of Vuorinen (2000) also revealed that the bulking material used for composting affected the potential capacity and property for mineralization of P in the manure composts. It is clear that further research is warranted to comparatively evaluate the true contributions of different organic amendments, particularly manure, on the soil enzyme pools. In addition, more information is needed to quantify and qualify the extent of alteration and resilience of the inherent soil microbial community and enzymatic pool over time following manure application.

6.3 Assay Protocols and Sampling to Determine the Response of Enzyme Activities to Manure

6.3.1 Protocols for Assay of Enzyme Activities in Soil

Most studies on the effects of manure on soil enzyme activities have predominantly used similar assay conditions (Nannipieri et al. 1978; Burns 1982; Sinsabaugh and Moorhead 1994; Tabatabai 1994) (Fig. 6.3, Table 6.1). Most assays for enzymes involved in C, P, and S cycling are based on determination of *p*-nitrophenol (PN) released during incubation of soil with buffered substrate solution (i.e., artificial substrate) under conditions determined to be optimal for the specific enzyme (Tabatabai 1994). The assays for enzymes involved in N cycling, such as the amidohydrolases (e.g. urease, aspartase, glutaminase, etc), are more commonly determined by back titration to quantify ammonium (NH_4^+) enzymatically released from specific amino acids or other substrates. Toluene has been commonly incorporated into enzyme assay solutions in order to inhibit microbial growth during the assay and stop further enzyme synthesis by living cells; however, some laboratories opt to omit toluene in order to reduce potential negative effects on the environment and human health following the handling and disposal of assay solutions (Acosta-Martínez and Tabatabai 2011). Among all enzyme assays, the assay for dehydrogenase activity, which is based on the colorimetric determination of the reduction of

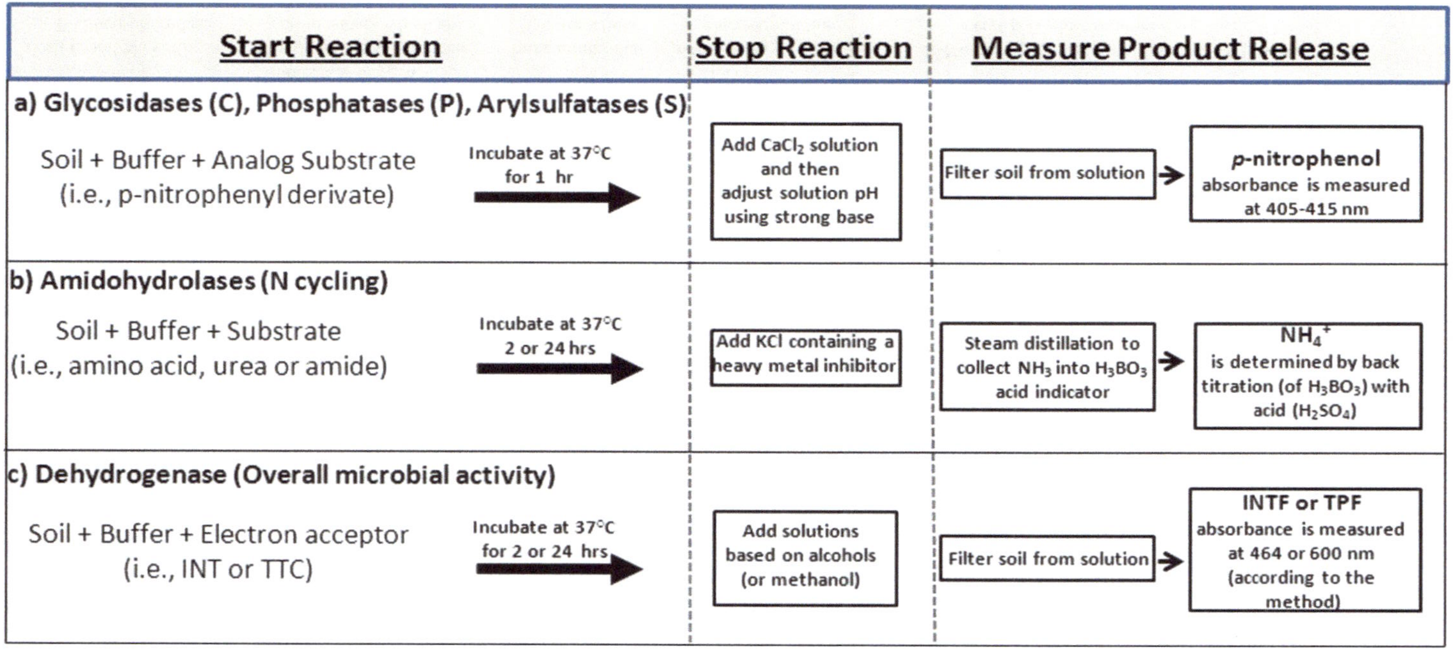

Fig. 6.3 General steps and conditions for determining the most commonly studied soil enzyme activities with: (**a**) colorimetric determination of p-nitrophenol (PN) based assays (Tabatabai 1994); (**b**) distillation and titration of product for amidohydrolases (Tabatabai 1994), and (**c**) colorimetric determination of the reduction of different electron acceptor-indicators (Casida et al. 1964; Prosser et al. 2011). Additional information on enzyme activities and assays are found in Table 6.1

Table 6.1 Description of assay conditions for most commonly assessed soil enzyme activities and their role in soil metabolic capacity and biogeochemical cycling (see assays described in Fig. 6.3)

			Assay conditions			
Recommended name and EC number	Role in soil metabolic function	Location in soils	Substrate	Reaction	Optimum pH	See Fig. 6.3
C cycling (Breakdown of C compounds)						
β-Glucosidase (3.2.1.21)	Glycosidase. Cellulose degradation, produce glucose required as energy source for plants and microorganisms	Intracellular and extracellular	*p*-Nitrophenyl-β-D-glucopyranoside (10 mM)	Glucoside-R + H_2O → Glucose + R-OH	6	a
Dehydrogenase (1.1.1.1)	Oxidoreductase. Catalyzes oxidation of various organic compounds during microbial respiration with the terminal acceptor being molecular oxygen	Strictly intracellular	2, 3, 5-triphenyltetrazollium chloride (TTC) or 2-(4-iodophenyl)-3-(4-nitrophenyl) 5-phenyl-2H-tetrazolium chloride (INT)	Organic matter + TTC + $2e^-$ + $2H^+$ → oxidized organic matter + TPF	7	c
Both C and N cycling β-Glucosaminidase (3.2.1.30)	Chitin degradation, produce amino sugars which are a major form of mineralizable N in soil	Intracellular and extracellular	*p*-Nitrophenyl-N-acetyl-β-D glucosaminidine (10 mM)	R-N-acetyl-β-D glucosaminide → R-OH + N-acetyl-β-D-glucosamine	5.5	a

N cycling (Amidohydrolases: mineralization of organic N compounds to release plant available N, i.e., NH_3 or NH_4^+)						
L-asparaginase (3.5.1.1)	Release of NH_4^+ from the amino acid asparagine	Intracellular and extracellular	Asparagine solution (0.5 M)	L-asparagine + $H_2O \rightarrow$ L-aspartic acid + NH_3	10	b
Amidase (3.5.1.4)	Catalyzes the hydrolysis of amides and produces NH_3 (C-N bonds other than peptide bonds in linear amides)	Intracellular and extracellular	Formamide solution (0.5 M)	$RCONH_2 + H_2O \rightarrow RCOOH + NH_3$	8.5	b
Urease (3.5.1.5)	Catalyzes the hydrolysis of urea to CO_2 and NH_3	Intracellular and extracellular	Urea solution (2 mg ml^{-1})	$NH_2CONH_2 + H_2O \rightarrow CO_2 + 2NH_3$	10	b
P cycling (Phosphatases: mineralization of organic P compounds into simpler inorganic P forms that can be taken up by plant roots, such as $H_2PO_4^-$ and HPO_4^{2-})						
Phosphodiesterase (3.1.4.1)	Produces phosphate monoesters	Intracellular and extracellular	*bis-p*-nitrophenyl phosphate (R-Na_2 $2PO_4$)	R-Na_2 $2PO_4$ + $\rightarrow$R-Na_2PO_4 + R-OH	8	a
Acid Phosphatase (3.1.3.2)	Produces plant available phosphates and is predominant in acid soils	Intracellular and extracellular	*p*-Nitrophenyl phosphate (R-Na_2PO_4)	R-Na_2PO_4 + $H_2O \rightarrow$ R-OH + Na_2HPO_4	6.5	a
Alkaline Phosphatase (3.1.3.1)	Produces plant available phosphates and is predominant in alkaline soils	More intracellular	*p*-Nitrophenyl phosphate (R-Na_2PO_4)	R-Na_2PO_4 + $H_2O \rightarrow$ R-OH + Na_2HPO_4	11	a
S cycling (Sulfatases: mineralization of organic S compounds)						
Arylsulfatase (3.1.6.1)	Produces plant available sulfates (SO_4)	Intracellular and extracellular	*p*-Nitrophenyl sulfate	R-OSO_3^- + $H_2O \rightarrow$ R-OH + SO_4^{2-}	5.8	a

2-(4-iodophenyl)-3-(4-nitrophenyl)-5-phenyl-2H-tetrazolium chloride (INT) (Prosser et al. 2011) or 2,3,5-triphenyltetrazolium chloride (TTC) (Casida et al. 1964), can provide information of overall microbial activity in soil.

To date, it does not appear that more recent approaches, such as microplate-fluorimetric assays, have yet been utilized to evaluate the response of enzyme activities to manure. Microplate-fluorimetric assays are based on detection of 4-methylumbelliferyl (MUF) released by enzymatic hydrolysis of specific substrates when incubated with soil at the enzyme optimal pH. The differences of the detection of MUF via microplate-fluorimetric assays vs. PN with traditional assays have been discussed thoroughly by Deng et al. (2011, 2013), and the use of these new approaches can be considered in future research to investigate the response of enzyme activities to manure applications.

Interpretation of results from current enzyme assay protocols is limited by the lack of approaches available to distinguish among the location of enzymes in soil, as different pools contribute to total activity and these vary for different enzymes and soils (Nannipieri et al. 2002). For example, enzymes more significantly linked to intracellular origin/sources will be more closely related to actual microbial community composition (structure and diversity) of soil. This complication limits researchers' ability to identify the effect of manure on specific enzyme pools and determine how these pools contribute to specific soil processes and nutrient cycling following manure application.

An additional factor to be considered for evaluating soil management effects, such as manure applications, on enzyme activities as indicators of biogeochemical cycling is the fact that enzyme assays only provide information on potential activity, as optimum conditions are typically set in order to achieve maximum enzyme activity (buffer with optimum pH, saturated substrate solution, temperature, and synthetic substrates). These controlled assay conditions described are considered necessary to allow for comparisons across regions and climatic conditions; however, in reality, optimum conditions are rarely found in the field. Further studies are required in order to identify relationships between enzyme activities derived under optimal conditions and actual enzyme activities under variable field conditions.

6.3.2 *Sampling Approaches to Evaluate Response of Enzyme Activities to Manure*

A chapter in an earlier published soil enzymology methodology book (Lorenz and Dick 2011) provides detailed procedural guidelines for soil sampling and pretreatment prior to enzyme analysis. Typically, enzyme activities are determined after air-drying soil subsamples. This use of air-dried soil is in contrast to microbiological assessments [e.g. DNA extraction, fatty acid methyl ester (FAME) analyses, chloroform fumigation methods for microbial biomass C or N], which are generally determined on field-moist samples. It has been reported that air-drying has more of an effect on intracellular enzymes associated with the active microbial community than on extracellular enzymes stabilized in the soil matrix (Tabatabai 1994; Lorenz and Dick 2011). Some enzymes, including urease, β-glucosidase, cellulase, invertase, acid phosphatase and arylsulfatase, do not appear to be affected by air-drying (Acosta-Martínez et al. 2011; Lorenz and Dick 2011). Furthermore, analyzing enzyme activity in air-dried soils may better reflect the soil management history (Lorenz and Dick 2011), as enzymes can become stabilized in soil over years. The choice of how to store and keep soil samples prior to analyzing for enzyme activities is important in order to maintain consistency between samples and allow comparisons over time (i.e., long-term assessment). In the literature, most studies that evaluated the effects of manure applications on enzyme activities have used soils that were air-dried. With air-drying as a common factor, comparisons could be made between studies conducted on soils from different regions, where manures from different livestock types were applied, or where different manure application rates were used, and under different land use and management activities as shown in Table 6.2.

Among biochemical analyses, soil enzyme activities can be determined with simple analyses that require low labor costs compared with other methodologies (Ndiaye et al. 2000). However, analyses of multiple enzyme-mediated reactions that transform different nutrients are required to best evaluate changes that represent overall biochemical cycling. Thus, the selection of soil sampling times and number of enzyme activities to evaluate is a difficult decision, as it is necessary to balance between cost/time involved in sampling and laboratory analyses and the value of the index of biochemical cycling obtained (Dick et al. 1996). The frequency in which soil samples are taken depends on the overall goals of the specific study and it is important to initially decide if the study will focus on short- or long-term effects of manure application.

If short-term effects are being investigated, there should be frequent samplings after beginning manure applications to assist in identification of a status-quo; however, the number of sampling times can be reduced if long-term effects are being addressed (Schinner et al. 1996; Lorenz and Dick 2011). Samples can be taken more often during the first year(s) of manure applications (with samples taken at the same times every year to reduce the influence of seasonal variation) to address changes in enzyme activities which occur within the first years of manure applications. However, this sampling regime is insufficient to assess long-term trends for the soil scenario being evaluated. For longer-term field studies we recommend that soil sampling not be conducted within the first few months following manure application (Schinner et al. 1996), or after tillage or other management practices are performed, in order to avoid confounding effects. Samples in long-term studies should be taken at least two different times within the same year to determine trends that include seasonal changes. Alternatively, sampling at the same time for two (or more) consecutive years can provide an overview of the longer-term effects of manure applications on enzyme activities at that point in time and could reveal year to year climatic variations. As an example, Lorenz and

Table 6.2 Literature review showing values of enzyme activities for manure-treated soils relative to untreated controls

Enzyme activities[a]									
β-glucosidase		Acid phosphatase		Alkaline phosphatase		Urease		Dehydrogenase	
Control	Treated	Control	Treated	Control	Treated	Control	Treated	Control	Treated
89	105			325	424				
89	112			325	415				
88	117	64	81	220	269				
88	143	64	143	220	386				
						65	70	5.4	7.1
						65	78	5.4	7.9
						65	90	5.4	9.2
42	68					59	130		
44	80	154	209	50	145	12	25		
		265	230	45	90			2.3	3.1
174	154			250	303	4.0	7.9	56	8.3
		740	879			42	122		
				110	190	34	63	3	7.5
154	273	304	310	nd					
		19	24	147	226			13.1	19.7
				416	198	42	65	1.60	1.60
				556	1,947			1.90	3.50
				556	1,391			1.90	2.80
				556	974			1.90	2.30
135	220			475	750				
135	195			475	625				
110	112			275	280				
110	145			275	350				
110	190			275	425				
110	188			275	400				
110	160			275	375				
84	232	120	188	295	589	11	45	0.2	0.4
154	281	304	393						
		633	915	273	425	0.1	0.2	1.8	4.2
						65	63	5.4	1.5
						65	58	5.4	4.6
						65	53	5.4	2.7
130	143	305	325	305	325	203	217	0.3	0.3
130	143	305	330	305	330	203	230	0.3	0.3
				430	512	12.5	12.5		
162	204	120	197	295	480	11	40	0.2	0.4

[a]Enzyme activities units vary: mg p-nitrophenol g^{-1} soil h^{-1} (for the first three enzymes), mg NH_4 g^{-1} soil h^{-1} (urease) and mg triphenyl formazan g^{-1} soil h^{-1} (for dehydrogenase)
[b]Manure properties are in Organic C/Total N/P unless preceded by OM (organic matter) or TC (total carbon); rates represent amendment amounts unless followed by N, which represents N application rate
[c]The section provides as much information possible given by each study/references. Goyal et al. (1993) is the only microcosms experiment, and Lalande et al. (2000) requires this clarification: *3.34 kg per m^3 total N, **90 m^3 per ha

Manure information[b]			Study information[c]			
				Soil pH		
Type as named	Organic C/N/P (% DM)	Rate (Mg ha^{-1} year^{-1})	Soil type	Control (treated)	Length (years)	References
Beef	TC30.8/1.3/0.3	5.2	Silt loam	8.3 (8.3)	3	Acosta-Martínez et al. (2011)
Beef	TC30.8/1.3/0.3	10.3	Silt loam	8.3 (8.1)	3	Acosta-Martínez et al. (2011)
Beef	TC36.6/2.3/0.5	1.5	Sandy loam	7.9 (8.0)	2	Acosta-Martínez et al. (2011)
Beef	TC36.6/2.3/0.6	4.2	Sandy loam	7.9 (7.8)	2	Acosta-Martínez et al. (2011)
Beef	nd	0.056 N	Clay loam	7.5 (7.4)	5	Deng et al. (2006)
Beef	nd	0.168 N	Clay loam	7.5 (7.4)	5	Deng et al. (2006)
Beef	nd	0.504 N	Clay loam	7.5 (7.5)	5	Deng et al. (2006)
Beef+Straw	nd	11.2	Silt loam	6.4 (7.0)	64	Bandick and Dick (1999)
Beef+Straw	nd	22.4	Silt loam	nd	55	Dick et al. (1988)
Cattle	nd	0.067 N	Silt loam	5.0 (5.3)	100	Parham et al. (2002)
Cattle	11.4/1.1/0.7	75	nd	8.2 (8.0)	29	Liu et al. (2010)
Cow	nd	5.2	Clay loam	5.5 (6.2)	3	Bhattacharyya et al. (2005)
Ovine	OM46/1.7/1.1	24	Sandy-silty loam	8.2	5	Albiach et al. (2000)
Dairy	TC12.1/0.8/1.2	50.4	Silty loam	6.8 (6.8)	3	Acosta-Martínez et al. (2011)
Farmyard	35/0.5/0.25	*	Sandy loam	8.3	34	Mandal et al. (2007)
Farmyard	32/1.6/0.8	5.6	Sandy loam	7.4 (7.3)	11	Goyal et al. (1999)
Farmyard	28/1.1/nd	90	Sandy loam	7.9	1	Goyal et al. (1993)
Farmyard	28/1.1/nd	45	Sandy loam	7.9	1	Goyal et al. (1993)
Farmyard	28/1.1/nd	15	Sandy loam	7.9	1	Goyal et al. (1993)
Poultry	29.2/4.0/3.3	6.7	Clay	8.4 (7.7)	4	Acosta-Martínez and Harmel (2006)
Poultry	29.2/4.0/3.3	13.4	Clay	8.4 (7.8)	4	Acosta-Martínez and Harmel (2006)
Poultry	29.2/4.0/3.3	4.5	Clay	7.9 (8.1)	4	Acosta-Martínez and Harmel (2006)
Poultry	29.2/4.0/3.3	6.7	Clay	7.9 (7.9)	4	Acosta-Martínez and Harmel (2006)
Poultry	29.2/4.0/3.3	9	Clay	7.9 (8.1)	4	Acosta-Martínez and Harmel (2006)
Poultry	29.2/4.0/3.3	11.2	Clay	7.9 (7.8)	4	Acosta-Martínez and Harmel (2006)
Poultry	29.2/4.0/3.3	13.4	Clay	7.9 (8.0)	4	Acosta-Martínez and Harmel (2006)
Poultry	nd	33.3	Sandy Clay loam	7.9	3	Martens et al. (1992)
Poultry	TC23.4/2.8/2.4	14.4	Silt loam	6.8 (6.5)	3	Acosta-Martínez et al. (2011)
Liquid Hog	*	**	Silt loam	6.5	18	Lalande et al. (2000)
Swine effluent	nd	0.056 N	Clay loam	7.5 (7.4)	5	Deng et al. (2006)
Swine effluent	nd	0.168 N	Clay loam	7.5 (7.4)	5	Deng et al. (2006)
Swine effluent	nd	0.504 N	Clay loam	7.5 (7.2)	5	Deng et al. (2006)
Municipal solid	2.8/2.1/nd	12	Clay loam	–	6	Crecchio et al. (2004)
Municipal solid	2.8/2.1/nd	24	Clay loam	–	6	Crecchio et al. (2004)
Sewage sludge	nd	7.6	nd	–	32	Marschner and Marschner (2003)
Sewage sludge	nd	33.3	Sandy Clay loam	7.9	3	Martens et al. (1992)

Dick (2011) suggested that mid-to-late spring or late fall might be optimal for sampling soil from cropping systems in temperate regions, as there will have been no recent fresh input of organic amendments or fertilizers. It is always important to avoid sampling after recent disturbances, such as tillage, as they can mask the effects of enzyme activities (Schinner et al. 1996; Lorenz and Dick 2011).

When the objective of a study is to evaluate the response of enzyme activities to different manure types and across different soils, it is important to use similar (or comparable) rates of applications and to measure soil enzyme activities at a time of the year when the climate is most stable and there have been no recent soil disturbances (e.g. tillage activities). Manure application rates can be based on target dry matter, N, or P additions. It is crucial to take into account the fact that manures can vary in moisture content, degree of decomposition, and concentrations of C and nutrients. Pagliari and Laboski (2012) analyzed physicochemical properties of 42 manure samples from seven livestock species that were collected from a variety of manure storages (lagoons, manure piles, bedded packs). These researchers found a wide range in concentrations of nutrients, C, moisture, and values for pH and EC, not only between the manures of different species, but also between manures from the same species. Thus, nutrient loads supplied to soils can differ between manures when applied based on specific nutrient requirements (e.g. N or P basis) or when applied on a mass basis (e.g. Mg ha^{-1}). These confounding effects can make it difficult to compare the effects different manure types have on enzyme activities. Manure properties can change soil conditions and influence environmental factors, particularly moisture and temperature, which affect soil chemical and biological properties (Aon and Colaneri 2001), especially under long-term manure applications. This is also very important when comparisons across soils and/or different manure types are intended. In addition to the factors already mentioned, other factors such as cropping history should be considered in order to produce a valid comparison of manure effects on soil enzyme activities. Lorenz and Dick (2011) suggest that for rhizosphere research it is important to identify key moments during the life cycle of the plant species of interest and sample accordingly. The same rationale holds true for research addressing the effects of manure, where care must be taken in sampling in order to reduce the number of factors that could confound analysis.

Lorenz and Dick (2011) also stressed the importance of characterizing a site before sampling in order to represent the horizontal and vertical spatial variability in soil physical and chemical properties. For a study on manure effects on soil properties, we also advise that the approach used for manure application (e.g., banding vs. broadcast) should be considered, as it is important in regard to appropriate depth for soil sampling and duration of the study (i.e., evaluation of long-term vs. short-term effects). We recommend reviewing literature on sampling approaches of soils under manure applications where manure is applied heterogeneously within the plot (e.g., liquid manure applied by banding) as compared to plots where manure solids or slurry are surface broadcast or applied by other methods that provide a more homogenous field coverage by manure (e.g., Tewolde et al. 2013). In order to fully evaluate the effects of manure application on enzyme activities, soil samples should also be taken from control plots that have not received manure (i.e., plots that were unfertilized or received inorganic fertilizer), but that were managed similarly (e.g., tillage, crop rotations, land use) as manure-amended plots with the same soil type.

6.4 Comparing Enzyme Activities as Affected by Different Manures

6.4.1 The Role of the Manures Chemical Composition

Among studies comparing the effects of manures and other organic amendments on enzyme activities, an important factor identified in enzyme response was the differences in amendment chemical composition in terms of the C:N ratios, quantity and quality of substrates, final products of reactions, cofactors, the presence of heavy metals or other inhibitory compounds, and other chemical characteristics (e.g., pH, etc). For example, a study with three different manure types (beef, poultry, and dairy) reported that beef manure applied to loamy soils promoted greater responses in activities of enzymes involved in C cycling (e.g., β-glucosidase, α-galactosidase) than did the other manure types (poultry and dairy) within the first 3 years of applications (Acosta-Martínez et al. 2011). This difference was likely due to the fact that the C content of beef manure was almost twice as high as the other manures. Additionally, when this study compared the enzyme activity response to poultry and dairy manure applications, acid phosphatase activity was greater in soil that received poultry manure than dairy manure. It is possible that this response could be explained by the lower pH of the poultry (pH < 7) and dairy (pH >7) manures that were applied. This study (Acosta-Martínez et al. 2011) also explained that the low pH of the poultry manure caused a decrease in soil pH and subsequent increase in acid phosphatase activity. In general, acid phosphatase activity may respond (increase) independently of soil organic matter content when soil pH is decreased within a given range (Eivazi and Tabatabai 1977; Acosta-Martínez and Tabatabai 2001). In contrast, application of layer hen manure to an acidic (pH 5.0) soil from Maine increased soil pH in a study by Waldrip et al. (2011), but had little effect on either acid or alkaline phosphatase activities in soils. However, application of organic dairy manures to the same soil type resulted in increased acid phosphatase activity over that of soils that received inorganic fertilizer N, and this increase in phosphatase activity in manure amended soil was correlated to manure C:N ratios (Waldrip et al. 2012).

Substrate quality in manure and other organic amendments has been identified as an important determinant of how enzymes will respond independent of C or N content. For example, in a greenhouse study, soil amended with pea vine had greater protease and β-glucosidase activities than soil amended with beef manure when both were applied on an equivalent N basis (Fauci and Dick 1994). These researchers (Fauci and Dick 1994) explained that the pea vine amendment contained less lignin than beef manure; therefore, the C compounds in pea vine were more readily metabolized by the microbial biota than beef manure and supported greater enzyme activities. Another study reported that an increase in β-glucosidase activity in soils amended with composted municipal solid wastes or uncomposted cow manure was not proportional to the quantity of C added with these two organic amendments (Marcote et al. 2001). The difference in response to these two amendments was likely due to organic matter quality or inhibitory effects of the high concentrations of heavy metals in municipal solid waste compost. These researchers (Marcote et al. 2001) also explained that the cellulose in the municipal solid waste (20 %) was derived primarily from paper and cardboard, while the cellulose in the manure (45 %) came mainly from more readily decomposable cereal straw that was used as animal bedding.

6.4.2 The Role of Soil Properties

The response of soil enzymes to organic amendments may vary among different soils due to their inherent soil properties (e.g., soil pH, texture). To date, few studies have simultaneously evaluated the response of different soils to manure (i.e., similar sampling times, manure types and/or regions); therefore, little is known about how specific soil properties influence enzyme activities in regard to manure application. Soil texture can be an important characteristic influencing the response of enzyme activities to manure applications, and in a multi-location study (Acosta-Martínez et al. 2011) reported a faster response of enzyme activities within the first years following application of beef manure to a fine sandy loam in Colorado than to a silt loam in Kansas. In this study (Acosta-Martínez et al. 2011), it was proposed that differences in enzyme activity responses were likely due to lower SOM content and greater sand content in the Colorado sandy loam than the Kansas silt loam.

In two unrelated studies (Albiach et al. 2000; Bhattacharyya et al. 2005), different responses in enzyme activities were observed following application of manure or municipal solid wastes, which may be partially explained by differences in soil pH. For example, one study (Albiach et al. 2000) evaluated the response of several enzyme activities (urease, alkaline phosphatase, phosphodiesterase, arylsulfatase, and dehydrogenase) to equal rates of different organic amendments (municipal solid waste, sewage sludge, sheep manure, vermicompost, and a commercial humic acid solution) added to a horticultural soil. These researchers (Albiach et al. 2000) reported that after 4 years, the highest activities occurred following application of municipal solid waste, followed by sheep manure and sewage. Among the enzyme activities, arylsulfatase and alkaline phosphatase showed the greatest increase (threefold) in response to the addition of municipal solid waste. However, another study (Bhattacharyya et al. 2005) reported higher enzyme activities (urease and acid phosphatase) in soils that had received decomposed cow manure than in soils that received municipal solid waste. An important difference between these two unrelated studies is that the study by Bhattacharyya et al. (2005), reporting lower activities under the municipal solid waste treatment, was conducted on an acidic soil (pH 5.5) in which the high metal concentration in the municipal solid waste may have become more accessible and inhibited enzyme activity or microbial growth. This same inhibitory effect was not observed with the municipal solid waste used in the study by Albiach et al. (2000), where enzyme activities were stimulated despite significantly higher concentrations of certain metals than found in sheep manure. In this study (Albiach et al. 2000), the soil pH was higher (pH 8.0), which would make metals more insoluble than in the lower pH soil of Bhattacharyya et al. (2005). Although there were other differences between these two studies that could have influenced the responses of enzyme activities to applied municipal solid wastes and manures (e.g., climate, soil genesis, texture, etc), it appears that soil pH was a major determinant controlling the response of enzyme activities to organic amendments.

6.5 Comparing the Response of Enzyme Activities to Various Rates of Manure

6.5.1 Studies Evaluating a Single Type of Manure Applied at Different Rates

Important ecosystem implications related to both water and soil quality can be elucidated from studies evaluating the response of soil enzyme activities to differing manure application rates. Recent studies have discussed how the leaching potential of manure-N is partly regulated by the activity of enzymes of soil involved in N transformations (Deng et al. 2000; Schimel and Bennett 2004). Similarly, studies conducted in three states where different manures were applied for 4 years reported that higher poultry manure application rates (13.5 Mg ha^{-1}) to a silt loam caused an increase in acid phosphatase activity and did not result in levels of residual soil test P or Cu and Zn that were considered harmful to surface water or cropping systems (Sistani et al. 2010; Acosta-Martínez et al. 2011). As emphasized in the previous section, trends in enzyme activities following manure applications can be very dependent on soil type. For example, a study on a cultivated high clay soil reported that poultry litter applications of $\geq$6.7 Mg ha^{-1} resulted in increased enzyme activities of C (β-glucosidase, α-galactosidase), C and N (β-glucosaminidase), P (alkaline phosphatase) and S (arylsulfatase) cycling after only four consecutive annual applications; however, these high rates also resulted in nutrient concentrations in excess of crop needs and created the potential for P loss in runoff (Harmel et al. 2004; Acosta-Martínez and Harmel 2006). These results were likely influenced by the impermeability (saturated hydraulic conductivity ~1.5 mm h^{-1}) and low infiltration capacity of the soil used in this study, which was a Texas Blackland Vertisol containing 55 % clay.

Evaluation of different rates of liquid pig amendments (liquid hog manure or pig slurry) on different soil types in two unrelated studies (Lalande et al. 2000; Plaza et al. 2004) revealed proportional increases in several enzyme activities up to a similar rate of organic amendment (90 m^3 ha^{-1}). Among the enzyme activities evaluated by Lalande et al. (2000), dehydrogenase, acid phosphatase and arylsulfatase showed the strongest response to liquid hog manure, while the lowest response was found for urease and alkaline phosphatase activities. However, both studies found a greater response of dehydrogenase activity to liquid pig manures than the other enzymes tested, and urease and phosphatase had the lowest levels of response. An important point raised by Lalande et al. (2000) about the response of enzyme activities to liquid manure applications is the possibility of anaerobic conditions created by higher rates of application (i.e., in their case 120 m^3 ha^{-1}). The points raised by Lalande et al. (2000) are very critical for soil quality and functioning because anaerobic conditions created by excess moisture from liquid manure could decrease microbial diversity and limit nutrient cycling and transformation over time in liquid manure treated soils.

6.5.2 Comparison of Different Types of Manure Applied at the Same Rate

The comparison of different manures applied at similar rates could allow better elucidation of the effects of substrate quality on enzyme activities. For example, Tejada et al. (2006) reported a proportional increase in enzyme activities when cotton gin compost or poultry manure were applied to soil at rates of 5, 8, and 10 Mg OM ha^{-1}; however, they also found higher activities (up to 30 %) of β-glucosidase, phosphatase, and arylsulfatase following application of poultry manure than cotton gin compost. Similarly, a study that compared the effects of swine effluent and beef manure applied to a semiarid soil at rates of 0, 56, 168 and 504 kg N ha^{-1} (Deng et al. 2006) reported that swine effluent caused a decrease, or no change, in the enzyme activities regardless of application rate. In contrast, there was a proportional response of some enzyme activities (L-glutaminase, L-asparaginase, urease) to beef manure applied at the same rates. In the study by Deng et al. (2006), dehydrogenase activity was increased only by the highest application rate (504 kg N ha^{-1}) of beef manure and there was no change in β-glucosaminidase activity.

It seems that it is possible to identify certain types of manure that, depending upon nutrient distribution, have more influence on specific enzyme activities and greater impact on a given nutrient cycle. For example, several studies have demonstrated that there is a higher response of phosphatases to poultry manure or litter application regardless of the soil type, likely due to the high concentrations of P in poultry manure (Acosta-Martínez and Harmel 2006; Deng et al. 2006).

6.5.3 Response of Enzyme Activities to Methods of Manure Application and Management Practices

In addition to the effects of manure types and application rates on enzyme activities, manure application practices could also have an impact on enzyme activities. There are many methods of manure application and incorporation, including irrigation, broadcast spreading, band spreading, and injection into the soil. However, few studies have evaluated the effects of manure application method or incorporation on soil enzyme activities. Some factors that influence the response of microbial communities and enzyme activities under different incorporation techniques could be due to variation in manure moisture content and infiltration between surface application of slurries vs. broadcast spreading of solid manure, soil disturbance with injection or disk incorporation, and compaction from machinery.

Tillage alone can have significant effects on microbial communities and the activities of enzymes. According to phospholipid fatty acids (PFLA) indicators, it appears that no-tillage practices can cause shifts in microbial community structure towards higher fungal populations, as compared to soils under tillage, and this can lead to increases in enzyme activities in no-till plots (Roldán et al. 2005; Kennedy and Schillinger 2006). However, in a study where enzyme activities were evaluated following application of three manure types to loamy soils under no-till and conventional tillage practices, there was no difference between the two management practices for the first 2 years of manure application (Acosta-Martínez et al. 2011). It is clear that additional studies are required to better address the effect of tillage practices and manure management on soil enzyme activities. However, methods of manure application can differ depending on the manure, which makes it difficult to address this factor. For example, liquid manures (pig slurries, dairy lagoon waste) are generally injected or sprayed on the soil surface, whereas solid manures (poultry manure, beef feedyard manure) are more often applied by broadcast spreading and left on the soil surface or incorporated by disking or tillage. In addition, more recent technologies are emerging that can also inject solid manure into the soil. In general, the United States Environmental Protection Agency (USEPA) encourages manure incorporation following application (USEPA 2012) and many states prohibit surface application of manure on frozen soils, during a rainfall event, or when conditions are such that a runoff event could occur. Therefore, there may be limitations on the number of published studies available to evaluate how manure application practices influence enzyme activities.

6.6 Future Research Needs for Comprehensive Knowledge of the Effects of Manure on Enzyme Activities and Soil Nutrient Cycling

Management practices that minimize the addition of organic amendments to soils diminish the potential for increasing enzyme activity, which could affect the ability of soils to cycle and provide nutrients for plant growth (Dick et al. 1988). Manure can also provide more benefits to soil biogeochemical cycling than inorganic fertilizers or other management practices. However, it is not possible to completely generalize a specific response of enzyme activities to manure under diverse soil types, as different factors can influence their response (climatic conditions, rates of application, soil properties, etc.). As new challenges emerge, particularly due to climate change, the selection of manure management practices that lead to a long and sustainable enhancement of biogeochemical cycling and SOM dynamics will be crucial. Thus, to increase our understanding of biogeochemical cycling and SOM dynamics in agroecosystems that receive manure applications, it will be important to investigate enzyme activities across larger temporal and spatial scales in order to take into account the effects of differing management practices, climatic conditions, soil properties, and manure characteristics. Additionally, information of the changes in enzyme activities and other soil processes should be coupled to characterization of the composition of the microbial community in order to establish linkages between the microbial communities associated to shifts in enzyme activities involved in specific reactions and biogeochemical cycling.

In the U.S., there is a need for more comprehensive assessments on how regional management and manure application practices, such as manure application methods, tillage (conventional vs. no-till) and application rates will influence soil enzyme activities. Regional comparisons with the most common manure types will lead to better understanding of the changes in enzyme activities in specific soil types and under certain climatic conditions. However, establishing comparisons across regions will only be possible if enzyme assays are performed with similar protocols for the determination of enzyme activities. We also recommend that analyses be conducted on the activities of several enzymes involved in different reactions in order to maintain the ability to compare results within the literature and within the same location over time. This will allow for a more mechanistic understanding of how soil and environmental properties influence biochemical cycling and nutrient availability. Currently, there is little information on the effects of manure applications on many enzyme activities related to biogeochemical cycling (e.g., N cycling enzymes such as protease, β-glucosaminidase). The use of a diverse range of enzymes can also provide an indication of the substrate quality applied with various type of manure, as they are substrate-specific.

We found it difficult to obtain details on the soil and manure properties within the existing literature. This type of information is important to be incorporated into existing process-level models for simulating soil nutrient transformations and predicting plant nutrient uptake and crop yields [e.g., the Denitrification-Decomposition model, DNDC (Li et al. 1999)]. Through the use of modeling approaches, there is potential for researchers to reach the next level of data interpretation and use, which is the integration of multiple enzyme activities and manure management scenarios into useful indexes for reaching a balance between desired improvements in soil biogeochemical cycling, crop yields, and environmental quality.

References

Acosta-Martínez V, Harmel RD (2006) Soil microbial communities and enzyme activities under various poultry litter application rates. J Environ Qual 35:1309–1318

Acosta-Martínez V, Klose S (2008) Soil enzymes. Encycl Soil Sci 1:1–5

Acosta-Martínez V, Tabatabai MA (2001) Enzyme activities in a limed agricultural soil. Biol Fertil Soils 31:85–91

Acosta-Martínez V, Tabatabai MA (2011) Phosphorus cycle enzymes. In: Dick RP (ed) Methods of soil enzymology, SSSA book series, no. 9. Soil Science Society of America, Madison, pp 161–183

Acosta-Martínez V, Mikha MM, Vigil MF (2007) Microbial communities and enzyme activities in soils under alternative crop rotations compared to wheat–fallow for the central Great Plains. Appl Soil Ecol 37:41–52

Acosta-Martínez V, Mikha MM, Sistani KR, Stahlman PW, Benjamin JG, Vigil MF, Erickson R (2011) Multi-location study of soil enzyme activities as affected by types and rates of manure application and tillage practices. Agriculture 1:4–21

Albiach R, Canet F, Pomares F, Ingelmo F (2000) Microbial biomass content and enzymatic activities after the application of organic amendments to a horticultural soil. Bioresour Technol 75:43–48

Aon MA, Colaneri AC (2001) Temporal and spatial evolution of enzymatic activities and physico-chemical properties in an agricultural soil. Appl Soil Ecol 18:255–270

Bandick AK, Dick RP (1999) Field management effect on soil enzyme activities. Soil Biol Biochem 31:1471–1479

Bhattacharyya P, Chakrabarti K, Chakraborty A (2005) Microbial biomass and enzyme activities in submerged rice soil amended with municipal solid waste compost and decomposed cow manure. Chemosphere 60:310–318

Bolton H Jr, Elliott LF, Papendick RI (1985) Soil microbial biomass and selected soil enzyme activities: effect of fertilization and cropping practices. Soil Biol Biochem 17:297–302

Burns RG (1978) Soil enzymes. Academic, New York/London, pp 1–340

Burns RG (1982) Enzyme activity in soil: location and a possible role in microbial ecology. Soil Biol Biochem 14(5):423–427

Casida LE Jr, Klein DA, Santoro T (1964) Soil dehydrogenase activity. Soil Sci 98:371–376

Crecchio C, Curci M, Pizzigallo M, Ricciuti P, Ruggiero P (2004) Effects of municipal solid waste compost amendments on soil enzyme activities and bacterial genetic diversity. Soil Biol Biochem 36:1595–1605

Deng SP, Moore JM, Tabatabai MA (2000) Characterization of active nitrogen pools in soils under different cropping systems. Biol Fertil Soils 32:302–309

Deng SP, Parham JA, Hattey JA, Babu D (2006) Animal manure and anhydrous ammonia amendment alter microbial carbon use efficiency, microbial biomass, and activities of dehydrogenase and amidohydrolases in semiarid agroecosystems. Appl Soil Ecol 33:258–268

Deng SP, Kang H, Freeman C (2011) Microplate fluorimetric assay of soil enzymes. In: Dick RP (ed) Methods of soil enzymology, SSSA book series, no. 9. Soil Science Society of America, Madison, pp 311–318

Deng SP, Popova IE, Dick L, Dick R (2013) Bench scale and microplate format assay of soil enzyme activities using spectroscopic and fluorometric approaches. Appl Soil Ecol 64:84–90
Dick RP (1992) A review: long-term effect of agricultural systems on soil biochemical and microbial parameters. Agric Ecosyst Environ 40:25–36
Dick RP (1994) Soil enzyme activities as indicators of soil quality. In: Doran JW, Coleman DC, Bezdicek DF, Stewart BA (eds) Defining soil quality for a sustainable environment. American Society of Agronomy, Madison, pp 107–124
Dick RP, Rasmussen PE, Kerle EA (1988) Influence of long-term residue management on soil enzyme activities in relation to soil chemical properties of a wheat-fallow system. Biol Fertil Soils 6:159–164
Dick RP, Breakwill D, Turco R (1996) Soil enzyme activities and biodiversity measurements as integrating biological indicators. In: Doran JW, Jones AJ (eds) Handbook of methods for assessment of soil quality. Soil Science Society of America, Madison, pp 247–271
Dixon M, Webb EC (1979) Enzymes, 3rd edn. Longman Group Ltd., London
Durso LA, Gilley JE, Marx DB, Woodbury BL (2011) Effects of animal diet, manure application rate, and tillage on transport of microorganisms from manure-amended fields. Appl Environ Microbiol 77:6715–6717
Eivazi F, Tabatabai MA (1977) Phosphatases in soils. Soil Biol Biochem 9:167–172
Fauci MF, Dick RP (1994) Soil microbial dynamics: short- and long-term effects of inorganic and organic nitrogen. Soil Sci Soc Am J 58:801–806
Goyal S, Mishra MM, Dhankar SS, Kapoor KK, Batra R (1993) Microbial biomass turnover and enzyme activities following the application of farmyard manure to field soils with and without previous long-term applications. Biol Fertil Soils 15:60–64
Goyal S, Chander K, Mundra MC, Kapoor KK (1999) Influence of inorganic fertilizers and organic amendments on soil organic matter and soil microbial properties under tropical conditions. Biol Fertil Soils 29:196–200
Gregorich EG, Beare MH, McKim UF, Skjemstad JO (2006) Chemical and biological characteristics of physically uncomplexed organic matter. Soil Sci Soc Am J 70:975–985
Harmel RD, Torbert HA, Haggard BE, Haney R, Dozier M (2004) Water quality impacts of converting to a poultry litter fertilization strategy. J Environ Qual 33:2229–2242
He Z, Griffin TS, Honeycutt CW (2004) Enzymatic hydrolysis of organic phosphorus in swine manure and soil. J Environ Qual 33:367–372
Kanchikerimath M, Singh D (2001) Soil organic matter and biological properties after 26 years of maize–wheat–cowpea cropping as affected by manure and fertilization in a Cambisol in semiarid region of India. Agric Ecosyst Environ 86:155–162
Kennedy AC, Schillinger WF (2006) Soil quality and water intake in traditional-till vs. no-till paired farms in Washington's Palouse region. Soil Sci Soc Am J 70:940–949
Khan SU (1970) Enzymatic activity in a gray wooded soil as influenced by cropping systems and fertilizers. Soil Biol Biochem 2:137–139
Kiss S, Drâgan-Bularda M, Radulescu D (1975) Biological significance of enzymes accumulated in soils. Adv Agron 27:25–87
Klose S (2003) Enzyme mediated reactions and microbial biomass of agricultural and fly ash influenced forest ecosystems. Habilitation dissertation. Department of Forest and Hydro Sciences, Dresden University of Technology. Dresden, Germany, 189 pp
Ladd JN, Butler JHA (1975) Humus- enzyme systems and synthetic, organic polymer-enzyme analogs. Soil Biochem 4:143–194
Lalande R, Gagnon B, Simard RR, Côte D (2000) Soil microbial biomass and enzyme activity following liquid hog manure application in a long-term trial. Can J Soil Sci 80:263–269
Larkin JE, Frank BC, Gavras H, Sultana R, Quackenbush J (2005) Independence and reproducibility across microarray platforms. Nat Methods 2:337–344
Li Y, Aneja VP, Arya SP, Rickman J, Brittig J, Roelle P, Kim DS (1999) Nitric oxide emission from intensively managed agricultural soil in North Carolina. J Geophys Res 104:26115–26123

Liang YC, Yang YF, Yang CG, Shen QQ, Zhou JM, Yang LZ (2003) Soil enzymatic activity and growth of rice and barley as influenced by organic matter in an anthropogenic soil. Geoderma 115:149–160

Liang Y, Nikolic M, Peng Y, Chen W, Jiang Y (2005) Organic manure stimulates biological activity and barley growth in soil subject to secondary salinization. Soil Biol Biochem 37:1185–1195

Liu E, Yan C, Mei X, He W, Bing SH, Ding L, Liu Q, Liu S, Fan T (2010) Longterm effect of chemical fertilizer, straw, and manure on soil chemical and biological properties in northwest China. Geoderma 158:173–180

Lorenz N, Dick RP (2011) Sampling and pretreatment of soil before enzyme analysis. In: Dick RP (ed) Methods of soil enzymology, SSSA book series, no. 9. Soil Science Society of America, Madison, pp 85–101

Mandal A, Patra AK, Singh D, Swarup A, Masto RE (2007) Effect of long-term application of manure and fertilizer on biological and biochemical activities in soil during crop development stages. Bioresour Technol 98:3585–3592

Marcote I, Hernández T, García C, Polo A (2001) Influence of one or two successive annual applications of organic fertilisers on the enzyme activity of a soil under barley cultivation. Bioresour Technol 79:147–154

Marschner P, Marschner EB (2003) Structure and function of the soil microbial community in a long-term fertilizer experiment. Soil Biol Biochem 35:453–461

Martens DA, Johanson JB, Frankenberger WT Jr (1992) Production and persistence of soil enzymes with repeated additions of organic residues. Soil Sci 153:53–61

Nannipieri P, Johnson RL, Paul EA (1978) Criteria for measurement of microbial growth and activity in soil. Soil Biol Biochem 10:223–229

Nannipieri P, Kandeler E, Ruggiero P (2002) Enzyme activity as monitors of soil microbial functional diversity. In: Burns R, Dick R (eds) Enzymes in the environment: activity, ecology and applications. Marcel Dekker, New York, pp 1–34

Ndiaye EL, Sandeno JM, McGrath D, Dick RP (2000) Integrative biological indicators for detecting change in soil quality. Am J Altern Agric 15:26–36

Pagliari PH, Laboski CAM (2012) Investigation of the inorganic and organic phosphorus forms in animal manure. J Environ Qual 41:901–910

Parham JA, Deng SP, Raun WR, Johnson GV (2002) Long-term cattle manure application in soil. I. Effect on soil phosphorus levels, microbial biomass C, and dehydrogenase and phosphatase activities. Biol Fertil Soils 35:328–337

Pérez-Piqueres A, Edel-Hermann V, Alabouvette C, Steinberg C (2006) Response of soil microbial communities to compost amendments. Soil Biol Biochem 38:460–470

Perucci P (1990) Effect of addition of municipal solid waste compost on microbial biomass and enzyme activities in soil. Biol Fertil Soils 10:221–226

Perucci P, Giusquiani PL (1990) Influence of municipal waste compost addition on chemical properties and soil-phosphatase activity. Zentralblatt für Mikrobiologie 145:615–620

Plaza C, Hernandez D, Garcia-Gil JC, Polo A (2004) Microbial activity in pig slurry-amended soils under semiarid conditions. Soil Biol Biochem 36:1577–1585

Prosser JA, Speir TW, Stott DE (2011) Soil oxidoreductases and FDA hydrolysis. In: Dick RP (ed) Methods of soil enzymology, SSSA book series, no. 9. Soil Science Society of America, Madison, pp 103–124

Roldán A, Salinas-García JR, Alguacil MM, Caravaca F (2005) Changes in soil enzyme activity, fertility, aggregation and C sequestration mediated by conservation tillage practices and water regime in a maize field. Appl Soil Ecol 30:11–20

Saison C, Degrange V, Oliver R, Millard P, Commeaux C, Montange D, Le Roux X (2006) Alteration and resilience of the soil microbial community following compost amendment: effects of compost level and compost-borne microbial community. Environ Microbiol 8:247–257

Schimel JP, Bennett J (2004) Nitrogen mineralization: challenges of a changing paradigm. Ecology 85:591–602
Schinner F, Ohlinger R, Kandeler E, Margesin R (1996) Methods in soil biology. Springer, Heidelberg/Berlin/New York
Senwo ZN, Ranatunga TD, Tazisong IA, Taylor RW, He Z (2007) Phosphatase activity of Ultisols and relationship to soil fertility indices. J Food Agric Environ 5(1):262–266
Sinsabaugh RL, Moorhead DL (1994) Resource allocation to extracellular enzyme production: a model for nitrogen and phosphorus control of litter decomposition. Soil Biol Biochem 26:1305–1311
Sistani KR, Bolster CH, Way TR, Tobert HA, Pote DH, Watts DB (2010) Influence of poultry litter application methods on the longevity of nutrient and E. coli in runoff from tall fescue pasture. Water Air Soil Pollut 206:3–12
Tabatabai MA (1994) Soil enzymes. In: Weaver RW, Angle S, Bottomley P, Bezdicek D, Smith S, Tabatabai MA, Wollum A, Weaver RW, Angle S, Bottomley P, Bezdicek D, Smith S, Tabatabai MA, Wollum A (eds) Methods of soil analysis. Part 2: Microbiological and biochemical properties. Soil Science Society of America, Madison, pp 775–833
Tejada M, Hernández MT, García C (2006) Application of two organic amendments on soil restoration: effects on the soil biological properties. J Environ Qual 35:1010–1017
Tewolde H, Way TR, Pote DH, Adeli A, Brooks JP, Shankle MW (2013) Method of soil sampling following subsurface banding of solid manures. Agron J 105:519–526
Tiquia SM (2005) Microbiological parameters as indicators of compost maturity. J Appl Microbiol 99:816–828
USEPA (2012) Protocols for land application of manure nutrients, NPDES permit writers' manual for CAFOs. United States Environmental Protection Agency, Washington, DC. Available at: http://www.epa.gov/npdes/pubs/cafo_permitmanual_chapter6.pdf. Accessed 17 June 2013
Verstraete W, Voets JP (1977) Soil microbial and biochemical characteristics in relation to soil management and fertility. Soil Biol Biochem 9:253–258
Vuorinen AH (2000) Effect of the bulking agent on acid and alkaline phosphomonoesterase and β-D-glucosidease activities during manure composting. Bioresour Technol 75:133–138
Waldrip HM, He Z, Erich MS (2011) Effects of poultry manure amendment on phosphorus uptake by ryegrass, soil phosphorus fractions and phosphatase activity. Biol Fertil Soils 47:407–418
Waldrip HM, He Z, Griffin TS (2012) Effects of organic dairy manure on soil phosphatase activity, available soil phosphorus and growth of sorghum-sudangrass. Soil Sci 177:629–637

Chapter 7
Phosphatase Activities and Their Effects on Phosphorus Availability in Soils Amended with Livestock Manures

Heidi M. Waldrip and Veronica Acosta-Martínez

Abstract The application of livestock manures can impact factors related to phosphorus (P) cycling and concentrations of plant-available P in soils. Specific manure physicochemical properties differ due to livestock species and management practices, which may result in differences in parameters related to soil fertility when manures are applied to soil. To date, no evaluation has been conducted on the differences among manure types on soil phosphatase activities and P availability. This chapter presents information on the most commonly studied soil phosphatases, acid and alkaline phosphomonoesterase and phosphodiesterase, and how manure application influences their activities and P cycling. In a case study, it is shown that soil application of manures from organically managed dairies increases acid phosphatase activity; however, concentrations of available P in soils that receive manure from organic dairies are similar to those that receive manure from conventional dairies. Depending upon specific research goals, different *in vitro* experimental approaches may be used prior to evaluation of phosphatase activity, but specific methodology influences phosphatase activities and can hinder among-study evaluation of effects of livestock manure on P dynamics. Research avenues are suggested to improve current understanding of the effects of livestock manure on soil quality and functioning related to P availability in soils.

H.M. Waldrip (✉)
USDA-ARS, Conservation and Production Research Laboratory, Bushland, TX 79012, USA
e-mail: heidi.waldrip@ars.usda.gov

V. Acosta-Martínez
USDA-ARS, Cropping Systems Research Laboratory, Lubbock, TX 79417, USA
e-mail: veronica.acostamart@ars.usda.gov

Z. He and H. Zhang (eds.), *Applied Manure and Nutrient Chemistry for Sustainable Agriculture and Environment*, DOI 10.1007/978-94-017-8807-6_7,

7.1 Introduction

Phosphorus (P) is a principal component of nucleic acids, phospholipids, adenosine triphosphate (ATP) and other essential metabolic molecules, and is a required nutrient for growing plants and all other living organisms. In soil, P exists in both inorganic and organic forms, and typically comprises 0.02–0.13 % of soil dry matter (Stevenson and Cole 1999). Despite its abundance in soil, most soil P is present in organic and relatively insoluble forms that are not immediately available for plant uptake (Hedley et al. 1995). Due to limitations of available P, synthetic fertilizers or organic residues, including livestock manure, are commonly applied to meet crop nutrient demands. However, care must be exercised when applying manure either as crop fertilizer or as a disposal mechanism for excess manure, as over-application or mismanagement can lead to environmental consequences in the event of runoff of soluble or particulate-associated P into sensitive surface waters (Dao et al. 2008; Kleinman et al. 2002).

Phosphorus cycling in soils occurs through a complex suite of reactions that includes immobilization by incorporation into microbial biomass, mineralization of organic P into simpler forms, sorption to mineral components, precipitation of mineral-P complexes, and mineral dissolution. A simplified version of the soil P cycle in agroecosystems, with relevant inputs, outputs, and P transformations is presented in Fig. 7.1. In cropping systems, P is often inaccessible to plants, as most is bound to mineral components or organic matter. Inputs of P into the system can be derived from inorganic fertilizers, plant residues, organic amendments, or direct deposition of animal manure.

The phosphatases are a diverse group of enzymes that are important in the mineralization of organic P. They catalyze the hydrolysis of different organic P moieties into simpler inorganic P forms that can be taken up directly by plants. The application of livestock manure changes the chemical, physical and microbial properties of soil, which may influence phosphatase activities and P availability (Acosta-Martinez and Harmel 2006; He et al. 2006a; Parham et al. 2002; Waldrip et al. 2011). However, manure-induced changes in soil properties are not easily predictable and may differ depending upon specific physicochemical characteristics of soil and manure.

The aims of this chapter are to: (1) provide a general overview of the enzymes responsible for organic P hydrolysis, (2) establish linkages between the role of soil phosphatase activities and P availability following application of livestock manure, and (3) reveal how specific methodologies can impede comparison of phosphatase activity results among studies.

7.2 Overview of Phosphatase Activities in Soil

In soil, enzymatic hydrolysis by extracellular phosphatases is the primary mechanism by which organic P forms are mineralized into bioavailable inorganic P ($H_2PO_4^-$ and HPO_4^{2-}) (Li et al. 2004; Tarafdar and Claassen 1998). Phosphatase activity is a key

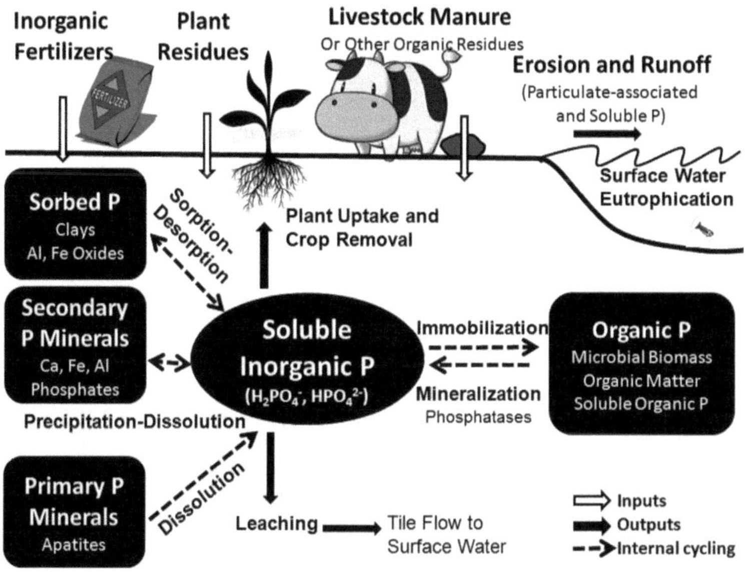

Fig. 7.1 Overview of the soil P cycle highlighting the physical, chemical, and microbial processes involved in transformation of P between soluble/insoluble and organic/inorganic forms (Original graphic by Emmylou Dail)

component of P cycling in soils and is one indicator of soil quality and microbial activity. Within the soil, there are five major groups of phosphatase enzymes: (1) phosphoric monoester hydrolases (EC 3.1.3, phosphomonoesterases), (2) phosphoric diester hydrolases (EC 3.1.4, phosphodiesterases), (3) phosphoric triester hydrolases (EC 3.1.5), (4) polyphosphate hydrolases (EC 3.6.1), and (5) enzymes acting on P–N bonds (EC 3.9). Among the phosphatase enzymes in soil, the most often investigated are the phosphodiesterases and phosphomonoesterases (Tabatabai 1994).

The breakdown of organic P begins with the action of the phosphodiesterases (PDE). Phosphodiesterases are responsible for the hydrolysis of phosphodiesters, including phospholipids and nucleic acids, to form simpler phosphomonoesters (Browman and Tabatabai 1978). Thus, the release of free phosphate from a diester requires the sequential action of PDE followed by phosphomonoesterase. The rate of PDE activity in soil tends to be an order of magnitude lower than that of phosphomonoesterase, and it has been suggested that PDE activity regulates organic P turnover in pasture soils (Turner and Haygarth 2005). Figure 7.2 illustrates how organic P inputs to soil are converted into free phosphate by the sequential action of PDE and phosphomonoesterase. Despite the apparent relationship between PDE activity and P availability, to date there has been far less study on the role of PDE in soil P cycling than the phosphomonoesterases.

Phosphomonoesterases are a group of relatively non-specific phosphohydrolases that are responsible for the degradation of a wide variety of phosphomonoesters, including inositol phosphates, β-glycerophosphate, phenylphosphate, sugar

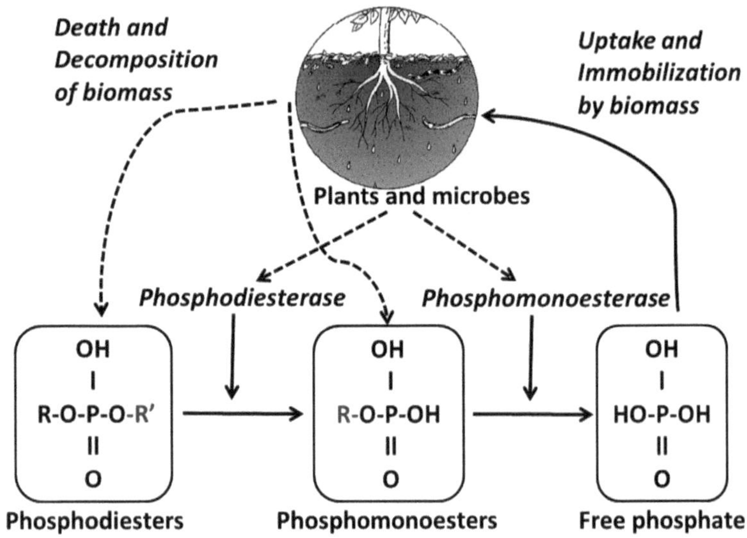

Fig. 7.2 A conceptual model of the turnover of organic phosphorus in the soil, outlining the sequential action of phosphodiesterase and phosphomonoesterase. *R* and *R′* represent organic moieties of P compounds. *Dashed arrows* indicate inputs from plants, residues, and microbial biomass (Original graphic by Heidi Waldrip)

phosphates, adenosine phosphates, β-napthyl phosphate, and *p*-nitrophenyl phosphate (He et al. 2004a). The most frequently investigated soil phosphomonoesterases are classified according to pH optima as acid phosphomonoesterase (pH 4.8, EC 3.1.3.2, ACP) and alkaline phosphomonoesterase (pH 8.5, EC 3.1.3.1, ALP). In addition to the less substrate-specific acid and alkaline phosphomonoesterases, phytases (EC 3.1.3) have been identified in soil and play a major role in mineralization of phytate and lower inositol phosphates (Lim et al. 2007; Richardson and Hadobas 1997).

7.2.1 Origins of Extracellular Soil Phosphatases

Extracellular soil phosphatases originate from a number of sources, including bacteria, fungi, and plant roots. Both plants and microorganisms produce ACP, and ACP can be excreted from the apoplast of lateral plant roots under conditions of P deficiency (Speir and Cowling 1991; Tarafdar and Claassen 1998). In contrast, it does not appear that plants can produce ALP (Li et al. 2004); therefore, extracellular ALP in soils is assumed to be microbial in origin. Pyrosequencing and other advanced techniques have recently been used to characterize the microbial diversity of soil, and have revealed significant positive correlations between ALP activity and several bacterial populations, including *Proteobacteria*, *Firmicutes*, *Chloroflexi*, *Verrucomicrobiae* and *Fibrobacteres* (Acosta-Martinez et al. 2010).

It appears that the primary source of soil PDE is bacteria or actinomycetes; however, some plants can secrete PDE under P-deficient conditions (Abel et al. 2000; George et al. 2008; Leake and Miles 1996). At present, it is unclear how much of a role root-derived PDE activity plays in P nutrition of plants, as no clear correlation has been made between P use efficiency and enzyme activities. However, Waldrip et al. (2011) identified a correlation between PDE activity and concentrations of labile inorganic P in the rhizosphere of perennial ryegrass.

7.2.2 Analysis of Soil Phosphatase Activities

Analyses of soil enzyme activities are usually conducted using standardized assays under controlled conditions. Specific details on methodologies can be found in Acosta-Martinez and Waldrip (Chapter 9). The general procedure involves the incubation of known amounts of soil and target substrate, followed by a colorimetric procedure to determine end-product concentration. Determination of activities of the phosphomonoesterases and PDE are generally conducted using the original method developed by Tabatabai and Bremner (1969). The method is simple and well-accepted, and has been widely used by researchers to investigate the impacts of various agricultural and environmental factors on soil P cycling (Acosta-Martinez and Harmel 2006; Chen et al. 2004; Colvan et al. 2001; Margesin and Schinner 1994; Waldrip et al. 2011, 2012). Although the procedure is often modified, the basic protocol involves the quantification of *p*-nitrophenol (*p*-NP) produced following incubation of soil with a buffered *p*-nitrophenol phosphate solution (or *bis-p*-NP-P in the case of PDE). The production of *p*-NP during a set incubation time, determined by measuring absorbance at 400 nm, is considered proportional to enzyme activity and is normally expressed as mass of *p*-NP produced per mass of soil per unit time (e.g. mg kg^{-1} h^{-1}).

Despite the simplicity and widespread acceptance of this methodology, full interpretation of the results is complicated by the fact that the standard phosphatase assay does not discriminate between enzymes of different origins. Thus, the assay detects activities of recently-produced root and microbial enzymes, as well as those from dead cells or immobilized on soil clay or humic colloids (Burns 1982; Sinsabaugh 1994).

7.2.3 Factors That Influence Soil Phosphatase Activities

There have been reports of a wide range of soil phosphatase activities and it can be difficult to make comparisons across soils, land uses, and climatic conditions. As an example, we found large differences in magnitude of phosphatase activities from two studies conducted in Texas in 2006 on soils with a high clay content (up to 38 %) and very similar chemical and physical characteristics (Acosta-Martinez and Harmel 2006; Bell et al. 2006). Bell et al. (2006) reported very low rates (<50 μg *p*-NP g^{-1} h^{-1})

of ALP activity in soil from grazed rangeland, while Acosta-Martinez and Harmel (2006) found much higher rates in grazed pasture soils (~400 to 500 μg *p*-NP g^{-1} h^{-1}). An overview of the literature revealed a wide range of reported phosphatase activities in unfertilized soils (Chap. 6, Table 6.1). Activities of ACP varied the most, ranging from 19 to over 740 μg *p*-NP g^{-1} h^{-1}, followed by ALP activities (1 to >550 μg *p*-NP g^{-1} h^{-1}). Activities of PDE also varied (30 to >450 μg *bis-p*-NP g^{-1} h^{-1}) among different soils and cropping systems, but were more consistent than ACP and ALP.

As organic P mineralization is a biotic process, it is influenced by any factor that affects biological activity, including soil pH, temperature, moisture content, and the presence of C, N, P and other nutrients required for microbial metabolism (Parham et al. 2002; Senwo et al. 2007; Huang et al. 2011). However, the extent to which these factors influence phosphatase activity can change with season and land management practices. As an example, Huang et al. (2011) reported that changes in ACP activity of forest soils in China coincided with seasonal climate pattern, with higher values observed during wet-warm periods (spring and summer) than during dry-cool periods (fall and winter). In contrast, Sardans et al. (2006) monitored phosphatase activities in soils from Mediterranean shrubland for five seasons, and found that moisture content did not influence either ACP or ALP activities and response to temperature was inconsistent.

According to Senwo et al. (2007), soil texture is the most significant property influencing the activity of ACP, accounting for 68 % of the variation observed between soils. These researchers (Senwo et al. 2007) regressed soil physiochemical properties against enzyme activities in soils from a range of land use types (pasture, mixed grasses, weeds, woodlands, and pine plantation) and found that ALP activity was most highly correlated to CEC, followed by Fe content and pH. For PDE, pH and SOM accounted for more than 58 % and 67 %, respectively, of the variation between different soil types.

In addition to soil properties, root exudates and other factors related to growing plants influence soil phosphatase activities and P availability (Hinsinger 2001; Waldrip et al. 2011). Tarafdar and Jungk (1987) reported that ACP and ALP activities were higher in the rhizosphere of four plant species than in the bulk soil, and this corresponded with greater fungal and bacterial populations. In the rhizosphere of barley, there is more diversity in the community of ALP-producing bacteria under P-limiting conditions (Chhabra et al. 2013). It is therefore not easy to determine if higher phosphatase activities are related to an increase in the production of root phosphatases or a result of root exudates enhancing the rhizosphere for microbial growth. As previously mentioned, the inability of the standard phosphatase assay to differentiate between root-derived activities from that of microbial phosphatase, or to separate activity of newly-produced enzymes from that of older, stabilized extracellular phosphatases can greatly complicate analysis.

While phosphomonoesterases and PDE can be produced in response to P deficient conditions, the presence of phosphate can also inhibit both activity and synthesis of phosphomonoesterases and phytase (Acosta-Martinez and Harmel 2006; Olander and Vitousek 2000; Oshima et al. 1996; Waldrip et al. 2011). Thus, the application of any material that increases the concentration of soluble orthophosphate could potentially affect the mineralization of organic P compounds and influence overall P cycling and plant P availability. For example, transcription of ACP and ALP in *Saccharomyces cerevisiae* is repressed by high concentrations of available P (Oshima et al. 1996).

7.3 Effects of Livestock Manures on Soil Phosphatase Activities and P Availability

The application of livestock manure to soils can change the P distribution in soils; however, these changes can differ depending upon specific soil and manure properties (He et al. 2006b; Waldrip-Dail et al. 2009; Waldrip et al. 2011). The response of soil phosphatase activities to livestock manure could be dependent on many factors, such as soil type, pH, specific P forms in soil and manure, C:N ratio of manure, history of fertilization, and other environmental conditions and management practices. The application of livestock manures is frequently reported to increase soil phosphatase activities (Garg and Bahl 2008; Leytem et al. 2011; Parham et al. 2002); however, little work has been conducted to compare the effects of different manures on phosphatase activities or P availability.

The chemical and physical composition of manures can differ dramatically due to species digestive system differences, bedding types, length of storage, and other management practices (Dou et al. 2001; Leinweber et al. 1997; Pagliari and Laboski 2012). Large differences also exist between the specific forms of P found in manure from different livestock production systems. One key difference between monogastric species, such as swine and poultry (non-ruminants), and ruminants (beef cattle, dairy cows, sheep, etc.) is that the phytase enzyme responsible for degrading phytate is lacking in non-ruminants; thus, manure from swine and poultry contains larger amounts of phytate than manure from beef cattle and dairy cows. The difference between P excreted by dairy cows and poultry is illustrated by comparison of work by He et al. (2008) and Dail et al. (2007) (Fig. 7.3). In these studies, manures were sequentially extracted with H_2O (soluble P), 0.5 M $NaHCO_3$ (labile P sorbed on crystalline surfaces), 0.1 M NaOH (moderately labile P associated with Al and Fe oxides or carbonates), and 1.0 M HCl (Stable P associated with Ca^{2+}). Clear differences exist between the P distributions in the manures, with 80 % of dairy manure P in labile inorganic forms. In contrast, poultry manure contained more recalcitrant mineral-associated inorganic P (NaOH- and HCl-Extractable P). The majority of organic P in poultry manure is Ca^{2+}-associated phytate (HCl-P), which has the potential to bioaccumulate in soils due to its large number of reactive P groups (He et al. 2008; Turner and Leytem 2004).

There have been conflicting reports on the effects of livestock manure on soil phosphatase activities. Most frequently, phosphatase activities are increased following application of organic fertilizers (Acosta-Martinez and Harmel 2006; Colvan et al. 2001; Pascual et al. 2002; Waldrip et al. 2011, 2012). The stimulatory effect of organic amendment may involve a complex combination of physical,

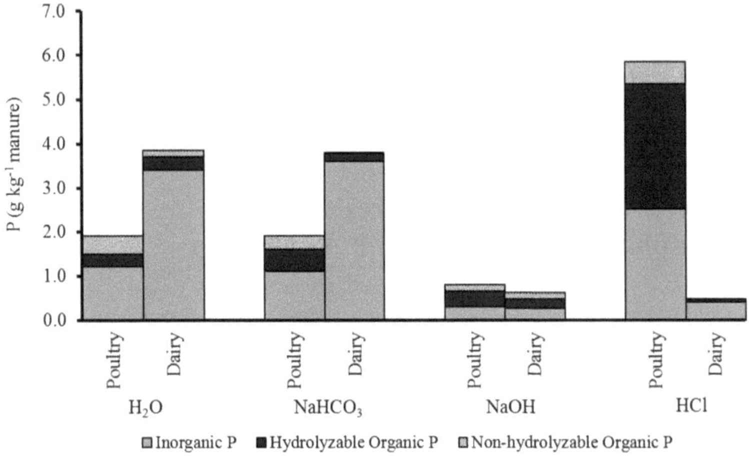

Fig. 7.3 Comparison of phosphorus distribution in sequential extracts of dairy manure and poultry manure (Poultry manure data adapted from Dail et al. (2007) and dairy manure from He et al. (2004b))

chemical, and biological soil properties and processes; however, it has been proposed that manure's contribution of organic matter and growth substrates for microorganisms is a major contributing factor (Garcia-Ruiz et al. 2008). Garcia-Ruiz et al. (2008) examined changes in phosphomonoesterase activity in 18 pairs of organic and conventionally managed olive orchards, and found higher activity in soils under organic management where there was a history of addition of animal manures or other organic amendments. In this study, there were no discernible differences in soil physicochemical properties between the two fertilization regimes, suggesting that differences in enzyme activity were related to manure addition rather than soil attributes.

Overall changes to phosphatase activities resulting from addition of manure may be short-lived due to adaptations by soil microorganisms to higher levels of soil organic matter and available growth substrates. Criquet et al. (2007) incubated soil with anaerobically digested sewage sludge and found higher phosphatase activities (ACP, ALP, and PDE), at 25 days, coinciding with higher microbial density and available P. However, after 85 days the enzyme activities were reduced 30–50 %, while microbial density and P levels remained elevated. Similar patterns in soil enzyme activities following organic amendment were reported in other studies (Kizilkaya and Bayrakli 2005; Pascual et al. 1998). Despite the trend towards adaptation of the soil microbial community to increased growth substrates and subsequent decreased enzyme activity over time, there have also been reports of long-term increases in ALP and PDE with manure addition (Acosta-Martinez and Harmel 2006; Bell et al. 2006; Parham et al. 2002). As an example, in a 3-year field study, Acosta-Martinez and Harmel (2006) found that addition of poultry manure to pasture increased both microbial biomass and ALP activity.

From the literature, it appears that poultry manure may have a more profound influence on phosphatase activities than manure from other livestock species. In a

90 day aerobic incubation study, Garg and Bahl (2008) compared the effects of different organic amendments (poultry manure, farmyard manure, green manure, and crop residue) on ALP activity and found that the greatest increases were observed with poultry manure, followed by farmyard manure, with activities that were up to 133 % and 104 % higher than unfertilized control, respectively. This increase in enzyme activity corresponded with increases in available (Olsen-extractable) P. These researchers (Garg and Bahl 2008) attributed these effects to low lignin content and high C, N and cellulose content in the poultry manure. The rationale for this is clear, as cellulose is comprised primarily of readily degradable monopolysaccharides, while lignin contains complex phenolic compounds that are more resistant to degradation; therefore, rapid decomposition of C-containing structural components of poultry manure led to faster nutrient turnover with poultry manure than other organic amendments.

The addition of composted manure to soils can result in longer-term increases in enzyme activities than when fresh materials are applied due to enzyme stabilization and a more favorable C:N ratio for microbial growth (He et al. 2010). However, there have also been reports of decreased soil phosphatase activity following the addition of both fresh manure and composted organic material (Garcia-Gil et al. 2000), leading to the conclusion that amendment effects are not easily predicted.

We recently measured ACP, ALP, and PDE activities in fresh and freeze-dried layer hen manure and found that activities increased in the order of ACP $<$ ALP $<$ PDE (unpublished data). All activities were significantly higher than in soil, where PDE activity of fresh poultry manure was 3,686 mg *p*-NP kg^{-1}: more than two orders of magnitude over what is commonly reported for soil. In addition, freeze-drying did not significantly change ACP or ALP activities in these manures, indicating that both fresh and older, stored manures could be a direct source of phosphatase activity.

7.3.1 Case Study: Effects of Application of Organic and Conventional Dairy Manures on Soil Phosphatase Activities

Organic dairy (OD) production is increasing in the Northeastern U.S. due to consumer demand for organic milk and other dairy products. Management of OD farms is necessarily reliant upon manure and compost as fertilizer sources for pastures and feedstuff crops. However, due to differences in diet composition and the use of more bedding on OD farms than conventional dairies (CD), the manure from organically managed dairies is different from that of CD dairies (He et al. 2009; He and Ohno 2012; He and Wang 2012). Specifically, OD manure contains more soluble inorganic P and Ca/Mg-associated P, and less monoester P and stable metal phytate species, than CD manure (He et al. 2009). Furthermore, the soluble organic matter in OD manure contains more stable humic- and lignin-related

Table 7.1 Selected properties of 13 organic dairy manures (OD) and a conventional dairy manure

		C	N		P	
Manure	DM (%)	% DM		C:N	g kg^{-1}	C:P
OD1	24	38.0	2.79	13.6	18.7	20.3
OD2	22	29.8	1.71	17.4	4.5	66.2
OD3	19	40.2	2.11	19.1	5.4	74.4
OD4	19	41.7	1.49	30.0	5.8	71.9
OD5	16	41.5	1.46	28.4	4.4	94.3
OD6	21	34.9	1.82	19.2	5.2	67.1
OD7	30	45.4	0.88	51.6	2.0	227.0
OD8	23	41.2	1.49	27.7	2.6	158.5
OD9	20	41.5	1.41	29.4	4.5	92.2
OD10	20	42.2	1.44	29.3	3.4	124.1
OD11	20	43.2	1.28	33.8	3.5	123.4
OD12	19	43.1	1.42	30.4	4.9	88.0
OD14	20	40.2	2.44	16.5	2.4	167.5
Mean (SD)[a]	21 (3.0)	40.2 (3.9)	1.67 (0.50)	26.5 (9.6)	5.2 (4.2)	106 (54)
Conventional dairy	15	36.8	1.34	27.5	6.9	53.3

Modified from Waldrip et al. (2012)
[a]SD, numbers in parentheses represent standard deviation from the mean

components and less amino/protein N-related components than CD manure (He and Ohno 2012). However, until recently it was unknown how these differences affect P availability following application of OD manure to soil.

Waldrip et al. (2012) examined the effects of OD manure on activities of ACP, ALP, PDE, available soil P, and growth of sorghum-sudangrass in a greenhouse study. In this study, soil was fertilized with manures taken from 13 organic dairies in Maine, CD manure, or ammonium nitrate. Soil phosphatase activities and modified Morgan P were determined at planting and after 16 weeks of plant growth. Selected manure properties are listed in Table 7.1.

Of the three enzymes tested, ACP activity was the highest, with initial rates averaging 150 mg *p*-NP kg^{-1} h^{-1} for soils that received OD manure (Fig. 7.4a). Both OD and CD manure amendments increased ACP activity over inorganic fertilizer by an average of 117 % and 101 %, respectively. The highest ACP activity was found following application of OD10, and corresponded to greater early plant growth. This manure (OD10) contained a relatively low concentration of total P (Table 7.1), suggesting that low levels of available P resulted in production of root-borne ACP or stimulation of extracellular soil ACP activity by rhizosphere processes. In general, there was little change in ACP over time following application of OD manure (Fig. 7.4a). Activity of ACP was positively related to the amount of manure P applied, and negatively related to manure C:P and C:N ratios.

These findings are in contrast to those of a previous study by Waldrip et al. (2011), where poultry manure amendment had little effect on ACP activity. However, in the previous study, soil ACP activity was higher than the current study, with an average activity of 323 mg *p*-NP kg^{-1} h^{-1}, which could have masked any manure-induced changes. It is difficult to make strong inferences about the relationship between

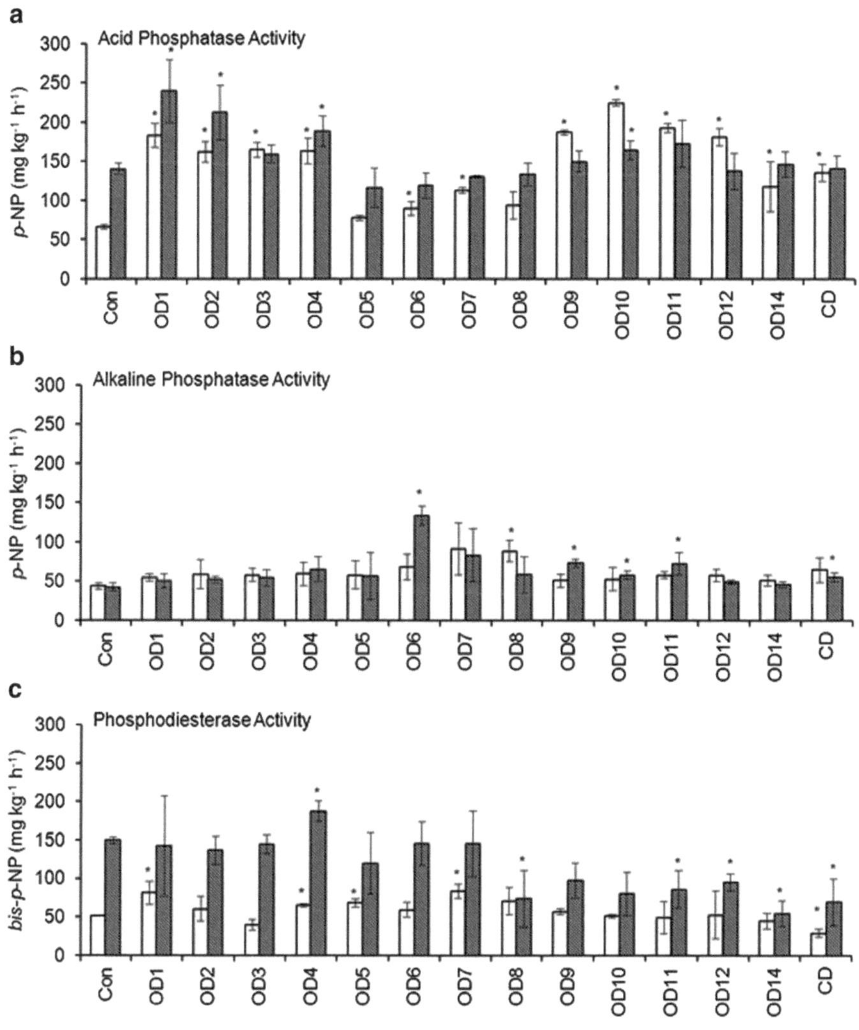

Fig. 7.4 Activities of (**a**) acid phosphomonoesterase, (**b**) alkaline phosphomonoesterase, and (**c**) phosphodiesterase in soil at planting (*open bars*) and after 16 weeks (*filled bars*) of sudangrass growth when soil was amended with inorganic fertilizer (Con), 13 organic dairy manures (OD), and manure from a conventional dairy (CD). *Asterisk* (*) indicates significant difference ($p < 0.05$) from inorganic fertilizer (Con) at a particular soil sampling time (Modified from Waldrip et al. 2012)

manure C:P ratios and ACP activities in the OD manure study by Waldrip et al. (2012). While there was a trend for lower ACP activity following application of OD manures with high C:P ratios, soils that received some manures (e.g. OD6, Table 7.1) had very low soil ACP activity (Fig. 7.4a).

In the study by Waldrip et al. (2012), OD manure had little effect on ALP activity (Fig. 7.4b). There was a relatively narrow range in ALP activity (44–87 mg *p*-NP kg^{-1} h^{-1}), with an average of 62 mg *p*-NP kg^{-1} h^{-1}. In general, ALP activity did not change significantly due to fertilizer treatment or over time. In addition, ALP activity was not well correlated to most of the OD manure variables, showing only a slight positive relationship with manure C:N ratio.

In this study (Waldrip et al. 2012), the initial activity of PDE following application of OD manure averaged 60 mg *bis-p*-NP kg^{-1} h^{-1}, and was higher than with CD manure, but not different from soils that received inorganic fertilizer (Fig. 7.4c). About half of the OD manure treatments resulted in initial PDE activities that were higher than with inorganic fertilizer. However, PDE activity following application of CD manure was lower than with inorganic fertilizer. The general trend for all treatments was an increase in PDE over time; however, there was a high degree of variability in measured final PDE activities in manured soils.

Similar to ALP, few relationships were noted between PDE activity and OD manure variables, with only a positive relationship between the amounts of manure P applied. The PDE activities measured in this study were similar to those reported in Waldrip et al. (2011). In addition, the findings of increases in PDE activity with plant growth were consistent with that of Waldrip et al. (2011). As PDE can be produced by plant roots as well as microorganisms and is affected by rhizosphere processes (Tarafdar and Claassen 1998), our results suggest that OD manure has little effect on soil PDE activity.

Based on this work, Waldrip et al. (2012) concluded that manure from organically managed dairies does not influence soil phosphatase activities, P cycling, or P availability any differently than does manure from conventional dairies when applied at equivalent N rates. This is despite large differences in the organic matter and P forms in the manures from the two systems. It appears that the hydrolysis of organic P following OD manure application is largely controlled by root- or microbial-borne ACP, in a manner similar to CD manure. Therefore, the information accumulated over years through research on conventionally managed dairies is relevant as a reference for predicting P availability and transformation rates for soils that receive organic dairy manure in the Northeastern U.S.

7.4 How Experimental Approach Influences Soil Phosphatase Activities

Evaluation of effects of manure application on soil properties, including phosphatase activities, has been conducted where soils are sampled under field conditions and from greenhouse and micro-scale incubation studies under controlled conditions. Unfortunately, comparison of the collected data among studies conducted using different experimental approaches can be complicated by this lack of standardized methodologies, and the results obtained by different researchers are often contradictory. Most commonly, phosphatase activities are determined on fresh soil samples taken directly from the field. This method can provide information on how many variables (e.g. fertilizer management, tillage practices, crops, soil type) influence soil quality and P cycling. However, conducting large field studies is not always practical, particularly when the objectives of the researcher are to compare a large number of soils or amendment types, or track short-term changes in soil properties. In these instances, an *in vitro* method may be more advantageous.

The two most commonly used *in vitro* soil incubation methods are aerobic incubation and greenhouse pot studies. In typical aerobic incubation studies, the soil and any amendment are mixed and placed in bottles, tubes or other containers, packed to a specific bulk density, and then brought to a predetermined % water-filled pore space (wfps). The containers are then loosely capped and put into a controlled-temperature/humidity incubator. Typically, incubation containers are vented periodically to ensure aerobic conditions, and water is added as needed to gravimetrically maintain desired % wfps throughout the experiment. Alternately, soil in the containers is allowed to dry during incubation in order to evaluate the effects of wet-dry cycling on enzyme activities. At designated time points during the incubation, a sub-sample is removed from the incubation containers (or the entire container is destructively sampled) for determination of enzyme activity. Variations of the aerobic incubation method have been used to study soil C, N, and P transformations and enzyme activities following application of inorganic fertilizer and manure to soil (Bell et al. 2006; Griffin et al. 2002, 2003; Waldrip-Dail et al. 2009). The highly controlled conditions of aerobic incubation allow for a detailed examination of treatment effects on soil enzyme activities. However, it is unknown if the results obtained via this method are truly representative of what occurs under greenhouse and field conditions, where soil temperature, moisture content, and degree of aeration are more variable.

Greenhouse experiments, where soil and any treatments are placed in pots (with or without growing plants) allows a soil to be exposed to airflow, humidity, and temperature changes that are more representative of natural environmental changes than is possible with the aerobic incubation method. In general, % wfps in the pots is maintained either gravimetrically or based on measurements from a soil moisture sensor. When growing plants are present, this type of incubation can also provide information about rhizosphere effects on phosphatase activities. While both aerobic incubation and greenhouse methods provide distinct benefits towards the investigation of factors that influence the mineralization of organic compounds, to our knowledge, no comparative investigation has been conducted to examine the effects of specific incubation method on phosphatase activity in soil amended with livestock manure.

As an example of the differences in phosphatase activities that can be obtained under different experimental methodologies, we compare two studies, an aerobic incubation and a greenhouse pot study conducted by Waldrip-Dail et al. (2009) and Waldrip et al. (2011). In these studies, activities of ACP, ALP, and PDE were determined following incubation of a Bangor silt loam with layer hen manure. The incubation study (Waldrip-Dail et al. 2009) was performed under aerobic conditions, where freeze-dried poultry manure was added at rates of 100 (Low-PM) or 200 (High-PM) mg P kg^{-1}, and held for 108 days at 25 °C in a controlled temperature incubator. In the greenhouse study (Waldrip et al. 2011), the same

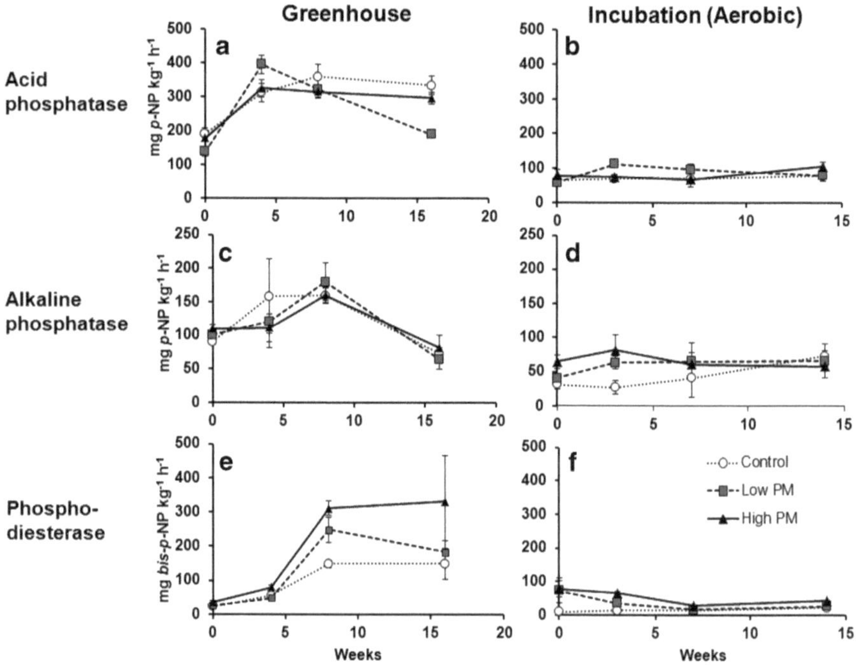

Fig. 7.5 Comparison showing how experimental approach (greenhouse pot study vs. aerobic incubation) influences activities of acid phosphomonoesterase (**a**, **b**), alkaline phosphomonoesterase (**c**, **d**), and phosphodiesterase (**e**, **f**) in soils amended with layer hen manure (Modified from Waldrip-Dail et al. 2009 and Waldrip et al. 2011)

manure was added to the same soil type as used in the aerobic incubation, placed in 6 in. diameter plastic pots, and kept on a greenhouse bench under natural lighting and ambient temperatures for 16 weeks. In both experiments, % wfps was maintained gravimetrically at 45 %.

Time course plots of the activities of ACP, ALP, and PDE are shown in Fig. 7.5, where the effects of methodology-based differences are clear. In general, the greenhouse-incubated soils had significantly higher activities of all enzymes and exhibited more dramatic changes over time and due to manure application. Most notably, PDE activity in the greenhouse study increased over time and was very responsive to manure addition (Fig. 7.5e). In contrast, ALP activity in the greenhouse study decreased over time and was not responsive to manure (Fig. 7.5c). In the aerobic incubation study (Waldrip-Dail et al. 2009), activities of all enzymes tested were low and remained relatively stable over time (Fig. 7.5b, d, f). Based on these findings, we recommend that extreme care be taken when comparing enzyme activities determined on soils following different experimental approaches.

7.5 Conclusions and Directions for Further Research

In conclusion, the application of livestock manures can have differential effects on soil phosphatase activities and P availability; however, these effects are not easily predicted and many factors, including manure physicochemical properties and experimental approach, can influence phosphatase activities and soil P cycling. Thus, there is a need for a better understanding of the effects of livestock manure on P cycling and availability. This type of information would assist with development of appropriate fertilization strategies for different soil-agroecosystem scenarios. In addition, this could lead to adoption of manure management practices that reduce the risk of negative environmental impacts due to leaching and runoff of P into sensitive surface waters following field application.

Further work is needed to better understand how and why livestock manures influence organic P mineralization and soil fertility. Specific areas of research needs include: (1) further study on soil PDE activities and their impact as a rate-limiting step in organic P hydrolysis, (2) analyses to determine the origin of specific enzyme activities, (3) better qualitative and quantitative understanding of the forms of organic P in manures and soils following application of manures, and (4) quantitative understanding of the direct contribution of manure-derived enzymes on soil phosphatase activities.

References

Abel S, Nürnberger T, Ahnert V, Krauss G-J, Glund K (2000) Induction of an extracellular cyclic nucleotide phosphodiesterases as an accessory ribonucleolytic activity during phosphate starvation of cultured tomato cells. Plant Physiol 2000:543–552

Acosta-Martinez V, Harmel RD (2006) Soil microbial communities and enzyme activities under various poultry litter application rates. J Environ Qual 35:1309–1318

Acosta-Martinez V, Dowd SE, Sun Y, Wester D, Allen V (2010) Pyrosequencing analysis for characterization of bacterial diversity in a soil as affected by integrated livestock-cotton production systems. Appl Soil Ecol 45:13–25

Bell JM, Robinson CA, Schwartz RC (2006) Changes in soil properties and enzymatic activities following manure applications to rangeland. Rangel Ecol Manag 59:314–320

Browman MG, Tabatabai MA (1978) Phosphodiesterase activity of soils. Soil Sci Soc Am J 42:284–290

Burns RG (1982) Enzyme activity in soil: location and a possible role in microbial ecology. Soil Biol Biochem 14:423–427

Chen CR, Condron LM, Davis MR, Sherlock RR (2004) Effects of plant species on microbial biomass phosphorus and phosphatase activity in a range of grassland soils. Biol Fertil Soils 40:313–322

Chhabra S, Brazil D, Morrissey J, Burke J, O'Gara F, Dowling DN (2013) Fertilization management affects the alkaline phosphatase bacterial community in barley rhizosphere soil. Biol Fertil Soils 49:31–39

Colvan SR, Syers JK, O'Donnell AG (2001) Effect of long-term fertilizer use on acid and alkaline phosphomonoesterase and phosphodiesterase activities in managed grassland. Biol Fertil Soils 34:258–263

Criquet S, Braud A, Nieble S (2007) Short-term effects of sewage sludge application on phosphatase activities and available P fractions in Mediterranean soils. Soil Biol Biochem 39:921–929
Dail HW, He Z, Erich MS, Honeycutt CW (2007) Effect of drying on phosphorus distribution in poultry manure. Commun Soil Sci Plant Anal 38:1879–1895
Dao TH, Guber AK, Sadeghi AM, Karns JS, van Kessel JS, Shelton DR, Pachepsky YA, McCarty G (2008) Loss of bioactive phosphorus and enteric bacteria in runoff from dairy manure applied to sod. Soil Sci 173:511–521
Dou Z, Galligan DT, Ramberg CF, Meadows C, Ferguson JD (2001) A survey of dairy farming in Pennsylvania: nutrient management practices and implications. J Dairy Sci 84:966–973
Garcia-Gil JC, Plaza C, Soler-Rovira P, Polo A (2000) Long-term effects of municipal solid waste compost application on soil enzyme activities and microbial biomass. Soil Biol Biochem 32:1907–1913
Garcia-Ruiz R, Ochoa V, Hinojosa MB, Carreira JA (2008) Suitability of enzyme activities for the monitoring of soil quality improvement in organic agricultural systems. Soil Biol Biochem 40:2137–2145
Garg S, Bahl GS (2008) Phosphorus availability to maize as influenced by organic manures and fertilizer P associated phosphatase activity in soils. Bioresour Technol 99:5773–5777
George TS, Gregory PJ, Hocking P, Richardson AE (2008) Variation in root-associated phosphatase activities in wheat contributes to the utilization of organic P substrates in vitro, but does not explain differences in the P-nutrition of plants when grown in soils. Environ Exp Bot 64(3):239–249
Griffin TS, Honeycutt CW, He Z (2002) Effects of temperature, soil water status, and soil type on swine slurry nitrogen transformations. Biol Fertil Soils 36:442–446
Griffin TS, Honeycutt CW, He Z (2003) Changes in soil phosphorus from manure application. Soil Sci Soc Am J 67:645–653
He Z, Ohno T (2012) Fourier transform infrared and fluorescence spectral features of organic matter in conventional and organic dairy manure. J Environ Qual 43:911–919
He Z, Wang JJ (2012) Characterization of plant nutrients and traceable marker components in dairy manure for organic dairy farming evaluation. In: He Z (ed) Applied research in animal manure management: challenges and opportunities beyond the adverse environmental concerns. Nova, New York, pp 1–19
He Z, Honeycutt CW, Griffin TS (2004a) Enzymatic hydrolysis of organic phosphorus in swine manure and soils. J Environ Qual 33:367–372
He Z, Honeycutt CW, Griffin TS (2004b) Phosphorus distribution in dairy manures. J Environ Qual 33:1528–1534
He Z, Griffin TS, Honeycutt CW (2006a) Soil phosphorus dynamics in response to dairy manure and fertilizer applications. Soil Sci 171:598–609
He Z, Fortuna AM, Senwo ZN, Tazisong IA, Honeycutt CW, Griffin TS (2006b) Hydrochloric fractions in Hedley fractionation may contain inorganic and organic phosphates. Soil Sci Soc Am J 70:893–899
He Z, Honeycutt CW, Cade-Menun BJ, Senwo ZN, Tazisong IA (2008) Phosphorus in poultry litter and soil: enzymatic and nuclear magnetic resonance characterization. Soil Sci Soc Am J 72:1425–1433
He Z, Honeycutt CW, Griffin TS, Cade-Menun BJ, Pellechia PJ, Dou Z (2009) Phosphorus forms in conventional and organic dairy manure identified by solution and solid state P-31 NMR spectroscopy. J Environ Qual 38:1909–1918
He Z, Honeycutt CW, Griffin TS, Larkin RP, Olanya M, Halloran JM (2010) Increases of soil phosphatase and urease activities in potato fields by cropping rotation practices. J Food Agric Environ 8:1112–1117
Hedley MJ, Mortvedt JJ, Bolan NS, Syers JK (1995) Phosphorus fertility management in agroecosystems. In: Tiessen H (ed) Phosphorus in the global environment. Wiley, West Sussex, pp 59–92

Hinsinger P (2001) Bioavailability of soil inorganic P in the rhizosphere as affected by root-induced chemical changes: a review. Plant Soil 237:173–195

Huang W, Liu J, Zhou G, Zhang D, Deng Q (2011) Effects of precipitation on soil acid phosphatase activity in three successional forests in southern China. Biogeosciences 8:1901–1910

Kizilkaya R, Bayrakli B (2005) Effect of N-enriched sewage sludge on soil enzyme activities. Appl Soil Ecol 30:192–202

Kleinman PJA, Sharpley AN, Wolf AM, Beegle DB, Moore PA (2002) Measuring water-extractable phosphorus in manure as an indicator of phosphorus in runoff. Soil Sci Soc Am J 66:2009–2015

Leake JR, Miles W (1996) Phosphodiesters as mycorrhizal P sources: I. Phosphodiesterase production and the utilization of DNA as a phosphorus source by the ericol mycorrhizal fungus *Hymenoscyphus ericae*. New Phytol 132:435–443

Leinweber P, Haumaier L, Zech W (1997) Sequential extractions and ^{31}P NMR spectroscopy of phosphorus forms in animal manures, whole soils and particle-size separates from a densely populated livestock area in northwest Germany. Biol Fertil Soils 25:89–94

Leytem AB, Dungan RS, Moore A (2011) Nutrient availability to corn from dairy manure and fertilizer in a calcareous soil. Soil Sci 176:426–434

Li SM, Li L, Zhang FS, Tang C (2004) Acid phosphatase role in chickpea/maize intercropping. Ann Bot 94:297–303

Lim BL, Yeung P, Cheng C, Hill JE (2007) Distribution and diversity of phytate-mineralizing bacteria. ISME J 1:321–330

Margesin R, Schinner F (1994) Phosphomonoesterase, phosphodiesterase, phosphotriesterase, and inorganic pyrophosphatase activities in forest soils in an alpine area: effect of pH on enzyme activity and extractability. Biol Fertil Soils 18:320–326

Olander LP, Vitousek PM (2000) Regulation of soil phosphatase and chitinase activity by N and P availability. Biogeochemistry 49:175–190

Oshima Y, Ogawa N, Harashima S (1996) Regulation of phosphatase synthesis in *Saccharomyces cerevisiae* – a review. Gene 179:171–177

Pagliari PH, Laboski CAM (2012) Investigation of the inorganic and organic phosphorus forms in animal manure. J Environ Qual 41:901–910

Parham JA, Deng SP, Raun WR, Johnson GV (2002) Long term cattle manure application in soil. I. Effect on soil phosphorus levels, microbial biomass C, and dehydrogenase and phosphatase activities. Biol Fertil Soils 35:328–337

Pascual JA, Garcia C, Hernandez T (1998) Comparison of fresh and composted organic waste in their efficacy for the improvement of arid soil quality. Bioresour Technol 68:255–265

Pascual JA, Moreno JL, Hernandez T, Garcia C (2002) Persistence of immobilized and total urease and phosphatase activities in a soil amended with organic wastes. Bioresour Technol 82:73–78

Richardson AE, Hadobas PA (1997) Soil isolates of *Pseudomonas* spp. that utilize inositol phosphates. Can J Microbiol 43:509–516

Sardans J, Peñuelas J, Estiarte M (2006) Warming and drought alter soil phosphatase activity and soil P availability in a Mediterranean shrubland. Plant Soil 289:227–238

Senwo ZN, Ranatunga TD, Tazisong IA, Taylor RW, He Z (2007) Phosphatase activity of Ultisols and relationship to soil fertility indices. J Food Agric Environ 5:262–266

Sinsabaugh RL (1994) Enzymatic analysis of microbial pattern and process. Biol Fertil Soils 17:69–74

Speir TW, Cowling JC (1991) Phosphatase activities of pasture plants and soils: relationship with plant productivity and soil P fertility indices. Biol Fertil Soils 12:189–194

Stevenson FJ, Cole MA (1999) The cycles of soil, 2nd edn. Wiley, New York, pp 279–379

Tabatabai MA (1994) Soil enzymes. In: Weaver RW, Angle JS, Bottomly PM (eds) Methods of soil analysis. Part 2: Microbiological and biochemical properties. Soil Science Society of America, Madison, pp 775–833

Tabatabai MA, Bremner JM (1969) Use of *p*-nitrophenyl phosphate for assay of soil phosphatase activity. Soil Biol Biochem 1:307–309

Tarafdar JC, Claassen N (1998) Organic phosphorus compounds as a phosphorus source for higher plants through the activity of phosphatases produced by plant roots and microorganisms. Biol Fertil Soils 5:308–312

Tarafdar JC, Jungk A (1987) Phosphatase activity in the rhizosphere and its relation to the depletion of soil organic phosphorus. Biol Fertil Soils 3:199–204

Turner BL, Haygarth PM (2005) Phosphatase activity in temperate pasture soils: potential regulation of labile organic phosphorus turnover by phosphodiesterase activity. Sci Total Environ 344:27–36

Turner BL, Leytem AB (2004) Phosphorus compounds in sequential extracts of animal manure: chemical speciation and a novel fractionation procedure. Environ Sci Technol 38:6101–6108

Waldrip HM, He Z, Erich MS (2011) Effects of poultry manure amendment on phosphorus uptake by ryegrass, soil phosphorus fractions and phosphatase activity. Biol Fertil Soils 47:407–418

Waldrip HM, He Z, Griffin TS (2012) Effects of organic dairy manure on soil phosphatase activity, available soil phosphorus and growth of sorghum-sudangrass. Soil Sci 177:629–637

Waldrip-Dail H, He Z, Erich MS, Honeycutt CW (2009) Soil phosphorus dynamics in response to poultry manure amendment. Soil Sci 174:195–201

Chapter 8
Variety and Solubility of Phosphorus Forms in Animal Manure and Their Effects on Soil Test Phosphorus

Paulo H. Pagliari

Abstract Animal manure has been recognized as one of the best sources of nutrients for crop production for centuries. However, only over the last few decades have we started to fully understand and recognize the negative environmental impacts of agricultural use of manure. Although all essential nutrients required for plant growth can be found in a manure sample, there are primarily two that are well known to negatively impact the environment the most, nitrogen (N) and phosphorus (P). Both of these nutrients have considerably high mobility while within the manure matrix; however, their mobility significantly decreases once manure is incorporated into the soil. Phosphorus is found in two forms in animal manure, inorganic (P_i) and organic (P_o). The P_o can be further categorized as enzymatically hydrolysable or non-hydrolysable P_o. The P_i and P_o can be found dissolved in solution and also precipitated as minerals or complexed with organic compounds and metals in the manure matrix. The relative amount of each P form is highly variable and depends among other factors on the animal species, animal age, growth stage, animal diet, and manure storage conditions. When manure is applied to soils, most often there is an increase in soil test phosphorus (STP) levels. However, the magnitude of the increase in STP will depend on soil and manure properties. Understanding how manure and soil properties interact is key to minimize the negative impact of manure on the environment. This chapter will discuss the variety and solubility of manure P and also how the different forms of P_i and P_o can affect the increase in STP.

P.H. Pagliari (✉)
Department of Soil, Water, and Climate, Southwest Research and Outreach Center, University of Minnesota, 23669 130th Street, Lamberton, MN 56152, USA
e-mail: pagli005@umn.edu

Z. He and H. Zhang (eds.), *Applied Manure and Nutrient Chemistry for Sustainable Agriculture and Environment*, DOI 10.1007/978-94-017-8807-6_8,

8.1 Advances in Manure Analysis Methods

Several methods and techniques such as solution and solid-state ^{31}P nuclear magnetic resonance (NMR), scanning electron microscopy with energy dispersive X-ray (SEM), X-ray absorption near-edge structure (XANES), sequential fractionation and others, have been developed or adapted to help determine which phosphorus (P) compounds are most predominant in manure from different animal species. Each of these techniques has its strengths and weaknesses. For example, the use of solution ^{31}P NMR gives less information about the specific mineral solid phases, but it provides information about the solubility and forms of organic compounds. The alkaline extraction required for sample preparation for solution ^{31}P NMR analysis favor the dissolution of P minerals present in manure which makes it impossible for solution NMR to identify the manure solid phases (Toor et al. 2006; Cade-Menun 2011). Solid-state ^{31}P NMR on the other hand does not require the alkaline extraction; however, the presence of paramagnetic metal cations in the sample presents challenges for precise solid-state NMR analysis (Toor et al. 2006; Akinremi et al. 2011). However, solid-state ^{31}P NMR spectroscopy is sensitive to inorganic forms of P that are associated with metals such as Ca and Mg (Akinremi et al. 2011). Therefore, solid-state ^{31}P NMR can be very useful to understanding manure P mineralogy with respect to Ca and Mg minerals (Akinremi et al. 2011). X-ray absorption near-edge structure spectroscopy has also been used to identify the organic and mineral P fractions of manure (Shober et al. 2006; Akinremi et al. 2011). The use of XANES has advantages over solution ^{31}P NMR because no extraction is required, and it has advantages over solid-state ^{31}P NMR because it is less affected by paramagnetic metals, such as Fe(III) (Akinremi et al. 2011). Sequential fractionation procedures have also been used to characterize P forms in animal manure (Dou et al. 2000; Haygarth and Sharpley 2000; Sharpley and Moyer 2000; He and Honeycutt 2001; He et al. 2004b; Toth et al. 2011). In these procedures, the most soluble inorganic (P_i) and organic (P_o) P forms are extracted with water and 0.5 M $NaHCO_3$ (pH = 8.5); iron (Fe) and Al associated P_i are extracted with 0.1 M NaOH; Ca phosphate stable minerals, such as hydroxyapatite, are extracted with 1.0 M HCl; and the Stable P_o and insoluble P_i (if any remains) left in the residue after HCl extraction is determined after concentrated H_2SO_4-H_2O_2 digestion (Hedley et al. 1982; Sui et al. 1999). The NaOH and HCl extractants can also hydrolyze some P_o, which can lead to an overestimation of P_i and underestimation of P_o in these fractions (He and Honeycutt 2001). Table 8.1 presents a range of total P, fractions of P_i, P_o, and residue P as quantified by sequential fractionation of animal manure from several animal species, as reported by several researchers (Dou et al. 2000; Sharpley and Moyer 2000; He and Honeycutt 2001; Ajiboye et al. 2004, 2007; He et al. 2004a, b, 2006, 2007; Maguire et al. 2004; McDowell and Stewart 2005; Pagliari and Laboski 2012).

More recently, many of the methods described above for manure analysis have been used together to provide a better characterization and distribution of the P forms in manure. For example, sequential fractionation method has been coupled

Table 8.1 Summary of total P, inorganic (P_i), organic (P_o), and residue P in sequential extraction of manures from several animal species[a]

		Total P	P_i	P_o	Residue P
P source	Number of samples	g kg^{-1}	% of total P		
Dairy	37	2.8–18.3	39–100	0–32	0–24
Swine	9	3.9–48.7	51–100	0–37	0–28
Beef	11	2.5–14.2	50–94	8–49	1–32
Turkey	7	10.1–28.2	49–90	21–50	0–5
Poultry	34	8.6–30.4	32–100	0–68	0–24
Sheep	3	7.2–10.7	70–81	25–27	4–8
Horse	2	5.4–12.4	56–74	15–30	2–10
Goat	2	10.3–13.1	65–68	21–37	3–4

[a]Summary from Dou et al. (2000), Sharpley and Moyer (2000), He and Honeycutt (2001), Ajiboye et al. (2004), He et al. (2004a, b, 2006, 2007), Maguire et al. (2004), McDowell and Stewart (2005), Ajiboye et al. (2007), and Pagliari and Laboski (2012)

with solution ^{31}P NMR to quantify/characterize the P_i and P_o forms extracted in each fractionation step (McDowell and Stewart 2005; Ajiboye et al. 2007; He et al. 2007). Sequential fractionation has also been coupled with enzyme hydrolysis to distinguish between several forms of labile and recalcitrant manure P (He and Honeycutt 2001; He et al. 2004a, b, 2010). By using certain enzymes alone or in combination, the hydrolysis of known P_o compounds can be achieved. Although the sequential fractionation methods seem very efficient at categorizing the P_o forms, it does not give any direct information on the mineralogy of the P_i species. However, it does give indirect information about the solubility of the mineral phases. Characterization of the mineral phase is still under progress and the best known methods used to date include sequential fractionation combined with chemical modeling software, scanning electron microscopy microprobe analysis (SEM), solid–state ^{31}P NMR, and XANES (Bril and Salomons 1990; Leinweber et al. 1997; Cooperband and Good 2002; Toor et al. 2006; Ajiboye et al. 2007; Güngör and Karthikeyan 2008; Toth et al. 2011). Cade-Menun (2011) and Akinremi et al. (2011) provided a more detailed description of each method as well as the principles and mechanisms involved in each method mentioned above.

8.2 Variety and Solubility of Manure Phosphorus

8.2.1 Identified Manure Inorganic Phosphorus Minerals

The complexity of the manure matrix leads to an interesting interaction between the inorganic and organic phases. Fordham and Schwertmann (1977) described manure as being in a sensitive and balanced dynamic equilibrium state where minor changes in any property (chemical, physical, or biological) could affect the whole matrix. Bril and Salomons (1990) reported that the majority of Ca and Mg are complexed with carbonates and organic compounds in manure, and that potassium (K^+), ammonium (NH_4^+), sodium (Na^+), and chloride (Cl^-) ions were the most important regulators of manure solution ionic strength. Therefore, these ions would have control over the mineral solid phases in the manure including those controlling P_i solubility. Understanding how the inorganic phases interact is the key in understanding P solubility.

Over the past few decades, significant amount of work has been done to try to better understand and characterize the mineral solid phases controlling P_i dissolution from animal manures. Most of those studies have shown that the forms and solubility of manure P_i change from sample to sample; however, some similarities are always found. Furthermore, computer modeling has allowed for a new way to identify minerals in manure that were previously unknown to be present in manure. Chemical modeling using chemical equilibrium software can be extremely helpful in understanding the probable solid phases of compounds with a complex matrix such as animal manure. Chemical modeling does not prove that a mineral phase exists as physical methods do, but it can establish that a substance may behave as if a particular mineral was present. Although chemical models are not a direct measurement of the mineral, when combined with other methods such as SEM, X-ray, or XANES it has been proven to identify the same minerals. In addition, under certain conditions some methods such as SEM and XANES may also indicate the presence of unknown minerals leaving the researchers wondering what they might be. Combining chemical models with other methods has shown to solve the problem of unknown mineral forms by providing a list of probable minerals that would be stable under the conditions in a manure sample. Although there is a large amount of mineral species that can be found in a manure sample (Akinremi et al. 2011), there are a few species that are commonly reported in the majority of the samples.

8.2.1.1 Swine Manure

Solid phase analysis of swine manure using SEM revealed the presence of $MgNH_4PO_4{\cdot}6H_2O$ (struvite) with small amounts of $MgKPO_4{\cdot}6H_2O$ (K-struvite) and also other magnesium and calcium minerals that could not be determined (Bril and Salomons 1990). Bril and Salomons (1990) then used chemical modeling and were able to report that the manure solutions were in dynamic equilibrium with struvite, β–$Ca_3(PO_4)_2$ (β-tricalcium phosphate, or β-TCP), and $CaHPO_4{\cdot}2H_2O$ (brushite). By using the chemical modeling Bril and Salomons (1990) were able to identify the unknown minerals. Fordham and Schwertmann (1977) reported that struvite, $Mg_3H(PO_4)_3{\cdot}3H_2O$ (trimagnesium phosphate), $Ca_8H_2(PO_4)_6{\cdot}5H_2O$ (octacalcium phosphate, or OCP), brushite, and $NH_4CaPO_4{\cdot}3H_2O$ were the probable mineral species in manure solutions based on chemical modeling. X-ray diffraction confirmed that struvite was present in their samples, which also confirmed the usefulness of the modeling. Ajiboye et al. (2007) used XANES to analyze different animal manures and reported that swine manure was dominated by brushite and $AlPO_4 \cdot 2H_2O$ (variscite). Although several minerals have bee reported in different manure samples, the majority of the analysis shows that struvite and brushite are almost always present in swine manure.

8.2.1.2 Beef and Dairy Manure

As in swine manure, analysis of the mineral solid phase in cattle manure has shown a large variety of phosphate minerals. Struvite and $CaHPO_4$ (dicalcium phosphate, or DCP) were found to be the main mineral forms of manure P_i determined by SEM and X-ray diffraction by Fordham and Schwertmann (1977) and Güngör et al. (2007). In contrast, Shober et al. (2006) reported that only $Ca_5(PO_4)_3 \cdot 2.5H_2O$ (hydroxyapatite) was present in dairy manure samples using XANES. In a different study Ajiboye et al. (2007) using XANES reported that the dairy manure was dominated by brushite and beef manure was dominated by struvite and brushite. Güngör and Karthikeyan (2005) used chemical modeling software and reported that the most probable phases controlling P solubility in the dairy manure samples investigated were struvite, OCP, and β-TCP. As reported in the swine manure section, the solid phase in beef and dairy manure is usually controlled by struvite and brushite.

8.2.1.3 Chicken and Turkey

Bril and Salomons (1990) reported that the solid phase in chicken manure was controlled by struvite and Ca-phosphate minerals including β-TCP, brushite, and DCP. Ajiboye et al. (2007) used XANES to molecularly characterize P forms in poultry manure and found that the solid phase was dominated by struvite and brushite. In contrast, Toor et al (2005) reported that DCP was the only mineral solid phase identified by XANES in chicken and turkey manure. Chemical modeling of chicken manure has also shown the probable presence of struvite, β-TCP, DCP, and brushite (Pagliari 2011).

8.2.1.4 Other Species

There has been limited amount of research that has investigated the mineralogy in manure from species such as horse, deer, sheep, goat and others. Solid-state ^{31}P NMR has shown that struvite and brushite were present in sheep manure (Shand et al. 2005). Others have reported that chemical modeling suggested the presence of struvite, β-TCP, and OCP (Pagliari 2011).

In summary, the mineral solid phase of most animal manure appear to have a mix of struvite and at least one form of Ca-phosphate mineral, usually brushite, with potentially other more stable Ca-phosphate forms. The amount of each form and the fraction that each form represents in a sample is one of the most important factors controlling P_i solubility in manures.

8.2.2 Solubility of Manure Inorganic Phosphorus

The solubility of manure P_i is a function of the mineral form, the amount of a particular mineral, the solution ionic strength, the solution pH used to dissolve the mineral, and the amount of time used in the extraction. A large fraction of manure Mg and Ca are associated with carbonates or organic matter; however, as these cations go into solution they are likely to affect the rate of P_i solubility. Several techniques can be employed to investigate the effects of Ca and Mg ionic activity on the solubility of manure P_i. Among those techniques this chapter will discuss manure solution pH titration, the use of an extract with a chelating agent such ethylenediaminetetraacetic acid (EDTA), and serial dilution where increasing manure dry matter to water ratio (MWR) is used.

Most of the P containing minerals in manure are extremely pH dependent and abrupt changes in solution pH will have a major impact on P solubility. Fordham and Schwertmann (1977) titrated manure solutions to pH values as low as 2.5 to study the effects of acidification on manure P solubility. Fordham and Schwertmann (1977) reported that during the initial titration struvite was likely dissolving, but as pH dropped below 6.7 no more struvite was apparent. Inorganic P continued to dissolve until the solution pH reached 5.5. Most of the P_i dissolved between pH 6.7 and 5.5 was attributed to Ca phosphate minerals (Fordham and Schwertmann 1977). Further acidification of those samples did not dissolve additional P_i indicating that at pH lower than 5.5, P containing minerals were no longer stable (Fordham and Schwertmann 1977). The compound EDTA is very effective at complexing Ca and Mg from solution, which result in low Ca and Mg activity. The lower Ca and Mg activity in solution will allow for more stable Ca- and Mg-phosphate compounds to dissolve and as a result more P_i will be dissolved. Figure 8.1 shows the results of a study that compared the total P_i extracted using the sequential fraction with the total P_i extracted using a 10 M EDTA solution buffered at pH 7.0 (Pagliari 2011). In this study 60 manure samples were collected from commercial beef, dairy, horse, swine, chicken, turkey, sheep, and dairy goat operations. The Fig. 8.1 shows that both methods extracted the same amount of total P_i from all manure samples. The P-value for the regression analysis is highly significant for the slope when compared to 0; however, when the slope was compared to 1 (the 1:1 relationship) the slope was not statistically different (P-value >0.05).

Other studies have reported that EDTA extraction recovered about 100 % of P from dairy manure and from 65 to 96 % of P from poultry manure (Toor et al. 2005; He et al. 2007; Turner 2004). Lower P_i recovery reported from EDTA extracts could be due to the fact that not enough EDTA was added to chelate 100 % of Ca and Mg from solution. In another study, goat, swine, chicken, and dairy manure

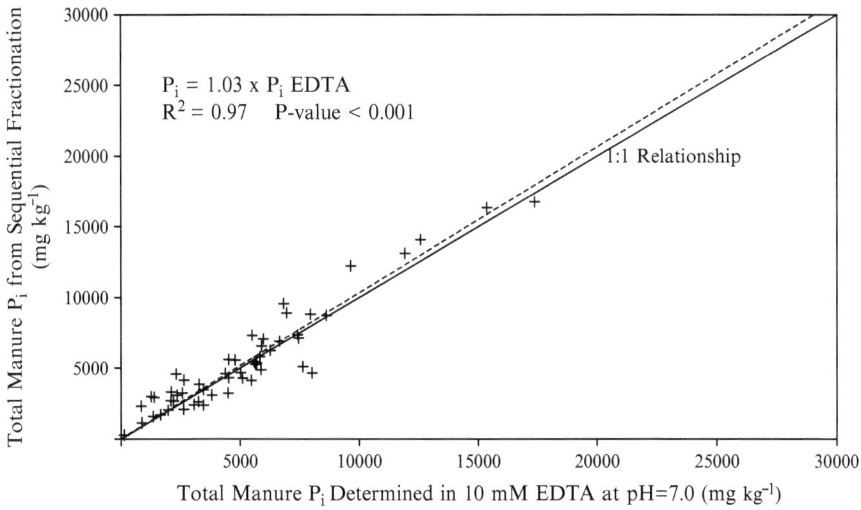

Fig. 8.1 Relationship between total inorganic P (P_i) determined using sequential fractionation and 10 mM EDTA at pH 7.0. *Dashed line* represents the regression equation and the *solid line* represents the 1:1 relationship (Pagliari 2011)

were extracted in solution with increasing EDTA concentrations of 1, 5, and 10 mM at an MWR of 1:100 (Pagliari 2011). It was found that at the 1 mmol L^{-1} EDTA, only a fraction of the total P_i was recovered from those manures. However, by adding 10 mmol L^{-1} EDTA 100 % of the P_i was recovered, but most importantly, adding 10 mmol L^{-1} EDTA at the MWR 1:100 kept the Mg and Ca ionic activity low enough in solution which allowed for all P_i to dissolve. Therefore, when using EDTA as an extracting solution it is important to know how much metal need to be chelated so that the EDTA solution can be as effective as needed.

In sequential fractionation studies, the manure dry matter to water ratio (MWR) is usually fixed, and in several cases the MWR used ranged from 1:10 to 1:200 (He and Honeycutt 2001; Kleinman et al. 2002; He et al. 2009a, b; Toor et al. 2005; Waldrip et al. 2011). However, in serial dilution studies the MWR used has, in most cases, ranged from 1:1 to 1:200. For example, Kleinman et al. (2002) used MWR ranging from 1:10 to 1:200 to study the solubility of P_i from dairy, poultry, and swine manure samples. At the 1:10 MWR, water was able to dissolve about 1.6, 2.1, and 2.2 g P_i kg^{-1} in dairy, poultry, and swine manures, respectively (Kleinman et al. 2002). However, at the 1:200 MWR, water was able to dissolve approximately 2.8, 5.8, and 7.7 g P_i kg^{-1} in dairy, poultry, and swine manures, respectively (Kleinman et al. 2002). In a different study, Studnicka et al. (2011) studied the effects of MWR at 1:100 and 1:1,000 on P_i solubility of dairy, chicken, and turkey manures. Studnicka et al. (2011) reported that at the 1:100 water was able to dissolve on average 2.3, 3.9, and 7.7 g P_i kg^{-1} in dairy, chicken, and turkey manures, respectively (Studnicka et al. 2011). Increasing the MWR to 1:1,000 increased the water extractable P_i to 3.3, 5.7, and 11.0 g P_i kg^{-1} in dairy, chicken,

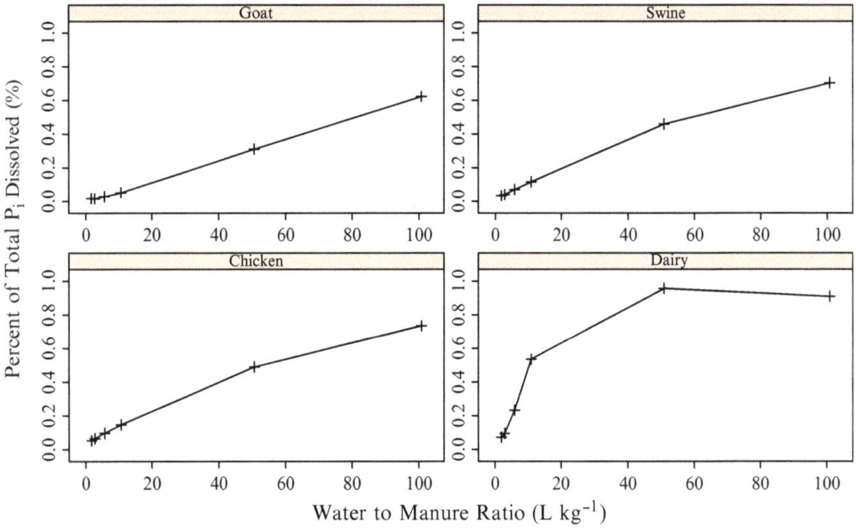

Fig. 8.2 Manure inorganic P (P_i) dissolved as the manure to water ratio increased from 1:1 to 1:100 in samples collected from goat, swine, chicken, and dairy manure (Pagliari 2011)

and turkey manures, respectively (Studnicka et al. 2011). In another study, the author of this chapter used MWR ranging from 1:1 to 1:100 to study the P_i solubility of goat, swine, chicken, and dairy manures (Pagliari 2011). Figure 8.2 shows the increasing amount of P_i that dissolved as the MWR increased from 1:1 to 1:100 for goat, swine, chicken, and dairy manures. The Fig. 8.2 also shows that for goat, swine, and chicken manure there were still more P_i that could potentially dissolve; unlike dairy manure, which appeared to have reached a maximum dissolution at the MWR 1:50. The reasons for the increase in P_i solubility with increased MWR reported in the literature are yet not fully understood. However, it is likely that increasing the MWR resulted in a decrease of the Ca and Mg ionic activity in solution. The decreased Ca and Mg ionic activity resulted in an increased P_i solubility. Increasing MWR is therefore analogous to using EDTA, which also minimizes the Ca and Mg ionic activity in solution allowing for P_i from manure to dissolve. However, the use of serial dilution is limited based on the volume of water that can be practically used. In contrast, the use of EDTA enables the use of rather small MWR ratios and can achieve more satisfactory results.

8.2.3 Identified Manure Organic Phosphorus Compounds

Organic P from animal manure has been studied for decades using several techniques including solution and solid-state ^{31}P NMR, SEM, XANES, and enzyme hydrolysis. However, the different methods provide information of varying degrees of usefulness. For example, ^{31}P NMR and XANES can identify a wide range of P_o in a sample; though those methods provide no information on the bioavailability of those forms to biological communities. In contrast, the methods that use enzyme hydrolysis provide what percent of P_o can actually become bioavailable to biological communities. He and Honeycutt (2011) provided a list of enzymes and how effective they were at catalyzing the hydrolysis of specific P_o compounds. Therefore, the objectives the researchers tried to achieve should be considered before deciding which method to use for P_o determination.

The amounts and forms of manure P_o reported in the literature vary widely within the same animal species and also among different species. The high variability of the manure P_o found in manures is a result of the different diets, differences in how the feed is digested inside the digestive tract, conditions and time of manure storage, type of bedding used, animal age, and other factors. Manure treatment such as composting, anaerobic digestion, or liquid-solid separation can also affect the amounts and forms of P_o in a manure sample.

8.2.3.1 Swine Manure

The P_o forms in swine manure reported in the literature vary dramatically among studies. This reflects the differences listed before, such as animal age, feed, manure age, type of storage, and also the method used for manure extraction. For example, Turner and Leytem (2004) used the sequential fractionation procedure to separate the P soluble in water, $NaHCO_3$, NaOH, and HCl from swine manure (total P 13,508 mg P kg^{-1}). Turner and Leytem (2004) then used ^{31}P NMR to determine what forms of P_o were present in each fraction. Although only small amounts of P_o were reported, monoester-P made up the majority of P_o in the water, $NaHCO_3$, and NaOH fractions (617 mg P kg^{-1}), while phytate was the major form of P_o in the HCl (401 mg P kg^{-1}) fraction with small amount in the NaOH (118 mg P kg^{-1}) fraction (Turner and Leytem 2004). Other P_o compounds found in the sample analyzed by Turner and Leytem (2004) included small amounts (<70 mg P kg^{-1}) of phospholipids and DNA in the water fraction only. In a different study, Leinweber et al. (1997) used 0.1 and 0.5 M NaOH to investigate the P_o forms in swine manure (total P 16,220 mg P kg^{-1}), and reported that mainly monoester-P (8,515 mg P kg^{-1}) and diester-P (3,973 mg P kg^{-1}) were present in the sample analyzed. He and Honeycutt (2001) also used the sequential fraction to characterize the P soluble in water, $NaHCO_3$, and NaOH from swine manure (total P 3,875 mg P kg^{-1}). In addition, He and Honeycutt (2001) used enzyme hydrolysis to determine the forms of P_o in each of the fractions. Monoester-P (356 mg P kg^{-1}) was the major form of P_o, followed by phytate (350 mg P kg^{-1}), DNA-P (61 mg P kg^{-1}), and organic pyrophosphate (35 mg P kg^{-1}) (He and Honeycutt 2001). Pagliari and Laboski (2012) also used the sequential fractionation and enzyme hydrolysis to investigate the hydrolysable P_o in several manures. The majority of the hydrolysable P_o was found as phytate (1,972 mg P kg^{-1}), followed by monoester-P (622 mg P kg^{-1}), and DNA-P (288 mg P kg^{-1}) (Pagliari and Laboski 2012). These studies show that although different P_o forms can be found in swine manure, monoester-P and phytate are the dominant forms. In addition, the concentration of each P_o form varies widely depending on the manure sample. All of the manure samples used in the studies reported in this section were from animals fed regular diets without feed additives, such as phytase.

8.2.3.2 Beef and Dairy Manure

The microflora present in the digestive tract of ruminant animals are known to produce the phytase enzymes that are responsible for breaking down the phytate from feed. Although phytase is present in the digestive tract of ruminant animals, there is still a significant amount of phytate reported in manures from ruminant animals. Pagliari and Laboski (2012) reported that phytate (471 mg P kg^{-1}) was the major P_o form in beef manure (average total P 8,500 mg P kg^{-1}), followed by monoester-P (301 mg P kg^{-1}), and DNA-P (182 mg P kg^{-1}). In contrast, Turner and Leytem (2004) used ^{31}P NMR to identify the forms of P_o in beef manure (total P 3,916 mg P kg^{-1}) and reported that monoester-P (1,100 mg P kg^{-1}) was the major form of P_o with small amounts of DNA-P (109 mg P kg^{-1}).

For dairy manure (total P 3,450 mg P kg^{-1}), He and Honeycutt (2001) reported that most of the hydrolysable P_o was DNA-P (135 mg P kg^{-1}), followed by phytate (121 mg P kg^{-1}), organic pyrophosphate (116 mg P kg^{-1}), and small amounts of monoester-P (48 mg P kg^{-1}). In a different study, He et al. (2004a) reported that phytate (average 457 mg P kg^{-1}) as the major form of hydrolysable P_o in 13 dairy manure samples (total P average 9,131 mg P kg^{-1}). In contrast Pagliari and Laboski (2012) reported similar amounts of monoester-P (295 mg P kg^{-1}) and phytate (254 mg P kg^{-1}) and smaller amounts of DNA-P (103 mg P kg^{-1}) in 18 dairy manure samples (total P average 8,600 mg P kg^{-1}). Nuclear magnetic resonance has also been used to determine the forms of P_o in dairy manure. The reports show that the majority of the P_o identified by ^{31}P NMR is phytate and monoester-P with similar amounts (about 250 mg P kg^{-1}) and small amounts of DNA-P and other diester-P forms (100 mg P kg^{-1}) (He et al. 2009b; McDowell and Stewart 2005). Others have used XANES to study the P_o forms in dairy manure and reported only phytate was present (Shober et al. 2006). He et al. (2007) compared the P_o forms identified by ^{31}P NMR and enzyme hydrolysis in dairy manure (total P 6,880 mg P kg^{-1}). He et al. (2007) reported that monoester-P (1,486 mg P kg^{-1}) was the highest form of P_o followed by phytate (413 mg P kg^{-1}) and DNA-P (89 mg P kg^{-1}) as indicated by ^{31}P NMR. In contrast, the enzyme hydrolysis showed similar amounts of monoester-P (608 mg P kg^{-1}) and phytate (678 mg P kg^{-1}), with greater amounts of DNA-P (434 mg P kg^{-1}) than ^{31}P NMR (He et al. 2007). Although the amounts of P_o are highly variable among the different studies, monoester-P and phytate are the dominant forms of P_o in beef and dairy manure.

8.2.3.3 Chicken and Turkey

The majority of P_o in poultry manure fed regular diets is in most cases reported to be phytate with smaller amounts of monoester-P and DNA-P. For example, Pagliari and Laboski (2012) reported that the amounts of phytate, monoester-P, and DNA-P were 3,195, 1,037, and 315 mg kg^{-1}, respectively. He et al. (2007) reported similar results to those of Pagliari and Laboski (2012), where phytate, monoester-P, and DNA-P were 2,060, 138, 407 mg kg^{-1}, respectively. When ^{31}P NMR or XANES were used for identification of P_o in poultry manure, higher amounts of phytate was observed compared with the enzyme hydrolysis method. For example, phytate has been reported to range from 3,638 to 8,208 mg kg^{-1} (Turner and Leytem 2004; Toor et al. 2005; Shober et al. 2006; He et al. 2007). Similar results have been observed for turkey manure where enzyme hydrolysis methods estimate lower concentration of phytate P_o in manures compared with ^{31}P NMR or XANES. Pagliari and Laboski (2012) reported that phytate in turkey manure determined with the enzyme method was 1,707 mg kg^{-1}, while Toor et al. (2005) reported that phytate was 4,000 mg kg^{-1} using XANES. Although very different results can be observed by the different methods used for P_o determination, the researcher should use the method that best suits their specific research goals.

8.2.3.4 Other Species

Limited research has been reported on P_o forms in manure from species such as horse, deer, sheep, goat and others. Pagliari and Laboski (2012) used the enzyme hydrolysis to determine the hydrolysable P_o in samples of goat, horse, and sheep manure and reported that Phytate was 508 mg kg^{-1} and 328 mg kg^{-1}; monoester-P was 305 and 243 mg kg^{-1}; and DNA-P was 209 and 157 mg kg^{-1} in sheep and horse respectively. In goat manure, 952 mg kg^{-1} was monoester-P, 717 mg kg^{-1} was DNA-P and 302 mg kg^{-1} was phytate. McDowell and Stewart (2005) used ^{31}P NMR to study the forms of P_o in deer and sheep manure and reported that monoester-P was 590 and 532 mg kg^{-1}, DNA-P was 140 and 215 mg kg^{-1}, and phospholipids-P was 2,147 and 105 mg kg^{-1} in deer and sheep manure, respectively. In their study, McDowell and Stewart (2005) grouped phytate into the same group as the other general forms of monoester-P.

In summary, the P_o compounds that are reported to be present in manure include more generally phytate, monoester-P and DNA-P. The amount of each form and the fraction that each form represents in a sample is highly variable. Therefore, testing each manure sample is the only way to know what form of P_o is present and at what concentrations that particular form is present in a sample. The next section examines the solubility of each P_o form in water, $NaHCO_3$, NaOH, and HCl.

8.2.4 Solubility of Organic Phosphorus

A diverse number of extractant solutions have been used to investigate the solubility of P in animal manure. Among the extractants employed the most used include water, $NaHCO_3$, NaOH, NaOH + EDTA, sodium acetate (NaOAc), NaOAc with EDTA or sodium dithionite (NaOAc-SD), and HCl. The concentration of those extracting solutions usually varies from study to study, with exception of the $NaHCO_3$, which is usually set as 0.5 M. Sodium hydroxide has been reported to be used at concentrations ranging from 0.1 to 0.5 mol L^{-1}. The EDTA is usually added at 0.05 M, which for most manures should be enough to complex most, if not all, of the Ca and Mg in the sample. Sodium acetate is usually used at 0.1 M, with the pH of the solution buffered at 5.0. Hydrochloric acid has been used in manure extractions at the concentration of 1.0 M. For a particular manure sample, the amount of P_o that is extracted with each of these different solutions varies widely. Water and NaOAc in general, extract the least amount of P_o compared with the other extractants. The addition of EDTA or sodium dithionite to NaOAc, however, makes NaOAc as effective as NaOH-EDTA or HCl (He et al. 2009a, b). For example He et al. (2009a) reported that the amount of P_o extracted in water was 928 and 170 mg P_o kg^{-1}, in 0.1 M NaOAc was 982 and 8,258 mg P_o kg^{-1}, and averaged 3,595 and 10,691 mg P_o kg^{-1} in 0.1 M NaOAc + 0.05 M EDTA, 0.25 and 0.5 M NaOH + 0.05 M EDTA, and 1.0 M HCl from poultry manure and poultry litter, respectively. In a different study, Turner and Leytem (2004) compared the effectiveness of 0.5 M $NaHCO_3$ in combination with either 1.0 M HCl or 0.5 M NaOH + 0.05 M EDTA with the sequential fractionation method to extract P from cattle, poultry, and swine manures. Turner and Leytem (2004) reported that, overall, combining $NaHCO_3$ with NaOH + EDTA (total P_o extracted 8,866, 809 and 1,229 mg P_o kg^{-1}, in poultry, cattle, and swine manure, respectively) was as effective as the sequential fractionation (total P_o extracted 10,232, 3,830, and 1,429 mg P_o kg^{-1}, in poultry, cattle, and swine manure respectively) with the exception of cattle manure. However the $NaHCO_3$ combined with NaOH + EDTA simplified the procedures, as only two extractions were needed as opposed to four in the sequential fractionation procedure.

Although a significant amount of work has been done to improve our understanding on manure P_o solubility by using different extractants, there has been no work that evaluated P_o solubility as a function of MWR. In the study of Turner and Leytem (2004) it was indicated that $NaHCO_3$ with NaOH + EDTA was as effective as the sequential fractionation to remove total P_o. However, Turner and Leytem (2004) also reported that the P_o extracted in the water + $NaHCO_3$ fraction was higher than the P_o extraction in the $NaHCO_3$ alone from the $NaHCO_3$ with NaOH-EDTA method. This suggests that even the more labile P_o pool in animal manure might be affected by the ionic activity of Ca and Mg in solution, as is the manure P_i. The fact that Ca and Mg might be affecting the rate of the more labile manure P_o solubility is also supported by the increased P_o recovery that is observed by adding EDTA to NaOH or NaOAc extracts (He et al. 2009a, b; Turner and Leytem 2004).

8.3 Increase in Soil Test Phosphorus After Manure Addition

Soil tests have been used to estimate how much nutrient in the soil may become available for plant uptake during the growing season. In the case of P, sol test phosphorus values are used to determine the P status in a soil, or the labile P. From the initial status, one can estimate how much P must be added to a soil to provide plants with an adequate amount of P for optimum growth and productivity. Several soil test methods have been developed over the past 60 years as an aid to best estimate of the amount of P required for crop growth. The diversity in soil tests used for P analysis reflects the differences in soil chemistry and also the ability of the different tests in extracting P from different binding sites within the soils. These methods include extraction of P in water (WSP), acidic solution (e.g. Bray-1 P, Mehlich-3, Morgan, Modified Morgan), alkaline solution (e.g. Olsen P), and also exchange reaction by using resins (e.g. iron oxide impregnated filter paper, namely FeO). Although those soil tests have been proven to correlate well with plant uptake, the best test to be used depends on the soil and also on the binding mechanisms between the soil particles and P to be determined.

The Mehlich-3 and Bray-1 P extracts are very effective at complexing Ca, Al, and Fe cations with fluoride (F) anions to release bound P (Kuo 1996). These tests use an acidic extracting solution to solubilize easily soluble P present as Al-phosphate, Fe-phosphate, and Ca-phosphate. However, the use of acidic extractants is appropriate for soils with pH ranging from slightly alkaline to acidic. Soils with high pH, with large concentration of hydroxyl anions (OH^-) or carbonates (HCO^{-3}) can neutralize the acidity of the extracting solution and subsequently underestimate extractable P.

The Olsen P test, on the other hand, works by replacing anions adsorbed on Ca (Kuo 1996). For example, the phosphate anions ($HPO_4{}^{-2}$) from sorbed as Ca-phosphate are replaced by bicarbonate (HCO^{-3}) anions and released into solution. In this test, the soil pH must be from slightly acid to alkaline, because low soil pH reduces HCO^{-3} activity by forming $H_2O + CO_2$ and subsequently underestimating the P status of that soil. The FeO test works differently from the other two extraction methods. Ferric oxide provides sites that specifically sorb P in solution (Kuo 1996).

The WSP test is the simplest of the three extraction methods presented above. In this test, the portion of P in solution that is in equilibrium with P sorbed on the soil is extracted. This portion, represents a very small fraction of available P and usually is not used as an estimate of plant available P nor for fertilizer recommendations. However, this test has been used to provide an index of the amount of P that may pose a risk to surface water quality (Bundy et al. 2001; Sharpley and Moyer 2000).

After samples are extracted several methods can be used for P determination in the extracts. For example, P can be determined by inductively coupled plasma optical emission spectroscopy (ICP-OES), or it can be determined by colorimetry, using methods proposed by Watanabe and Olsen (1965), or more recently the method of He and Honeycutt (2005). Extensive work has been done to show that the ICP-OES and the colorimetry methods quantify different amounts of P when the same samples are analyzed by both methods (Pittman et al. 2005; He et al. 2012). This indicates that both methods are measuring different pools of P in the extracts. Pittman et al. (2005) and He et al. (2012) have suggested that the differences between ICP-OES P and the P measured by colorimetry was mainly due to the P_o fraction that was extracted together with the P_i when using the Mehlich-3 and modified Morgan method for P extraction. Understanding that different P pools are extracted when different soil extractants are used and that different P pools are quantified when different methods are used is crucial to optimizing fertilizer efficiency and minimizing potential environmental problems. Furthermore, understanding how manure affects the distribution of P in the soil, may help developing management plans that can minimize the negative impacts of manure application to agricultural fields.

8.3.1 The Inorganic Fraction

Our understanding of how manure affects soil test P (STP) has significantly improved over the last couple of years (Leconte et al. 2011; Waldrip et al. 2011, 2012; Roboredo et al. 2012; Pagliari and Laboski 2013). The better understanding of the manure solid phases, such as what minerals control P solubility, and also the solubility and of the P_o forms in manure has helped us better understand the effects of manure on STP. Up until recently it was believed that the labile P_i from manure (P_i soluble in the water and $NaHCO_3$ fraction) was the fraction that had the most impact on STP. However, the work of Waldrip et al. (2011) and Pagliari and Laboski (2013), have shown that not only the total P_i in a manure sample is bioavailable, but also the total hydrolysable P_o. We will now discuss the effects of manure P_i on STP, while the effects of the hydrolysable P_o on STP will be discussed on the next section (Sect. 8.3.2).

A linear increase in STP is usually observed when increasing amounts of inorganic P from inorganic fertilizer such as triple superphosphate, or monoammonium phosphate, or diammonium phosphate are applied to a given soil (Griffin et al. 2003; Leytem et al. 2005; Pagliari et al. 2010). Figure 8.3 exemplifies this theoretical relationship for a typical Mollisol soil from Wisconsin, based on results reported by Pagliari and Laboski 2013. The Fig. 8.3 shows that the increase in STP measured with the Bray-1 P after inorganic P addition follows a normal distribution and 95 % of the points fall within the 95 % confidence limit (95 % CI) represented by the dotted lines.

In the case of manure, we can evaluate what the effects of P_i soluble in water, $NaHCO_3$, NaOH, and HCl is on the increase in STP. For example, if one would consider that the P_i soluble in water is the only fraction of manure P_i that interacts with the soil to increase STP, then plotting the increase in STP as a function of manure applied P_i should result in at least 95 % of the points falling within the 95 % CI for inorganic fertilizer. However, this is not what is observed when the water P_i is assumed to be the only fraction of manure P that is highly reactive in the soil.

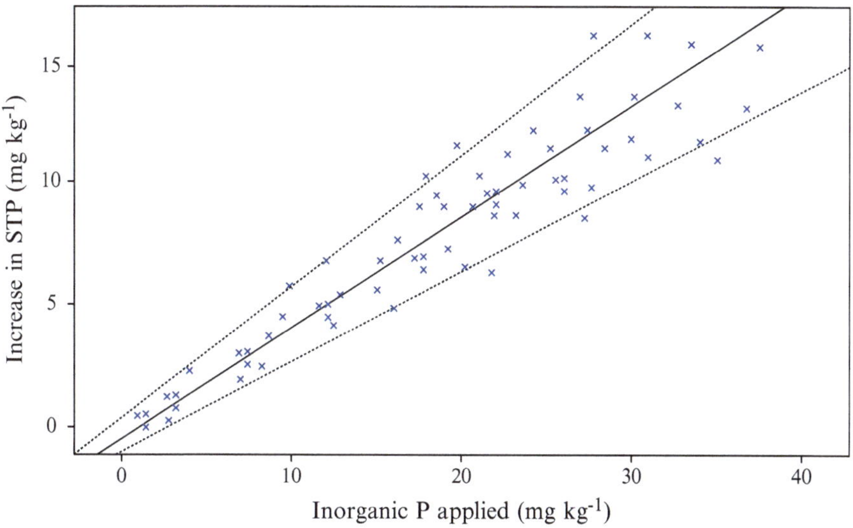

Fig. 8.3 Increase in soil test P (STP) measured with the Bray-1 P as a function of inorganic P from inorganic fertilizer applied. Solid line represents the regression equation for the increase in STP as a function of inorganic P applied; *dotted lines* are the 95 % confidence interval around the regression line (Pagliari and Laboski 2013)

Figure 8.4 shows the increase in STP measured with the Bray-1 P as a function of P_i soluble in water for 40 different manures applied to a soil from Wisconsin (Pagliari 2011). In this figure, the solid line represents the regression equation obtained by applied inorganic P fertilizer and the dotted lines represent the 95 % CI for the inorganic fertilizer.

If the manure P_i soluble in water was the only fraction of manure P that would react with soil to increase STP, then all points would be expected to fall within the 95 % CI for the fertilizer. However, the majority of the points fell above the 95 % CI. This suggests that the amount of manure P_i that can react with soil and increase STP is greater than the amount that is water-soluble. Figure 8.1 showed that 100 % of the manure P_i was soluble in a 0.05 mol L^{-1} EDTA solution buffered at pH 5. In addition, it was mentioned earlier that if enough water is added to a manure sample then most of the P_i from manure could dissolve. Pagliari and Laboski (2013) have shown that 100 % of the manure P_i can in fact dissolve in the soil and participate in reactions that are responsible for increasing STP after manure addition. Figure 8.5 shows the increase in STP measured with the Bray-1 P for the soil and manures showed in Fig 8.4 when the total manure P_i is used as the amount of P applied. In this example, 88 % of the points for the increase in STP after manure addition fell within the 95 % CI for the inorganic fertilizer. The same behavior has been observed for over 20 soils amended with more than 40 different manures (Pagliari and Laboski 2013).

As explained earlier in Sect. 8.2.1, the solid phase of animal manure is composed of a mixture of Mg- and Ca-phosphate minerals and the solubility of those minerals

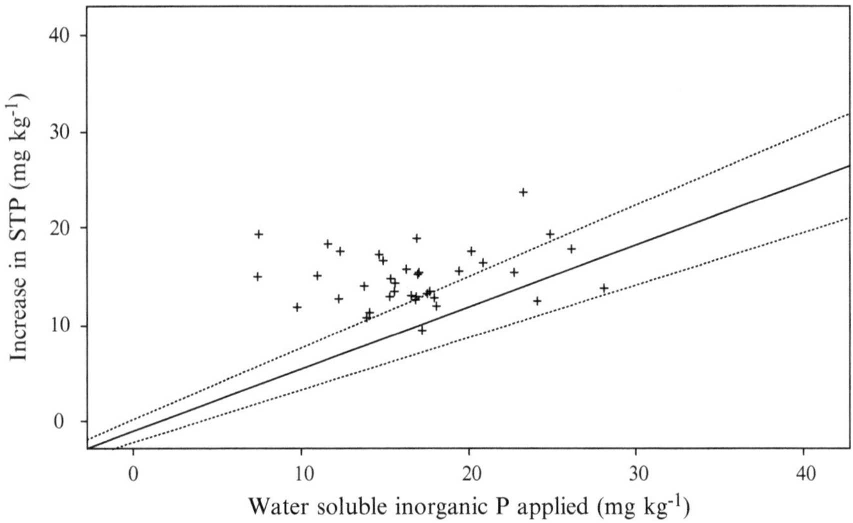

Fig. 8.4 Increase in soil test P (STP) measured with the Bray-1 P as a function of water-soluble manure inorganic P from 42 different manures. Each point represents one manure sample applied at 40 mg total P kg^{-1}. *Solid line* represents the regression equation for the increase in STP as a function of inorganic fertilizer P applied; *dotted lines* are the 95 % confidence interval around the regression line for the inorganic fertilizer (Pagliari 2011)

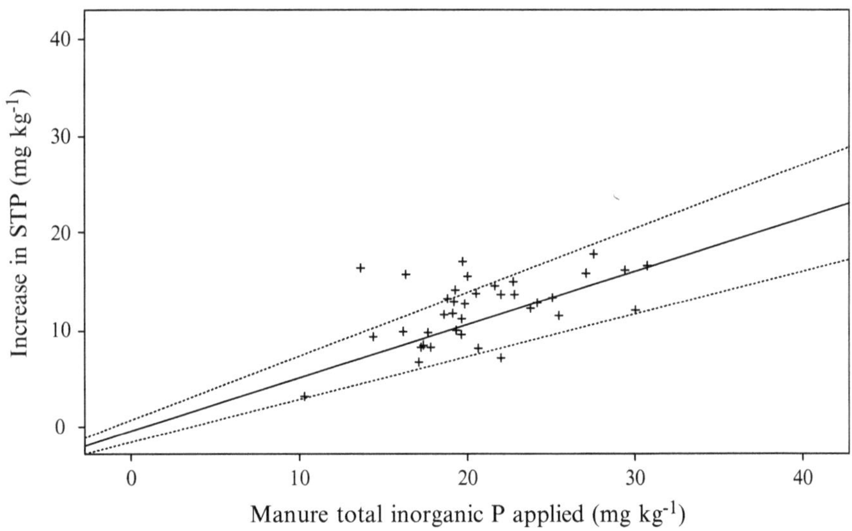

Fig. 8.5 Increase in soil test P (STP) measured with the Bray-1 P as a function of total manure P_i applied from 42 different manures. Each point represents one manure sample applied at 40 mg total P kg^{-1}. *Solid line* represents the regression equation for the increase in STP as a function of inorganic fertilizer P applied; *dotted lines* are the 95 % confidence interval around the regression line for the inorganic fertilizer (Pagliari 2011)

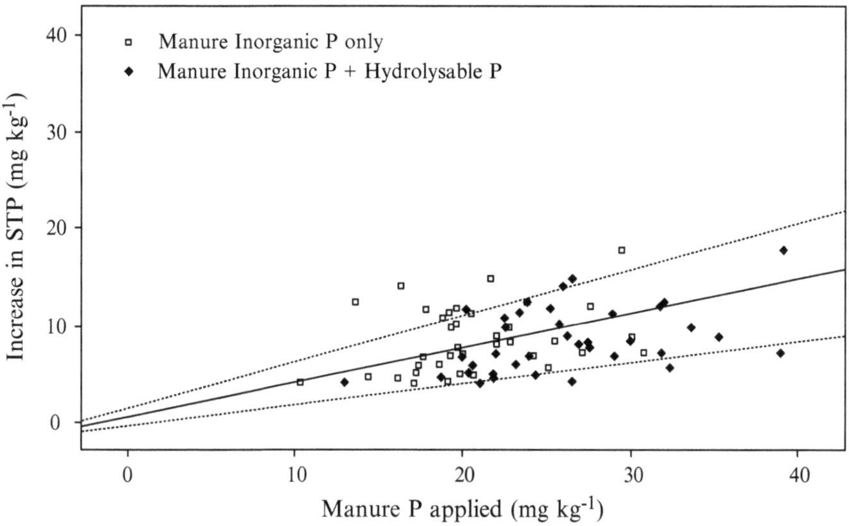

Fig. 8.6 Increase in soil test P (STP) measured with the Bray-1 P as a function of total manure P_i (*open squares*) and total manure P_i+ hydrolysable P_o (*black diamond*) applied from 42 different manures. Each point represents one manure sample applied at 40 mg total P kg^{-1}. *Solid line* represents the regression equation for the increase in STP as a function of inorganic fertilizer P applied; *dotted lines* are the 95 % confidence interval around the regression line for the inorganic fertilizer (Pagliari 2011)

is pH dependent. Therefore, when applied to soils with pH below the solubility constant for those minerals, they will tend to dissolve, and this is what recent research has started to observe (Pagliari and Laboski 2013). However, under calcareous soils, where the soil pH is higher than the solubility constant for the minerals present in the manure, less available P might be extracted than the total manure P_i applied.

8.3.2 The Organic Fraction

Although there has been increased research that has categorized manure P_o into different fractions and forms, there was limited amount of work that evaluated the availability of those forms to enzyme hydrolysis in soils. Waldrip et al. (2011) reported that ryegrass root biomass was highly correlated with labile hydrolysable P_o on poultry manure amended soil. These authors also reported that the addition of poultry manure to soil promoted mineralization of P_o. In a different study, Pagliari and Laboski (2013) showed evidences that the hydrolysable forms of P_o from manure were hydrolyzed in certain soils and affected the increase in STP. Figure 8.6 illustrates the potential effect that the hydrolysable P_o has on the increase in STP. Figure 8.6 is a plot of the increase in STP measured with the Bray-1 P as a function of total P_i and total P_i+ total hydrolysable P_o applied with 42 different manures for a soil from Wisconsin.

The open squares show the increase in STP observed as a function of total P_i applied. About 21 % of the manures increased STP more (points above the 95 % CI line) than would be expected by assuming that the manure P_i is the only fraction that increases STP. Adding the total hydrolysable P_o decreased the number of points that were above the 95 % CI from 21 to 5 %. Adding the manure hydrolysable P_o also increased the number of points that fell below the 95 % CI to 7 %. Enzyme hydrolysis of sequential extracts from these same manure samples showed that the hydrolysable P_o ranged from 4 to 29 % of the total P and were classified as phytate, monoester-P, and DNA-P (Pagliari and Laboski 2012). Waldrip et al. (2012) showed that the addition of organic dairy manure and conventional dairy manure increased acid phosphomonoesterase enzyme activity compared with soil that received inorganic fertilizer. The results of Waldrip et al. (2012) provided further evidences that the hydrolysable P_o from manure is potentially bioavailable to soil microbes and plant roots. Chapter 10 provides more information on phosphatase activity in soils.

Although there have been evidences that the manure hydrolysable P_o can affect how manure increases STP, there is still more information needed to fully understand the entire process. The work of Pagliari showed that only a certain group of soils had signs of P_o hydrolysis, while a different set of soils showed no signs of P_o hydrolysis. Pagliari reported that soil clay content was the determinant soil property that dictated whether or not P_o would hydrolyze in the soils studied. However, Pagliari were not able to provide conclusive evidences neither did the authors provided what were the mechanisms controlling the hydrolysis reactions. Phytate is known to adsorb strongly onto clay particles (Celi et al. 1999; Karathanasis and Shumaker 2009), which can decrease the availability of phytate for enzyme hydrolysis (Giaveno et al. 2010). Furthermore, certain types of clay are more effective at protecting phytate against hydrolysis than others. For example, Giaveno et al. (2010) reported that phytate adsorbed onto kaolinite types of clay were available for hydrolysis, though only small amounts of P_i could be recovered. In contrast, phytate adsorbed onto montmorillonite, goethite, or hematite is unavailable for hydrolysis (Giaveno et al. 2010).

8.4 Future Research

As reported in this chapter the understanding of the effects of various forms of P from manure has significantly improved over the last decade. Being able to identify that 100 % of manure P_i is bioavailable as opposed to only the fraction that is soluble in water and $NaHCO_3$ will help in the future development of manure management strategies that focus on reducing the unnecessary use of manure as a P source for crop production. However, much research is still needed.

Most of the research that has used the sequential fraction to investigate the effects of the various P_i fractions on STP has been done in the laboratory and also used soils with acidic pH. Soils with acid pH are more likely to provide the right conditions for the mineral solid phases present in the manure to dissolve. Soils with high pH values, such as >7.0, have not been thoroughly investigated. Therefore, it is extremely important to assess whether the results observed in the laboratory would also be observed in the field under unpredictable and uncontrollable weather variations. In addition, the presence of growing plants in the field can also provide interactions that could not happen in incubation studies. Understanding the effects of those potential environmental differences on how manure and soil interact is needed before significant improvements on current practices can be achieved.

Although evidences for the availability of P_o for enzyme hydrolysis in soils have been reported, much more work is still needed to fully understand the P_o behavior in soils. Clay content has been identified as one of the soil properties that have the most impact on P_o hydrolysis under laboratory conditions. However, there has been no field research studies that have confirmed the results reported under laboratory conditions. In summary, field research is needed so that the laboratory research is confirmed.

References

Ajiboye B, Akinremi OO, Racz GJ (2004) Laboratory characterization of phosphorus in fresh and oven-dried organic amendments. J Environ Qual 33:1062–1069

Ajiboye B, Akinremi OO, Hu Y, Flaten DN (2007) Phosphorus speciation of sequential extracts of organic amendments using nuclear magnetic resonance and X-ray absorption near-edge structure spectroscopies. J Environ Qual 36:1563–1576

Akinremi OO, Ajiboye B, He Z (2011) Metal speciation of phosphorus derived from solid state spectroscopic analysis. In: He Z (ed) Environmental chemistry of animal manure. Nova, New York

Bril J, Salomons W (1990) Chemical composition of animal manure: a modelling approach. Neth J Agric Sci 38:333–349

Bundy LG, Andraski TW, Powell JM (2001) Management practice effects on phosphorus losses in runoff in corn production systems. J Environ Qual 30:1822–1828

Cade-Menun BJ (2011) Characterizing phosphorus in animal waste with solution ^{31}P NMR spectroscopy. In: He Z (ed) Environmental chemistry of animal manure. Nova, New York, pp 275–299

Celi L, Lamacchia S, Marsan FA, Barberis E (1999) Interaction of inositol hexaphosphate on clays: adsorption and charging phenomena. Soil Sci 164:574–585

Cooperband LR, Good LW (2002) Biogenic phosphate minerals in manure: implications for phosphorus loss to surface waters. Environ Sci Technol 36:5075–5082

Dou Z, Toth JD, Galligan DT, Ramberg CF Jr, Ferguson JD (2000) Laboratory procedures for characterizing manure phosphorus. J Environ Qual 29:508–514

Fordham AW, Schwertmann U (1977) Composition and reactions of liquid manure (gülle), with particular reference to phosphate: II. solid phase Components1. J Environ Qual 6:136–140

Giaveno C, Celi L, Richardson AE, Simpson RJ, Barberis E (2010) Interaction of phytases with minerals and availability of substrate affect the hydrolysis of inositol phosphates. Soil Biol Biochem 42:491–498

Griffin TS, Honeycutt CW, He Z (2003) Changes in soil phosphorus from manure application. Soil Sci Soc Am J 67:645–653

Güngör K, Karthikeyan KG (2005) Probable phosphorus solid phases and their stability in anaerobically digested dairy manure. Trans ASAE 48:1509–1520
Güngör K, Karthikeyan KG (2008) Phosphorus forms and extractability in dairy manure: a case study for Wisconsin on-farm anaerobic digesters. Bioresour Technol 99:425–436
Güngör K, Jürgensen A, Karthikeyan KG (2007) Determination of phosphorus speciation in dairy manure using XRD and XANES spectroscopy. J Environ Qual 36:1856–1863
Haygarth PM, Sharpley AN (2000) Terminology for phosphorus transfer. J Environ Qual 29:10–15
He Z, Honeycutt CW (2001) Enzymatic characterization of organic phosphorus in animal manure. J Environ Qual 30:1685–1692
He Z, Honeycutt CW (2005) A modified molybdenum blue method for orthophosphate determination suitable for investigating enzymatic hydrolysis of organic phosphates. Commun Soil Sci Plant Anal 36:1373–1383
He Z, Honeycutt CW (2011) Enzymatic hydrolysis of organic phosphorus. In: He Z (ed) Environmental chemistry of animal manure. Nova, New York, pp 253–274
He Z, Griffin TS, Honeycutt CW (2004a) Phosphorus distribution in dairy manures. J Environ Qual 33:1528–1534
He Z, Griffin TS, Honeycutt CW (2004b) Enzymatic hydrolysis of organic phosphorus in swine manure and soil. J Environ Qual 33:367–372
He Z, Senwo ZS, Mankolo RN, Honeycutt CW (2006) Phosphorus fractions in poultry litter characterized by sequential fraction coupled with phosphate hydrolysis. J Food Agric Environ 4:304–312
He Z, Cade-Menun BJ, Toor GS, Fortuna A, Honeycutt CW, Sims JT (2007) Comparison of phosphorus forms in wet and dried animal manures by solution phosphorus-31 nuclear magnetic resonance spectroscopy and enzymatic hydrolysis. J Environ Qual 36:1086–1095
He Z, Honeycutt CW, Griffin TS, Cade-Menun B, Pellechia PJ, Dou Z (2009a) Phosphorus forms in conventional and organic dairy manure identified by solution and solid state P-31 NMR spectroscopy. J Environ Qual 38:1909–1918
He Z, Waldrip H, Honeycutt CW, Erich MS, Senwo Z (2009b) Enzymatic quantification of phytate in animal manure. Commun Soil Sci Plant Anal 40:566–575
He Z, Zhang H, Toor GS, Dou Z, Honeycutt CW, Haggard BE, Reiter MS (2010) Phosphorus distribution in sequentially extracted fractions of biosolids, poultry litter, and granulated products. Soil Sci 175:154–161
He Z, Zhang H, Olanya OM, Frantz JM, Larkin RP (2012) Differences in modified Morgan phosphorus levels determined by colorimetric and inductively coupled plasma methods. Open J Soil Sci 2(3):256–262
Hedley MJ, Stewart JWB, Chauhan BS (1982) Changes in inorganic and organic soil phosphorus fractions induced by cultivation practices and by laboratory incubations. Soil Sci Soc Am J 46:970–976
Karathanasis AD, Shumaker PD (2009) Preferential sorption and desorption of organic and inorganic phosphates by soil hydroxyinterlayered minerals. Soil Sci 174:417–423
Kleinman PJA, Sharpley AN, Wolf AM, Beegle DB, Moore PA (2002) Measuring water-extractable phosphorus in manure as an indicator of phosphorus in runoff. Soil Sci Soc Am J 66:2009–2015
Kuo S (1996) Phosphorous. In: Sparks DL (ed) Methods of soil analysis: Part 3, Chemical methods. American Society of Agronomy, Madison
Leconte M, Mazzarino M, Satti P, Crego M (2011) Nitrogen and phosphorus release from poultry manure composts: the role of carbonaceous bulking agents and compost particle sizes. Biol Fertil Soils 47:897–906
Leinweber P, Haumaier L, Zech W (1997) Sequential extractions and ^{31}P NMR spectroscopy of phosphorus forms in animal manures, whole soils and particle-size separates from a densely populated livestock area in northwest Germany. Biol Fertil Soils 25:89–94
Leytem AB, Turner BL, Raboy V, Peterson KL (2005) Linking manure properties to phosphorus solubility in calcareous soils: importance of the manure carbon to phosphorus ratio. Soil Sci Soc Am J 69:1516–1524

Maguire RO, Sims JT, Saylor WW, Turner BL, Angel R, Applegate TJ (2004) Influence of phytase addition to poultry diets on phosphorus forms and solubility in litters and amended soils. J Environ Qual 33:2306–2316

McDowell RW, Stewart I (2005) Phosphorus in fresh and dry dung of grazing dairy cattle, deer, and sheep: sequential fraction and phosphorus-31 nuclear magnetic resonance analyses. J Environ Qual 34:598–607

Pagliari PH (2011) Investigation of inorganic and organic phosphorus in animal manure and their effects on soil test phosphorus. Dissertation, University of Wisconsin-Madison

Pagliari PH, Laboski CAM (2012) Investigation of the inorganic and organic phosphorus forms in animal manure. J Environ Qual 41:901–910

Pagliari P, Laboski CAM (2013) Dairy manure treatment effects on manure phosphorus fractionation and changes in soil test phosphorus. Biol Fertil Soils. doi:10.1007/s00374-013-0798-2

Pagliari PH, Rosen CJ, Strock JS, Russelle MP (2010) Phosphorus availability and early corn growth response in soil amended with turkey manure ash. Commun Soil Sci Plant Anal 41:1369–1382

Pittman JJ, Zhang H, Schroder JL, Payton ME (2005) Differences of phosphorus in mehlich 3 extracts determined by colorimetric and spectroscopic methods. Commun Soil Sci Plant Anal 36:1641–1659

Roboredo M, Fangueiro D, Lage S, Coutinho J (2012) Phosphorus dynamics in soils amended with acidified pig slurry and derived solid fraction. Geoderma 190:328–333

Shand CA, Coutts G, Hillier S, Lumsdon DG, Chudek A, Eubeler J (2005) Phosphorus composition of sheep feces and changes in the field determined by 31P NMR spectroscopy and XRPD. Environ Sci Technol 39:9205–9210

Sharpley A, Moyer B (2000) Phosphorus forms in manure and compost and their release during simulated rainfall. J Environ Qual 29:1462–1469

Shober AL, Hesterberg DL, Sims JT, Gardner S (2006) Characterization of phosphorus species in biosolids and manures using XANES spectroscopy. J Environ Qual 35:1983–1993

Studnicka JS, Bundy LG, Andraski TW, Good LW, Powell JM (2011) Measuring water-extractable phosphorus in manures to predict phosphorus concentrations in runoff. Commun Soil Sci Plant Anal 42:1071–1084

Sui Y, Thompson ML, Shang C (1999) Fractionation of phosphorus in a mollisol amended with biosolids. Soil Sci Soc Am J 63:1174–1180

Toor GS, Cade-Menun BJ, Sims JT (2005) Establishing a linkage between phosphorus forms in dairy diets, feces, and manures. J Environ Qual 34:1380–1391

Toor GS, Hunger S, Peak JD, Sims JT, Sparks DL (2006) Advances in the characterization of phosphorus in organic wastes: environmental and agronomic applications. Adv Agron 89:1–72

Toth JD, Dou Z, He Z (2011) Solubility of manure phosphorus characterized by selective and sequential extractions. In: He Z (ed) Environmental chemistry of animal manure. Nova, New York

Turner BL (2004) Optimizing phosphorus characterization in animal manures by solution phosphorus-31 nuclear magnetic resonance spectroscopy. J Environ Qual 33:757–766

Turner BL, Leytem AB (2004) Phosphorus compounds in sequential extracts of animal manures: chemical speciation and a novel fractionation procedure. Environ Sci Technol 38:6101–6108

Waldrip H, He Z, Erich MS (2011) Effects of poultry manure amendment on phosphorus uptake by ryegrass, soil phosphorus fractions and phosphatase activity. Biol Fertil Soils 47:407–418

Waldrip HM, He Z, Griffin TS (2012) Effects of organic dairy manure on soil phosphatase activity, available soil phosphorus, and growth of sorghum-sudangrass. Soil Sci 177:629–637

Watanabe FS, Olsen SR (1965) Test of an ascorbic acid method for determining phosphorus in water and NaHCO3 extracts from Soil1. Soil Sci Soc Am J 29:677–678

Chapter 9
Phytate in Animal Manure and Soils: Abundance, Cycling and Bioavailability

Courtney D. Giles and Barbara J. Cade-Menun

Abstract The importance of phytate in phosphorus (P) cycling in soil and manure has long been recognized. Phytate is a storage compound in seeds that cannot be fully digested by many animal species, resulting in the accumulation of phytate in manures. It can enter the soil either directly from plants, or from application of manures. In this chapter we will discuss the abiotic and biotic factors that control the cycling and bioavailability of phytate in soil and manure. An understanding of these processes is key to enhancing the availability of P to plants and animals, minimizing the losses of P from soil to water, and sustainably managing the use of P in agricultural systems.

9.1 Introduction

Phosphorus (P) research has increased in recent years. One reason for this is the concern for water quality and eutrophication, which can be exacerbated by the loss of P from land to water, particularly in areas of high soil P concentration (e.g. Sharpley et al. 1994). In contrast, research has also become focused on the potential for P deficiencies and sustainable ways to utilize this finite resource in agricultural systems (e.g. Cordell et al. 2009). In a global analysis of agricultural P budgets, Sattari et al. (2012) calculated the difference between inputs (e.g., fertilization) and outputs (e.g., crop uptake, erosional loss), and estimated that the

C.D. Giles
Experimental Program to Stimulate Competitive Research – Adaptation to Climate Change, University of Vermont, 529 Cook Physical Science Bldg., 82 University Place, Burlington, VT 05405, USA

B.J. Cade-Menun (✉)
Semiarid Prairie Agricultural Research Centre, Agriculture and Agri-Food Canada, Swift Current, SK S9H 3X2, Canada
e-mail: Barbara.Cade-Menun@agr.gc.ca

Z. He and H. Zhang (eds.), *Applied Manure and Nutrient Chemistry for Sustainable Agriculture and Environment*, DOI 10.1007/978-94-017-8807-6_9,

residual P fraction (e.g., native, legacy) accounts for 10–30 % of P in crop systems, becoming available as labile-P over variable time-lags. As such, improving the P efficiency of crops, and their ability to access residual soil P, is a priority for the management of P resources world-wide to mitigate agricultural nutrient pollution and enhance crop and animal productivity (e.g. Richardson et al. 2011).

Historically, P research has focused on inorganic P (P_i). However, there is a growing awareness that organic P (P_o) is an important contributor to both soil fertility and water quality problems (Condron et al. 2005). The best studied P_o form is phytate (e.g. Turner et al. 2002b; Giles et al. 2011). The objective of this chapter is to review our current knowledge of the abundance, cycling and bioavailability of phytate in soils and manure, and to suggest areas where knowledge is limited, and thus where further research is needed.

9.2 Phytate Definition

Inositol phosphate (IP_x) is a 6-carbon cyclohexanehexol (inositol) ring that can have between one and six possible phosphate moieties linked via an ester bond. Inositol hexakisphosphate (IP_6) can occur in nine stereoisomeric forms (*myo, scyllo, neo,* L-*chiro*-(−), D-*chiro*-(+), *epi, muco, allo,* or *cis*-IP_6), differing only in the axial or equatorial orientation of the six phosphate groups (Turner et al. 2002b; Giles et al. 2011). Phytic acid is the term commonly used for *myo*-inositol hexakisphosphate (*myo*-IP_6). However, phytic acid does not exist in nature; instead, it is found as a salt, which is called phytate. The term "phytate" generally refers to any salt of phytic acid. Specific salts such as calcium (Ca)/magnesium (Mg)/potassium (K)-*myo*-IP_6 can be referred to as 'phytin' (Shears and Turner 2007). However, the term "phytin" is less commonly used than the more general "phytate" for soils and manures.

Under environmental conditions where the pH ranges between 6 and 9, phytate will have one axial and five equatorial phosphate groups (Fig. 9.1), while at pH $>$10 the opposite will be true (Barrientos and Murthy 1996).

9.3 Methods to Identify and Quantify Phytate

As shown in Tables 9.1 and 9.2, a variety of methods have been used to quantify phytate in soil and manure. Ideally, phytate should be quantified directly in soil or manure samples without the need for extraction to minimize potential artifacts, and by a method that distinguishes among the various phytate salts (e.g. Ca-phytate vs. Mg-phytate). However, techniques for direct measurement may not be sensitive enough to quantify P in all samples. Various methods that have been used to identify and quantify phytate in soil and manure are discussed in this section.

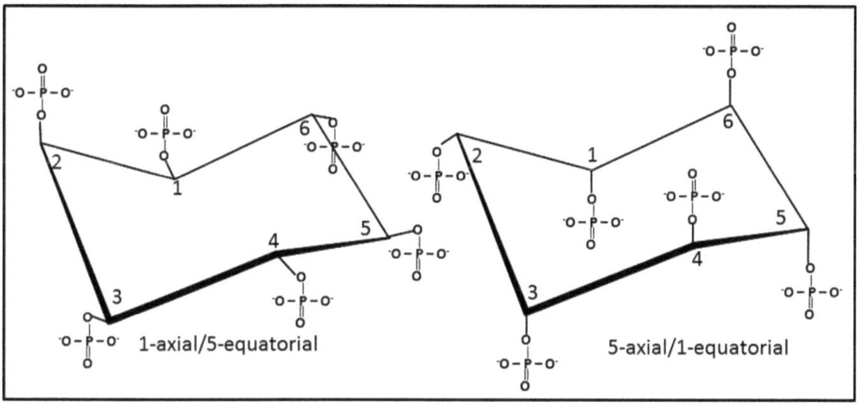

Fig. 9.1 The two configurations of phytic acid. In nature, these molecules exist as salts with cations, termed phytate

9.3.1 Solid-State Techniques

He and colleagues have shown that it is possible to identify and quantify various phytate salts in prepared mixtures of phytate compounds using the solid-state techniques of Fourier transform infrared (FT-IR), P-K-edge X-ray Adsorption Near-Edge Structure (XANES), solid-state 1D ^{31}P and ^{13}C nuclear magnetic resonance (NMR) and 2D ^{1}H-^{13}C NMR spectroscopies (He et al. 2006a, 2007b, c, 2013). However, these techniques have been less successful for manure and soil samples. Phytate has not been identified in FT-IR or solid-state ^{31}P NMR spectra of soils or manure samples, which are generally broad and poorly resolved due to the random orientation of molecules in the matrix, the presence of paramagnetic ions, and the wide range of P compounds with overlapping P resonances (Shand et al. 1999; Hunger et al. 2004, 2008; He et al. 2007b; Conte et al. 2008). In contrast, P-K-edge XANES spectroscopy has been more widely used to characterize P in manure (Toor et al. 2005a; Shober et al. 2006; Ajiboye et al. 2007) and soil (Ajiboye et al. 2008; Prietzel et al. 2013) samples. However, as noted by Prietzel et al. (2013), the quantification of phytate by XANES must be treated with caution. The P K-edge XANES spectra of different organic P species, including phytate, are almost identical (Kruse and Leinweber 2008). For most XANES studies of soils and manure, phytate is the only organic P standard used, which may result in an overestimation of phytate if other organic P species are present. Additionally, most studies use Na-phytate as a standard. However, different salts of phytate will have subtle differences in their XANES spectra (He et al. 2007c), which may be missed if only one phytate form is used as a standard.

Table 9.1 Concentrations of phytate in soil from selected studies, including the methods used to quantify phytate

References	Location	n[a]	Total P (mg kg^{-1})	Phytate (mg kg^{-1})	Manure[b]	Extractant	Analysis
Caldwell and Black (1958)	USA	49	n.a.[c]	2–62	N	HCl, 0.5 M NaOH	Ion exchange chromatography
McKercher and Anderson (1968b)	Canada	10	n.a.	18–71	N	HCl, 3 M NaOH	Ion exchange chromatography
Islam and Ahmed (1973)	Bangladesh	10	284–792	19–130	N	HCl, 3 M NaOH	Ion exchange chromatography
Appiah and Thomas (1982)	Canada, Ghana	17	144–1,094	15–31	N	HCl, 3 M NaOH	Paper chromatography
Borie et al. (1989)	Chile	9	1,107–3,121	499–987	N	NaOH, NaOBr	Ion exchange chromatography
Turner et al. (2003a)	USA	18	220–1,210	1–8	N	Bicarbonate	Phytase hydrolysis
Turner et al. (2003b)	Great Britain	29	376–1,981	26–189	N	NaOH-EDTA	Solution P-NMR
Hansen et al. (2004)	USA	4	440–1,000	36–157	Y	NaOH-EDTA	Solution P-NMR
Chen et al. (2004)	New Zealand	21	357–2,746	17–142	N	NaOH-EDTA	Solution P-NMR
McDowell et al. (2005)	New Zealand	24	116–2,746	0–220	N	NaOH-EDTA	Solution P-NMR
Turner (2006)	Madagascar	13	133–1,378	0–33	N	NaOH-EDTA	Solution P-NMR
Ajiboye et al. (2008)	Canada	1	1,320	22	N	None	XANES
		1	1,833	231	Y	None	XANES
He et al. (2008)	USA	1	87	11	N	NaOH	Solution P-NMR
		1	172	12	Y	NaOH	Solution P-NMR
Dou et al. (2009)	USA	5	304–866	31–111	N	NaOH-EDTA	Solution P-NMR
		10	808–4,866	53–106	Y	NaOH-EDTA	Solution P-NMR
Hill and Cade-Menun (2009)	USA	2	538–1,094	42–53	Y	NaOH-EDTA	Solution P-NMR
Murphy et al. (2009)	Ireland	25	606–2,580	97–185	N	NaOH-EDTA	Solution P-NMR
Cade-Menun et al. (2010)	Canada	24	320–1,096	23–137	N	NaOH-EDTA	Solution P-NMR
Doolette et al. (2011)	Australia	16	239–2,293	0–55	N	NaOH-EDTA	Solution P-NMR
		2	1,214–1,511	65–86	Y	NaOH-EDTA	Solution P-NMR
Baknäs et al. (2012)	Finland	16	74–313	0–105	N	NaOH-EDTA	Solution P-NMR
Vincent et al. (2012)	Sweden	6	927–1,433	57–494	N	NaOH-EDTA	Solution P-NMR
Prietzel et al. (2013)	Switzerland	7	190–850	0–576	N	None	XANES

[a]Number of soil samples analyzed
[b]Manure application to soil
[c]Information not available in published paper

Table 9.2 Concentrations of phytate in manure from selected studies, including the methods used to quantify phytate

References	n[a]	Sample	Total P (g kg^{-1})	Phytate (g kg^{-1})	Extractant	Analysis
Hansen et al. (2004)	1	Dairy solid	11	1.7	NaOH-EDTA	Solution P-NMR
	1	Dairy lagoon	0.08	0.01	NaOH-EDTA	Solution P-NMR
Turner (2004)	1	Broiler litter	0.02	0.01	NaOH-EDTA	Solution P-NMR
	1	Swine manure	0.01	0.001	NaOH-EDTA	Solution P-NMR
Toor et al. (2005a)	2	Poultry litter	20.1–22.4	2.4–4.5	None	XANES
	2	Poultry litter +phytase[b]	11.3–19.0	0–2.3	None	XANES
Shober et al. (2006)	5	Dairy slurry	4.3–7.3	1.0–1.8	None	XANES
	1	Dairy solid	2.5	0.6	None	XANES
	2	Poultry litter	20.5–21.4	3.8–6.2	None	XANES
Ajiboye et al. (2007)	1	Hog liquid	120	38.4	None	XANES
	1	Hog liquid[c]	2.8	0.3	NaOH	Solution P-NMR
	1	Hog liquid[c]	11.2	0.5	HCl	Solution P-NMR
	1	Dairy solid	240	42	None	XANES
	1	Dairy solid[c]	1.4	0.3	NaOH	Solution P-NMR
	1	Dairy solid[c]	1.0	0.3	HCl	Solution P-NMR
	1	Beef solid	310	60	None	XANES
	1	Beef solid[c]	0.2	0.4	NaOH	Solution P-NMR
	1	Poultry litter	290	203	None	XANES
	1	Poultry litter[c]	1.3	0.4	Bicarbonate	Solution P-NMR
	1	Poultry litter[c]	1.7	1.0	NaOH	Solution P-NMR
	1	Poultry litter[c]	5.2	3.1	HCl	Solution P-NMR
He et al. (2008)	1	Poultry litter[c]	3.0	0.7	Water	Phytase hydrolysis
	1	Poultry litter[c]	3.0	0.2	Water	Solution P-NMR
	1	Poultry litter[c]	1.0	0.05	Bicarbonate	Phytase hydrolysis
	1	Poultry litter[c]	1.0	0	Bicarbonate	Solution P-NMR
	1	Poultry litter[c]	2.5	1.8	NaOH	Phytase hydrolysis
	1	Poultry litter[c]	2.5	1.5	NaOH	Solution P-NMR

(continued)

Table 9.2 (continued)

References	n[a]	Sample	Total P (g kg^{-1})	Phytate (g kg^{-1})	Extractant	Analysis
	1	Poultry litter[c]	6.8	4.0	HCl	Phytase hydrolysis
	1	Poultry litter[c]	6.8	5.3	HCl	Solution P-NMR
Leytem and Thacker (2008)	4	Swine – corn[d]	11.4	0	NaOH-EDTA	Solution P-NMR
	4	Swine – normal barley[d]	8.8	0.26	NaOH-EDTA	Solution P-NMR
	4	Swine – low phytate barley[d]	8.2	0	NaOH-EDTA	Solution P-NMR
	4	Swine – oat[d]	7.2	0.43	NaOH-EDTA	Solution P-NMR
	4	Swine – wheat[d]	11.2	0.22	NaOH-EDTA	Solution P-NMR
Leytem et al. (2008)	9	Poultry litter	7.0–11.3	4.2–6.3	NaOH-EDTA	Solution P-NMR
	9	Poultry litter	7.0–11.3	4.9–6.7	HCl	HPLC
	9	Poultry litter +phytase[b]	5.9–8.3	2.8–3.5	NaOH-EDTA	Solution P-NMR
	9	Poultry litter +phytase[b]	5.9–8.3	2.6–4.3	HCl	HPLC
Hill and Cade-Menun (2009)	2	Poultry litter	14.3–17.9	5.1–6.0	NaOH-EDTA	Solution P-NMR
	2	Poultry litter compost	12.6–18.5	8.4	NaOH-EDTA	Solution P-NMR
He et al. (2009a)	2	Dairy Manure	2.94–7.01	2.6–6.6	Na-Acetate, dithionite	Solution P-NMR
	2	Dairy Manure	2.94–7.01	2.4–6.8	NaOH-EDTA	Solution P-NMR
He et al. (2009b)		Poultry manure	13.9	0.22	Water	Phytase hydrolysis
		Poultry manure	13.9	0.46	Na Acetate, pH 5	Phytase hydrolysis
		Poultry manure	13.9	2.1	Na Acetate +EDTA	Phytase hydrolysis
		Poultry manure	13.9	3.5	HCl	Phytase hydrolysis
		Poultry manure	13.9	2.1	0.25 M NaOH +EDTA	Phytase hydrolysis
		Poultry manure	13.9	2.2	0.5 M NaOH +EDTA	Phytase hydrolysis
		Poultry litter	24.6	0.01	Water	Phytase hydrolysis
		Poultry litter	24.6	8.0	Na Acetate, pH 5	Phytase hydrolysis

(continued)

Table 9.2 (continued)

References	n[a]	Sample	Total P (g kg^{-1})	Phytate (g kg^{-1})	Extractant	Analysis
		Poultry litter	24.6	9.8	Na Acetate +EDTA	Phytase hydrolysis
		Poultry litter	24.6	9.1	HCl	Phytase hydrolysis
		Poultry litter	24.6	10.0	0.25 M NaOH +EDTA	Phytase hydrolysis
		Poultry litter	24.6	10.2	0.5 M NaOH +EDTA	Phytase hydrolysis
		Dairy manure	6.9	0.29	Water	Phytase hydrolysis
		Dairy manure	6.9	0.77	Na Acetate, pH 5	Phytase hydrolysis
		Dairy manure	6.9	0.39	Na Acetate +EDTA	Phytase hydrolysis
		Dairy manure	6.9	0.97	HCl	Phytase hydrolysis
		Dairy manure	6.9	0.72	0.25 M NaOH +EDTA	Phytase hydrolysis
		Dairy manure	6.9	0.86	0.5 M NaOH +EDTA	Phytase hydrolysis

[a]Number of samples analyzed
[b]Phytase added to diet
[c]Sequentially-extracted
[d]Diet modification

9.3.2 *Extraction Techniques*

The majority of studies quantifying phytate in soil and manure use an extraction step followed by some form of analysis. Studies of phytate and other inositol phosphates prior to 2000 predominantly relied on a two-step extraction: pretreatment with HCl to remove carbonates followed by extraction with hot 3 M NaOH (McKercher and Anderson 1968a, b). Inositol phosphates were then precipitated as barium salts and were analyzed by ion-exchange chromatography. There were minor variations in this general protocol, some of which are shown in Table 9.1. However, as noted by Turner et al. (2002b), this procedure had the potential to overestimate phytate by also extracting other organic P forms, which could be overcome with alkaline bromination (Irving and Cosgrove 1981, 1982). However, it could still be difficult to separate phytate from its stereoisomers. In more recent years, there have been improvements in these separation techniques, such as the use of chromatographic separation and mass spectroscopy. These are described in detail by Cooper et al. (2007).

Some authors have used phytase hydrolysis to quantify phytate in soil and manure extracts (Tables 9.1 and 9.2). Phytase is a specialized enzyme in the phosphomonoesterase class of phosphatases (Eivazi and Tabatabai 1977) that is produced by plants and microorganisms and displays high hydrolysis efficiencies towards phytate (Menezes-Blackburn et al. 2013). Phytase-hydrolysable P is estimated by adding a commercially available phytase to soil, manure, or an extract of these, and then measuring the orthophosphate released colorimetrically. The first use of phytase to estimate phytate concentration was applied in soils (Pant et al. 1994; Shand and Smith 1997) and later in manures (He and Honeycutt 2001, 2005; He et al. 2004, 2006b, 2007a, 2008, 2009b). As shown by He et al. (2009b) and in Table 9.1, the choice of extractant will influence the phytate concentration determined by phytase hydrolysis. Others have coupled phytase hydrolysis to sequential extraction (e.g. He et al. 2007a; Table 9.1). It should be noted, however, that phytase may hydrolyze IP_6 stereoisomers in addition to, or instead of, phytate (He et al. 2011). As such, phytate determined by phytase hydrolysis should be identified as phytate-like, rather than phytate (He et al. 2011).

The most widely used technique at present to quantify phytate in soils and manures is solution ^{31}P NMR spectroscopy after a one-step extraction with 0.25 M NaOH plus 0.05 M Na_2EDTA (Cade-Menun and Preston 1996). Turner et al. (2003b) used spectral deconvolution to clearly identify phytate peaks in a number of soil samples, which could be clearly distinguished in many samples even without alkaline bromination. Subsequent analysis has shown that phytate can be clearly distinguished from its stereoisomers with solution P-NMR (Turner and Richardson 2004; Turner et al. 2012). However, careful spiking experiments are required to confirm the identification of the four phytate peaks (McDowell et al. 2007; Smernik and Dougherty 2007; Doolette et al. 2009). To obtain more information about phytate in soil and manure, solution P-NMR has been coupled with sequential extraction (e.g. Turner and Leytem 2004; Ajiboye et al. 2007; He et al. 2008) or with enzyme hydrolysis (e.g. He et al. 2008; Johnson and Hill 2010). Different extractants will influence the concentration of phytate determined by P-NMR (Tables 9.1 and 9.2).

9.4 Abundance of Phytate in Soil and Manure

9.4.1 Source of Phytate in Soil and Manure

Plants and manures are the main sources of phytate to soils (Leytem and Maguire 2007; Leytem et al. 2007; Raboy 2007). Plants accumulate phytate as a storage compound, which can represent 60–89 % of mature seed total P (Raboy 2007). In plants, the majority of phytate is stored in seeds, with far lower concentrations in other parts of the plant (Noack et al. 2012). The concentration of phytate varies with plant species as well (Schlemmer et al. 2009). Phytate concentrations of seeds (in g per 100 g dry weight) include (from Schlemmer et al. 2009): 0.35–9.42 (almonds), 1.44–5.36 (sesame seeds), 0.61–2.38 (kidney beans), 0.72–2.22 (maize, soybeans), 0.38–1.16 (barley, oats), 0.06–1.08 (rice) and 0.23–0.92 (hazelnuts).

Monogastric animals, particularly poultry and humans, are limited in their ability to hydrolyze phytate-P (Giles et al. 2011). As such, phytate from undigested plant material can be added to soil from sewage sludge (e.g. Smith et al. 2006) or manure (discussed in more detail below).

9.4.2 Abundance of Phytate in Manure

The concentration of phytate in animal waste will depend on the animal species, animal diet, the nature of the waste (feces versus manure, which is feces plus bedding material), and the storage of the waste (Table 9.2).

In general, the feces of poultry fed a normal, grain-based diet will have the highest phytate content (Table 2, Ajiboye et al. 2007; Leytem and Maguire 2007). This is in part due to physiology: monogastric animals, particularly poultry, have low gut phytase activity compared to ruminants such as cows. Although swine are also monogastric, they can digest phytate in their hindgut (Leytem and Thacker 2008). As such, their feces and manure contain little phytate. Ruminants, or rather their associated rumen microflora, produce phytase to digest dietary phytate (Toor et al. 2005b; Dao 2007; Menezes-Blackburn et al. 2013). However, some phytate may persist in ruminant feces, possibly due to precipitation with cations or rapid passage through the gut (Menezes-Blackburn et al. 2013).

The phytate content of feces and manure can also be controlled through diet, through low-phytate grain varieties (Raboy 2007) or dietary phytase additions (Leytem and Maguire 2007). Plumstead et al. (2008) reported decreased digestibility of phytate in broilers with increasing dietary Ca levels, while water soluble P in manures also increased (Leytem et al. 2008), suggesting an important link between Ca, phytate digestibility, and P excretion.

Combining bedding material with animal feces has been shown to reduce the concentration of phytate in manures. Toor et al. (2005b) reported a decreasing phytate concentration of 1.4–0.2 g P kg^{-1} dry matter from dairy feces to manure. Manure storage will also affect phytate concentrations. Solid dairy manure contained much higher phytate concentrations than dairy lagoons (Hansen et al. 2004; Table 9.2). For poultry litter, McGrath et al. (2005) reported that wet storage decreased the phytate concentration of poultry litter by 22 %, while Hill and Cade-Menun (2009) reported little difference in phytate concentration in fresh versus composted litter (Table 9.2). Drying of dairy and poultry manures also promoted the conversion of phytate to orthophosphate (He et al. 2007a).

9.4.3 Abundance of Phytate in Soil

The range of phytate concentrations in soils can vary greatly, from as low as 0 to as much as 987 mg P kg^{-1} soil (Table 9.1). There is no clear pattern to phytate concentration in soils, although Table 9.1 suggests that analytical technique may play a role. Phytate concentrations were not higher in soils with high total P concentrations, or in soils from temperate regions such as the USA (e.g. Hansen et al. 2004; Hill and Cade-Menun 2009) than from warmer climates such as Bangladesh (Islam and Ahmed 1973) or Australia (Doolette et al. 2011). Additionally, while manured soils had higher total P concentrations, their phytate concentrations were not elevated to the same degree (e.g. He et al. 2008; Dou et al. 2009). Phytate concentration has not been shown to correlate with factors associated with soil organic matter (OM) content, such as organic carbon, nitrogen, clay or microbial biomass (Turner 2007), but instead is correlated with factors linked to P stabilization in soil. These are discussed in more detail in the next section on chemical cycling of phytate (Sect. 9.5).

Although phytate was long thought to be very stable in soil (e.g. Turner et al. 2002b), recent studies suggest that phytate may mineralize rapidly in soil. Leytem et al. (2006) and Doolette et al. (2010) both added phytate to calcareous soils, either as manure (Leytem et al. 2006) or as Na-phytate (Doolette et al. 2010) and then incubated the soils. Over time, the phytate concentration decreased while orthophosphate increased. Although some phytate may have precipitated as Ca salts in these calcareous soils, becoming unextractable for ^{31}P NMR analysis, it appears from these studies that phytate was mineralized during incubation. The enzymatic conversion of phytate to orthophosphate, referred to here as hydrolysis or mineralization, may explain the lack of elevated phytate concentrations in other studies of manured soils (e.g. Hill and Cade-Menun 2009; Dou et al. 2009). The factors influencing phytate mineralization are discussed later in Sect. 9.6 (Bioavailability of Phytate in Manure and Soil). Plant species can also influence soil phytate concentration. For example, Chen et al. (2004) showed a decrease in phytate in soils planted with pine seedlings compared with the same soils planted with ryegrass. Management practices will also influence soil phytate concentrations. Cade-Menun et al. (2010) showed a significant increase in phytate concentration under zero-tillage compared to conventional tillage. However, for both of these studies, it is not clear whether these differences in phytate were due to enhanced mineralization of phytate or decreased phytate inputs from the different vegetation or management practices.

9.5 Chemical Cycling of Phytate

Phytate cycling in soils is influenced by physical and chemical processes that are a function of soil properties and management practices. Soil properties that control phytate cycling include pH, surface charge, mineralogy, organic matter and texture (Borie et al. 1989; Celi et al. 1999; Celi and Barberis 2007).

Phytate is sensitive to solution pH. Below pH 9, phytate occurs in a 1-axial/5-equatorial conformation, which shifts to 5-axial/1-equatorial above pH 9 (Fig. 9.1). Phytate acid dissociation constants are (Celi and Barberis 2007): 1.1 (pK_1), 1.5 (p$K_{2,3}$), 1.8 (p$K_{4\text{–}6}$), 5.7 (pK_7), 6.9 (pK_8), 7.6 (pK_9), 10.0 (p$K_{10,11}$) and 12.0 (pK_{12}).

Thus, at the typical soil pH range of 5–8, phytate will have 6–9 hydroxyl groups with negative charge that can react with the mineral surface or cations in solution (Celi and Barberis 2007). In addition to affecting the charge of the phytate molecule, solution pH will also influence the surface charge of the soil substrate (Barrow et al. 1980, 1993; Bolan et al. 1986). As pH increases, the number of negative charges on both the P_o compound and soil components (minerals and OM) will increase, leading to mounting repulsive forces (Celi et al. 2001) that may lead to dispersion and/or flocculation of soil particles (Celi and Barberis 2007). In studies using amorphous aluminum (Al) hydroxide (Guan et al. 2006), the amount of phytate sorbed was constant at pH 4.0–7.0, but decreased considerably above pH 7.0. The sorption of phytate was also shown to increase solution pH by increasing –OH release (Guan et al. 2006). In studies using goethite, phytate sorption was also highest above pH 7.0 (Johnson et al. 2012), but nearly 50 % of the added phytate remained sorbed at pH 10.0.

Surface charge is also strongly influenced by the concentration of polyvalent cations in solution that can diminish negative charge concentration at pH values as high as 9–10 (Celi et al. 2001). Recently, He et al. (2013) showed that cations bonded to phytate could induce the same conformational change in phytate as solution pH. Phytate adsorption is related to the availability of –OH groups at the mineral or organic-solution interface. The phosphate groups of phytate will interact with surface –OH groups until steric or charge effects occur due to surface saturation (Anderson and Arlidge 1962; Anderson and Malcolm 1974; Celi et al. 2001). An example of this is the binding of phytate with goethite via four of six phosphate groups. The remaining two phosphate groups that are not bound to surface hydroxyls will impact the surface-binding characteristics of other ionic species in solution through steric hindrance or altered charge (Ognalaga et al. 1994; Celi et al. 1999, 2000; Giaveno et al. 2010a). The stabilization of phytate in soils has been linked to precipitation with metals [copper (Cu)II > zinc (Zn) > nickel (Ni) II > cobalt (Co)II > manganese Mn(II) > Fe III > Ca; Martin and Evans 1987; Celi et al. 2001; He et al. 2013]. Amorphous iron (Fe) and Al oxides have been shown to sorb phytate more strongly than crystalline forms in acid soils (Ognalaga et al. 1994; Celi et al. 1999, 2003), while phytate sorption in neutral and alkaline soils is controlled by calcite, clays, and OM (McKercher and Anderson 1989).

The capacity of soils (acid, neutral, alkaline) to sorb phytate tended to increase as the number of phosphate moieties increased in non-competitive (single IP_x) experiments. For example, McKercher and Anderson (1989) studied neutral and alkaline soils and showed the sorption capacities for glucose 1-phosphate < glycerophosphate < phytate. Similarly, Berg and Joern (2006) demonstrated that the magnitude of sorption for P_o compounds in slightly acidic soils occurred in the order of *myo*-IHP > glucose 6-phosphate > adenosine triphosphate (ATP) > phosphate. When more than one organic P species is present, as is usually the case in natural systems, the P species will compete for substrate sorption sites. Competitive isotherm studies have shown preferential binding of organic P relative to inorganic P species, while sorption of the organic P species increased with the number of phosphate moieties (Anderson and Malcolm 1974; Berg and Joern 2006;

Fransson and Jones 2007). Phytate inhibited orthophosphate sorption to soil and mineral surfaces (Anderson and Malcolm 1974; De Groot and Golterman 1993) and was the least displaced orthophosphate monoester upon addition of orthophosphate in a competitive isotherm study of soils (glucose 6-phosphate > ATP > IP_6; Berg and Joern 2006). Increases in P loss from manured soils (Mozaffari and Sims 1994, 1996; Dou et al. 2009) may be related to the competitive displacement of organic or inorganic P by phytate from manure (Berg and Joern 2006). Recent research suggests that this preferential binding is due to the sorption of phytate through outer-sphere complex formation, while other P forms bind through inner-sphere complexation (Johnson et al. 2012).

9.6 Bioavailability of Phytate in Manure and Soil

The extent to which a nutrient is bioavailable is defined by its ability to be consumed by a primary producer. Orthophosphate can be directly assimilated by plants and algae; however, in order for phytate to be converted to orthophosphate and consumed it must first be hydrolyzed by phytases, which are produced by plants and microorganisms in the environment. There are no examples of direct assimilation of phytate by primary producers, and few examples exist for plant uptake of other soluble organic P forms (Hayes et al. 2000a). If the appropriate biogeochemical conditions exist to support the production and activity of phytases, the enzymatic conversion of phytate to orthophosphate may only be a matter of time (Makoi and Ndakidemi 2008) (Fig. 9.2).

9.6.1 The Role of Phytases in the Bioavailability of Phytate

Early studies of phytate bioavailability were focused on understanding the role of enzymatic hydrolysis in the conversion of phytate to orthophosphate by phytases. There are four sub-classes of phytase (Giles et al. 2011), namely: (1) beta-propeller phytase (βPP), (2) histidine-acid-phosphatase (HAP), (3) cysteine phytase (CPhy), and (4) purple-acid-phosphatase (PAP). Each phytase type has a specific mechanism of action, related to pH optima, requirements for co-factor cations (e.g., βPP with Ca^{2+}), and the pattern of dephosphorylation (Greiner 2007).

In manures, phytase originates from microbes that are endogenous to the animal digestive system or through direct secretion of enzymes in the digestive tract (Leytem et al. 2007). Phytase is actively produced in soils by plants and microorganisms, or may be introduced in lesser amounts during fertilization with manures. *In silico* analyses of phytase gene fragments in soils and manures indicate the role of soil and rhizosphere microorganisms in its production (Greiner 2007; Lim et al. 2007), and is reviewed further by Menezes-Blackburn et al. (2013). Microbial culturing techniques have also been used to isolate phytate-mineralizing soil bacteria for

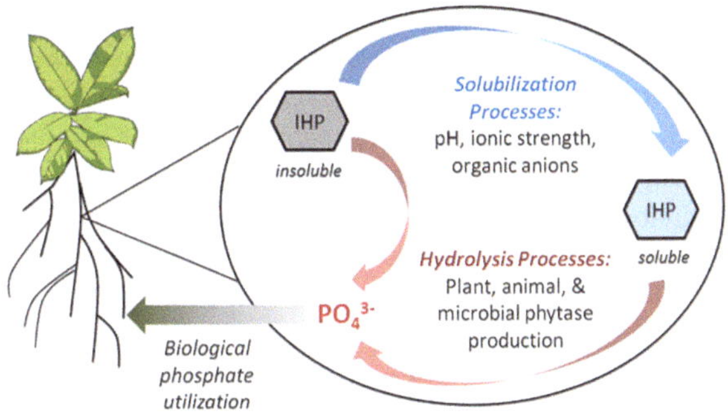

Fig. 9.2 A model of phytate cycling in soil and manure. The figure illustrates the abiotic and biotic solubilization and hydrolysis processes that lead to the generation of orthophosphate from phytate

identification of specific organisms (e.g., *Pseudomonas* spp., *Bacillus* spp.) and identification of functional genes (Hill and Richardson 2007; Jorquera et al. 2008; Patel et al. 2010a; Richardson et al. 2011; Richardson and Simpson 2011). Finally, exudation of phytases by crop plants (e.g., tobacco, wheat, maize) or residual amounts of phytase activity originating from plant seeds also contribute to the abundance of phytase enzymes in agricultural systems (George et al. 2007a; Lung et al. 2008).

The enzymatic hydrolysis of phytate in manures and soils is strongly influenced by the solubility of the enzyme and substrate, the solid and aqueous phase enzyme activity, and the relative abundance of both substrate and enzyme in a given medium (George et al. 2007a). Upon introduction to the soil environment, phytase enzymes may become immobilized on soil mineral surfaces or organic matter phases through electrostatic interactions (Mullaney and Ulla 2007). The sorption of phytase to solid-phase surfaces is primarily controlled by the specific isoelectric points (pI) of the enzyme and soil, as well as the aqueous phase pH (George et al. 2005b, 2007b). Giaveno et al. (2010b) reported a strong affinity of phytase (*Aspergillus niger, Peniophora lycii*, *PhyA*, HAP) for clay and oxy-hydroxide minerals in 24 h incubation experiments, and noted an overall decrease in phytase activity as well as a partitioning of activity from the aqueous to solid phases. The influence of surface interactions was further illustrated in experiments with phytate-saturated minerals, in which phytase partitioning to the solid phase was inhibited, but hydrolysis of sorbed phytate was also minimal (Giaveno et al. 2010b). The inability of *A. niger PhyA* to hydrolyze phytate-saturated goethite minerals was in contrast to that reported for the simple orthophosphate monoester, glucose-1-phosphate (Olsson et al. 2012), and is likely related to the high affinity that phytate has for goethite relative to less-phosphorylated compounds. The hydrolysis of metal-precipitated phytate by three classes of phytase (i.e., HAP, βPP, PAP) only occurred when phytate was complexed with Ca^{2+}, Mg^{2+}, and Mn^{2+}, whereas hydrolysis was negligible for Al^{3+}, Cu^{2+}, Zn^{2+}, Fe^{2+} and Fe^{3+} complexes (Tang et al. 2006).

For phytases remaining in solution, pH, dissolved metal concentrations, and temperature will affect enzyme activity (George et al. 2007a). Beta-propeller-phytases require two Ca^{2+} cations for proper activation (Kerovuo et al. 1998; Oh et al. 2001; Fu et al. 2008), whereas the activity of some fungal phytases (e.g., *A. ficuum*

PhyA, HAP) will be enhanced in the presence of arsenic, barium, bromide, and iodide (Tazisong et al. 2008). Certain dissolved cations [i.e., cadmium (II), Zn^{2+}, cobalt (Co^{2+}), Al^{3+}, Fe^{2+}, Fe^{3+}, Cu^{2+}; Yamada et al. 1968; Hayes et al. 1999; Tang et al. 2006] and anions (i.e., phosphate, fluoride, arsenate; Nagai and Funahashi 1962; Chang 1966; Mandal et al. 1972; Lolas and Markakis 1977) will also inhibit the activity of purified and dissolved phytases.

9.6.2 The Role of Solubility in the Bioavailability of Phytate

The solubility of phytate and its accessibility to the active site of phytase is a further requirement for dephosphorylation to occur. The abiotic factors controlling the solubility of phytate in soil and manure have been described in Sect. 9.5. Here we will describe additional biotic mechanisms controlling the solubility of phytate. The presence of organic ligands (Jones and Darrah 1994; Jones et al. 2003; Palomo et al. 2006; Wang et al. 2008) may enhance the solubility of phytate through the dissolution of phytate-metal/mineral complexes (Dao 2003, 2007; Oburger et al. 2011; Giles et al. 2012). Plant exudation and microbial secretion are the major sources of organic anions (OAs) in soils and manures. Plant and microbial production of OAs is often coupled to the release of protons, leading to localized changes in pH, which affect the interaction of phytate with metal and mineral phases (Richardson et al. 2009). Specifically, organic anions such as mono- (e.g., acetate, oxalate), poly- (e.g., maleate), and hydroxy-carboxylates (e.g., citrate) as well as redox-active OAs (e.g., ascorbate) increase phytate solubility in soils and manures through reductive dissolution, the direct chelation of metals, or the saturation of sorption sites on minerals. Bolan et al. (1994) reported a phosphate solubilization series for organic anions (tricarboxylates > dicarboxylates > monocarboxylates) that was positively correlated with the affinity of the various OAs for trivalent metal cations. Simple *in vitro* experiments comparing phosphate and phytate desorption by citrate have demonstrated the relative resistance of phytate to desorption (tenfold less than phosphate; Martin et al. 2002, 2004) due to the affinity of its multiple charged phosphate groups for the mineral surface. The improved recovery of organic P from soils extracted with citrate is also frequently reported (Palomo et al. 2006; Wang et al. 2008; Wei et al. 2009, 2010). The recovery of organic P from forest soils was approximately 30 % greater for citrate compared to other OAs (e.g., maleate, oxalate; Martin et al. 2002; Wei et al. 2010). The release of phytate from goethite particles followed a similar order: ascorbate > citrate > oxalate > pyruvate > acetate (Giles et al. 2012). Single mineral-single compound incubations such as those carried out with goethite-bound phytate are useful for identifying the likely mechanisms of desorption (i.e., reductive dissolution, chelation), however further work is needed to demonstrate the influence of single and multiple OAs on the solubilization of phytate from soils of varying chemistries.

Recent studies of phytate bioavailability have become focused on the role of organic anions in enabling the conversion of phytate to orthophosphate by phytases. For example, OA-driven solubilization of precipitated phytate led to the improved hydrolysis of phytate by phytases *in vitro*, in the order: oxalate > citrate > malate (Tang et al. 2006). Incubations of Ca-phytate with common bacterial-produced organic anions (e.g., gluconate, citrate, acetate) also led to the solubilization and subsequent hydrolysis of phytate by bacterial and fungal phytases (Patel et al. 2010a; Giles et al. 2014).

Animal feeds are often supplemented with metals, rendering the phytate in manure fertilizers less soluble and therefore less available to dephosphorylation by native soil or manure enzymes (Dao 2007). For example, Dao (2003) reported a decrease in the hydrolysis of manure phytate with increasing counter ion concentrations in the order $Fe^{3+} > Al^{3+} > Ca^{2+}$ due to the formation of insoluble precipitates. This is particularly relevant in cases where animal feeds are supplemented with metals, as the resulting dephosphorylation of phytate salts may depend on the local pH and ionic conditions or the biological production of OAs (Dao 2007).

9.6.3 Methods for Assessing the Bioavailability of Phytate in Soils and Manures

Phosphorus bioavailability was traditionally assessed based on the solubility of manure or soil P in water or weak salt solution. Sequential extraction procedures have been developed to characterize organic P in soils (Hedley et al. 1982) and manures (Leytem and Maguire 2007), and are defined based on the relative solubility of inorganic and organic P compounds in extractants that range from neutral (e.g. water) to extreme acid (e.g., HCl) or base (e.g., NaOH). Phosphorus retained in the water-soluble fraction is considered immediately available for plant uptake, whereas refractory and non-soluble fractions (e.g., strong base or acid-extractable P) are sparingly bioavailable, and associated with organic P (such as phytate) that has precipitated, sorbed, or become bound to solid phase metals and minerals (Turner et al. 2005). As demonstrated in Table 9.2, various studies have applied a sequential extraction scheme to manures to assess the P species composition of the readily-soluble (i.e., water, sodium bicarbonate) and stable (i.e., NaOH, HCl) fractions using solution ^{31}P NMR spectroscopy (e.g. Turner and Leytem 2004; Ajiboye et al. 2007; He et al. 2008). Generally, these studies found the majority of phytate to be associated with the stable fraction, smaller proportions detected in the readily soluble fractions, and some variation of this noted across animal species. The relative solubility of phytate does not guarantee its availability in agricultural systems, as additional hydrolysis steps are required to convert phytate to phosphate. However, the hydrolysis of phytate is diminished due to precipitation and sorption reactions, making estimates of phytate solubility of critical importance to our understanding and measurement of bioavailable phytate pools.

As previously described, hydrolysis by phytase has been used to quantify phytate in soils and manures (Menezes-Blackburn et al. 2013). Manure and soil can be treated through the direct addition of phosphatases (Bünemann 2008; George et al. 2005a; Keller et al. 2012), or alternatively, the enzymes may be added to an extract of the solid material (Hayes et al. 2000a; Turner et al. 2002a; He et al. 2007a, 2008; Johnson and Hill 2010; Keller et al. 2012). Following an incubation period, the phosphate released from the targeted organic P class is quantified using molybdate colorimetry. Phytase may be applied individually or in combination with other phosphatases. For example, Hayes et al. (2000a) incubated citrate and bicarbonate extracts of soil with individual preparations of a commercially available phosphatase and a phytase isolated from *Aspergillus niger*. Alternatively, He et al. (2008) estimated simple-monoester-P, phytate-like-P, and DNA-like-P in manures through the single addition of acid-phosphatase from potato (PP; simple-monoester-P) or the combination of PP with wheat germ phosphatase (PP+GP; phytate-like-P), and nuclease-P1 from *Penicillium citrinum* (DNA-like-P). When enzymes are applied in combination, the calculation of P associated with each enzyme-labile class is determined by difference, with DNA-like-P being equivalent to the total P released by all three enzymes, less the phosphate released by PP+GP enzyme incubations.

Phytate bioavailability has been measured directly using bioassay organisms (e.g., plant, microorganism) that metabolically facilitate the transformation of phytate to orthophosphate and serve as the ultimate sink of P. Manure slurry (Patel et al. 2010a, b), soluble (Richardson et al. 2000; Hayes et al. 2000b), and insoluble phytate sources (Giles et al. 2012, 2014) are provided as the sole source of P. The amount of phytate-P that is incorporated into the organism can be estimated based on biomass P (e.g., plant root/shoots) in cases where growth is carried out in a closed system. Alternatively, phytate utilization by microorganisms grown in liquid culture can be assayed by measuring the depletion of total P (Patel et al. 2010a), or generation of orthophosphate in the media. The total P incorporated into the biomass or depleted from the culture media is proportional to the bioavailable fraction of phytate that has been utilized during the growth period.

Direct measurements of phytate utilization by wild-type plants have been carried out for numerous species (e.g., tobacco (*Nicotiana tabacum*), wheat (*Triticum aestivum* L.), and mustard (*Arabidopsis thaliana*); Hayes et al. 2000a; Richardson et al. 2001a; Sepehr et al. 2012). Wild-type and transgenic plant bioassays have been particularly useful for elucidating the mechanisms of phytate utilization in the rhizosphere (George et al. 2005a, 2009; Lung et al. 2005; Chan et al. 2006; Lung and Lim 2006; Giles et al. 2012). For example, phosphatase and phytase activity originating from plant roots is associated with the ability of wild-type wheat (Richardson et al. 2000; Hayes et al. 2000b) and tobacco (George et al. 2005b; Lung and Lim 2006) to utilize soluble and insoluble phytate sources in agar. In goethite-containing media, transgenic tobacco expressing the *A. ficuum* fungal phytase could access tenfold more phytate than wild-type plants when phytate was added at twice the equilibrium concentration needed to saturate the goethite particles (Giles et al. 2012). A high citrate-exuding plant-line incorporated nearly twice as much P into shoots when phosphate was sorbed to synthetic goethite particles in agar (Giles et al. 2012). A transgenic plant-line (*N. tabacum ex::phyA*) with high levels of root phytase and organic anion exudation could access greater than twofold more P from Ca-phytate salts than wild-type plants under similar growth conditions. In the same study, the inoculation of two low phytase-exuding tobacco lines with a *Pseudomonas* sp. soil isolate resulted in a six to tenfold improvement in the assimilation of Ca-phytate compared to uninoculated plants (Giles et al. 2014), supporting a symbiotic mechanism of plant organic anion and bacterial phytase production in the bioavailability of phytate.

Bioassays have also been employed to investigate the role of soil microorganisms in phytate availability to plants. Plants inoculated with phytate-solubilizing and -hydrolyzing rhizosphere bacteria displayed improvements in growth over uninoculated controls (Richardson and Hadobas 1997; Richardson et al. 2001b; Patel et al. 2010a, b; Giles et al. 2014). For example, phytate-mineralizing rhizobacteria (e.g., *Pseudomonas* spp., *Bacillus* spp.) have promoted the hydrolysis of insoluble phytate *in vitro* (Hill and Richardson 2007; Jorquera et al. 2008) as well as the utilization of phytate by pasture plants grown in soil (Richardson and Hadobas 1997). Specific mechanisms of phytate solubilization by plant inocula are linked to the production of organic anions and the coupled transport of protons across the outer membrane of bacterial cells. Gluconate is a common bacterial metabolite that is produced through the pentose-phosphate pathway and is associated with Ca-phosphate (Goldstein 1995; Browne et al. 2009; Miller et al. 2010) and Ca-phytate solubilization and hydrolysis by plant-growth promoting bacteria (e.g., *Pseudomonas* spp., Giles et al. 2014; *Citrobacter* spp., *Burkholderia* spp., Patel et al. 2010b). Several *Pseudomonas* spp. soil isolates produce other organic anions such as ascorbate, citrate, oxalate, and acetate, which improve the availability of phytate to tobacco plants in synthetic soil (Richardson and Hadobas 1997), calcareous-, and goethite-containing growth systems (Giles et al. 2012, 2014).

Finally, we can garner some understanding of the rate at which phytate becomes bioavailability in soils and manures through the direct measurement of phytate flux and turnover rates. Estimates of phytate mineralization and transport into and out of environmental compartments (e.g., microorganisms, plants, soil pore water, etc.) are based on a suite of analyses, which provide information on (1) time-series changes in phytate abundance (e.g. Leytem et al. 2006; Doolette et al. 2010); (2) the lability of phytate to phytases; (3) *in situ* phytase activity measurements, which are currently limited to a small collection of fluorescent-tagged phytate analogs (Berry and Berry 2005; Berry et al. 2009); (4) radiolabeling techniques to track phytate transformations and uptake by plants or microorganisms (Caldwell and Black 1958; Cosgrove 1964; Bünemann 2008); and (5) gross estimates of P flux using isotopic dilution methods (Bünemann et al. 2012).

In combination, these methods provide information on how quickly phytate is converted to orthophosphate in a given system, and therefore the rate at which phytate-derived orthophosphate or carbon will be taken up by plants and microorganisms. As previously noted, estimates of phytate and orthophosphate abundance in soil and manure are currently limited to extract-based isolation of P compounds and subsequent analysis by colorimetry or spectroscopic and chromatography-based technologies. The development of compound-specific methodologies for estimating P abundance would allow for *in situ* assessment of phytate turnover and flux in a range of environments. The time-dependence of phytate bioavailability may be most important from a land-management perspective. The design of nutrient-efficient farm systems will depend, in-part, on the ability of crops to access native soil P on seasonally relevant time-scales. The ability to estimate the timing of phytate conversion to orthophosphate will therefore be needed to enhance phytate bioavailability at specific times of the year.

9.6.4 Plant and Microbial Strategies for Improving the Bioavailability of Phytate in Agricultural Systems

The specific role of phytate in P utilization efficiency has been studied for various crop plants as an approach to localize the utilization of soil P in the root-zone (rhizosphere) of plants and improve crop acquisition of soil P (Richardson et al. 2007; Ramaekers et al. 2010). Specific strategies for this have included the genetic modification of plants for enhanced exudation of organic anions (Giles et al. 2012) and phytases from roots (Richardson et al. 2001b; George et al. 2005b, 2009; Kong et al. 2005; Lung et al. 2005; Lung and Lim 2006; Giles et al. 2012; Liu et al. 2012; Wang et al. 2013) or rhizosphere inoculation with phytate-solubilizing or hydrolyzing microorganisms (Patel et al. 2010a, b; Richardson et al. 2011; Richardson and Simpson 2011). The utilization of phytate by plants has been improved *in vitro* due to bacterial inoculation, genetic modification of plant exudation patterns or a combination of these strategies (Richardson and Hadobas 1997; Giles et al. 2012). However, limited studies have been carried out *in vivo* (i.e., in soils; George et al. 2009) or at the field-scale in order to determine the impact of these treatments on agricultural P loss (e.g., leaching). Several groups have also reported the importance of microbial community structure and function in controlling the transformation and bioavailability of phytate in soils (Unno et al. 2005; Richardson et al. 2011; Hayatsu 2013; Unno and Shinano 2013).

9.7 Conclusions

As a ubiquitous organic P compound in soils and manures, phytate represents an ecologically significant component of P cycles in a variety of ecosystems, including agriculture. As described in this chapter, numerous chemical and biological controls influence the availability of phytate to plants and microorganisms, as well as its potential to contribute to nutrient enrichment of surface waters and eutrophication. A number of techniques are available that can be used to quantify total phytate in samples, although many require the use of complex instrumentation. Various phytate salts can be distinguished in model compounds; however, less is known about the relative abundances of these salts in natural samples. Phytate concentrations in animal manures will vary with animal species and diet; animal species vary in their ability to digest dietary phytate, which can be enhanced by supplementation with phytase. Soil phytate concentrations vary with inputs from plants and manure, and also with soil mineralogy and pH. Phytate sorption is highest below pH 7.0, and in soils containing more amorphous minerals than crystalline minerals. Because it forms outer sphere complexes, phytate will preferentially sorb relative to other organic P forms that form inner-sphere complexes.

Phytate is hydrolyzed by phytase, which occurs as four subclasses: beta-propeller phytase, histidine-acid-phosphatase, cysteine phytase and purple-acid-phosphatase. Phytate bioavailability is controlled by both abiotic and biotic factors, which influence phytate solubility and phytase production and activity. Agricultural practices (e.g., manure storage) and biotechnologies (e.g., transgenic plants and microorganisms) that promote the efficient use of soil and manure phytate by plants could be used to reduce demands on external or inorganic P sources for crop fertilization, thereby improving the sustainable use of P at the field scale. A deeper understanding of phytate chemistry and bioavailability will be needed to further expand the suite of sustainable management options that are available to farmers and land managers. The following section addresses some specific research needs pertaining to phytate in manures and soils.

9.8 Future Research Needs Related to Phytate in Manures and Soils

A further understanding of the chemical and biological controls on phytate availability could be used to leverage phytate as a sustainable source of P in modern agricultural systems. The use of advanced chemical techniques (e.g. NMR, XANES) will further elucidate chemical controls, particularly when multiple techniques are used together (e.g. Liu et al. 2013). Linking these spectroscopic techniques to other techniques such as radioisotope tracer studies (e.g. L'Annunziata 2013) will provide information on chemical and biological cycling. However, there is also a need to develop simple field tests for phytate concentration in soils and manures for more effective management of phytate at the farm level. It should also be noted that the majority of the research on phytate chemistry has been conducted using model systems; in light of this, tests need to be conduct in natural systems. In particular, there is a need to understand phytate movement in the environment, including whether it exists in truly dissolved forms in runoff and receiving waters, and how it binds to natural colloids.

Further research on phytate bioavailability should include the development of *in situ* techniques for P species measurement to improve estimates of flux, transport, and turnover of phytate in the environment. Enzyme-based methods can be used to determine the relative or potential bioavailability of phytate; however these methods should be standardized to enable the comparison of estimates across studies. More work is needed to produce suitable substrates for phytase activity estimates in soils and manures. Studies of microbial community structure should continue to characterize the organic anion and phytase production profiles of the major resident organisms, which may be driving phytate transformations. These studies should also be tied to chemical studies, to assess the effects of anion and phytase production on phytate chemistry in soil. Site-specific assessments of phytate turnover and flux may be needed at both the rhizosphere and field-scale. Finally, although pot- and *in vitro* experiments have demonstrated the efficacy of plant and microbial strategies in improving phytate bioavailability, field-scale trials will be needed to determine the net impact of these practices on crop yields and phosphate loss reductions from agricultural systems.

References

Ajiboye B, Akinremi OO, Hu Y, Flaten DN (2007) Phosphorus speciation of sequential extracts of organic amendments using nuclear magnetic resonance and X-ray absorption near-edge structure spectroscopies. J Environ Qual 36:1563–1576

Ajiboye B, Akinremi OO, Hu Y, Jürgensen A (2008) XANES speciation of phosphorus in organically amended and fertilized vertisol and mollisol. Soil Sci Soc Am J 72:1256–1262

Anderson G, Arlidge EZ (1962) The adsorption of inositol phosphates and glycerophosphate by soil clays, clay minerals and hydrated sesquioxides in acid media. J Soil Sci 25:51–62

Anderson G, Malcolm RE (1974) The nature of alkaline soluble soil organic phosphates. J Soil Sci 25:282–297

Appiah MR, Thomas RL (1982) Inositol phosphate and organic phosphorus contents and phosphatase activity of some Canadian and Ghanian soil. Can J Soil Sci 62:31–38

Baknäs S, Laine-Kauio H, Kløve B (2012) Phosphorus forms and related soil chemistry in preferential flowpaths and the soil matrix of a forested podzolic till soil profile. Geoderma 189–190:50–64

Barrientos LG, Murthy PPN (1996) Conformational studies of *myo*-inositol phosphates. Carbohydr Res 296:39–54

Barrow NJ, Bowdan JW, Posner AM, Quirk JP (1980) Describing the effects of electrolyte on adsorption of phosphate by variable charge surface. Aust J Soil Res 18:395–404

Barrow NJ, Brummer GW, Strauss R (1993) Effects of surface heterogeneity on ion adsorption by metal-oxides and by soils. Langmuir 9:2606–2611

Berg AS, Joern BC (2006) Sorption dynamics of organic and inorganic phosphorus compounds in soil. J Environ Qual 35:1855–1862

Berry DF, Berry DA (2005) Tethered phytic acid as a probe for measuring phytase activity. Bioorg Med Chem Lett 15:3157–3161

Berry DF, Shang C, Zelazny LW (2009) Measurement of phytase activity in soil using a chromophoric tethered phytic acid probe. Soil Biol Biochem 41:192–200

Bolan NS, Syers JK, Tillman RW (1986) Ionic-strength effects on surface charge and adsorption of phosphate and sulfate by soils. J Soil Sci 37:379–388

Bolan NS, Naidu R, Mahimairaja S, Baskaran S (1994) Influence of low-molecular-weight organic acids on the solubilization of phosphate. Biol Fertil Soils 18:311–319

Borie F, Zunino H, Martinez L (1989) Macromolecule-P associations and inositol phosphates in some Chilean volcanic soils of temperate regions. Commun Soil Sci Plant Anal 20:1881–1894
Browne P, Rice O, Miller SH, Burke J, Dowling DN, Morrissey JP, O'Gara F (2009) Superior inorganic phosphate solubilization is linked to phylogeny within the *Pseudomonas fluorescens* complex. Appl Soil Ecol 43:131–138
Bünemann EK (2008) Enzyme additions as a tool to assess the potential bioavailability of organically bound nutrients. Soil Biol Biochem 40:2116–2129
Bünemann EK, Oberson A, Liebisch F, Keller F, Annaheim KE, Huguenin-Elie O, Frossard E (2012) Rapid microbial phosphorus immobilization dominates gross phosphorus fluxes in a grassland soil with low inorganic phosphorus availability. Soil Biol Biochem 51:84–95
Cade-Menun BJ, Preston CM (1996) A comparison of soil extraction procedures for P-31 NMR spectroscopy. Soil Sci 161:770–785
Cade-Menun BJ, Carter MR, James DC, Liu CW (2010) Phosphorus forms and chemistry in the soil profile under long-term conservation tillage: a phosphorus-31 nuclear magnetic resonance study. J Environ Qual 39:1647–1656
Caldwell AG, Black CA (1958) Inositol hexaphosphate. II. Synthesis by soil microorganisms. Soil Sci Soc Am J 22:293–296
Celi L, Barberis E (2007) Abiotic reactions of inositol phosphates in soils. In: Turner BL, Richardson AE, Mullaney EJ (eds) Inositol phosphates: linking agriculture and the environment. CAB International, Oxfordshire, pp 207–220
Celi L, Lamacchia S, Marsan FA, Barberis E (1999) Interaction of inositol hexaphosphate on clays: adsorption and charging phenomena. Soil Sci 164:574–585
Celi L, Barberis E, Ajmone-Marsan F (2000) Sorption of phosphate on goethite at high concentrations. Soil Sci 165:657–664
Celi L, Presta M, Ajmore-Marsan F, Barberis E (2001) Effects of pH and electrolytes on inositol hexaphosphate interaction with goethite. Soil Sci Soc Am J 65:753–760
Celi L, De Luca G, Barberis E (2003) Effects of interaction of organic and inorganic P with ferrihydrite and kaolinite-iron oxide systems on iron release. Soil Sci 168:479–488
Chan WL, Lung SC, Lim BL (2006) Properties of beta-propeller phytase expressed in transgenic tobacco. Protein Expr Purif 46:100–106
Chang CW (1966) Study of phytase and fluoride effects in germinating corn seeds. Cereal Chem 43:129–142
Chen CR, Condron LM, Turner BL, Mahieu N, Davies MR, Xu ZH, Sherlock RR (2004) Mineralisation of soil orthophosphate monoesters under pine seedlings and ryegrass. Aust J Soil Res 42:189–196
Condron LM, Turner BL, Cade-Menun BJ (2005) Chemistry and dynamics of soil organic phosphorus. In: Sims JT, Sharpley AN (eds) Phosphorus, agriculture and the environment, Monograph no 46. Soil Science Society of America, Madison, pp 87–121
Conte P, Šmejkalová D, Piccolo A, Spaccini R (2008) Evaluation of the factors affecting direct polarization solid-state ^{31}P NMR spectroscopy of bulk soils. Eur J Soil Sci 59:584–591
Cooper WT, Heerboth M, Salters VJM (2007) High-performance chromatographic separations of inositol phosphates and their detection by mass spectrometry. In: Turner BL, Richardson AE, Mullaney EJ (eds) Inositol phosphates: linking agriculture and the environment. CAB International, Oxfordshire, pp 23–40
Cordell D, Drangert JO, White S (2009) The story of phosphorus: global food security and food for thought. Glob Environ Change Hum Policy Dim 19:292–305
Cosgrove DJ (1964) An examination of some possible sources of soil inositol phosphates. Plant Soil 21:137–141
Dao TH (2003) Polyvalent cation effects on *myo*-inositol Hexakis dihydrogenphosphate enzymatic dephosphorylation in dairy wastewater. J Environ Qual 32:694–701
Dao TH (2007) Ligand effects on inositol phosphate solubility and bioavailability in animal manures. In: Turner BL, Richardson AE, Mullaney EJ (eds) Inositol phosphates: linking agriculture and the environment. CAB International, Oxfordshire, pp 169–185

De Groot CJ, Golterman HL (1993) On the presence of organic phosphate in some Camargue sediments – evidence for the importance of phytate. Hydrobiologia 252:117–126
Doolette AL, Smernik RJ, Dougherty WJ (2009) Spiking improved solution phosphorus-31 nuclear magnetic resonance identification of soil phosphorus compounds. Soil Sci Soc Am J 73:919–927
Doolette AL, Smernik RJ, Dougherty WJ (2010) Rapid decomposition of phytate applied to a calcareous soil demonstrated by a solution ^{31}P NMR study. Eur J Soil Sci 61:563–575
Doolette AL, Smernik RJ, Dougherty WJ (2011) A quantitative assessment of phosphorus forms in some Australian soils. Soil Res 49:152–165
Dou ZX, Ramberg CF, Toth JD, Wang Y, Sharpley AN, Boyd SE, Chen CR, Williams D, Xu ZH (2009) Phosphorus speciation and sorption–desorption characteristics in heavily manured soils. Soil Sci Soc Am J 73:93–101
Eivazi F, Tabatabai MA (1977) Phosphatases in soils. Soil Biol Biochem 9:167–172
Fransson AM, Jones DL (2007) Phosphatase activity does not limit the microbial use of low molecular weight organic-P substrates in soil. Soil Biol Biochem 39:1213–1217
Fu DW, Huang HQ, Luo HY, Wang YR, Yang PL, Meng K, Bai YG, Wu NF, Yao B (2008) A highly pH-stable phytase from *Yersinia kristeensenii*: cloning, expression, and characterization. Enzym Microb Technol 42:499–505
George TS, Richardson AE, Simpson RJ (2005a) Behaviour of plant-derived extracellular phytase upon addition to soil. Soil Biol Biochem 37:977–988
George TS, Simpson RJ, Hadobas PA, Richardson AE (2005b) Expression of a fungal phytase gene in *Nicotiana tabacum* improves phosphorus nutrition of plants grown in amended soils. Plant Biotechnol J 3:129–140
George TS, Quiquampoix H, Simpson RJ, Richardson AE (2007a) Interactions between phytases and soil constituents: implications for the hydrolysis of inositol phosphates. In: Turner BL, Richardson AE, Mullaney EJ (eds) Inositol phosphates: linking agriculture and the environment. CAB International, Oxfordshire, pp 221–241
George TS, Simpson RJ, Gregory PJ, Richardson AE (2007b) Differential interaction of *Aspergillus niger* and *Peniophora lycii* phytases with soil particles affects the hydrolysis of inositol phosphates. Soil Biol Biochem 39:793–803
George TS, Richardson AE, Li SS, Gregory PJ, Daniell TJ (2009) Extracellular release of a heterologous phytase from roots of transgenic plants: does manipulation of rhizosphere biochemistry impact microbial community structure? FEMS Microbiol Ecol 70:433–445
Giaveno C, Celi L, Maja Aveiro Cessa R, Prati M, Bonifacio E, Barberis E (2010a) Interaction of organic phosphorus with clays extracted from Oxisols. Soil Sci 173:694–706
Giaveno C, Celi L, Richardson A, Simpson RJ, Barberis E (2010b) Interaction of phytases with minerals and availability of substrate affect the hydrolysis of inositol phosphates. Soil Biol Biochem 42:491–498
Giles CD, Cade-Menun BJ, Hill JE (2011) The inositol phosphates in soils and manures: abundance, cycling and measurements. Can J Soil Sci 91:397–416
Giles CD, Richardson AE, Druschel GK, Hill JE (2012) Organic anion-driven solubilization of precipitated and sorbed phytate improves hydrolysis by phytases and bioavailability to *Nicotiana tabacum*. Soil Sci 177:591–598
Giles CD, Hsu L, Richardson AE, Hurst M, Hill JE (2014) Plant assimilation of phosphorus from an insoluble organic form is improved by addition of an organic anion producing Pseudomonas sp. Soil Biol Biochem 68:263–269
Goldstein AH (1995) Recent progress in understanding the molecular genetics and biochemistry of calcium-phosphate solubilization by Gram-negative bacteria. Biol Agric Hortic 12:185–193
Greiner R (2007) Phytate-degrading enzymes: regulation of synthesis in microorganisms and plants. In: Turner BL, Richardson AE, Mullaney EJ (eds) Inositol phosphates: linking agriculture and the environment. CAB International, Oxfordshire, pp 78–96
Guan X-H, Shang C, Zhu J, Chen G-H (2006) ATR-FTIR investigation on the complexation of myo-inositol hexaphosphate with aluminum hydroxide. J Colloid Interface Sci 293:296–302

Hansen JC, Cade-Menun BJ, Strawn DG (2004) Phosphorus speciation in manure-amended alkaline soils. J Environ Qual 33:1521–1527
Hayatsu M (2013) Utilization of phytic acid by cooperative interaction in rhizosphere. Microbes Environ 28:1–2
Hayes JE, Richardson AE, Simpson RJ (1999) Phytase and acid phosphatase activities in extracts from roots of temperate pasture grass and legume seedlings. Aust J Plant Physiol 26:801–809
Hayes JE, Richardson AE, Simpson RJ (2000a) Components of organic phosphorus in soil extracts that are hydrolysed by phytase and acid phosphatase. Biol Fertil Soils 32:279–286
Hayes JE, Simpson RJ, Richardson AE (2000b) The growth and phosphorus utilisation of plants in sterile media when supplied with inositol hexaphosphate, glucose 1-phosphate or inorganic phosphate. Plant Soil 220:165–174
He ZQ, Honeycutt CW (2001) Enzymatic characterization of organic phosphorus in animal manure. J Environ Qual 30:1685–1692
He ZQ, Honeycutt CW (2005) A modified molybdenum blue method for orthophosphate determination suitable for investigating enzymatic hydrolysis of organic phosphates. Commun Soil Sci Plant Anal 36:1373–1383
He ZQ, Griffin TS, Honeycutt CW (2004) Enzymatic hydrolysis of organic phosphorus in swine manure and soil. J Environ Qual 33:367–372
He ZQ, Honeycutt CW, Zhang TQ, Bertsch PM (2006a) Preparation and FT-IR characterization of metal phytate compounds. J Environ Qual 35:1319–1328
He ZQ, Toor GS, Honeycutt CW, Sims JT (2006b) An enzymatic hydrolysis approach for characterizing labile phosphorus forms in dairy manure under mild assay conditions. Bioresour Technol 97:1660–1668
He ZQ, Cade-Menun BJ, Toor GS, Fortuna AM, Honeycutt CW, Sims JT (2007a) Comparison of phosphorus forms in wet and dried animal manures by solution phosphorus-31 nuclear magnetic resonance spectroscopy and enzymatic hydrolysis. J Environ Qual 36:1086–1095
He ZQ, Honeycutt CW, Xing B, McDowell RW, Pellechia PJ, Zhang TQ (2007b) Solid-state fourier transform infrared and P-31 nuclear magnetic resonance spectral features of phosphate compounds. Soil Sci 172:501–515
He ZQ, Honeycutt CW, Zhang TQ, Pellechia PJ, Caliebe WA (2007c) Distinction of metal species of phytate by solid-state spectroscopic techniques. Soil Sci Soc Am J 71:940–943
He ZQ, Honeycutt CW, Cade-Menun BJ, Senwo ZN, Tazisong IA (2008) Phosphorus in poultry litter and soil: enzymatic and nuclear magnetic resonance characterization. Soil Sci Soc Am J 72:1425–1433
He ZQ, Honeycutt CW, Griffin TS, Cade-Menun BJ, Pellechia PJ, Dou ZX (2009a) Phosphorus forms in conventional and organic dairy manure identified by solution and solid state P-31 NMR spectroscopy. J Environ Qual 38:1909–1918
He ZQ, Waldrip HW, Honeycutt CW, Erich MS, Senwo ZN (2009b) Enzymatic quantification of phytate in animal manure. Commun Soil Sci Plant Anal 40:566–575
He ZQ, Olk DC, Cade-Menun BJ (2011) Forms and lability of phosphorus in humic acid fractions of Hord silt loam soil. Soil Sci Soc Am J 75:1712–1722
He ZQ, Zhong J, Cheng HN (2013) Conformational change of metal phytates: solid state 1D ^{13}C and 2D ^{1}H–^{13}C NMR spectroscopic investigations. J Food Agric Environ 11:965–970
Hedley MJ, Stewart JWB, Chauhan BS (1982) Changes in inorganic and organic soil phosphorus fractions induced by cultivation practices and by laboratory incubations. Soil Sci Soc Am J 46:970–976
Hill JE, Cade-Menun BJ (2009) Phosphorus-31 nuclear magnetic resonance spectroscopy transect study of poultry operations on the Delmarva Peninsula. J Environ Qual 38:130–138
Hill JE, Richardson AE (2007) Isolation and assessment of microorganisms that utilize phytate. In: Turner BL, Richardson AE, Mullaney EJ (eds) Inositol phosphates: linking agriculture and the environment. CAB International, Oxfordshire, pp 61–77
Hunger S, Cho H, Sims JT, Spark DL (2004) Direct speciation of phosphorus in alum-amended poultry litter: solid-state P-31 NMR investigation. Environ Sci Technol 38:674–681

Hunger S, Sims JT, Sparks DL (2008) Evidence for struvite in poultry litter: effects of storage and drying. J Environ Qual 37:1617–1625
Irving GCJ, Cosgrove DJ (1981) The use of hypobromite oxidation to evaluate two current methods for the estimation of inositol polyphosphates in alkaline extracts of soils. Commun Soil Sci Plant Anal 12:495–509
Irving GCJ, Cosgrove DJ (1982) The use of gas–liquid-chromatography to determine the proportions of inositol isomers present as pentakisphosphates and hexakisphosphates in alkaline extracts of soils. Commun Soil Sci Plant Anal 13:957–967
Islam A, Ahmed B (1973) Distribution of inositol phosphates, phospholipids and nucleic acids and mineralization of inositol phosphates in some Bangladesh soils. J Soil Sci 24:193–198
Johnson NR, Hill JE (2010) Phosphorus species composition of poultry manure-amended soil using high-throughput enzymatic hydrolysis. Soil Sci Soc Am J 74:1786–1791
Johnson BB, Quill E, Angove MJ (2012) An investigation of the mode of sorption of inositol hexaphosphate to goethite. J Colloid Interface Sci 367:436–442
Jones DL, Darrah PR (1994) Role of root derived organic acids in the mobilization of nutrients from the rhizosphere. Plant Soil 166:247–257
Jones DL, Dennis PG, Owen AG, van Hees PAW (2003) Organic acid behavior in soils – misconceptions and knowledge gaps. Plant Soil 248:31–41
Jorquera MA, Hernandez MT, Rengel Z, Marschner P, Mora MD (2008) Isolation of culturable phosphobacteria with both phytate-mineralization and phosphate-solubilization activity from the rhizosphere of plants grown in a volcanic soil. Biol Fertil Soils 44:1025–1034
Keller M, Oberson A, Annaheim KE, Tamburini F, Mader P, Mayer J, Frossard E, Bünemann EK (2012) Phosphorus forms and enzymatic hydrolyzability of organic phosphorus in soils after 30 years of organic and conventional farming. J Plant Nutr Soil Sci 175:385–393
Kerovuo J, Lauraeus M, Nurminen P, Kalkkinen N, Apajalahti J (1998) Isolation, characterization, molecular gene cloning, and sequencing of a novel phytase from *Bacillus subtilis*. Appl Environ Microbiol 64:2079–2085
Kong F, Lin W, Yan X, Liao H (2005) Phytate-phosphorus uptake and utilization by transgenic tobacco carrying *Bacillus subtilis* phytase gene. Yingyong Shengtai Xuebao 16:2389–2393
Kruse J, Leinweber P (2008) Phosphorus in sequentially extracted fen peat soils: a K-edte X-ray absorption near-edge structure (XANES) spectroscopy study. J Plant Nutr Soil Sci 171:613–620
L'Annunziata MF (2013) Soil inositol phosphate biochemistry: a call to link radiotracer applications with spectroscopic structural analysis. Soil Sci Soc Am J 77:1107–1116
Leytem AB, Maguire RO (2007) Environmental implications of inositol phosphates in animal manure. In: Turner BL, Richardson AE, Mullaney EJ (eds) Inositol phosphates: linking agriculture and the environment. CAB International, Oxfordshire, pp 150–168
Leytem AB, Thacker PA (2008) Fecal phosphorus excretion and characterization from swine fed diets containing a variety of cereal grains. J Anim Vet Adv 7:113–120
Leytem AB, Smith DR, Applegate TJ, Thacker PA (2006) The influence of manure phytic acid on phosphorus solubility in calcareous soils. Soil Sci Soc Am J 70:1629–1638
Leytem AB, Plumstead PW, Maguire RO, Kwanyuen P, Brake J (2007) What aspect of dietary modification in broilers controls litter water-soluble phosphorus: dietary phosphorus, phytase, or calcium? J Environ Qual 36:453–463
Leytem AB, Kwanyuen P, Plumstead PW, Maguire RO, Brake J (2008) Evaluation of phosphorus characterization in broiler ileal digesta, manure, and litter samples: P-31-NMR vs. HPLC. J Environ Qual 37:494–500
Lim BL, Yeung P, Cheng C, Hill JE (2007) Distribution and diversity of phytate-mineralizing bacteria. ISME J 1:321–330
Liu J, Wang X, Huang H, Want J, Li Z, Wu L, Zhang G, Ma Z (2012) Efficiency of phosphorus utilization in *phyA*-expressing cotton lines. Indian J Biochem Biophys 49:250–256
Liu J, Yang J, Cade-Menun BJ, Liang X, Hu Y, Liu CW, Zhao Y, Li L, Shi J (2013) Complementary phosphorus speciation in agricultural soils by sequential fractionation, solution 31P NMR, and P K-edge XANES spectroscopy. J Environ Qual 42:1763–1770

Lolas GM, Markakis P (1977) The phytase of navy beans (*Phaseolus vulgaris*). J Food Sci 42:1094–1097
Lung SC, Lim BL (2006) Assimilation of phytate-phosphorus by the extracellular phytase activity of tobacco (*Nicotiana tabacum*) is affected by the availability of soluble phytate. Plant Soil 279:187–199
Lung SC, Chan WL, Yip W, Wang LJ, Yeung EC, Lim BL (2005) Secretion of beta-propeller phytase from tobacco and Arabidopsis roots enhances phosphorus utilization. Plant Sci 169:341–349
Lung SC, Leung A, Kuang R, Wang Y, Leung P, Lim BL (2008) Phytase activity in tobacco (*Nicotiana tabacum*) root exudates is exhibited by a purple acid phosphatase. Phytochemistry 69:365–373
Makoi J, Ndakidemi PA (2008) Selected soil enzymes: examples of their potential roles in the ecosystem. Afr J Biotechnol 7:181–191
Mandal NC, Burman S, Biwas BB (1972) Isolation, purification and characterisation of phytase from germinating mung beans. Phytochemistry 11:495–502
Martin CJ, Evans WJ (1987) Phytic acid-divalent-cation interactions. V. Titrimetric, colorimetric, and binding studies with cobalt (II) and nickel (II) and their comparison with other metal ions. J Inorg Biochem 30:101–119
Martin M, Celi L, Barberis E (2002) Extractability and plant availability of phosphate from P-goethite complexes. Commun Soil Sci Plant Anal 33:143–153
Martin M, Celi L, Barberis E (2004) Desorption and plant availability of *myo*-inositol hexaphosphate adsorbed on goethite. Soil Sci 169:115–124
McDowell RW, Condron LM, Stewart I, Cave V (2005) Chemical nature and diversity of phosphorus in New Zealand pasture soils using P-31 nuclear magnetic resonance spectroscopy and sequential fractionation. Nutr Cycl Agroecosyst 72:241–254
McDowell RW, Cade-Menun B, Stewart I (2007) Organic phosphorus speciation and pedogenesis: analysis by solution ^{31}P nuclear magnetic resonance spectroscopy. Eur J Soil Sci 58:1348–1357
McGrath JM, Sims JT, Maguire RO, Saylor WW, Angel CR, Turner BL (2005) Broiler diet modification and litter storage: impacts on phosphorus in litters, soils, and runoff. J Environ Qual 34:1896–1909
McKercher RB, Anderson G (1968a) Characterization of inositol penta- and hexaphosphate fractions of a number of Canadian and Scottish soils. J Soil Sci 19:303–310
McKercher RB, Anderson G (1968b) Content of inositol penta- and hexaphosphates in some Canadian soils. J Soil Sci 19:47–55
McKercher RB, Anderson G (1989) Organic phosphate sorption by neutral and basic soils. Commun Soil Sci Plant Anal 20:723–732
Menezes-Blackburn D, Jorquera MA, Greiner R, Gianfreda L, Mora M (2013) Phytases and phytase-labile organic phosphorus in manures and soils. Crit Rev Environ Sci Technol 43:916–954
Miller SH, Browne P, Prigent-Combaret C, Combes-Meyne E, Morrissey JP, O'Gara F (2010) Biochemical and genomic comparison of inorganic phosphate solubilization in Pseudomonas species. Environ Microbiol Rep 2:403–411
Mozaffari M, Sims JT (1994) Phosphorus availability and sorption in an Atlantic coastal-plain watershed dominated by animal-based agriculture. Soil Sci 157:97–107
Mozaffari M, Sims JT (1996) Phosphorus transformations in poultry litter-amended soils of the Atlantic coastal plain. J Environ Qual 25:1357–1365
Mullaney EJ, Ulla AHJ (2007) Phytases: attributes, catalytic mechanisms and applications. In: Turner BL, Richardson AE, Mullaney EJ (eds) Inositol phosphates: linking agriculture and the environment. CAB International, Oxfordshire, pp 97–110
Murphy PNC, Bell A, Turner BL (2009) Phosphorus speciation in temperate basaltic grassland soils by solution 31P NMR spectroscopy. Eur J Soil Sci 60:638–651
Nagai Y, Funahashi S (1962) Phytase from wheat bran. I. Purification and substrate specificity. Agric Biol Chem 26:794–803

Noack SR, McLaughlin MJ, Smernik RJ, McBeath TM, Armstrong RD (2012) Crop residue phosphorus: speciation and potential bio-availability. Plant Soil 359:375–385

Oburger E, Jones DL, Wenzel WW (2011) Phosphorus saturation and pH differentially regulate the efficiency of organic acid anion-mediated P solubilization mechanisms in soil. Plant Soil 341:363–382

Ognalaga M, Frossard E, Thomas F (1994) Glucose-1-phosphate and *myo*-inositol hexaphosphate adsorption mechanisms on goethite. Soil Sci Soc Am J 58:332–337

Oh BC, Chang BS, Park KH, Ha NC, Kim HK, Oh BH, Oh TK (2001) Calcium-dependent catalytic activity of a novel phytase from *Bacillus amyloliquefaciens* DS11. Biochemistry 40:9669–9676

Olsson R, Giesler R, Loring JS, Persson P (2012) Enzymatic hydrolysis of organic phosphates adsorbed on mineral surfaces. Environ Sci Technol 46:285–291

Palomo L, Claassen N, Jones DL (2006) Differential mobilization of P in the maize rhizosphere by citric acid and potassium citrate. Soil Biol Biochem 38:683–692

Pant HK, Edwards AC, Vaughan D (1994) Extraction, molecular fractionation and enzyme degradation of organically associated phosphorus in soil solutions. Biol Fertil Soils 17:196–200

Patel KJ, Singh AK, Nareshkumar G, Archana G (2010a) Organic-acid-producing, phytate-mineralizing rhizobacteria and their effect on growth of pigeon pea (*Cajanus cajan*). Appl Soil Ecol 44:252–261

Patel KJ, Vig S, Nareshkumar G, Archana G (2010b) Effect of transgenic rhizobacteria overexpressing *Citrobacter braakii* appA on phytate-P availability to mung bean plants. J Microbiol Biotechnol 20:1491–1499

Plumstead PW, Leytem AB, Maguire RO, Spears JW, Kwanyuen P, Brake J (2008) Interaction of calcium and phytate in broiler diets. 1. Effects on apparent prececal digestibility and retention of phosphorus. Poult Sci 87:449–458

Prietzel J, Dümig A, Wu Y, Zhou J, Klysubun W (2013) Synchrotron-based P K-edge XANES spectroscopy reveals rapid changes of phosphorus speciation in the topsoil of two glacier foreland chronosequences. Geochim Cosmochim Acta 108:154–171

Raboy V (2007) Seed phosphorus and the development of low-phytate crops. In: Turner BL, Richardson AE, Mullaney EJ (eds) Inositol phosphates: linking agriculture and the environment. CAB International, Oxfordshire, pp 111–132

Ramaekers L, Remans R, Rao IM, Blair MW, Vanderleyden J (2010) Strategies for improving phosphorus acquisition efficiency of crop plants. Field Crops Res 117:169–176

Richardson AE, Hadobas PA (1997) Soil isolates of *Pseudomonas* spp. that utilize inositol phosphates. Can J Microbiol 43:509–516

Richardson AE, Simpson RJ (2011) Soil microorganisms mediating phosphorus availability. Plant Physiol 156:989–996

Richardson AE, Hadobas PA, Hayes JE (2000) Acid phosphomonoesterase and phytase activities of wheat (*Triticum aestivum* L.) roots and utilization of organic phosphorus substrates by seedlings grown in sterile culture. Plant Cell Environ 23:397–405

Richardson AE, Hadobas PA, Hayes JE (2001a) Extracellular secretion of *Aspergillus* phytase from *Arabidopsis* roots enables plants to obtain phosphorus from phytate. Plant J 25:641–649

Richardson AE, Hadobas PA, Hayes JE, O'Hara CP, Simpson RJ (2001b) Utilization of phosphorus by pasture plants supplied with *myo*-inositol hexaphosphate is enhanced by the presence of soil micro-organisms. Plant Soil 229:47–56

Richardson AE, George TS, Jakobsen I, Simpson RJ (2007) Plant utilization of inositol phosphates. In: Turner BL, Richardson AE, Mullaney EJ (eds) Inositol phosphates: linking agriculture and the environment. CAB International, Oxfordshire, pp 242–260

Richardson AE, Hocking PJ, Simpson RJ, George TS (2009) Plant mechanisms to optimize access to soil phosphorus. Crop Pasture Sci 60:124–143

Richardson AE, Lynch JP, Ryan PR, Delhaize E, Smith FA, Smith SE, Harvey PR, Ryan MH, Veneklaas EJ, Lambers H, Oberson A, Culveno RA, Simpson RJ (2011) Plant and microbial strategies to improve the phosphorus efficiency of agriculture. Plant Soil 349:121–156

Sattari SZ, Bouwman AF, Giller KE, van Ittersum MK (2012) Residual soil phosphorus as the missing piece in the global phosphorus crisis puzzle. Proc Natl Acad Sci U S A 109:6348–6353
Schlemmer U, Frølich W, Prieto RM, Grases F (2009) Phytate in foods and significance for humans: food sources, intake, processing, bioavailability, protective role and analysis. Mol Nutr Food Res 53:S330–S375
Sepehr E, Rengel Z, Fateh E, Sadaghiani MR (2012) Differential capacity of wheat cultivars and white lupin to acquire phosphorus from rock phosphate, phytate and soluble phosphorus sources. J Plant Nutr 35:1180–1191
Shand CA, Smith S (1997) Enzymatic release of phosphate from model substrates and P compounds in soil solution from a peaty podzol. Biol Fertil Soils 24:183–187
Shand CA, Cheshire MV, Bedrock CN, Chapman PJ, Frase AR, Chudek JA (1999) Solid-phase ^{31}P NMR spectra of peat and mineral soils, humic acids and soil solution components: influence of iron and manganese. Plant Soil 214:153–163
Sharpley AN, Chapra SC, Wedepohl R, Sims JT, Daniel TC, Reddy KR (1994) Managing agricultural phosphorus for protection of surface waters – issues and options. J Environ Qual 23:437–451
Shears SB, Turner BL (2007) Nomenclature and terminology of inositol phosphates: clarification and a glossary of terms. In: Turner BL, Richardson AE, Mullaney EJ (eds) Inositol phosphates: linking agriculture and the environment. CAB International, Oxfordshire, pp 1–6
Shober AL, Hesterberg DL, Sims JT, Gardner S (2006) Characterization of phosphorus species in biosolids and manures using XANES spectroscopy. J Environ Qual 35:1983–1993
Smernik RJ, Dougherty WJ (2007) Identification of phytate in phosphorus-31 nuclear magnetic resonance spectra: the need for spiking. Soil Sci Soc Am J 71:1045–1050
Smith MTE, Cade-Menun BJ, Tibbett M (2006) Soil phosphorus dynamics from sewage sludge at different stages in a treatment stream. Biol Fertil Soils 42:186–197
Tang J, Leung A, Leung C, Lim BL (2006) Hydrolysis of precipitated phytate by three distinct families of phytases. Soil Biol Biochem 38:1316–1324
Tazisong IA, Senwo ZN, Taylor RW, He ZQ (2008) Hydrolysis of organic phosphates by commercially available phytases: biocatalytic potentials and effects of ions on their enzymatic activities. J Food Agric Environ 6:500–505
Toor GS, Peak JD, Sims JT (2005a) Phosphorus speciation in broiler litter and turkey manure produced from modified diets. J Environ Qual 34:687–697
Toor GS, Cade-Menun BJ, Sims JT (2005b) Establishing a linkage between phosphorus forms in dairy diets, feces and manures. J Environ Qual 34:1380–1391
Turner BL (2004) Optimizing phosphorus characterization in animal manures by solution phosphorus-31 nuclear magnetic resonance spectroscopy. J Environ Qual 33:757–766
Turner BL (2006) Organic phosphorus in Madagascan rice soils. Geoderma 136:279–288
Turner BL (2007) Inositol phosphates in soil: amounts, forms and significance of the phosphorylated inositol stereoisomers. In: Turner BL, Richardson AE, Mullaney EJ (eds) Inositol phosphates: linking agriculture and the environment. CAB International, Oxfordshire, pp 186–206
Turner BL, Leytem AB (2004) Phosphorus compounds in sequential extracts of animal manures: chemical speciation and a novel fractionation procedure. Environ Sci Technol 38:6101–6108
Turner BL, Richardson AE (2004) Identification of *scyllo*-inositol phosphates in soil by solution phosphorus-31 nuclear magnetic resonance spectroscopy. Soil Sci Soc Am J 68:802–808
Turner BL, McKelvie ID, Haygarth PM (2002a) Characterisation of water-extractable soil organic phosphorus by phosphatase hydrolysis. Soil Biol Biochem 34:27–35
Turner BL, Paphazy MJ, Haygarth PM, McKelvie ID (2002b) Inositol phosphates in the environment. Phil Trans R Soc Lond B 357:449–469
Turner BL, Cade-Menun BJ, Westermann DT (2003a) Organic phosphorus composition and potential bioavailability in semi-arid arable soils of the western United States. Soil Sci Soc Am J 67:1168–1179

Turner BL, Mahieu N, Condron LM (2003b) Quantification of *myo*-inositol hexakisphosphate in alkaline soil extracts by solution P-31 NMR spectroscopy and spectral deconvolution. Soil Sci 168:469–478

Turner BL, Cade-Menun BJ, Condron LM, Newman S (2005) Extraction of soil organic phosphorus. Talanta 66:294–306

Turner BL, Cheesman AW, Godage HY, Riley AM, Potter BVL (2012) Determination of *neo*- and D-*chiro*-inositol hexakisphosphate in soils by solution ^{31}P NMR spectroscopy. Environ Sci Technol 46:4994–5002

Unno Y, Shinano T (2013) Metagenomic analysis of the rhizosphere soil microbiome with respect to phytic acid utilization. Microbes Environ 28:120–127

Unno Y, Okubo K, Wasaki J, Shinano T, Osaki M (2005) Plant growth promotion abilities and microscale bacterial dynamics in the rhizosphere of Lupin analysed by phytate utilization ability. Environ Microbiol 7:396–404

Vincent AG, Schleucher J, Gröbner G, Vestergren J, Persson P, Jansson M, Giesler R (2012) Changes in organic phopshorus composition in boral forest humus soils: the role of iron and aluminium. Biogeochemistry 108:485–499

Wang Y, He Y, Zhang H, Schroder J, Li C, Zhou D (2008) Phosphate mobilization by citric, tartaric, and oxalic acids in a clay loam Ultisol. Soil Sci Soc Am J 72:1263–1268

Wang Y, Ye X, Ding G, Xu F (2013) Overexpression of *phyA* and *appA* genes improves soil organic phosphorus utilisation and seed phytase activity in *Brassica napus*. PLoS One 8:e60801

Wei LL, Chen CR, Xu ZH (2009) The effect of low-molecular-weight organic acids and inorganic phosphorus concentration on the determination of soil phosphorus by the molybdenum blue reaction. Biol Fertil Soils 45:775–779

Wei LL, Chen CR, Xu ZH (2010) Citric acid enhances the mobilization of organic phosphorus in subtropical and tropical forest soils. Biol Fertil Soils 46:765–769

Yamada K, Minoda Y, Yamamoto S (1968) Phytase from *Aspergillus terreus*. Part I. Production, purification and some general properties of the enzyme. Agric Biol Chem 32:1275–1282

Chapter 10
Phosphorus Forms and Mineralization Potentials of Alabama Upland Cotton Production Soils Amended with Poultry Litter

Irenus A. Tazisong, Zachary N. Senwo, Barbara J. Cade-Menun, and Zhongqi He

Abstract The essential role of phosphorus (P) in almost all biological processes has led to its extensive studies. Phosphorus in its inorganic form (P_i) is required for metabolic reactions and energy transfer. In contrast, organic P (P_o) forms become bioavailable usually after hydrolysis to P_i. Organic P dephosphorylation in the soil environment is largely dependent on the stereochemical and stereoisomeric structures of the compounds in question. In addition, P_o dephosphorylation also depends on the types and enzymes sources involved in such catalytic activities. This chapter focuses on P fractions and mineralization potentials of different P forms in poultry litter amended soils of northern Alabama cropped with cotton. Phosphorus forms characterized by solution ^{31}P nuclear magnetic resonance (NMR) spectroscopy showed orthophosphate to be the dominant form, accounting for 63.6–76.1 % of the total P forms in these soils. A sequential fractionation was used to separate soil P_i and P_o into several fractions. The fractionation revealed that soil P_i was mainly associated with Fe and Al oxides and was not readily bioavailable. Water-, Fe- and Al- associated P_i increased in soils treated with poultry litter compared with the control. The Ca-Al bound P_o was the most abundant fraction with more than 40 % of the total P_o but the Ca-Al bound P_o values was lower for the control and no-tilled soils. Phosphatase enzymes were able to hydrolyze 7–62 % water, 17–53 % Ca-EDTA, 18–88 % Na-EDTA, 40–77 % H_2SO_4, and 50–75 % NaOH extractable P_o. Information obtained from this study indicates that P dynamics

I.A. Tazisong (✉) • Z.N. Senwo
Department of Biological and Environmental Sciences, College of Agricultural, Life and Natural Sciences, Alabama A&M University, Normal, AL 35762, USA
e-mail: irenus.tazisong@aamu.edu

B.J. Cade-Menun
Agriculture and Agri-Food Canada/Agriculture et Agroalimentaire Canada SPARC, Swift Current, SK, Canada

Z. He
Southern Regional Research Center, USDA-ARS, New Orleans, LA 70124, USA

Z. He and H. Zhang (eds.), *Applied Manure and Nutrient Chemistry for Sustainable Agriculture and Environment*, DOI 10.1007/978-94-017-8807-6_10,

and uptake in cotton soils amended with poultry litter is not well understood. More research is needed to better understand the impact of poultry litter application on P forms and mineralization potentials in cotton production soils.

10.1 Introduction

Phosphorus (P) is an essential nutrient for life in both aquatic and terrestrial systems. In soils, too little P can limit plant growth in many regions, requiring the addition of P fertilizers and strategies to improve the P efficiency of agriculture (e.g. Richardson et al. 2011). However, addition of P to soils in excess of plant needs can result in P being transported out of the field into water bodies, triggering eutrophication (e.g. Elser and Bennett 2011).

Soils can be considered a sink for both aquatic and atmospheric P. Plants usually take up P from soil solutions in inorganic forms. Although P_i equilibrium concentrations in soil solutions is low, there is a continual replenishment for plants and microbial uptake from soil solid phases (Wild and Oke 1966). There are several P_i forms of soils, which include phosphate and anhydrides (complex P_i forms such as pyrophosphate and polyphosphate) (Condron et al. 2005; Pierzynski et al. 2005). Organic P in soils is classified based on the type of bond between the P and C atoms, and is divided into orthophosphate monoesters, orthophosphate diesters and phosphonates (Condron et al. 2005). Orthophosphate monoesters have the general structure $ROPO_3^{2-}$ (where R is an organic moiety), with one orthophosphate per C group. This is the predominant group of P_o compounds in most soils, and includes inositol phosphates, sugar phosphates, phosphoproteins and mononucleotides. Orthophosphate diesters ($R_1OROPO_2^-$, where R and R_1 are C moieties), have two C groups per orthophosphate, and include nucleic acids, phospholipids and teichoic acid. Phosphonates differ from other P_o forms because they have a direct C-P bond (not an ester bond through O). These have the structure $[RP(O)(OH)_2]$, and include 2-aminoethyl phosphonic acid (AEP), and agricultural chemicals such as the herbicide glyphosate.

The forms and amounts of P in soils are governed by a combination of chemical, physical and biological parameters, which in turn are controlled by such environmental conditions as temperature, moisture, salinity, pH, land use intensity, and fertilization history. Other investigators (Condron et al. 1990; Ottabong et al. 1997; Chen et al. 2002) have reported that long-term P fertilization or animal manure applications increase P_o concentrations in soils, while long-term cultivation and afforestation of grasslands significantly decrease soil P_o. Stevenson (1986), classified soil P into seven categories: (1) soluble inorganic and organic compounds in soil solutions, (2) weakly sorbed inorganic phosphate, (3) insoluble phosphate-calcium bound, (4) iron and aluminum bound, (5) strongly adsorbed phosphate or occluded hydrous oxides of Fe and Al, (6) fixed phosphate of silicate minerals, and (7) insoluble organic forms (undecomposed plant and animal residues).

Phosphatases are diverse groups of metalloenzymes that deserve special attention because they are involved in P_i and P_o transformations in the environment.

They hydrolyze P_o and anhydrides to release phosphate. These enzymes in the environment are derived from plants and microorganisms and are known to hydrolyze a variety of P compounds (Condron et al. 2005). These enzymes are categorized into general groups such as diesterases, which generally hydrolyze orthophosphate diesters. Alkaline and acid phosphatases degrade orthophosphate monoesters, operating at alkaline and acid pH, respectively. They can also be more specific in their hydrolysis. Phytase catalyze the hydrolysis of phytic acid to less phosphorylated inositol phosphate and inorganic phosphates, all utilizing a phosphohistidine intermediate in their phosphoryl transfer reactions (Vats et al. 2005). Phytases are specific for phytic acid and it is possible to distinguish them from acid and alkaline phosphatases, which are incapable of degrading phytic acid. However, the hydrolytic activity of phytase may not be only limited to phytic acid, for it has also been noted to hydrolyze (not effectively) various forms of phosphomonoesters compounds (Tazisong et al. 2008).

Soil tillage and management practices, coupled with poultry litter applications, greatly affect the fate of P_i and P_o in soils. In the southeastern United States where the soils are mostly highly weathered Ultisols, P has been a key nutrient limiting soil productivity and health (Senwo et al. 2003). However, opinions differ on strategies to best manage P in these highly weathered acid soils that typify these systems. Investigators have long believed that a very high initial addition of P was needed to maximize crop yields and that the excess P applied had a long-term residual value. The advantages for the initially large P applications do not only lead to P deficiency corrections for crops, but also affect soil pH, CEC, organic matter content, and biological activities (Kingery et al. 1994; Dobermann et al. 2002; Acosta-Martinez and Harmel 2006).

In Ultisols and Oxisols where P fixation is high, the risk of water contamination is low in the short run, but in soils with less sorbing capacity, negative impacts to water quality are possible due to mismanaged land applied animal manure. Poultry litter applications have not only created P crises in soils, but have also lead to a shift in P fractions. Dobermann et al. (2002), found an increase of P_i in the resin, NaOH, and H_2SO_4 fractions due to manure application. In contrast, Dobermann et al. (2002) reported no change in NaOH-P_o fraction, which accounted for over 80 % of the total P in soils without manure application. Eghball et al. (1996) reported a significant movement of P from manure applications through high P adsorbing calcium carbonate soil layers than P from chemical fertilization.

Alabama is one of the leading cotton and poultry producing states in the USA. To optimize cotton yield in Alabama Ultisols and other soils, poultry litter is usually applied to cotton fields (Tewolde et al. 2010, 2011). There is currently little or no comprehensive information available on the P forms, fractions, and their potential mineralization in Alabama Ultisols amended with poultry litter and cropped with cotton despite numerous studies on its carbon, nitrogen, macro and micro elements contents (Tazisong et al. 2005; Tazisong and Senwo 2009; Sainju et al. 2008a, b, 2010; Ohno et al. 2009). This information is needed to understand the biogeochemistry and potential agroecosystem responses to complex organic compounds in these soils. Therefore, this chapter presents and discusses published

Table 10.1 Poultry litter and P amount applied to cotton soils (Belle Mina-Alabama) from 1997 to 2004

Year	Amount of poultry litter applied to supply 100 kg N ha^{-1} (Mg ha^{-1})	Amount of P applied to inorganic fertilizer plots[a] (Kg ha^{-1})
1997	3.70	67
1998	3.33	67
2000	2.94	67
2001	2.86	67
2003	3.13	67
2004	3.57	67
Avg.	3.23	67

Adapted from Sainju et al. (2010)
[a]The rate of P applied is similar to that supplied by poultry litter

Table 10.2 Cropping and fertilization management practices used in Belle Mina-Alabama cotton soils

Management practice	Description of treatment
BF	Bare fallow-no crop planted
CTAN	Conventional-till, cotton-winter fallow + 100 kg N ha^{-1} from ammonium nitrate
CTR	Conventional-till, cotton/rye cropping, 0 kg N ha^{-1}
CTRAN	Conventional-till, cotton/rye cropping + 100 kg N ha^{-1} from ammonium nitrate
CTRP	Conventional-till, cotton/rye cropping + 100 kg N ha^{-1} from poultry litter
MTRAN	Mulch-till, cotton/rye cropping + 100 kg N ha^{-1} from ammonium nitrate
MTRP	Mulch-till. cotton/rye cropping + 100 kg N ha^{-1} from poultry litter
NT	No-till, cotton-winter fallow, 0 kg n ha^{-1}
NTAN	No-till, cotton-winter fallow + 100 kg N ha^{-1} from ammonium nitrate
NTRAN	No-till, cotton/rye cropping + 100 kg N ha^{-1} from ammonium nitrate
NTRP	No-till, cotton/rye cropping + 100 kg N ha^{-1} from poultry litter
NTRPP	No-till, cotton/rye cropping + 200 kg N ha^{-1} form poultry litter

Adapted from Tazisong et al. (2013)

(Tazisong et al. 2013) and unpublished data on P forms and mineralization potential in Alabama cotton soils amended with poultry litter (Table 10.1) and managed as no-tilled, tilled, and mulch-tilled practices (Table 10.2).

10.2 P Forms in Alabama Cotton Soil

The soils used for this study is a Decatur silt loam (clayey, kaolinitic, thermic, typic paleudults) collected from an upland cotton production site at the Alabama Agricultural Experimental Station at Belle Mina, AL (34°41′ N, 86°52′ W).

The site had been continuously cropped with cotton (*Gossypium hirsutum* L.) for 5 years prior to sampling (Nyakatawa and Reddy 2000). Crop management practices consisted of various combinations of tillage systems (conventional till, mulch till, and no-till), cropping systems and management practices [cotton in summer and rye (*Secale cereale* L.) cover crop in winter], inorganic nitrogen fertilization [ammonium nitrate (0 and 100 kg N ha^{-1})], and poultry litter (0, 100, and 200 kg N ha^{-1}) (Table 10.2). The soils used in this study were collected in spring of 2005 (no crops were in the field) with a hand probe (10 cm i.d.) from the 0 to 10 cm depth in four replicates in the central rows of the plots after removing the surface soil residues. The soils were air-dried, ground, and sieved to a 2 mm size into plastic bags and stored until use. Selected soil properties of these soils have been listed in other studies (Ohno et al. 2009; Sainju et al. 2008b; Tazisong et al. 2013). Solution ^{31}P NMR spectroscopy was used to study the P forms of these soils. Phosphorus was extracted for solution ^{31}P NMR spectroscopy using a modified version of the Cade-Menun and Preston (1996) procedures. The NMR spectra for the conventional till, mulch-till and No-till were visually similar among the 12 soil treatments; however, only the representative spectra of the conventionally-tilled soils are presented in Fig. 10.1. The spectra are visually similar for soil samples reported in other literature (Cade-Menun et al. 2010; Doolette et al. 2010; He et al. 2011; Ohno et al. 2011). The distributions of the P compound classes in the soils used are listed in Table 10.3. Phosphonates and polyphosphates were not detected in these soil samples. The dominant P species in the investigated soil samples irrespective of tillage and amendments was orthophosphate, accounting for 63.6–76.1 % of the total P forms in the soils. The high concentrations of orthophosphate is not due to the hydrolysis of organic compounds during extraction, but the orthophosphates associated with Fe, Al and Mn oxides (Bishop et al. 1994). Orthophosphate monoesters (phytate, glucose-6-phosphate, choline phosphate, adenosine-5-phosphate etc.) were the predominant P_o forms. The orthophosphate monoesters accounted for 20.1–29.5 % of the total P in the soils. Phytate was the dominant monoester species in the soils accounting for 5–10 % of the total orthophosphate monoesters. However, the accumulation of phytate P in the poultry litter-amended soils was not obvious, which is consistent with previous observations in studies using Alabama soils under pasture management (He et al. 2008) although poultry litter contains higher amounts of phytate (He et al. 2006, 2007). The predominance of monoester P_o forms in these soils is in agreement with those reported in several NMR studies of organic P forms in various soil types (He et al. 2008; Cade-Menun et al. 2010; Cade-Menun 2011; Johnson and Hill 2010; Giles et al. 2011). The orthophosphate diesters (DNA, Teichoic acid, RNA, phospholipids, phosphatidyl choline, etc.) accounted for less than 4 % of the total P in soils (Table 10.3). Similar findings were reported by Rheinheimer et al. (2002) in Brazilian Oxisols and Ultisols.

Tillage and soil amendments can greatly influence the quantity of P forms in soils and thus affect P quality. Although, all the spectra were visually similar irrespective of the treatments and tillage practices, the relative similarity suggests

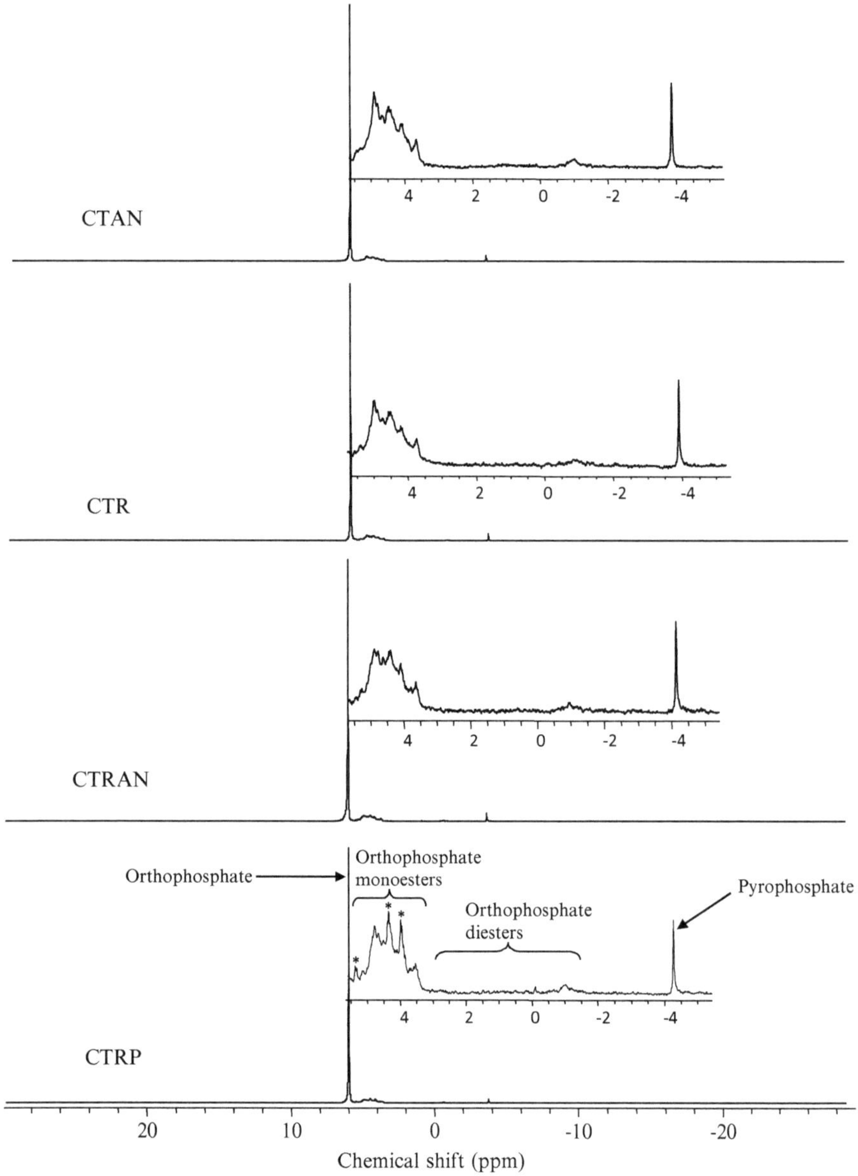

Fig. 10.1 NMR spectra of NaOH-EDTA ^{31}P extracts from conventionally-tilled (CT) soils amended with poultry litter and ammonium nitrate. The *inserts* show expanded orthophosphate monoester and diester regions (Refer to Table 10.2 for treatment abbreviations)

that tillage and soil amendments did not significantly alter the P forms in these soils. However, the average amounts of monoesters in the bare fallow (BF) and no-tilled samples was slightly higher (one unit higher) than in the conventionally and mulch tilled samples (Table 10.3).

Table 10.3 Relative amount (%) of P within each P form assessed by ^{31}P NMR spectroscopy

Treatment	Orthophosphate	Monophosphate	Other diesters (%)	DNA	Pyrophosphate	Recovery[a]
BF	69.1	25.5	0.8	1.6	2.9	42.2
CTAN	72.5	23.4	0.8	1.5	1.9	26.3
CTR	66.0	28.2	1.2	2.3	2.3	25.9
CTRAN	66.7	27.2	0.8	2.4	2.8	29.3
CTRP	74.7	21.6	0.7	1.3	1.7	31.7
MTRAN	66.0	27.2	1.5	2.3	3.0	28.9
MTRP	73.0	23.0	0.8	1.6	1.6	33.1
NT	67.5	27.6	0.8	1.6	2.4	28.4
NTAN	65.8	27.7	1.3	2.6	2.6	27.2
NTRAN	63.6	29.5	0.8	2.7	3.4	29.4
NTRP	72.8	22.9	0.7	1.8	1.8	34.5
NTRPP	76.1	20.1	0.9	1.1	1.7	31.8

[a]Percent of total P that was extracted for NMR analysis. Refer to Table 10.2 for treatment abbreviations

Cultivated soils are usually but not always lower in P_o content than their uncultivated counterparts, because tillage increases aeration and exposes organic matter to microbial decomposition. On the other hand, cultivation increases the rate of orthophosphate monoesters and diesters adsorptions on Fe and Al oxides, thereby preventing subsequent degradations. This may explain the reason why the differences (average) in P categories between conventional, conservation and mulch tillage was not much (0.66 %). Similar trends were observed for diesters except that those in mulch-tilled soils were slightly higher (0.25 %). Soils amended with poultry litter were lower in monoesters and diesters and higher in orthophosphates contents than soils that have received inorganic fertilizations. Faster mineralization rates of the orthophosphates monoesters and diesters due to high microbial populations in the poultry litter may have been responsible for the low P forms in the poultry amended soils. Further, correlation analysis revealed that the orthophosphates positively correlated with microbial biomass N, soil organic C, particulate organic C, and negatively correlated with microbial biomass C and the potential C mineralized (Table 10.4). Pyrophosphates negatively correlated with all the biological properties except with microbial biomass C and the potential C mineralized which showed a positive correlation. The lack of correlation between the organic P forms and the soil biological, physical and chemical properties is in line with that of Sumann et al. (1998).

They explained that the lack of such correlations between the P_o forms and the soil properties suggest that climate was the primary determinant for the soil P_o compositions, due to the strong correlation between climatic factors (mean annual temperature and precipitation) and the P_o forms in the Great Plains soils.

Table 10.4 Correlation matrix for relationships between P forms and soil biological properties (n = 12)

	POC:PON	MBN	MBC	PON	SOC	POC	PCM
Ortho	−0.32 ns	0.69*	−0.61*	0.54 ns	0.60*	0.63*	−0.72**
DNA	0.30 ns	−0.54 ns	0.41 ns	−0.26 ns	−0.37 ns	−0.35 ns	0.54 ns
Pyro	0.05 ns	−0.65*	0.58*	−0.58*	−0.59*	−0.61*	0.62*
Other diest	0.28 ns	0.01 ns	−0.07 ns	−0.17 ns	−0.11 ns	−0.15 ns	0.14 ns
IHP	0.01 ns	0.18 ns	0.24 ns	−0.05 ns	−0.10 ns	−0.13 ns	−0.15 ns
Scyllo	0.04 ns	0.29 ns	0.19 ns	0.24 ns	0.10 ns	0.05 ns	−0.24 ns
Other mono	0.23 ns	−0.63*	0.23 ns	−0.34 ns	−0.34 ns	−0.33 ns	0.62*

MBN microbial biomass nitrogen, *MBC* microbial biomass carbon, *PON* particulate organic carbon, *SOC* soil organic carbon, *POC* particulate organic carbon, *PCM* potential carbon mineralization

*, **, significant at $P \leq 0.05$, 0.01, respectively; *ns* not significant at $P = 0.05$

10.3 Inorganic P Fractions in Alabama Cotton Soils

In this chapter, the sequential procedure developed by Golterman et al. (1998) was used to elucidate the various P fractions in the soils (Table 10.5). Golterman (1988, 1996) suggested the use of the chelating agent ethylenediaminetetraacetic acid (EDTA) at neutral pH. The Ca-EDTA step in the method renders Fe soluble while at the same time the phosphate and pH prevents re-precipitation. The Na-EDTA step removes calcium and aluminum bound phosphate, thus minimizing extracting organic materials. In brief, one gram of soil was used via shaking with 30 mL of each extraction solution, after which the remaining P was digested with persulfate. Following extraction, each soil suspension was centrifuged at 4,000 X g for 10 min, filtered under suction using a 0.45 μm filter paper. Portion of the extract from each fractionation step was immediately frozen at −20 °C after adding 10 μL of sodium azide [NaN_3 (to prevent P uptake and microbial growth)] until used for enzyme hydrolysis determination. An aliquot of the remaining extract was taken for inorganic P determination. To determine the inorganic P from the Ca- and Na-EDTA extracts, a 25-time dilution was performed to prevent the interference of EDTA with the Mo-P colorimetric reaction.

Due to the incomplete factorial treatments arrangements used in this study, contrast analysis was used for specific treatments means comparisons. Water extractable, Ca-EDTA, and Na-EDTA extractable P_i were significantly different from the control whereas; little or no significant differences existed for P_i between various tillage systems (Table 10.6). The water soluble fraction was among the least distributed fraction and accounted for less than 5 % of the total soil P_i (Table 10.6). This fraction is considered to be the most important fraction because it is readily available for plant and microbial uptake. Most studies on soil P fractions (Sui et al. 1999; Gilbin et al. 2000; He et al. 2004a, b; McDowell and Stewart 2005) have reported the water fraction to have the lowest P concentration. In contrast, water soluble P in animal manure has been reported to be the dominant form of soluble P (He and Honeycutt 2001; He et al. 2004b, c; Wienhold and Miller 2004).

Table 10.5 Sequential extraction procedure of organic phosphorus (Golterman et al. 1998)

1.	Deionzed water (2 h)	Water soluble P_i, P_o
2.	0.05 M Ca-EDTA + 1 % Na-dithionite pH 7.5 (2 h twice)	Fe associated P_i, P_o
3.	0.1 M Na-EDTA pH 4.2 (16 h twice)	Ca-Al associated P_i, P_o
4.	0.5 M H_2SO_4 (2 h)	Acid soluble P_i, P_o
5.	0.5 M TCA (0 °C, 2 h)	Sugar bound P_o
6.	0.5 M TCA (95 °C, 30 min)	Nucleic acid P_o and polyphosphate
7.	2 M NaOH (90 °C, 1 h)	Humic bound P_o
8.	$K_2S_2O_8$ + 0.5 M H_2SO_4 (150 °C 2 h)	Residual P_o

Table 10.6 P_i fractions as affected by poultry litter amendments and tillage system

Contrast	Water	CaEDTA	NaEDTA	H_2SO_4	NaOH
BF vs MTRP	9.92 vs 37.6***	303 vs 489***	158 vs 271**	89.0 vs 113 ns	61.3 vs 70.1 ns
BF vs NTRP	9.92 vs 36.7***	303 vs 386 ns	158 vs 244 ns	89.0 vs 112 ns	61.3 vs 68.5 ns
BF vs NTRPP	9.92 vs 50.7***	303 vs 486***	158 vs 371***	89.0 vs 113 ns	61.3 vs 68.7 ns
CTRP vs MTRP	39.4 vs 37.6 ns	537 vs 489 ns	283 vs 271 ns	117 vs 113 ns	69.4 vs 70.1 ns
MTRP vs NTRP	37.6 vs 36.5 ns	489 vs 386 ns	271 vs 244 ns	113 vs 112 ns	70.1 vs 68.5 ns
MTRP vs NTRPP	37.6 vs 50.7*	489 vs 486 ns	271 vs 371*	113 vs 113 ns	70.1 vs 68.7 ns
CTRP vs NTRP	39.4 vs 36.5 ns	537 vs 386**	282 vs 244 ns	117 vs 112 ns	69.4 vs 68.5 ns
CTRP vs NTRPP	39.4 vs 50.7 ns	537 vs 486 ns	283 vs 371 ns	117 vs 113 ns	69.4 vs 68.7 ns

*, **, ***, significant at $P \leq 0.05$, 0.01, 0.001; respectively. *ns* not significant at $P < 0.05$

A decline in water extracted P_i within 21 days after amending a coarse-loamy, mixed, frigid Typic Haplorthod soil at a rate of 100 mg total P kg^{-1} has been reported (Griffin et al. 2003; He et al. 2004c). Such decline is attributed to the high levels of exchangeable Al and Fe in the soils.

The P extracted with Ca-EDTA was found to have the greatest amounts of P in the soils irrespective of management practices (Table 10.6). This fraction represents Fe associated P_i and accounted for more than 43 % of the total P_i in the soils. The high association of P_i with Fe is expected due to the high Fe content of the soils (Tazisong et al. 2004). Reported data by others (Gilbin et al. 2000; Saavedra et al. 2007) did not agree with our results probable due to the low Fe contents in the soils analyzed in their studies. Approximately 25–30 % was Ca and Al bound P (i.e. Na-EDTA fraction) (Table 10.6), which was the second most abundant fraction after the Fe associated fraction. Soils with low Fe contents tend to show strong P_i association with Al (Saavedra et al. 2007; Gilbin et al. 2000). Dobermann et al. (2002) reported an increase in Fe and Al bound P_i fractions in Ultisols and Oxisols of Philippines and Indonesia fertilized with single and triple superphosphate to a depth of 15 cm. Similar results were reported by Cox et al. (1997), which showed surface applied sewage sludge increased Ca-Al and Fe associated P_i in the 0–7.5 cm soil depth. The Fe and Ca-Al associated P_i make up about 70 % of the total P_i in the soils studied. The P_i associated with these two fractions is tightly bound and relatively resistant to leaching thereby reducing the risk of ground and surface water contamination. Andraski et al. (2003) observed that soil test P levels were linearly related to dissolved P and bioavailable P loads in runoff from soils with long-term applied manure under NT and CT systems. The P_i tightly bound to these two fractions may not be available for plant or microbial uptake in the short term; however, in the long term it may be slowly released for uptake in times of high P_i deficiency. The acid soluble P_i (H_2SO_4 fraction) was about 10 % of the total P_i while organic matter (NaOH fraction) associated P_i accounted for about 9 % of the total P. The NaOH extractable P represents the P fraction associated with humic materials and phytate.

The low organic matter associated with P_i in the soils is because the soils are highly weathered with usually less than 3 % organic matter content (Senwo et al. 2003). Phosphorus studies in highly weathered tropical soils have indicated that organic matter is the most important source that provides labile P to plants (Tiessen et al. 1994; Selles et al. 1997). In conclusion, the amount of P_i extracted in the soils studied was in the order: Ca-EDTA > Na-EDTA > H_2SO4 > NaOH > water > residual.

10.4 Organic P Fractions in Alabama Cotton Soils

Results from this study showed that the water extractable P_o was significantly (contrast analysis) affected by tillage and manure applications compared to the control (Table 10.7). Neither tillage nor applied manure significantly affected the acid soluble, cold TCA, nor residual P_o fractions compared to the control (Table 10.7). Differences in P_o among treatments were only observed between CTRP vs NTRP for the water fraction and CTRP vs MTRP, MTRP vs NTRPP, and CTRP vs NTRP, for the Ca-EDTA P_o. Most of the P_o fractions were not significantly different amongst the treatments.

The water extractable P_o was among the least extractable fraction and accounted for less than 5 % of the total soil P_o (Table 10.7). The water extractable P_o fraction is an important source of P for microbes and plants in terrestrial and aquatic environments. McDowell and Koopmans (2006) suggested that the water soluble P fraction to be the least abundant fraction in some pastured and cultivated New Zealand soils. In our study, the Na-EDTA extractable P_o was the most abundant fraction with more than 40 % of the total P_o extracted in all fractions combined. McDowell and Koopmans (2006) showed that Na-EDTA extracted the second highest amount of P_o in some cultivated New Zealand soils.

Table 10.7 P_o fractions as affected by poultry litter amendments and tillage systems

Contrast	Water	CaEDTA	NaEDTA	H_2SO_4	TCA 0 °C	TCA 95 °C	NaOH	Residual
BF vs CTRP	8.70 vs 26***	46 vs 22 ns	72.1 vs 279 ns	20 vs 5.86 ns	15.8 vs 9.01 ns	18.5 vs 34.3***	73.1 vs 84.7 ns	18.5 vs 17.3 ns
BF vs MTRP	8.70 vs 20.2**	46 vs 103**	72 vs 504***	20 vs 14.0 ns	15.8 vs 11.7 ns	18.5 vs 29.3**	73.1 vs 107*	18.5 vs 16.6 ns
BF vs NTRP	8.70 vs 14.5 ns	46 vs 131***	72.1 vs 392**	20 vs 8.80 ns	15.8 vs 9.01 ns	18.5 vs 29.8**	73.1 vs 76.6 ns	18.5 vs 16.4 ns
BF vs NTRPP	8.70 vs 25***	46 vs 31 ns	72.1 vs 329*	20 vs 8.34 ns	15.8 vs 9.50 ns	18.5 vs 37.0***	73.1 vs 98.9 ns	18.5 vs 18.6 ns
CTRP vs MTRP	25.7 vs 20.2 ns	22 vs 103***	279 vs 505 ns	5.86 vs 14 ns	9.1 vs 11.7 ns	34.3 vs 29.3 ns	84.7 vs 107 ns	17.3 vs 16.6 ns
MTRP vs NTRP	20.2 vs 14.5 ns	103 vs 131 ns	505 vs 392 ns	14 vs 8.79 ns	11.7 vs 9.01 ns	29.3 vs 29.8 ns	107 vs 76.6 ns	16.6 vs 16.4 ns
MTRP vs NTRPP	20.2 vs 24.8 ns	103 vs 31**	505 vs 329 ns	14 vs 8.34 ns	11.7 vs 9.47 ns	29.3 vs 37.0 ns	107 vs 98.9 ns	16.6 vs 18.6 ns
CTRP vs NTRP	25.7 vs 14.5*	22 vs 131***	279 vs 392 ns	5.86 vs 8.9 ns	9.01 vs 9.01 ns	34.3 vs 29.8 ns	84.7 vs 76.6 ns	17.3 vs 16.4 ns
CTRP vs NTRPP	25.7 vs 24.8 ns	22 vs 310 ns	279 vs 329 ns	5.86 vs 8.0 ns	9.01 vs 9.47 ns	34.3 vs 37.0 ns	84.7 vs 98.9 ns	17.3 vs 18.6 ns

*, **, ***, significant at $P \leq 0.05, 0.01, 0.001$; respectively. *ns* not significant at $P = 0.05$

Approximately 15–26 % of the total P_o was extractable with NaOH in most treatments except for the NT soils, which had 40 % of its total P_o in this fraction. This fraction also accounted for 27 % of total P_o in the control (BF), making it the dominant fraction in both the NT and the control. The NaOH-P_o fraction was the second most abundant after the Ca-Al bound P_o. The NaOH-P_o fraction has been reported by several investigators (Daroub et al. 2000; Dobermann et al. 2002; Lehmann et al. 2005; McDowell and Koopmans 2006) to constitute the dominant fractions, especially in soils with significant amounts of organic matter contents.

The hot-TCA fraction (nucleic acid-P_o and polyphosphate) contributed >5 % of the total P_o with the exception of the MTRP treatments where it was <4 %. Sugar bound (cold TCA) P_o ranged from 2 to 8 % while acid soluble (H_2SO_4) P_o ranged from 1 to 11 % depending on the treatments. The residual fractions accounted for 2–5 % of the total P_o. The ranking of the various P_o fractions in all treatments was in the order: Na-EDTA-P_o > NaOH-P_o > Ca-EDTA-P_o > hot TCA-P_o ≥ Residual-P_o ≥ H_2SO_4-P_o ≥ water ≥ cold TCA-P_o.

10.5 Mineralization Potentials of Organic P in Soils Under Cotton Production

Adding specific enzymes with known activities to extracts from a fractionation step enables researchers characterize various P_o forms in the respective extracts. To estimate the labile P_o forms in sequential extracts of soils in the studies reported in this chapter, each of the extract was treated with phytase from two sources [wheat (PHW), *Aspergillus ficuum* (APP)], acid phosphatase from various sources [wheat germ (WGP), sweet potato (SPP), potato (PP)] and alkaline phosphatase (ALK P). In the water fraction, wheat phytase (PHW) hydrolysis of P_o was higher in most of the treatments compared to acid phosphatase and alkaline phosphatase hydrolysis of the water soluble P_o; although statistical analysis (means separations) revealed that the hydrolysis of the water soluble P_o by each enzyme within each treatment was not significantly different at $P < 0.05$ (Fig. 10.2). More water soluble P_o from the NT and MT treatments were hydrolyzed than from the CT treatment (Fig. 10.2).

The results reported in this study are consistent with earlier reports (Pant et al. 1994; Hayes et al. 2000; He and Honeycutt 2001, 2011; He et al. 2004a) and suggest that phytic acid was the major form of hydrolyzable P_o extracted in water. Surprisingly, *Aspergillus ficuum* phytase (APP) activity was far lower than that of PHW in most of the treatments. The low APP activity may suggest this enzyme is easily inhibited by metal ions (such as Pb, Hg, Cd), degraded or complexed in soils. Among the acid and alkaline phosphatase, the enzyme hydrolytic potentials were in the order of WGP > PP > Alk P ≥ SPP. Acid and alkaline phosphatase activities suggest the hydrolysis of simple liable monoester P compounds (He and Honeycutt 2001; He et al. 2004a). Overall, less than 13 % of the water extractable P_o from the

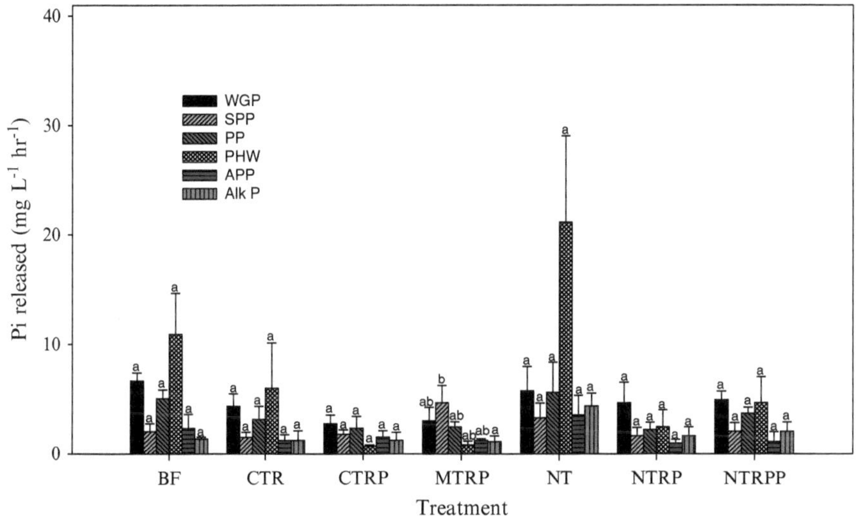

Fig. 10.2 Enzymes activities on water soluble organic P extracted from soils. *Vertical bars* represent the standard deviations of four field replicates. The incubation pH was 5.0 for wheat phytase (*PHW*), wheat germ acid phosphatase (*WGP*), sweet potato acid phosphatase (*SPP*), and potato acid phosphatase (*PP*); pH 2.5 for phytase from *Aspergillus ficuum* (*APP*); and pH 10.4 for alkaline phosphatase (*Alk P*). *Bars* with different *letters* are significant at $P < 0.05$

treatments that received poultry litter was hydrolyzed. Wheat phytase activity was higher than WGP, PP, and SPP activities in the water and Ca-EDTA extracts, and the Ca-EDTA fraction showed higher P_o concentrations than the water fraction (Data not shown). The high phytase values in these fractions suggest that phytic acid and other polyphosphates may be soluble and liable. The APP activity was lower than that of PHW in all the treatments suggesting that this enzyme may be easily inhibited in soils or unable to hydrolyze phosphates bond attached to colloidal materials. The WGP, PP, and SPP activity was similar to each other in the water extracted fraction, with 17–42 % of the P_o in the water extracted fraction hydrolyzed compared to 20–53 % for the phytase and 17–36 % for alkaline phosphatase.

The activities of PHW and WGP far exceeded that of SPP and PP in the Na-EDTA fraction (Fig. 10.3). The high PHW and WGP activities suggest that Na-EDTA extracted a range of P compounds that include phytic acid, simple monoesters, complex diesters, linear polyphosphates, and cyclic polyphosphates. However, the Na-EDTA extracted P_o is considered less available. The hydrolysis of P_o bound to Ca-Al suggests that this fraction may serve as a source of P_i after the depletion of the more readily available P_o. Wheat phytase and WGP hydrolyzed approximately 50–88 % of the P_o extracted in Na-EDTA compared to about 33–61 % for SPP and PP, and 18–53 % by APP.

Wheat germ acid phosphatase hydrolyzed more H_2SO_4 extracted P_o than PHW, PP, SPP, APP, and ALK (data not shown). Unlike in the Na-EDTA fraction where PHW hydrolyzed the greatest amounts of P_o extracted, the low PHW activity in the H_2SO_4 fraction suggests low concentrations of phytic acid and complex

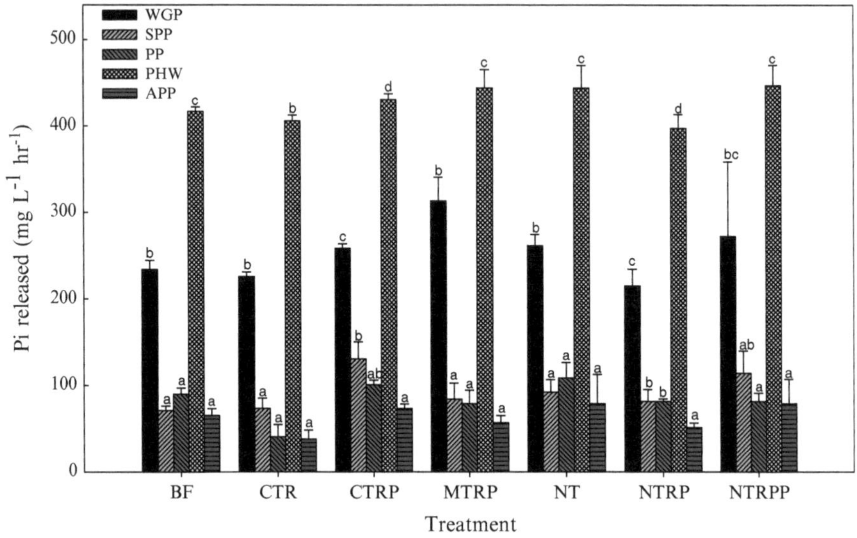

Fig. 10.3 Enzymes activities on Na-EDTA soluble organic P extracted from soil. *Vertical bars* represent the standard deviations of four field replicates. *Bars* with different *letters* are significant at $P < 0.05$ (Refer to Fig. 12.3 for enzyme abbreviations)

monophosphates and high concentrations of simple monophosphate in this fraction. Published data by Hayes et al. (2000) showed less than 13 % HCl extractable P_o was hydrolyzed by commercial and purified phytase. In contrast, 40–77 % of H_2SO_4-extractable P_o in this study was enzymatically hydrolysable (data not shown). Differences in the acid, enzyme preparation, and incubation techniques used might have contributed to such differences.

The NaOH-extractable organic P is the P_o mostly attached to organic matter and humic substances. The enzymes hydrolyzed 50–75 % of the NaOH-extractable P_o with the PHW activity being the highest (data not shown). The findings are in agreement with that of Pant and Warman (2000) and Bishop et al. (1994), but contrary to those reported by He et al. (2004a) where only 29–49 % of NaOH-extractable P_o was hydrolysable. Among all the fractions, NaOH P_o was the most sensitive fraction to all the enzymes (data not shown) which may suggest that this fraction is the primary source for plant or microbial available P_i after mineralization (Beck and Sanchez 1994; Zheng et al. 2002). Enzymes hydrolysis of NaOH extracted P_o was in the order: PHW > APP > SPP $\geq$ WGP $\geq$ PP > Alk P and statistically different.

10.6 Conclusion

By using NaOH-EDTA extraction and ^{31}P NMR analysis, typical soil P forms i.e. orthophosphate, pyrophosphate, monoesters including phytate, and diesters were observed in Alabama Decatur silt loam soils under cotton production.

The bulk of P in these soils was in the inorganic forms. The P_o extracted was mainly orthophosphate monoesters including phytate, with smaller concentrations of orthophosphate diesters and pyrophosphates. Deoxyribonucleic acid (DNA) was the major orthophosphates diester (ranged from 1.3 to 2.6 %) suggesting their resistance to biodegradation than RNA and phospholipids.

Sequential fractionation revealed that soil P_i in these cotton fields was mainly associated with Fe and Al oxides and is not readily bioavailable in the short run but may become available under appropriate environmental conditions. Soils treated with poultry litter showed elevated levels of water soluble, and Fe, and Al associated P_i compared with the control. The elevated water soluble P_i may be of environmental concern because of possible runoff with rain water and leaching into ground water leading to eutrophication of aquatic systems. Irrespective of treatments, the concentrations of P_i extracted followed the order: Ca-EDTA > Na-EDTA > H_2SO_4 > NaOH > water > residual. Chemical fractionation revealed that most soil P_o was associated with Al oxide (>40 %) except in the NT and BF treatments where the bulk of the soil P_o was organic matter bound. The NaOH-P_o was the next most abundant, and it has also been used to estimate short and long-term P availability of P to algae in lake sediments. As expected, water soluble P_o increased after poultry litter applications. The concentrations of P_o extracted followed the order: Na-EDTA > NaOH > Ca-EDTA > hot TCA $\geq$ Residual $\geq$ cold TCA $\geq$ H_2SO_4.

The quantity of water soluble P_o hydrolyzed by phytase, acid and alkaline phosphatases were minimal. This may suggest that only low levels of potential biomineralizable P_o occurred in soil solutions. Less than half of the P_o in Ca-EDTA was hydrolysable by the tested enzymes. Wheat phytase and WGP hydrolyzed 50–88 % of P_o in Na-EDTA compared to 33–61 % for SPP and PP. Simple monoester P was the major hydrolysable P in the H_2SO_4 fraction. Greater than 50 % of P_o in NaOH was hydrolysable by the various enzymes, which suggest that NaOH-extractable P_o is liable. Any portion of P_o extracted in the fractions that was not hydrolyzed by the enzymes is considered to be in more complex forms tightly bound to colloidal materials.

10.7 Future Research

It is reported that poultry litter can serve as a sole nutrient source for cotton production (Tewolde et al. 2010, 2011). Recently, He et al. (2013) reported that the K, Mg, and P contents of cotton seed harvested from a Falkner silt loam soil in Mississippi were higher in plots that received poultry litter compared with synthetic fertilizer and control treatments. The authors further observed that the contents of P, Mg, and K were highly correlated ($P < 0.01$), indicating these three elements were co-accumulated in cottonseed, probably in the form of the mixture of K/Mg phytate compounds. The results of our research showed that poultry litter applied to soils affected many of the soil P fractions. However, this effect is not well understood as the content of major P_o form, phytate, was not remarkably different between poultry litter applied soils and controls shown by data in the cotton soil in this work and other Alabama soils under pasture management (He et al. 2008). Thus, further research should be focused on poultry litter P supplying potential in soils used for cotton production and the impact of P supply on mineral contents in cotton seeds, or vice versa. Such research would shed light on more efficient use of poultry litter for environmentally-friendly and sustainable cotton productions in Alabama and other US southern states.

References

Acosta-Martinez V, Harmel RD (2006) Soil microbial communities and enzyme activities under various poultry litter application rates. J Environ Qual 35:1309–1318

Andraski TW, Bundy LG, Kilian KC (2003) Manure history and long-term tillage effects on soil properties and phosphorus losses in runoff. J Environ Qual 32:1782–1789

Beck MA, Sanchez A (1994) Soil phosphorus fraction dynamics during 18 years of cultivation on a Typic Paleudult. Soil Sci 34:1424–1431

Bishop ML, Chang AC, Lee RWK (1994) Enzymatic mineralization of organic phosphorus in a volcanic soil in Chile. Soil Sci 157:238–242

Cade-Menun BJ (2011) Characterizing phosphorus in animal waste with solution ^{31}P NMR spectroscopy. In: He Z (ed) Environmental chemistry of animal manure. Nova, New York, pp 275–299

Cade-Menun BJ, Preston CM (1996) A comparison of soil extraction procedures for ^{31}P NMR spectroscopy. Soil Sci 161:770–785

Cade-Menun BJ, Carter MR, James DC, Liu CW (2010) Phosphorus forms and chemistry in the soil profile under long-term conservation tillage: a phosphorus-31 nuclear magnetic resonance study. J Environ Qual 39:1647–1656

Chen CR, Condron LM, Davis MR, Sherlock RR (2002) Phosphorus dynamics in the rhizosphere of perennial ryegrass (*Lolium perenne* L.) and radiata pine (*Pinus radiata* D. Don.). Soil Biol Biochem 34:487–499

Condron LM, Frossard E, Tiessen H, Newman RH, Stewart JWB (1990) Chemical nature of organic phosphorus in cultivated and uncultivated soils under different environmental conditions. J Soil Sci 41:41–50

Condron LM, Turner BL, Cade-Menun BJ (2005) Chapter 4: Chemistry and dynamics of soil organic phosphorus. In: Sims JT, Sharpley AN (eds) Phosphorus, agriculture and the environment, Monograph no. 46. Soil Science Society of America, Madison, pp 87–121

Cox AE, Camberato JJ, Smith BR (1997) Phosphate availability and inorganic transformation in an alum sludge-affected soil. J Environ Qual 26:1393–1398

Daroub SH, Pierce FJ, Ellis BG (2000) Phosphorus fractions and fate of phosphorus-33 in soils under plowing and no-tillage. Soil Sci Soc Am J 64:170–176

Dobermann A, George T, Thevs N (2002) Phosphorus fertilizer effects on soil phosphorus pools in acid upland soils. Soil Sci Soc Am J 66:652–660

Doolette AL, Smernik RJ, Dougherty WJ (2010) Rapid decomposition of phytate applied to a calcareous soil demonstrated by a solution ^{31}P NMR study. Eur J Soil Sci 61:563–575

Eghball B, Binford GD, Baltensperger DD (1996) Phosphorus movement and adsorption in a soil receiving long-term manure and fertilizer application. J Environ Qual 25:1339–1343

Elser J, Bennett E (2011) A broken biogeochemical cycle. Nature 478:29–31

Gilbin R, Gomez E, Picot B (2000) Phosphorus and organic matter in wetland sediments: analysis through gel permeation chromatography (GPC). Agronomie 20:567–576

Giles CD, Cade-Menun BJ, Hill JE (2011) The inositol phosphates in soils and manures: abundance, cycling, and measurement. Can J Soil Sci 91:397–416

Golterman HL (1988) Reflections on fractionation and bioavailability of sediments bound phosphorus. Arch fur Hydrobiologie 30:1–4

Golterman HL (1996) Fractionation of sediments phosphate with chelating compounds: a simplification, and comparison with other methods. Hydrobiologia 335:87–95

Golterman HL, Paing J, Serrano L, Gomez E (1998) Presence of and phosphate release from polyphosphate or phytate phosphate in lake sediments. Hydrobiologia 364:99–104

Griffin TS, Honeycutt CW, He Z (2003) Changes in soil phosphorus from manure applications. Soil Sci Soc Am J 67:645–653

Hayes JE, Richardson AE, Simpson RJ (2000) Components of organic phosphorus in soil extracts that are hydrolysed by phytase and acid phosphatase. Biol Fertil Soils 32:279–286

He Z, Honeycutt CW (2001) Enzymatic characterization of organic phosphorus in animal manure. J Environ Qual 30:1685–1692

He Z, Honeycutt CW (2011) Enzymatic hydrolysis of organic phosphorus. In: He Z (ed) Environmental chemistry of animal manure. Nova, New York, pp 253–274

He Z, Griffin TS, Honeycutt CW (2004a) Enzymatic hydrolysis of organic phosphorus in swine manure and soil. J Environ Qual 33:367–372

He Z, Griffin TS, Honeycutt CW (2004b) Evaluation of soil phosphorus transformations by sequential fractionation and phosphatase hydrolysis. Soil Sci 169:515–527

He Z, Griffin TS, Honeycutt CW (2004c) Phosphorus distribution in dairy manures. J Environ Qual 33:1528–1534

He Z, Senwo ZN, Mankolo RN, Honeycutt CW (2006) Phosphorus fractions in poultry litter characterized by sequential fractionation coupled with phosphatase hydrolysis. J Food Agric Environ 4(1):304–312

He Z, Cade-Menun BJ, Toor GS, Fortuna A, Honeycutt CW, Sims JT (2007) Comparison of phosphorus forms in wet and dried animal manures by solution phosphorus-31 nuclear magnetic resonance spectroscopy and enzymatic hydrolysis. J Environ Qual 36:1086–1095

He Z, Honeycutt CW, Cade-Menun BJ, Senwo ZN, Tazisong IA (2008) Phosphorus in poultry litter and soil: enzymatic and nuclear magnetic resonance characterization. Soil Sci Soc Am J 72:1425–1433

He Z, Olk DC, Cade-Menun BJ (2011) Forms and lability of phosphorus in humic acid fractions of Hord silt loam soil. Soil Sci Soc Am J 75:1712–1722

He Z, Shankle M, Zhang H, Way TR, Tewolde H, Uchimiya M (2013) Mineral composition of cottonseed is affected by fertilization management practices. Agron J 105:341–350

Johnson NR, Hill JE (2010) Phosphorus species composition of poultry manure-amended soil using high-throughput enzymatic hydrolysis. Soil Sci Soc Am J 74:1786–1791

Kingery WL, Wood CW, Delaney DP, Williams JC, Mullins GL (1994) Impact of long-term application of broiler litter on environmentally related soil properties. J Environ Qual 23:139–147

Lehmann J, Lan Z, Hyland C, Sato S, Solomon D, Ketterings QM (2005) Long-term dynamics of phosphorus forms and retention in manure-amended soils. Environ Sci Technol 39:6672–6680

McDowell RW, Koopmans GF (2006) Assessing the bioavailability of dissolved organic phosphorus in pasture and cultivated soils treated with different rates of nitrogen fertilizer. Soil Biol Biochem 32:61–70

McDowell RW, Stewart I (2005) Phosphorus in fresh and dry dung of grazing dairy cattle, deer, and sheep: sequential fraction and phosphorus-31 nuclear magnetic resonance analysis. J Environ Qual 34:598–607

Nyakatawa EZ, Reddy KC (2000) Tillage, cover cropping, and poultry litter effects on cotton: 1. Germination and seedling growth. Agron J 92:992–999

Ohno T, He Z, Tazisong IA, Senwo ZN (2009) Influence of tillage, cropping, and nitrogen source on the chemical characteristics of humic acid, fulvic acid, and water-soluble soil organic matter fractions of a long-term cropping system study. Soil Sci 174:652–660

Ohno T, Hiradate S, He Z (2011) Phosphorus solubility of agricultural soils: a surface charge and phosphorus-31 NMR speciation study. Soil Sci Soc Am J 75:1704–1711

Ottabong E, Persson J, Iakimenko O, Sadovnikova L (1997) The Ultuna long-term soil organic matter experiment. 2. Phosphorus status and distribution in soils. Plant Soil 195:17–23

Pant HK, Warman PR (2000) Phosphorus release from soils upon exposure to ultra-violet light. Commun Soil Sci Plant Anal 31:321–329

Pant HK, Vaughan D, Edwards AC (1994) Molecular size distribution and enzymatic degradation of organic phosphorus in root exudates of spring barley. Biol Fertil Soils 18:285–290

Pierzynski GM, McDowell RW, Sims JT (2005) Chapter 3: Chemistry, cycling and potential movement of inorganic phosphorus in soils. In: Sims JT, Sharpley AN (eds) Phosphorus, agriculture and the environment, Monograph no. 46. Soil Science Society of America, Madison, pp 53–86

Rheinheimer DS, Anghinoni I, Flores AF (2002) Organic and inorganic phosphorus as characterized by phsophorus-31 nuclear magnetic resonance in subtropical soils under management systems. Commun Soil Sci Plant Anal 33:1853–1871

Richardson AE, Lynch JP, Ryan PR, Delhaize E, Smith FA, Smith SE, Harvey PR, Ryan MH, Veneklaas EJ, Lambers H, Oberson A, Culvenor RA, Simpson RJ (2011) Plant and microbial strategies to improve the phosphorus efficiency of agriculture. Plant Soil 349:121–156

Saavedra C, Velasco J, Pajuelo P, Perea F, Delgado A (2007) Effects of tillage on phosphorus release potential in a Spanish vertisol. Soil Sci Soc Am J 71:56–63

Sainju UM, Senwo ZN, Nyakatawa EZ, Tazisong IA, Reddy KC (2008a) Soil carbon and nitrogen sequestration as affected by long-term tillage, cropping systems, and nitrogen fertilizer sources. Agric Ecosyst Environ 127:234–240

Sainju UM, Senwo ZN, Nyakatawa EZ, Tazisong IA, Reddy KC (2008b) Tillage, cropping systems, and nitrogen fertilizer source effects on soil carbon sequestration and fractions. J Environ Qual 37:880–888

Sainju UM, Senwo ZN, Nyakatawa EZ, Tazisong IA, Reddy KC (2010) Poultry litter application increases nitrogen cycling compared with inorganic nitrogen fertilization. Agron J 102:917–925

Selles F, Kochhann RA, Denardin JE, Zentner RP, Faganello A (1997) Distribution of phosphorus fraction in a Brazilian Oxisols under different tillage systems. Soil Tillage Res 44:23–34

Senwo ZN, Taylor RW, Sistani KR (2003) Phosphorus distribution in five highly weathered soils. Commun Soil Sci Plant Anal 34:97–109

Stevenson FJ (1986) Cycles of soil (carbon, nitrogen, phosphorus, sulfur, micronutrients). Wiley, New York, pp 231–284

Sui Y, Thompson ML, Shang C (1999) Fractionation of phosphorus in a Mollisol with biosolids. Soil Sci Soc Am J 63:1174–1180

Sumann M, Amelung W, Haumaier L, Zech W (1998) Climate effects on soil organic phosphorus in the North America great plains identified by phosphorus-31 nuclear magnetic resonance. Soil Sci Soc Am J 62:1580–1586

Tazisong IA, Senwo ZN (2009) Mercury concentration and distribution in soils impacted by long-term applied broiler litter. Bull Environ Contam Toxicol 83:291–294

Tazisong IA, Senwo ZN, Taylor RW, Mbila MO, Wang Y (2004) Concentration and distribution of iron and manganese fractions in Alabama ultisols. Soil Sci 169:489–496

Tazisong IA, Senwo ZN, Taylor RW (2005) Trends in trace elements in an Ultisol impacted by long-term applied broiler litter. Bull Environ Contam Toxicol 75:975–981

Tazisong IA, Senwo ZN, Taylor RW, He Z (2008) Hydrolysis of organic phosphates by commercially available phytases: biocatalytic potentials and effects of ions on their enzymatic activities. J Food Agric Environ 6(3&4):500–505

Tazisong IA, He Z, Senwo ZN (2013) Inorganic and enzymatically hydrolysable organic phosphorus in Alabama Decatur silt loam soil cropped with upland cotton. Soil Sci 178:231–239

Tewolde H, Buehring N, Adeli A, Sistani KR, Rowe DE, Pratt RG (2010) Equivalency of broiler litter to ammonium nitrate as a cotton fertilizer in an upland soil. Agron J 102:251–257

Tewolde H, Adeli A, Sistani KR, Rowe DE (2011) Mineral nutrition of cotton fertilized with poultry litter or ammonium nitrate. Agron J 103:1704–1711

Tiessen H, Cuevas E, Chacon P (1994) The role of soil organic matter in sustaining soil fertility. Nature 371:783–785

Vats P, Bhattacharyya MS, Banerjee UC (2005) Use of phytases (*myo*-inositol hexakisphosphate phosphohydrolases) for combating environmental pollution: a biological approach. Crit Rev Environ Sci Technol 35:469–486

Wienhold BJ, Miller PS (2004) Phosphorus fractionation in manure from swine fed traditional and low-phytate corn diets. J Environ Qual 33:389–393

Wild A, Oke OL (1966) Organic phosphate compounds in calcium chloride extracts of soils: identification and availability to plants. J Soil Sci 17:356–371

Zheng Z, Simard RR, Lafond J, Parent LE (2002) Pathways of soil phosphorus transformations after 8 years of cultivation under contrasting cropping practices. Soil Sci Soc Am J 66:999–1007

Chapter 11
Chemistry and Application of Industrial By-products to Animal Manure for Reducing Phosphorus Losses to Surface Waters

Chad J. Penn and Joshua M. McGrath

Abstract Several industries produce by-products capable of sorbing phosphorus (P) that are typically placed in a landfill. These P sorbing materials (PSMs) can reduce soluble P concentrations when added to animal manure, thereby reducing the potential for non-point P transport to surface waters after land application of the manure. Addition of PSMs to manure also provides a beneficial re-use for the by-product material. This chapter reviews and discusses the use of PSMs as manure amendments to reduce soluble P concentrations. The by-product PSMs can generally be separated chemically into two groups: iron/aluminum and calcium/magnesium. However, the ability of a PSM to reduce manure soluble P concentrations is a function of the chemistry of both the PSM and the receiving manure.

11.1 Phosphorus Transport to Surface Waters and Animal Waste Management

One of the major challenges facing animal agriculture is the loss of nutrients, especially phosphorus (P), from manured fields to surface waters. Phosphorus is typically the most limiting nutrient in surface waters and therefore small inputs have the potential to cause a water body to become eutrophic. Transport of P to surface waters can occur by both subsurface movement and overland flow. Animal manure can potentially become a source of non-point P to surface waters directly or indirectly. Direct loss of P can occur from rainfall/runoff that interact directly with

C.J. Penn (✉)
Department of Plant and Soil Science, Oklahoma State University,
367 Agricultural Hall, Stillwater OK 47078-1020, USA
e-mail: chad.penn@okstate.edu

J.M. McGrath
Department of Environmental Science and Technology, University of Maryland,
H.J. Patterson Hall, College Park MD 20742, USA
e-mail: mcgrathj@umd.edu

Z. He and H. Zhang (eds.), *Applied Manure and Nutrient Chemistry for Sustainable Agriculture and Environment*, DOI 10.1007/978-94-017-8807-6_11,

stockpiled or spilled manure (Hively et al. 2005), or recently land applied manure (Warren et al. 2006; Penn et al. 2004; Withers et al. 2003). Indirect loss of manure P occurs from soils that have not received a recent manure application, but possess elevated soil test P concentrations due to historic manure applications (Schroeder et al. 2004; McDowell et al. 2001).

There are two broad categories of P that is transported in runoff: particulate and dissolved. Particulate P is P that is sorbed onto the surfaces of minerals and organic matter, and is therefore transported with eroded sediment or incidental manure particles. Particulate P transport is mostly controlled through best management practices (BMPs) that reduce erosion such as no-till agriculture and vegetated buffer strips (Sharpley and Smith 1991). Dissolved P is simply P that is already dissolved in solution, which makes it much more difficult to control its transport to surface waters. The source of dissolved P in runoff can be from recently applied chemical fertilizers and manure or from soils that are elevated in soil test P concentrations (Penn et al. 2005, 2006; Schroeder et al. 2004; Gaston et al. 2003; Pote et al. 1999). Certain BMPs such as vegetated filter strips that can reduce runoff volume can reduce some dissolved P transport, but the reductions are often minimal (Carlyle and Hill 2001; Osborne and Kovacic 1993). A study by Deng et al. (2011) examined the effectiveness of Hippophae rhamnoides (grass) as a vegetative filter strip for reducing pollutants in runoff. The authors found that while total P (which includes particulate P) was reduced by more than 70 %, the dissolved P losses actually increased; this increase was attributed to vegetative biomass at the surface. A big challenge with regard to dissolved P transport from high P soils is the longevity of P in such soils. In other words, soils that are elevated in soil test P will remain elevated for many years, even if all P applications cease. For example, Four years of corn production without P application decreased Bray-P soil test levels from 265 to 171 mg kg^{-1} (Eghball et al. 2003). The authors concluded that it would require 10 years of continuous drawdown in that soil to achieve a soil test level of 69 mg kg^{-1}. The P lost from high P testing soils is known as "legacy phosphorus".

Phosphorus transport from agricultural fields is a function of the landscape's hydrologic connectivity to surface waters (transport factors) and the amount and forms of P in the soils (source factors; Gburek et al. 2000). Transport factors controlling P loss are typically inherent to a site and more difficult to manage; whereas, P source characteristics can more easily be controlled by management decisions. For example, limiting P applications to meet crop needs and not exceed crop removal rates is an effective strategy to maintain soil P concentrations below levels of environmental concern. As expected, a mass balance approach where P additions are restricted to below P crop removal rates best achieves the goal of managing soil P to control environmental P losses (Daniel et al. 1994). However, P forms and soil characteristics may be just as critical as actual P concentrations in controlling source factors related to P loss. Numerous studies have showed that soil water Extractable P and forms of P added in manures or fertilizers, is strongly related to runoff dissolved P losses (McGrath et al. 2005; Penn et al. 2005; Vadas et al. 2005). In addition, when P sources are surface applied, such as is often the case with manures, the forms of P in the manures dominate P losses from those soils

(Penn et al. 2004; Warren et al. 2006). If organic sources of P are incorporated by tillage a combination of manure characteristics and soil chemistry together determine the amount of P available for loss (Shober and Sims 2007).

11.2 Treatment of Animal Manure with Chemicals to Reduce Phosphorus Solubility

Due to the importance of soil and manure chemistry in determining the potential for source contributions to environmental P losses, several soil and manure amendments have been studied in regard to their ability to sorb (adsorb and precipitate) P. These materials act to reduce water soluble P concentrations and potential dissolved P losses in runoff. Direct addition of P sorbing materials (PSMs) that are rich in Fe ($FeCl_3$) or Al ($Al_2(SO_4)_3$) are currently used in poultry production to suppress ammonia loss and have been shown effective at reducing P solubility in the manure and amended soils and thereby reducing the potential for dissolved P losses in runoff (Smith et al. 2001, 2004a, b; Shreve et al. 1995; Moore 1999; Moore et al. 2000; Moore and Edwards 2007; Warren et al. 2006; Kalbasi and Karthikeyan 2004). For example, Smith et al. (2004a) showed a 60 % reduction in runoff P losses when poultry litter was treated with aluminum sulfate compared to non-treated litter. The authors noted that the most important factor in determining P losses in runoff was dissolved P concentrations in the applied litter. Brennan et al. (2011a) significantly reduced dissolved P losses in runoff through addition of alum, poly-aluminum chloride hydroxide, ferric chloride, and calcium hydroxide to dairy slurry, in comparison to non-treated slurry. The authors also conducted an economic analysis and showed that alum and ferric chloride were the most cost efficient (i.e. lowest cost per kg P reduction). With regard to swine manure, Smith et al. (2001) also noted nearly 100 % reductions in manure dissolved P concentrations by amending with aluminum sulfate and aluminum chloride, which resulted in a decrease in runoff dissolved P concentrations from 5.5 to less than 1 mg L^{-1} after surface application. Several different chemical amendments can clearly decrease dissolved P concentrations in manure and therefore runoff after materials have been land applied. However, these studies used commercially produced Fe or Al materials. As an alternative, the use of industrial byproducts that have a strong affinity for P can reduce P solubility in manures and therefore decrease environmental losses of P. In addition, the beneficial reuse of byproducts could reduce the cost of their disposal.

11.3 By-product Phosphorus Sorbing Materials (PSMs)

By-products from several industries are rich in different forms of Fe, Al, Ca, and Mg that are able to sorb P from solutions, soils, and manures. Such industries produce by-products from mining, coal-fired power plants, metal casting, oil and

Table 11.1 Examples of several by-product phosphorus sorbing materials (PSMs) that may serve as manure amendments

By-product PSM	Main P sorbing elements	Studies involving application to manure
Fly-ash	Ca, Mg, Al, and Fe; varies	O'Flynn et al. (2012), Brennan et al. (2011b), Dao (1999), Dou et al. (2003), and Dao and Daniel (2002)
Blast furnace slag	Ca, Mg, Al, and Fe; varies	
Electric arc furnace slag	Ca, Mg, Al, and Fe; varies	
Melter slag	Ca, Mg, Al, and Fe; varies	
Bauxite waste (red mud)	Ca, Mg, Al, and Fe; varies	Wang et al. (2011)
Drinking water treatment residuals (WTRs)	Ca, Mg, Al, and Fe; varies	Brennan et al. (2011b), Dao et al. (2001), Codling et al. (2000), and Gallimore et al. (1999)
Titanium oxide manufacturing by-product	Fe and Al	Dao et al. (2001), and Codling et al. (2000)
Foundry sand	Mg, Fe, and Al	
Fe-coated sand	Fe	
Flue gas desulfurization (FGD) gypsum	Ca	O'Flynn et al. (2012), Brennan et al. (2011b), and Dou et al. (2003)
Mg fertilizer by-product	Mg	
Sandblast grit	Fe and Al	
Acid mine drainage residuals (AMDRs)	Ca, Al, and Fe; varies	Han et al. (2005), Fenton et al. (2009), and Adler and Sibrell (2003)

gas exploration, steel production, drinking water treatment, wall board production, fertilizer manufacture, and others. In general, one can split PSMs into two different categories based on P sorption mechanisms; Fe + Al and Ca + Mg based materials. Table 11.1 provides some examples of several by-products that may serve as a manure amendment PSM. Several of these materials are pictured in Fig. 11.1.

The Al and Fe rich PSMs often contain these elements in the form of oxides and hydroxides. Such minerals can strongly adsorb P by ligand exchange directly onto the surface; depending on the mineralogy and pH, these minerals may precipitate Al and Fe phosphates. At low pH, precipitation of P with Al and Fe may occur, but as pH increases, soluble Al and Fe precipitate into oxides and hydroxides that adsorb P via ligand exchange. On the other hand, PSMs rich in Ca and Mg dominantly sorb P by precipitation of Ca and Mg phosphates rather than ligand exchange; the degree of such precipitation generally increases with pH. One of the most important differences between Al + Fe and Ca + Mg rich materials is that Ca and Mg must first dissolve from the material (mineral) before they can precipitate P.

Example characteristics, including mineralogy, of several PSMs are shown in Table 11.2. Leader et al. (2008) characterized several PSM materials and reported

Fig. 11.1 Pictures of example phosphorus sorbing materials (*PSMs*) that are industrial by-products. From *top left* moving clockwise: fly ash, gypsum, acid mine drainage treatment residuals (*AMDRs*), and drinking water treatment residuals (*WTRs*)

ammonium oxalate extractable Al and Fe, and HCl extractable Ca and Mg. A P fractionation was also conducted on treated PSMs and showed that materials rich in Ca and Mg (such as a SuperMag byproduct) were dominated with HCl-Extractable P, which is indicative of Ca and Mg related phosphates. Similarly, treated PSMs that were elevated in Fe and Al (ore processing waste, Fe drinking water treatment residuals, and coated sand) were dominated with NaOH Extractable P, which is indicative of Fe and Al related P.

Acid mine drainage treatment residuals (AMDRs) have been shown to effectively remove P from solutions. Sibrell et al. (2009) used acid mine drainage residuals to remove dissolved P from solutions at several contact times and found that a 5 min or less was sufficient for removal of 60–90 % of solution P. Another study with AMDR materials showed that when the solid dosing rate was increased above 1 g L^{-1} for a 20 mg P L^{-1} effluent solution, P removal increased above 95 % (Wei et al. 2008). For both of those studies, the materials were dominated with Fe and Al, suggesting the P removal was achieved by either ligand with metal oxides/hydroxides or precipitation with Fe and Al. Other materials enriched with mostly Ca are likely to remove P via Ca phosphate precipitation. Ugurlu and Salman (1998) removed >99 % P in batch studies with fly ash. The fly ash used in their study contained both calcite (Ca carbonate) and gypsum (Ca sulfate), and after P treatment the samples were found to contain Ca phosphate as identified by X-ray

Table 11.2 Chemical characteristics of example by-product phosphorus sorbing materials (PSMs) amended to manure samples described in Table 11.3 and Fig. 11.2

	Total[a]					Amorphous[b]	Water soluble				EC[c]	
	Ca	Mg	Fe	Al	Mn	Fe + Al	Ca	Mg	Na			
By-product	$g\ kg^{-1}$				$mg\ kg^{-1}$	$g\ kg^{-1}$	$mg\ kg^{-1}$			pH	($mS\ cm^{-1}$)	Crystalline minerals[d]
AMDR1	301.0	6.9	88.1	1.6	1,715	6.1	411	608	2,964	8.2	3.3	Calcite
AMDR2	22.6	38.1	102.8	55.5	600	17.3	66	117	3,757	9.1	3.4	ND[e]
WTR	5.2	4.6	31.6	76.4	1,868	13.7	1,485	173	162	7.5	2.2	Muscovite
Fly ash	38.6	2.7	12.0	26.5	102	2.0	4,364	56	129	8.2	1.5	Calcite
Bauxite residual	12.7	2.2	55.6	77.8	126	9.0	10	3	4,418	9.8	4.2	Hematite
FGD	198.4	0.6	1.3	0.5	11	0.1	5,890	5	22	7.9	2.1	Gypsum
LSD[f]	8.9	1.6	6.8	6.1	38	1.0	356	10	153	0.12	0.17	–

[a]Determined by EPA 3051 digestion method
[b]Determined by ammonium oxalate extraction
[c]Electrical conductivity
[d]Determined by X-ray diffraction
[e]None detected
[f]Least significant difference at $p = 0.05$

diffraction. Electric arc furnace (EAF) slag was shown by Drizo et al. (2002) to sorb 1.35 g P kg^{-1} over 278 days, with most of the P sorption occurring in the Fe, Al, and Ca fractions of the slag. Cucarella and Renman (2009) compiled a review of several different PSMs, including their capacity to sorb P directly from solution. The authors noted that pH and particle size of the PSMs had a great impact on their ability to remove P from solution.

Penn et al. (2011) investigated the mechanisms of P sorption onto six different PSMs (Table 11.2) by adding solution P to them and conducting a P fractionation before and after sequentially desorbing the previously P treated PSM four times with anion exchange membranes. This analysis also showed what P pools were most susceptible to desorption. The results illustrated that the capacity and mechanism of a PSM to sorb P was not simply a function of the total Ca, Mg, Al, and Fe. Instead, P sorption was a function of the chemical environment (pH, buffering capacity, ionic strength, common ion effects), forms (mineralogy) that the P sorbing elements were in (for example, calcite vs. gypsum), and the "exposure" of the P sorbing elements to solution (i.e. occluded vs. non-occluded). In regard to P retention, the results showed the P bound to Fe was more resistant to desorption compared to Ca and Al-bound P.

In addition to the main P sorbing elements, other properties of the PSMs may influence potential uses. For example, excessive EC, soluble Na, F, trace metals and extreme pH buffering could have adverse environmental impacts when applied or utilized in soil systems or natural waters (Udeigwe et al. 2011). The AMDR2 and Bauxite residual shown in Table 11.2 both possess elevated soluble Na concentrations resulting from the use of NaOH in producing these materials. Bauxite residuals are known to have excessive pH, Na, and EC; as a result they are typically diluted with other materials before land application (Summers et al. 1996; Ippolito et al. 2005). In addition, contaminants such as heavy metals and certain organic chemicals can also have a negative impact on the environment and the ability of a soil to grow crops. When compared to EPA limits for biosolids applications (EPA 2009), all materials shown in Table 11.2 would be considered "exceptional quality" due to low trace metals concentrations (metals concentrations from Penn et al. 2011). Although total trace metal concentrations are often correlated with bioavailability, total trace metal concentration does not necessarily provide information on bioavailability upon addition to soils or manures. Therefore, caution should be exercised in use of such materials. Fenton et al. (2009) used an AMDR material to remove P from dairy effluent and found that several trace metals were released to solution at levels exceeding drinking water standards. A P-enriched AMDR material applied to Scottish soils, however, did not introduce trace metals beyond safe levels (Dobbie et al. 2005). The same author (Dobbie et al. 2009) also found no detectable trace metals in wastewater that was treated with an AMDR. Although not measured in this study, coal combustion byproducts such as fly ash and FGD have been shown to possess some radioactivity due to decay of Ra^{226}, Ac^{228}, K^{40}, U^{238}, and Th^{234} (Jala and Goyal 2006; Zielinski and Budahn 1998; Baba 2002). All PSMs should be characterized for environmental safety prior to use in the field.

11.4 Use of By-product PSMs for Reducing Soluble Phosphorus in Manure

As expected, since PSMs rich in Al, Fe, Ca, or Mg can sorb P, application of these materials to manure can reduce soluble P concentrations in manure-PSM mixtures. Compared to pure chemicals, less research has been conducted on the use of by-products for reducing P solubility in animal manures. Table 11.1 lists several studies in which by-products have been tested for reducing manure soluble P concentrations. Through application of a drinking water treatment residual (WTR) and a Fe-rich residual from titanium processing, Codling et al. (2000) was able to reduce water soluble P concentrations in poultry litter from 13 to 88 % depending on application rate and incubation time. Dao and Daniel (2002) reported a reduction in dairy manure dissolved P concentration from 1,164 to 652 mg kg^{-1} after addition of 10 g kg^{-1} of fly ash. Similarly, application of fly ash to cattle feedlot manure at a rate of 100 g kg^{-1} reduced water soluble P concentrations 85 and 93 % for stockpiled and composted manure, respectively (Dao 1999). Other studies have shown that AMDRs can be very effective at reducing P solubility in manure (Fenton et al. 2009; Adler and Sibrell 2003). Dou et al. (2003) found that aluminum sulfate consistently reduced soluble P concentrations in swine, dairy, and broiler manure, while flue gas desulfurization (FGD) gypsum and fly-ash varied in P reductions depending on application rate and manure type. Results suggested that P sorption was a function of both PSM chemical characteristics and manure chemistry.

Figure 11.2 shows the change in manure (poultry, swine, and dairy) P solubility with the application of various by-products.

Table 11.3 provides the equation parameters for the relationship between PSM application rate and the final water soluble P concentration shown in Fig. 11.2. Increasing the rate of byproduct added to each manure generally resulted in a greater amount of P sorbed (Fig. 11.2 and Table 11.3). However, some PSMs reached their maximum effectiveness at relatively low application rates (Fig. 11.2 and Table 11.3). This approach provides a means of deriving a recommended rate of addition for a specific PSM for a specific manure type.

11.5 Impact of PSM Additions on Manure Organic Phosphorus Forms, Stability, and Potential Plant Availability

The addition of by-products could have a profound impact on the stability of manure organic P, mainly through the addition of Al and Fe. This is especially true for phytic acid (inositol hexaphosphoric acid, IP6), which is considered to be the predominant form of organic P in animal manure (He et al. 2006a). Phytic acid reacts with metal cations to form metal-phytate compounds, which can have reduced solubility depending on the metal cation (Dao 2003; He et al. 2006b).

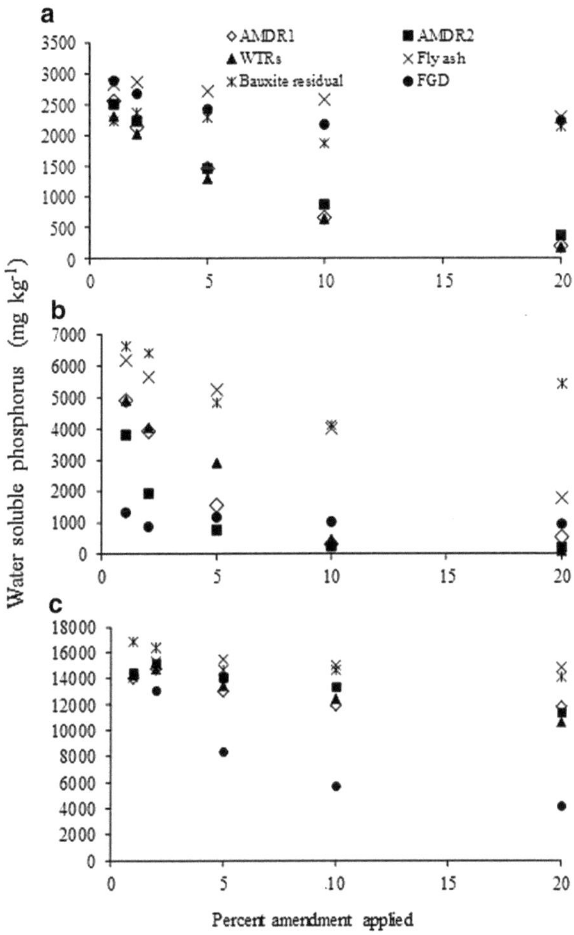

Fig. 11.2 Water soluble phosphorus (*WSP*) concentrations among dairy (**a**), swine (**b**), and poultry (**c**) manure samples amended with P sorbing materials (*PSMs*) shown in Table 11.2, as a function of PSM amendment rate

The six phosphates held onto the ring structure of IP6 can be released to solution, and therefore available for transport or plant uptake, only after being cleaved by phytase enzyme (Dao 2004a). Several organic ligands such as oxalate and citrate can also contribute to the release of P from these metal-phytate compounds (Dao 2004a, b). Organic ligands and phytase enzyme occur naturally in soils as they are excreted by microorganisms and plant roots. Certain organic ligands that are not produced in nature, such as EDTA, are extremely effective at contributing to the release of P from metal-phytate compounds (Dao 2004a).

Dao (2003) found that polyanionic IP6 had a strong affinity for cations and formed insoluble precipitants with them. Dephosphorylation was increasingly inhibited as counterion concentration increased the cation: IP6-P mole ratios to 6:6 in the order of Fe^{3+}, Al^{3+}, and Ca^{2+} (least). In a study involving the addition of several metal cations to solution of IP6, Dao (2004a) showed that the dephosphorylation of IP6 is inhibited as the solution Al and Fe concentration increases. The authors also noted that in

Table 11.3 Statistical comparison of the slope and intercepts obtained from linear regression of material amendment rate with soluble P concentration for each manure type shown in Fig. 11.2

Material	Slope	Intercept	R^2
Dairy			
AMDR1	−118 a	2299 bcd	0.87
AMDR2	−108 a	2309 bcd	0.88
WTRs	−109 a	2115 cd	0.88
Fly-ash	−29 b	2879 a	0.99
Bauxite residual	−12 b	2272 cd	0.23
FGD	−30 b	2711 ab	0.62
Swine			
AMDR1	−212 ab	3863 bc	0.63
AMDR2	−146 abc	2504 cd	0.55
WTRs	−250 a	4375 bc	0.82
Fly-ash	−226 a	6291 a	0.99
Bauxite residual	−62 bc	5951 ab	0.2
FGD	−11 c	1147 d	0.21
Poultry			
AMDR1	−168 b	14166 b	0.72
AMDR2	−179 b	15079 b	0.95
WTRs	−208 b	14742 b	0.97
Fly-ash	5 c	14931 b	0.006
Bauxite residual	−129 b	16311 a	0.68
FGD	−516 a	13061 b	0.81

Materials with a different letter indicate a significant difference at $P = 0.05$

addition to complexation of Al^{3+} and Fe^{3+} with P of IP6, amorphous metal hydroxides can sorb and protect IP6 from enzymatic hydrolysis. Such a reaction would be especially important through the addition of by-products to manures since many industrial by-products are rich in Al and Fe oxides/hydroxides. While almost all IP6-P (90 %) was released by addition of 3-phytase in 30 h, only 50 % was released when the ratio of IP6-P:Fe was 6:2; less than 3 % was hydrolyzed when the ratio was increased to 6:6 (He et al. 2006a). A similar effect was observed for Al.

Although not a by-product, Warren et al. (2008) found that aluminum sulfate application inhibited the hydrolysis of decomposing poultry litter compared to un-treated litter. Through nuclear magnetic resonance (NMR) spectroscopy and wet fractionation techniques, the authors showed that while IP6 in un-treated litter degraded and released P to the water soluble pool, the alum treatment transformed 30 % of the organic P from the HCl to NaOH extractable organic pool and that both P fractions were more persistent after 93 days of incubation. Note that the NaOH extractable pool represents the P bound with Al and Fe. After both the alum treated and un-treated degraded litter samples were amended to soils, the NaOH extractable organic P pool was found to be recalcitrant after 63 days of incubation. This suggests that the newly formed Al-IP6 compounds are not only less soluble than normal IP6 found in poultry manure, but may be less plant available as well.

Little information is available with regard to the plant availability of P from amended manure samples, and we are not aware of any studies regarding plant P availability from by-product PSM treated manures after land application. Although not a by-product, Warren et al. (2006) found that while alum treatment of poultry litter appreciably reduced water soluble P concentrations in the litter, land application to tall fescue did not decrease P uptake or forage production compared to non-treated litter. Similarly, Moore and Edwards (2007) noted that alum-treated litter land applied each year since 1995 did not decrease P uptake or forage yield of tall fescue.

11.6 Variation in Efficacy of Manure Soluble Phosphorus Reductions as a Function of Manure and PSM Properties

While information on P sorption by PSMs in pure inorganic solutions can provide useful information in regard to prediction of P removal performance when amended to manure samples, manure chemistry can have an appreciable impact on the efficacy of PSMs.

For example, although the PSMs shown in Table 11.2 were found to be highly effective at removing P directly from pure P solutions (Penn et al. 2011) manure type influenced the effectiveness of certain PSMs to sorb and retain P (Fig. 11.2). For example, FGD gypsum was one of the least effective materials in sorbing P from dairy manure, yet it was by far the most effective PSM for poultry manure and was more effective at lower rates than other PSMs when added to swine manure. In regard to dairy manure, AMDR1, AMDR2, and WTR were the most effective at P sorption. For swine manure, all materials were relatively effective, with the exception of the bauxite residual. The FGD gypsum, which sorbs P primarily by forming Ca phosphates (i.e. HCl extractable), was most effective for poultry manure. Acid mine drainage residual 1, AMDR2, and WTR, all of which demonstrated moderate degrees of formation of Ca/Mg phosphates with inorganic P in solution (Penn et al. 2011), were only somewhat effective in poultry manure when applied at higher rates (Fig. 11.2). Thus, the sorption of P from manure onto PSMs is a function of the properties of the PSMs and the P forms of the non-amended manures. The forms of P in three manure types before ("control") and after PSM additions are presented in Fig. 11.3. Swine manure typically contains an appreciable proportion of water soluble P (Leytem et al. 2004; Kleinman et al. 2005). The remaining P in the un-amended swine manure shown in Fig. 11.3 was relatively evenly distributed among the other P fractions. The dairy manure contained an intermediate amount of water soluble P, but the Fe/Al bound P dominated the P fractions.

Poultry manure was the driest manure and had the lowest water soluble P concentration; in contrast the Ca/Mg bound P fraction dominated the poultry manure P fractions. This is due to the large Ca and Mg concentrations in poultry manure derived from inorganic supplements provided in poultry feed and the favorable pH for Ca/Mg phosphate formation (Maguire et al. 2004; McGrath et al. 2005). The elevated Ca and Mg of the poultry manure can also inhibit P solubility in water. A general characterization of the three manure samples amended with PSMs presented in Figs. 11.2 and 11.3 is shown in Table 11.4.

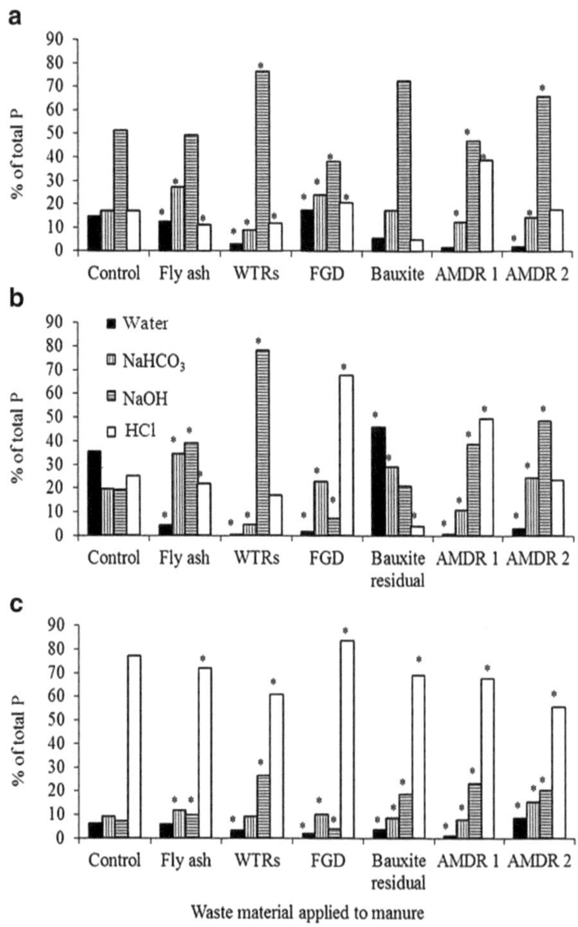

Fig. 11.3 Changes in phosphorus (P) distribution for dairy (**a**), swine (**b**), and poultry (**c**) manure after addition of 20 g 100 g^{-1} (20 % by weight) of P sorbing materials (PSMs) to manure, shown in Fig. 11.2. PSM and manure characteristics are shown in Tables 11.2 and 11.4, respectively (*Indicates a significant decrease or increase in P form compared to un-amended manure (*control*) at P = 0.05)

Phosphorus forms are dynamic in the manure chemical environment (Fig. 11.3). From the perspective of water quality protection, the ideal shift would be from the water Extractable P and $NaHCO_3$ Extractable P (considered weakly bound Ca, Mg, Fe, and Al related P) fractions into the NaOH extractable (Fe/Al bound P) and HCl extractable (Ca/Mg bound P) fractions. The NaOH and HCl extractions of the Hedley fractionation also remove organic P associated with Fe/Al and Ca/Mg (Turner and Leytem 2004; Warren et al. 2008; Seiter et al. 2008). However, as previously discussed, such a shift to a less soluble P pool will vary with the chemistry of the PSM and the manure.

While chemical properties of the PSM-manure mixture will dictate changes in P solubility, especially pH, one should also consider nitrogen. If the manure contains appreciable nitrogen in the form of ammonium, then an excessive pH in the PSM-manure mixture could result in loss of nitrogen through ammonia volatilization. This would decrease the monetary value of the manure since nitrogen is required in large amounts by growing plants. On the other hand, acidifying

Table 11.4 Total elemental analysis, pH, and solids contents of the animal manures amended with phosphorus sorbing materials (PSMs) shown in Figs. 11.2 and 11.3, and Tables 11.2 and 11.3

	Total N	Organic N	P	K	Ca	Mg	S	Na	Cu	Zn	Mn	Fe	Al	pH	solids
Manure	$g\ kg^{-1}$								$mg\ kg^{-1}$					%	
Dairy	26.4	17.1	4.2	1.45	13.9	4.9	2.8	3.3	26	103	129	388	0	7.4	19.3
Swine	122.1	28.4	36.3	4.91	47.1	16.9	10.4	11.5	521	1,302	521	3,906	781	7.3	1.9
Poultry	44.0	38.2	45.5	3.17	70.9	16.0	7.2	9.8	95	856	1,394	998	0	7.5	32

Results are presented on a dry weight basis

amendments have been shown to stabilize ammonium-nitrogen thereby preventing volatilization (Sims and Luka-McCafferty 2002; Moore et al. 2000).

11.6.1 Reducing Manure Phosphorus Solubility with Calcium-Rich Materials

With regard to Ca-rich PSMs, the effectiveness of PSMs at reducing soluble P concentrations in manure by Ca phosphate precipitation will be a function of several manure and PSM characteristics. For manure, this includes P and Ca content and forms, pH, and pH buffer capacity. For PSMs, factors such as Ca content and forms, pH, and pH buffer capacity will be influential. For a Ca-rich PSM to be effective at reducing soluble P concentrations, it must be able to dissolve Ca into solution and maintain a pH of 7 or above (Stoner et al. 2012; Vohla et al. 2011). However, as more Ca is released to solution from the PSM, a lower pH will be required to precipitate Ca phosphate minerals (Lindsay 1979). If the manure-PSM mixture results in an elevated pH, this could prevent the Ca in the PSM from being released, depending on the type of Ca minerals present, therefore reducing the PSM effectiveness. This will be a function of the pH buffer capacity of both materials and Ca forms in the PSM. Three common Ca bearing minerals found in several PSMs include the following:

$$CaCO_3 + 2\,H^+ \leftrightarrow Ca^{2+} + CO_{2(g)} \tag{11.1}$$

$$CaSO_4 \leftrightarrow Ca^{2+} + {SO_4}^{2-} \tag{11.2}$$

$$CaO + 2H^+ \leftrightarrow Ca^{2+} + H_2O \tag{11.3}$$

The Ca^{2+} produced from these dissolution reactions can then precipitate with manure P. Consider the following reaction:

$$Ca^{2+} + H_2PO_{4^-} + 2H_2O \leftrightarrow CaHPO_4 2H_2O_{(solid)} + H^+ \tag{11.4}$$

These Ca minerals shown in reactions 11.1, 11.2, and 11.3 differ from each other in regard to solubility, pH buffering capacity, and the impact of pH on their solubility. $CaCO_3$ minerals are generally less soluble than gypsum at pH 7 and above, while $CaSO_4$ minerals on the other hand, have minimal impact on increasing pH compared to $CaCO_3$. The CaO and $Ca(OH)_2$ minerals are highly soluble at pH levels of 9 and below, which not only release Ca^{2+} to the solution for precipitation of Ca phosphates, but it also maintains an elevated pH that is favorable for Ca phosphate precipitation. Equation 11.4 illustrates the fact that Ca phosphate minerals are generally able to form more readily at higher pH compared to low pH. The notion of increased solution Ca concentrations and pH on precipitation

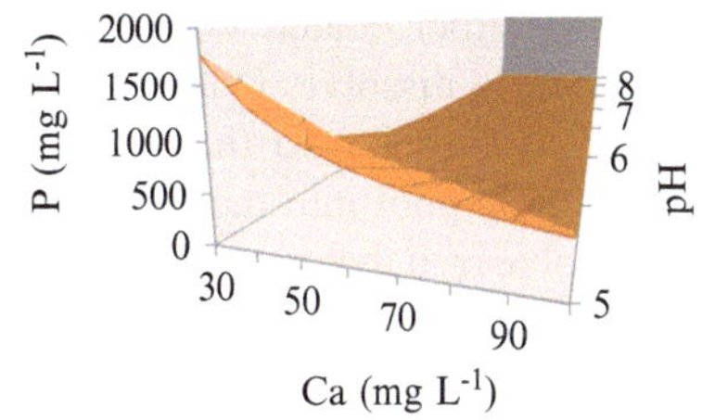

Fig. 11.4 The dependence of calcium phosphate on pH and equilibrium calcium concentrations. The calcium phosphate mineral in this example is brushite

of Ca phosphate minerals is shown visually in Fig. 11.4. Although not using manure as the P source, Johansson (1999) found that Ca-rich PSMs containing Ca in the form of CaO were more effective at P sorption compared to $CaCO_3$ rich materials.

For example, Moore and Miller (1994) found that the addition of Ca in the form of CaO and $Ca(OH)_2$ were successful at decreasing soluble P concentrations in poultry litter, while $CaCO_3$ and $CaMg(CO_3)_2$ additions had little impact. The difference in effectiveness between the different Ca minerals is due to the much higher solubility of CaO and $Ca(OH)_2$ compared to $CaCO_3$ and $CaMg(CO_3)_2$ at the pH level of the poultry litter (8.2). However, it should be noted that the stability of freshly formed Ca phosphates at a pH >8.2 is unknown; it is possible that P could be slowly released from these Ca phosphate minerals since carbonate is an effective competitor with phosphate for Ca at this high pH level.

Figures 11.2 and 11.3 provide several illustrations of the importance of manure and PSM characteristics on reducing manure soluble P concentrations with Ca-rich PSMs. For amended dairy manure, acid mine drainage residual 1, which contained calcite ($CaCO_3$), appeared to have sorbed P mainly by reaction with Ca (HCl extractable; Fig. 11.3), whereas the other Ca rich PSMs (fly ash and FGD gypsum; Table 11.2) were not as effective at sorbing P with Ca and Mg. This lack of P sorption by the Ca-rich FGD gypsum and fly ash may be due to differences in pH buffering; pH of the PSM-amended manures was 6.7, 6.7, and 7.3 for fly ash, FGD, and AMDR1, respectively. Acid mine drainage residual 1 was able to supply Ca/Mg to solution and maintain a suitable elevated pH, thereby promoting greater precipitation of Ca phosphate. Kostura et al. (2005) found a high correlation between acid-neutralizing capacity and P-sorption capacity of Ca-rich PSMs. Although AMDR materials are typically rich in Al and Fe, they can also contain significant amounts of Ca in the forms of either $CaCO_3$ or $CaSO_4$, depending on the source AMD water and the treatment process used to neutralize the AMD. For example, Adler and Sibrell (2003) created AMDRs by neutralizing AMD with either NaOH or Ca materials (CaO and $Ca(OH)_2$). They found that AMDRs that were produced using Ca were more effective at reducing soluble P concentrations in rainbow trout (*Oncorhynchus mykiss*) manure.

Similar to dairy manure shown in Figs. 11.2 and 11.3, the Ca rich fly ash did not sorb P from the swine manure by reaction with Ca/Mg even though the pH of the mixture (7.7) was sufficient to form Ca/Mg phosphates. This suggests that the

Ca was either coated with Al minerals or the pH was too high for calcite to sufficiently dissolve. Dou et al. (2003) also found that in swine manure, fly ash failed to sorb P into the Ca/Mg-P fraction; instead, P shifted from the water extractable pool into the $NaHCO_3$ extractable, weakly bound fraction. It should be emphasized that not all fly ash is the same, and the results described are specific to that particular sample only. Several authors have shown that fly ash is able to sorb P by forming Ca-phosphate minerals (Ugurlu and Salman 1998; Stout et al. 1998).

The results in Figs. 11.2 and 11.3 illustrate the importance of manure P forms on the mechanism for PSMs to sorb P. For example, all of the PSMs except for the high rate of bauxite residual were able to sorb high amounts of P from the swine manure (Fig. 11.2 and Table 11.3). Since much of the swine manure P was water soluble and the pH was not strongly buffered, both of the two main groups of P sorption elements (Fe/Al and Ca/Mg) could sorb significant amounts of P (Fig. 11.2). In fact, results from the swine manure amendments were similar to those of Penn et al. (2011) where inorganic P was reacted with these PSMs. As expected, the fly ash, WTR, AMDR1, and AMDR2 sequestered P into the Fe/Al-P fraction (NaOH extractable; Fig. 11.3). Calcium and Mg-rich materials, FGD gypsum and AMDR1, sorbed P into the Ca/Mg-P fraction (Fig 11.3).

It is interesting to contrast the effectiveness of the PSMs (Table 11.2) in different manure types, particularly between poultry litter, and swine and dairy since the poultry manure sample was dominated by Ca phosphates (HCl extractable pool; Fig. 11.3). Although Penn et al. (2011) found that FGD gypsum sorbed the least amount of P from inorganic solutions, it was the only PSM tested among Table 11.2 that effectively reduced water Extractable P in poultry manure to levels that may be expected to reduce P losses in runoff (Fig. 11.2 and Table 11.3). When amended with FGD gypsum, P shifted from the water extractable fraction to the Ca/Mg bound P fraction (HCl extractable; Fig. 11.3). Because poultry manure contained elevated Ca and Mg concentrations compared to the other manure types (Table 11.4), the other Ca/Mg-rich PSMs (fly ash and AMDR1; Table 11.2) were probably not able to sorb P by this mechanism due to the common ion effect and likely a buffered pH near 8 that prevented the dissolution of calcite. The FGD gypsum was therefore more soluble in the high Ca/Mg environment of the poultry manure compared to other PSMs and provided Ca/Mg to solution that reacted with P (gypsum is more soluble than calcite and its solubility is not as dependent on pH). This is similar to the results of Moore and Miller (1994), where poultry litter was amended with either $CaCO_3$ (calcite), $CaSO_4$ (gypsum), CaO, and $Ca(OH)_2$.

Beyond the initial formation of Ca phosphates, Maurer et al. (1999) found that the Ca phosphates precipitated at a pH below 8 were more soluble than those formed at a pH above 8. The authors precipitated P from domestic wastewater through simple manipulation of pH. The more "reversibly" formed Ca phosphates that formed at pH <8 were hydroxydicalcium phosphate, while hydroxyl apatite was considered to be less reversible in solubility.

11.6.2 Reducing Manure Phosphorus Solubility with Iron and Aluminum-Rich Materials

Phosphorus sorbing materials dominated with Fe and Al typically sorb P by ligand exchange reactions onto the surface of the metal oxides and hydroxides. While precipitation of Al and Fe phosphate minerals is possible, it is uncommon at the typical pH values of most animal manure-PSM mixtures (i.e. near neutral or above). This is due to the fact that Fe and Al are only in the free cation form (Al^{3+} and Fe^{3+}) at relatively acidic pH levels. For example, Peak et al. (2002) showed that aluminum sulfate additions to poultry litter precipitated amorphous Al hydroxide minerals which were then able to sorb P by ligand exchange.

The effectiveness of Al and Fe-rich PSMs are dependent on pH because Al and Fe oxides and hydroxides possess variable charges. Specifically, the point of zero charge (PZC) for most Al and Fe oxides and hydroxides is typically between pH 7 and 9 (Zelazny et al. 1996). Due to the decreasing positive charge with increasing pH, and also because of increased competition with solution hydroxyls, excessive pH levels can greatly reduce P sorption by PSMs rich in Al and Fe. For example, while increased application of a Fe-rich by-product, ferrihydrite, increased the P sorption capacity of soils, P removal decreased as the pH increased (Rhoton and Bigham 2005). The authors attributed this to the impact of pH on the variable charged ferrihydrite mineral. Bowden et al. (2009) used basic oxygen steel slag to remove P from simulated wastewaters. Although the slag contained 42–44 % CaO and 27–31 % FeO/Fe_2O_2, there was little evidence of P removal via Fe sorption ascertained through XRD, E-SEM, and EDX analysis, which showed only Ca phosphate minerals. Authors also noted that since P sorption only increased with increasing pH, P sorption was mostly due to the Ca portion of the slag. This was likely due to the elevated pH of the slag materials. Similar results were noted by Claveau-Mallet et al. (2011) in a study that involved sorption of P with an electric arc furnace steel slag that contained nearly equal amounts of CaO and Fe_2O_3.

The potential impact of a PSM on the resulting manure-PSM mixture pH is extremely important with regard to animal manures dominated with P in the form of Ca phosphate. For the normal pH range (4.5 to 8), Ca phosphates solubility increases with decrease in pH; however, as pH increases above 8 the solubility increases again due to increased competition of carbonate with phosphate for Ca (Lindsay 1979). Figure 11.5 shows a seven-times increase in solution P concentration for a Ca rich dairy manure when the pH is decreased from 7.4 to 4.5. This illustrates the importance of the potential impact of PSM pH buffering on soluble P in manure-PSM mixtures.

If the pH of manure is not strongly buffered, then an added PSM will remove P mainly as a function of the PSM characteristics (i.e. reactive elements and pH). For example, all of the PSMs except for the high rate of bauxite residual were able to sorb high amounts of P from the swine manure illustrated in Fig. 11.2 and Table 11.3. Since much of the swine manure P was water soluble and the pH was not strongly buffered, both of the two main groups of P sorption elements (Fe/Al and Ca/Mg) could sorb significant amounts of P (Fig. 11.2). In fact, results from the PSM amended swine manure were similar to those of Penn et al. (2011) where inorganic P was reacted with

Fig. 11.5 Changes in solution phosphorus (P) concentrations of a dairy slurry (3 % solids) after acidification with HCl

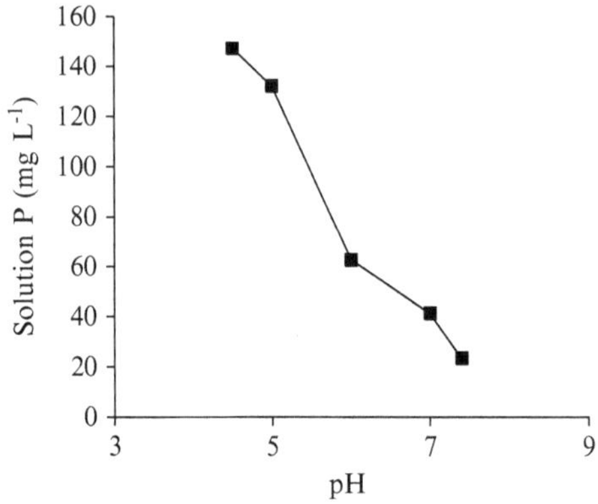

these PSMs. As expected, the fly ash, WTR, AMDR1, and AMDR2 sequestered P into the Fe/Al-P fraction (NaOH extractable; Fig. 11.3). Calcium and Mg-rich materials, FGD and AMDR1, sorbed P into the Ca/Mg-P fraction (HCl extractable; Fig. 11.3). One exception in Fig. 11.2 was the bauxite residual, which is a good example of how total Fe and Al alone is not sufficient to determine if a material will be an effective PSM. Figure 11.3 shows that although bauxite residual contained the highest amount of total Al and Fe, the swine effluent Fe/Al-P fraction actually decreased with addition of the material, likely due in part to the excessive pH (9.8; Table 11.2). In addition, bauxite residual may have been less effective at the 20 % addition rate due to excessive Na and ionic strength (Table 11.2; EC $= 4.2$ mS cm^{-1}), which likely hindered ligand exchange of P, as observed in all three manure types (Fig. 11.2). Another hypothesis for the poor P sorbing ability of the bauxite residual is that fluoride (F) contained in the material may have been released as the material weathered during the reaction period (i.e. shaking). Although this material was not analyzed for F, previous studies have shown that bauxite residual typically contains appreciable concentrations of F that may leach with additions of rainfall (Summers et al. 1996; Ho et al. 1989). Fluoride released from weathering minerals during the shaking/reaction process could have prevented P from sorbing with Fe and Al since F is a more effective competitor for such surfaces. Previous studies have shown that bauxite residuals can be "pre-washed" to remove such potential contaminants prior to use (Wang et al. 2011; McConchie et al. 2002).

When a Fe/Al-rich PSM is first added to a manure, it can immediately sorb P by ligand exchange:

$$FeOH_2{}^{+0.5} + H_2PO_4{}^{-} \leftrightarrow FeOPO_3H_2{}^{-0.5} + H_2O \quad (11.5)$$

In addition, an Fe/Al-rich PSM can also inhibit P release through the stabilization of organic phytate as described in a previous section. In Figs. 11.2 and 11.3, the dairy manure P that sorbed onto WTR, bauxite waste, and AMDR2 was primarily shifted from the water and $NaHCO_3$ fraction into the Fe/Al-P fraction, as evidenced by a significant increase in the percent NaOH extractable pool (Fig. 11.3). This is expected based on the fact that these three materials contained the highest amount of amorphous Al and Fe, which is considered indicative of the most reactive forms

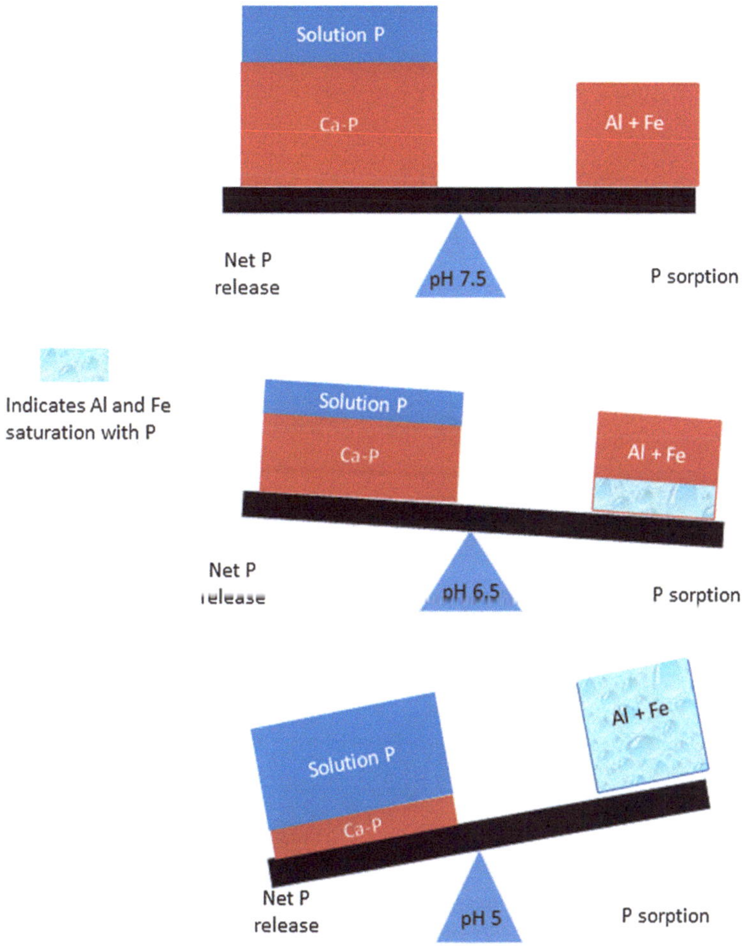

Fig. 11.6 Phosphorus (P) pool dynamics with changes in pH (fulcrum) for a hypothetical system where the initial calcium phosphate (Ca-P) pool is larger than the amount of P-unsaturated aluminum (Al) and iron (Fe). The direction of the lever pointing downward is indicative of either net P release or P sorption

of Al and Fe in by-products with regard to P sorption (Penn et al. 2011; Dayton and Basta 2005; Sakadevan and Bavor 1998). Total Al and Fe may overestimate P sorption capacity of PSMs not only because of the impact of pH on Al and Fe oxide/hydroxide effectiveness, but also because such minerals must have exposed surface area in order to efficiently remove P by ligand exchange reactions.

If a PSM decreases the pH of a manure dominated by Ca phosphate minerals, then the result can be dissolution of such minerals, i.e. the reverse of reaction 11.4, which is demonstrated in Fig. 11.5. If the dissolution of manure Ca phosphate minerals exceeds the P sorption capacity of the added Fe and Al from the PSM, then the result will be an increase in solution P concentrations; this scenario is illustrated in Fig. 11.6.

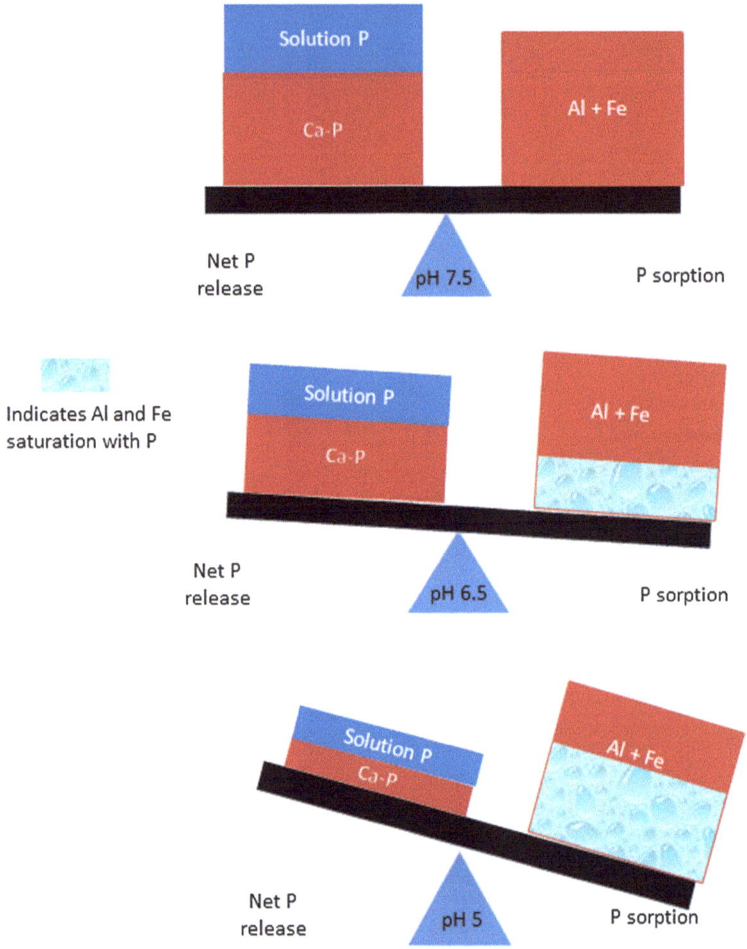

Fig. 11.7 Phosphorus (P) pool dynamics with changes in pH (fulcrum) for a hypothetical system where the initial P-unsaturated aluminum (Al) and iron (Fe) is larger than the calcium phosphate (Ca-P) pool. The direction of the lever pointing *downward* is indicative of either net P release or P sorption

Thus, there is a balance between the size of the Ca phosphate pool, available or "unsaturated" Al + Fe, and P sorption status as a function of pH. For example, Fig. 11.7 illustrates changes in P solution status with acidification in the case where the size of the initially unsaturated Al + Fe pool is larger than the Ca phosphate pool. Penn and Bryant (2008) acidified a manure-impacted and Ca-P dominated soil to various pH values down to about 5 and found that with acidification, soil water soluble P (WSP) initially decreased until about pH 7 and then increased with further acidification. However, further 2 w of incubation time of the soils with pH <7 allowed dissolved Ca-P to react with the P unsaturated soil Al and Fe oxides/hydroxides and decrease soil WSP from 40 to about 15 mg kg^{-1}. Such a scenario is illustrated in Fig. 11.7.

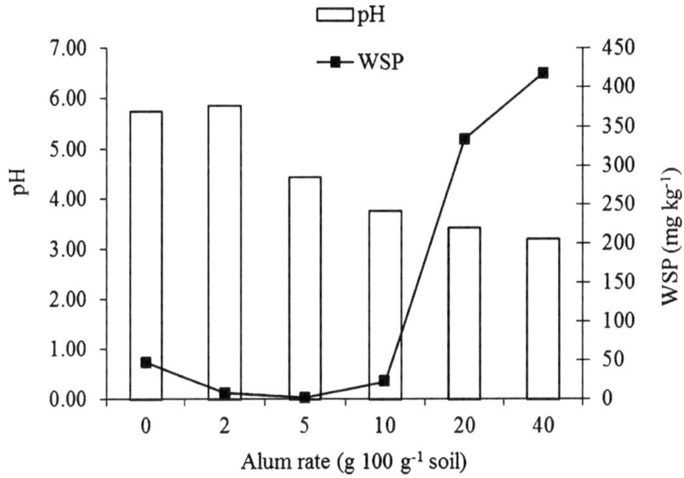

Fig. 11.8 Changes in water soluble phosphorus (*WSP*) and pH with application of alum (aluminum sulfate) to a manure-impacted soil. The soil properties are presented in Penn and Bryant (2006)

Although not a by-product, Fig. 11.8 shows that increased application of aluminum sulfate (alum) to a manure-impacted soil initially decreased WSP concentrations until the pH decreased below 4 at an application rate of 50 g kg^{-1}. After the pH decreased below 4, increased alum application appreciably increased WSP concentrations as pH continued to decrease. This scenario is illustrated in Fig. 11.6. Similarly, Moore and Miller (1994) showed that increased application of alum, ferric chloride, and ferric sulfate initially decreased soluble P concentrations in poultry litter, further applications beyond 20, 40, and 60 g kg^{-1}, respectively, increased soluble P concentrations. Huang and Shenker (2004) also noted the same phenomenon of Ca phosphate dissolution in biosolids with excessive ferric sulfate application.

In Fig. 11.2, P sorption by WTR, AMDR1, and AMDR2 in poultry manure is a result of shifting the water Extractable P into the more stable Fe/Al-P fraction (NaOH fraction; Fig. 11.3). The ability of Fe and Al to effectively sorb P in poultry manure has been well documented (Moore and Miller 1994; Sims and Luka-McCafferty 2002; Staats et al. 2004). However under the pH conditions in this study, these Fe/Al-rich PSMs did not appreciably reduce water Extractable P to levels in poultry manure that would be expected to significantly decrease soluble P losses from surface applied manure in runoff (Fig. 11.2 and Table 11.3). Each of those materials would likely have sorbed more P if the pH of the PSM-manure mixture was lower.

The interaction between Fe/Al from added PSMs with manure is somewhat complicated and contradictory in regard to the impact on P solubility. If the Fe and Al are soluble, then they can potentially stabilize organic phytate P as discussed in a previous section. However, if the added Al/Fe are not soluble, as is the most probable case considering the pH of most animal manure and by-product PSMs, then the presence of organic compounds in the manure can decrease P sorption of the PSM. Another consideration in regard to the use of Fe and Al-rich PSMs is that organic compounds can potentially block some of the P sorption sites (Ohno and Crannell 1996; Traina et al. 1986). For example, Pratt et al. (2007) found that a large portion of the oxide/hydroxide surfaces in wastewater-treated slag sorbed little P. The authors speculated that this was a result of organic compounds blocking sorption sites and preventing the release of cations that could precipitate with P.

11.7 Treatment of Manure-Impacted Soils and Runoff Phosphorus Filters

In addition to direct treatment of animal manure with PSMs prior to land application, application of PSMs to soils that have been heavily impacted by excessive manure applications can also reduce dissolved P losses. In fact, there have been more studies on the application of by-product PSMs to soils than manure (Penn et al. 2006; Callahan et al. 2002; Rhoton and Bigham 2005; Stout et al. 1998; Elliott et al. 2002; Brauer et al. 2005; Ippolito et al. 2003; Torbert et al. 2005; Peters and Basta 1996). The chemistry of direct application of PSMs to manure-impacted soils is the same as described in previous sections; however, P solubility is likely to be controlled by the properties of the soil instead of the applied manure. Elliott et al. (2002) reduced dissolved P losses in leachate using Ca, Al, and Fe-based WTRs co-applied with P on bahiagrass. Application of the WTRs reduced dissolved P losses to less than 3.5 % of the applied P. Penn and Bryant (2006) conducted a study in which several by-products (fly ash, FGD gypsum, WTRs) and aluminum sulfate were applied to streamside cattle loafing areas that were extremely high in soil test P (408–878 mg kg^{-1} of Mehlich-3 P). Simulated rainfall was applied before PSM application to determine background runoff P concentrations. Rainfall simulations were conducted 7 and 28 days after PSM application and incorporation with a simulated cow hoof, and P was measured in the runoff. While dissolved P reductions at day 7 were significantly different from the pre-application levels, P concentrations after 28 days were not different from the pre-application levels. This suggests that for high P soils, direct PSM applications will only reduce dissolved P losses temporarily.

A potential solution to this problem is the construction of landscape runoff P removal structures to directly remove P from runoff before it reaches a surface water body. In this context, the PSMs are physically contained to avoid loss, and water is channeled directly into the structure containing the PSMs. After the materials become saturated with P, they can be removed and replace with new materials, thereby allowing for true P removal (Penn et al. 2007, 2010). Although this new BMP is still being developed (Stoner et al. 2012), P-removal structures using PSMs has been demonstrated to be effective in some situations (Penn et al. 2006, 2012; Penn and McGrath 2011). A modeling tool is being developed to aid in design of such structures. A complete economic analysis is currently needed.

11.8 Conclusion and Future Research

By-product PSMs can clearly reduce soluble P concentrations in animal manure if they are applied at the proper rates and if the chemistry of the manure and PSM are compatible. Important manure properties include P forms and amounts, pH, pH buffering capacity, and relevant cations. For the PSM, the most relevant factors are forms, solubility, and amount of P sorbing elements, pH, pH buffering capacity, and safety. The maximum benefit of reducing soluble P concentrations in runoff is realized in surface applications of the PSM-manure mixture with no incorporation. This is particularly important in no-till agriculture and manure applications made to pasture and hay fields. With incorporation, the increased contact of soil to manure immediately reduces soluble P concentrations as manure P is sorbed onto the soil (Sharpley 2003; Daverede et al. 2003). In such a case, the added benefit of the PSM to the manure is little. However, a surface manure application with no incorporation into the soil causes minimal manure to soil contact, and is thereby the worst case scenario for P loss. In such a scenario, dissolved P losses will be dictated directly by the manure P solubility (Elliott et al. 2006), and therefore treatment of manure with PSMs prior to land application will have the most benefit to reducing P losses.

There is a great need for research on the impact of PSM treated manure on crop growth after the materials have been land applied. To our knowledge, there is not a single published study on this topic. However, there are several studies that examined the use of P saturated PSMs as potential fertilizers (Dobbie et al. 2005; Bird and Drizo 2009; Grubb et al. 2011, 2012). For example, Dobbie et al. (2005) applied a P saturated AMDR to soils and tested the growth of grass, barley, birch seedlings, and spruce tree seedlings. The authors concluded that the P saturated AMDR served as a suitable slow release fertilizer and was as effect as chemical P fertilizer with regard to biomass production. It is likely that a PSM-manure mixture will produce similar results when land applied.

Information on the economics of using by-products as manure PSMs is scarce. A study by O'Flynn et al. (2012) calculated the cost of P reductions in pig slurry using chemical and by-product amendments and found that fly ash was among the least expensive materials compared to chemicals such as alum. On the other hand, FGD gypsum was the most costly material for P reductions. The same group of scientists also examined the economics of P reductions in dairy slurry (Brennan et al. 2011b). The authors showed that in the context of dairy slurry, WTRs (dominated with Al) were the most cost effective materials while FGD gypsum and fly ash were the least economical.

References

Adler PR, Sibrell PL (2003) Sequestration of phosphorus by acid mine drainage floc. J Environ Qual 32:1122–1129

Baba A (2002) Assessments of radioactive contaminants in by-products from Yatagan (Mugla, Turkey) coal-fired power plant. Environ Geol 41:916–921

Bird SC, Drizo A (2009) Investigation on phosphorus recovery and reuse as soil amendment from electric arc furnace slag filters. J Environ Sci Health A 44:1476–1483

Bowden LI, Jarvis AP, Younger PL, Johnson KL (2009) Phosphorus removal from waste waters using basic oxygen steel slag. Environ Sci Technol 43:2476–2481

Brauer D, Aiken GE, Pote DH, Livingston SJ, Norton LD, Way TR, Edwards JH (2005) Amendment effects of soil test P. J Environ Qual 34:1682–1686

Brennan RB, Fenton O, Grant J, Healy MG (2011a) Impact of chemical amendment of dairy cattle slurry on phosphorus, suspended sediment and metal loss to runoff from a grassland soil. Sci Total Environ 409:5111–5118

Brennan RB, Fenton O, Rodgers M, Healy MG (2011b) Evaluation of chemical amendments to control phosphorus losses from dairy slurry. Soil Use Manage 27:238–246

Callahan MP, Kleinman PJA, Sharpley AN, Stout WL (2002) Assessing the efficacy of alternative phosphorus sorbing soil amendments. Soil Sci 167:539–547

Carlyle GC, Hill AR (2001) Groundwater phosphate dynamics in a river riparian zone: effects of hydrologic flow paths, lithology, and redox chemistry. J Hydrol 247:151–168

Claveau-Mallet D, Wallace S, Comeau Y (2011) Model of phosphorus precipitation and crystal formation in electric arc furnace steel slag filters. Environ Sci Tech 46:1465–1470

Codling EE, Chaney RL, Mulchi CL (2000) Use of aluminum- and iron-rich residues to immobilize phosphorus in poultry litter and litter-amended soils. J Environ Qual 29:1924–1931

Cucarella V, Renman G (2009) Phosphorus sorption capacity of filter materials used for on-site wastewater treatment determined in batch experiments – a comparative study. J Environ Qual 38:381–392

Daniel TC, Sharpley AN, Edwards DR, Wedepohl R, Lemunyon JL (1994) Minimizing surface water eutrophication from agriculture by phosphorus management. J Soil Water Conserv Suppl 49:30–38

Dao TH (1999) Coamendments to modify phosphorus extractability and nitrogen/phosphorus ratio in feedlot manure and composted manure. J Environ Qual 28:1114–1121

Dao TH (2003) Polyvalent cation effects on myo-inositol hexakis dihydrogenphosphate enzymatic dephosphorylation in dairy wastewater. J Environ Qual 32:694–701

Dao TH (2004a) Organic ligand effects on enzymatic dephosphorylation of myo-inositol hexakis dihydrogenphosphate in dairy wastewater. J Environ Qual 33:349–357

Dao TH (2004b) Ligands and phytase hydrolysis of organic phosphorus in soils amended with dairy manure. Agron J 96:1188–1195

Dao TH, Daniel TC (2002) Particulate and dissolved phosphorus chemical separation and phosphorus release from treated dairy manure. J Environ Qual 31:1388–1398

Dao TH, Sikora LJ, Hamasaki A, Chaney RL (2001) Manure phosphorus extractability as affected by aluminum- and iron by-products and aerobic composting. J Environ Qual 30:1693–1698

Daverede IC, Kravchenko AN, Hoeft RG, Nafziger ED, Bullock DG, Warren JJ, Gonzini LC (2003) Phosphorus runoff: effect of tillage and soil phosphorus. J Environ Qual 32:1436–1444

Dayton EA, Basta NT (2005) A method for determining the phosphorus sorption capacity and amorphous aluminum of aluminum-based drinking water treatment residuals. J Environ Qual 34:1112–1118

Deng N, Huaien L, Shi D (2011) Preliminary experimental study on effectiveness of vegetative filter strip to pollutants in surface runoff. J Water Res Prot 3(4):222–227. http://www.scirp.org/journal/jwarp

Dobbie KE, Heal KV, Smith KA (2005) Assessing the performance of phosphorus-saturated ochre as a fertiliser and its environmental acceptability. Soil Use Manage 21:231–239

Dobbie KE, Heal KV, Aumonier J, Smith KA, Johnston A, Younger PL (2009) Evaluation of iron ochre from mine drainage treatment for removal of phosphorus from wastewater. Chemosphere 75:795–800

Dou Z, Zhang GY, Stout WL, Toth JD, Ferguson JD (2003) Efficacy of alum and coal combustion by-products in stabilizing manure phosphorus. J Environ Qual 32:1490–1497

Drizo A, Comeau Y, Forget C, Chapuis RP (2002) Phosphorus saturation potential: a parameter for estimating the longevity of constructed wetland systems. Environ Sci Technol 36:4642–4648

Eghball B, Shanahan JV, Varvel GE, Gilley JE (2003) Reduction of high soil test phosphorus by corn and soybean varieties. Agron J 95:1233–1239

Elliott HA, O'Connor GA, Lu P, Brinton S (2002) Influence of water treatment residuals on phosphorus solubility and leaching. J Environ Qual 31:1362–1369

Elliott HA, Brandt RC, Kleinman PJA, Sharpley AN, Beegle DB (2006) Estimating source coefficients for phosphorus site indices. J Environ Qual 35:2195–2201

Fenton O, Healy MG, Rodgers M (2009) Use of ochre from an abandoned metal mine in the south east of Ireland for phosphorus sequestration from dairy dirty water. J Environ Qual 38:1120–1125

Gallimore LE, Basta NT, Storm DE, Payton ME, Huhnke RH, Smolen MD (1999) Water treatment residual to reduce nutrients in surface runoff from agricultural land. J Environ Qual 28:1474–1478

Gaston LA, Drapcho CM, Tapadar S, Kovar JL (2003) Phosphorus runoff relationships for Louisiana Coastal Plain soils amended with poultry litter. J Environ Qual 32:1422–1429

Gburek WJ, Sharpley AN, Heathwaite AL, Folmar GJ (2000) Phosphorus management at the watershed scale: a modification of the phosphorus index. J Environ Qual 29:130–144

Grubb KL, McGrath JM, Penn CJ, Bryant RB (2011) Land application of spent gypsum from ditch filters: phosphorus source or sink? Agric Sci 2:364–374

Grubb KL, McGrath JM, Penn CJ, Bryant RB (2012) Effect of land application of phosphorus saturated gypsum on soil phosphorus. Appl Environ Soil Sci. doi:10.1155.2012.5069.1

Han JS, Min SH, Kim YK (2005) Removal of phosphorus using AMD-treated lignocellulosic material. For Prod J 55:48–53

He Z, Dao TH, Honeycutt CW (2006a) Insoluble Fe-associated inorganic and organic phosphates in animal manure and soil. Soil Sci 171:117–126

He Z, Honeycutt CW, Zhang T, Bertsch PM (2006b) Preparation and FT-IR characterization of metal phytate compounds. J Environ Qual 35:1319–1328

Hively WD, Bryant RB, Fahey TJ (2005) Phosphorus concentrations in overland flow from diverse locations on a New York dairy farm. J Environ Qual 34:1224–1233

Ho GE, Mathew K, Newman PWG (1989) Leachate quality from gypsum neutralized red mud applied to sandy soils. Water Air Soil Pollut 47:1–18

Huang XL, Shenker M (2004) Water-soluble and solid-state speciation of phosphorus in stabilized sewage sludge. J Environ Qual 33:1085–1093

Ippolito JA, Barbarick KA, Heil DM, Chandler JP, Redente EF (2003) Phosphorus retention mechanisms of a water treatment residual. J Environ Qual 32:1857–1864

Ippolito JA, Redente EF, Barbarick KA (2005) Amendment effects on pH and salt content of bauxite residue. Soil Sci 170:832–841

Jala S, Goyal D (2006) Fly ash as a soil ameliorant for improving crop production-a review. Bio Res Technol 97:1136–1147

Johansson L (1999) Industrial by-products and natural substrata as phosphorus sorbents. Environ Technol 20:309–316

Kalbasi M, Karthikeyan KG (2004) Phosphorus dynamics in soils receiving chemically treated dairy manure. J Environ Qual 33:2296–2305

Kleinman PJA, Wolf AM, Sharpley AN, Beegle DB, Saporito LS (2005) Survey of water-extractable phosphorus in livestock manures. Soil Sci Soc Am J 69:701–708

Kostura B, Kulveitová H, Leško J (2005) Blast furnace slags as sorbents of phosphate from water solutions. Water Res 39:1795–1802

Leader JW, Dunne EJ, Reddy KR (2008) Phosphorus sorbing materials: sorption dynamics and physicochemical characteristics. J Environ Qual 37:174–181
Leytem AB, Sims JT, Coale FJ (2004) Determination of phosphorus source coefficients for organic phosphorus sources: laboratory studies. J Environ Qual 33:380–388
Lindsay WL (1979) Chemical equilibria in soils. Wiley, New York
Maguire RO, Sims JT, Saylor WW, Turner BL, Angel R, Applegate TJ (2004) Influence of phytase addition to poultry diets on phosphorus forms and solubility in litters and amended soils. J Environ Qual 33:1431–1439
Maurer M, Abramomovich D, Siegrist H, Gujer W (1999) Kinetics of biologically induced phosphorus precipitation in waste-water treatment. Water Res 33:484–493
McConchie D, Clark MW, Davies-McConchie F (2002) New strategies for the management of bauxite refinery residues (red mud). In: Proceedings of the 6th international alumina quality workshop, 8–13 September, Brisbane, Australia, pp 327–332
McDowell RW, Sharpley AN, Folmar G (2001) Phosphorus export from an agricultural watershed: linking source and transport mechanisms. J Environ Qual 32:1587–1595
McGrath JM, Sims JT, Maguire RO, Saylor WW, Angel CR, Turner BL (2005) Broiler diet modification and litter storage: impacts on phosphorus in litters, soils, and runoff. J Environ Qual 34:1896–1909
Moore PA (1999) Reducing phosphorus runoff and improving poultry production with alum. Poult Sci 78:692–698
Moore PA, Daniel TC, Edwards DR (2000) Reducing phosphorus runoff and inhibiting ammonia loss from poultry manure with aluminum sulfate. J Environ Qual 29:37–49
Moore PA, Edwards DR (2007) Long-term effects of poultry litter, alum-treated litter, and ammonium nitrate on phosphorus availability in soils. J Environ Qual 36:163–174
Moore PA Jr, Miller DM (1994) Decreasing phosphorus solubility in poultry litter with aluminum, calcium, and iron amendments. J Environ Qual 23:325–330
O'Flynn CJ, Fenton O, Healy MG (2012) Evaluation of amendments to control phosphorus losses in runoff from pig slurry applications to land. Clean Soil Air Water 40:164–170
Ohno T, Crannell BS (1996) Green and animal manure-derived dissolved organic matter effects on phosphorus sorption. J Environ Qual 25:1137–1143
Osborne LL, Kovacic DA (1993) Riparian vegetated buffer strips in water-quality restoration and stream management. Freshw Biol 29:243–258
Peak D, Sims JT, Sparks DL (2002) Solid-state speciation of natural and alum-amended poultry litter using XANES spectroscopy. Environ Sci Technol 36:4253–4261
Penn CJ, Bryant R (2006) Application of phosphorus sorbing materials to streamside cattle loafing areas. J Soil Water Conserv 61:303–310
Penn CJ, Bryant RB (2008) Phosphorus solubility in response to acidification of dairy manure impacted soils. Soil Sci Soc Am J 72:238–243
Penn CJ, McGrath JM (2011) Predicting phosphorus sorption onto normal and modified slag using a flow-through approach. J Wat Res Prot 3:235–244
Penn CJ, Mullins GL, Zelazny LW, Warren JG, McGrath JM (2004) Surface runoff losses of phosphorus from Virginia soils amended with turkey manure using phytase and HAP corn diets. J Environ Qual 33:1431–1439
Penn CJ, Mullins GL, Zelazny LW (2005) Mineralogy in relation to phosphorus sorption and dissolved phosphorus losses in runoff. Soil Sci Soc Am J 69:1532–1540
Penn CJ, Mullins GL, Zelazny LW, Sharpley AN (2006) Estimating dissolved phosphorus concentrations in runoff from three physiographic regions of Virginia. Soil Sci Soc Am J 70:1967–1974
Penn CJ, Bryant RB, Kleinman PA, Allen A (2007) Removing dissolved phosphorus from drainage ditch water with phosphorus sorbing materials. J Soil Water Conserv 62:269–276
Penn CJ, McGrath JM, Bryant RB (2010). Ditch drainage management for water quality improvement. In: Moore MT (ed) Agricultural drainage ditches: mitigation wetlands for the 21st century. Research Signpost, Kerala, India

Penn CJ, Bryant RB, Callahan MA, McGrath JM (2011) Use of industrial byproducts to sorb and retain phosphorus. Commun Soil Sci Plant Anal 42:633–644
Penn CJ, McGrath JM, Rounds E, Fox G, Heeren D (2012) Trapping phosphorus in runoff with a phosphorus removal structure. J Environ Qual 41:672–679
Peters JM, Basta NT (1996) Reduction of excessive bioavailable phosphorus in soils by using municipal and industrial wastes. J Environ Qual 25:1236–1241
Pote DH, Daniel TC, Nichols DJ, Sharpley AN, Moore PA, Miller DM, Edwards DR (1999) Relationship between phosphorus levels in three ultisols and phosphorus concentrations in runoff. J Environ Qual 28:170–175
Pratt C, Shilton A, Pratt S, Haverkamp RG, Bolan NS (2007) Phosphorus removal mechanisms in active slag filters treating waste stabilization pond effluent. Environ Sci Technol 41:3296–3301
Rhoton FE, Bigham JM (2005) Phosphate adsorption by ferrihydrite-amended soils. J Environ Qual 34:890–896
Sakadevan K, Bavor HJ (1998) Phosphate adsorption characteristics of soils, slags and zeolite to be used as substrates in constructed wetland systems. Water Res 32:393–399
Schroeder PD, Radcliffe DE, Cabrera ML, Belew CD (2004) Relationship between soil test phosphorus and phosphorus in runoff: effects of soil series variability. J Environ Qual 33:1452–1463
Seiter JM, Staats-Borda KE, Ginder-Vogel M, Sparks DL (2008) XANES spectroscopic analysis of phosphorus speciation in alum-amended poultry litter. J Environ Qual 37:477–485
Sharpley AN (2003) Soil mixing to decrease surface stratification of phosphorus in manured soils. J Environ Qual 32:1375–1384
Sharpley AN, Smith SJ (1991) Effect of cover crops on surface water quality. In: Hargrove WL (ed) Cover crops for clean water. Soil and Water Conservation Society, Ankeny, pp 41–50
Shober AL, Sims JT (2007) Integrating phosphorus source and soil properties into risk assessments for phosphorus loss. Soil Sci Soc Am J 71:551–560
Shreve BR, Moore PA Jr, Daniel TC, Edwards DR, Miller DM (1995) Reduction of phosphorus in runoff from field-applied poultry litter using chemical amendments. J Environ Qual 24:106–111
Sibrell PL, Montgomery GA, Ritenour KL, Tucker TW (2009) Removal of phosphorus from agricultural wastewaters using adsorption media prepared from acid mine drainage sludge. Water Res 43:2240–2250
Sims JT, Luka-McCafferty NJ (2002) On-farm evaluation of aluminum sulfate (alum) as a poultry litter amendment: effects on litter properties. J Environ Qual 31:2066–2073
Smith DR, Moore PA Jr, Griffis CL, Daniel TC, Edwards DR, Boothe DL (2001) Effects of alum and aluminum chloride on phosphorus runoff from swine manure. J Environ Qual 30:992–998
Smith DR, Moore PA, Miles DM, Haggard BE, Daniel TC (2004a) Decreasing phosphorus runoff losses from land-applied poultry litter with dietary modifications and alum addition. J Environ Qual 33:2210–2216
Smith DR, Moore PA Jr, Maxwell CV, Haggard BE, Daniel TC (2004b) Reducing phosphorus runoff from swine manure with dietary phytase and aluminum chloride. J Environ Qual 33:1048–1054
Staats KE, Arai Y, Sparks DL (2004) Alum amendment effects on phosphorus release and distribution in poultry litter–amended sandy soils. J Environ Qual 33:1904–1911
Stoner D, Penn CJ, McGrath JM, Warren JG (2012) Phosphorus removal with by-products in a flow-through setting. J Environ Qual 41:654–663
Stout WL, Sharpley AN, Pionke HB (1998) Reducing soil phosphorus solubility with coal combustion by-products. J Environ Qual 27:111–118
Summers RN, Smirk DD, Karafilis D (1996) Phosphorus retention and leachages from sandy soil amended with bauxite residue (red mud). Aust J Soil Res 34:555–567
Torbert HA, King KW, Harmel RD (2005) Impact of soil amendments on reducing phosphorus losses from runoff in sod. J Environ Qual 34:1415–1421

Traina SJ, Sposito G, Hesterberg D, Kafkafi U (1986) Effects of organic acids on orthophosphate solubility in an acidic, montmorillonitic soil. Soil Sci Soc Am J 50:45–52

Turner BL, Leytem AB (2004) Phosphorus compounds in sequential extracts of animal manures: chemical speciation and a novel fractionation procedure. Environ Sci Technol 38:6101–6108

Udeigwe TK, Eze PN, Teboh JM, Stietiya MH (2011) Application, chemistry, and environmental implications of contaminant-immobilisation amendments on agricultural soil and water quality. Environ Int 37:258–267

Ugurlu A, Salman B (1998) Phosphate removal from water by fly ash. Environ Int 24:911–918

USEPA (2009) A plain English guide to the EPA part 503 biosolids rule. http://www.epa.gov/owm/mtb/biosolids/503pe/. Accessed 20 June 2013

Vadas PA, Kleinman PJ, Sharpley AN, Turner BL (2005) Relating soil phosphorus to dissolved phosphorus in runoff: a single extraction coefficient for water quality modeling. J Environ Qual 34:572–580

Vohla C, Kõiv M, Bavor HJ, Chazarenc F, Mander Ü (2011) Filter materials for phosphorus removal from wastewater in treatment wetlands—a review. Ecol Eng 37:70–89

Wang JJ, Zhang H, Schroder JL, Udeigwe TK, Zhang Z, Dodla SK, Stietiya MH (2011) Reducing potential leaching of phosphorus, heavy metals, and fecal coliform from animal wastes using bauxite residues. Water Air Soil Pollut 214:241–252

Warren JG, Phillips SB, Mullins GL, Keahey DA, Penn CJ (2006) Environmental and production consequences of using alum-amended poultry litter as a nutrient source for corn. J Environ Qual 35:303–311

Warren JG, Penn CJ, McGrath JM, Sistani K (2008) The impact of alum additions on organic P transformations in poultry litter and soils receiving alum-treated poultry litter. J Environ Qual 37:469–476

Wei X, Viadero RC, Bhojappa S (2008) Phosphorus removal by acid mine drainage sludge from secondary effluents of municipal wastewater treatment plants. Water Res 42:3275–3284

Withers PJA, Ulen B, Stamm C, Bechmann M (2003) Incidental phosphorus loss—is it significant and can it be predicted? J Soil Sci Plant Nutr 166:459–468

Zelazny LW, He L, Vanwormhoudt A (1996) Charge analysis of soils and anion exchange. In: Sparks DL (ed) Chemical methods—methods of soil analysis. Part 3. Soil Science Society of America/American Society of Agronomy, Madison

Zielinski RA, Budahn JR (1998) Radionuclides in fly ash and bottom ash: improved characterization based on radiography and low energy gamma-ray spectrometry. Fuel 77:259–267

Chapter 12
Nutrient Chemistry of Manure and Manure-Impacted Soils as Influenced by Application of Bauxite Residue

Jim J. Wang and Lewis A. Gaston

Abstract Adequate management of animal manure and manure-affected soils is important for sustainable animal production as nutrient runoff from agricultural fields with long-term application of animal manure has caused great concerns over its impacts on the environment. This chapter demonstrates the effects of using bauxite residue, a by-product from the aluminum refinery industry, to modify nutrient characteristics of animal manure and manure-affected soils. The chapter begins with an introduction of bauxite residue properties followed with different aspects of treatment effects including pH, phosphorus (P) chemistry, mobility of nitrogen (N), organic carbon (OC), trace metals with focus on zinc (Zn), copper (Cu) and arsenic (As) as well as fecal coliforms in animal manure and manure-affected soils. Bauxite residue generally improves the nutrient characteristics of animal manure by suppressing the leaching potential of water-soluble P, NH_4-N, OC, Zn, Cu and As. It also decreases populations of fecal coliform in manure. The high organic nature of animal manures appears to be able to buffer the alkalinity contributed from the bauxite residues, preventing a substantial increase in pH of the treated manures and manure-affected soils although the soil sodium adsorption ratio (SAR) is elevated. Sorption and transport of phosphate in soil amended with bauxite residue indicate that bauxite residue can be used as possible reactive permeable barriers. Seawater neutralized bauxite residues, in general, exhibit greater retention of P and trace metals than un-neutralized ones. Bauxite residue addition up to 10 % (w/w) had no effect on the growth of transplanted switchgrass seedlings. Overall, bauxite residues could be used as a potential amendment for reducing P and other contaminant leaching in animal manures and manure-affected soils, and manure-bauxite residue treatment improves the application of both wastes for beneficial use.

J.J. Wang (✉) • L.A. Gaston
School of Plant, Environmental and Soil Sciences, Louisiana State University,
104 Sturgis Hall, Baton Rouge, LA 70803, USA
e-mail: jjwang@agcenter.lsu.edu

Z. He and H. Zhang (eds.), *Applied Manure and Nutrient Chemistry for Sustainable Agriculture and Environment*, DOI 10.1007/978-94-017-8807-6_12,

12.1 Introduction

Animal manure is rich in nutrients and its use as fertilizer amendments helps reduce the cost of crop production as well as provides waste disposal for the concentrated animal production industry. Long-term and extensive land application of animal manures to agricultural fields has, however, caused concerns over the runoff losses of nutrients such as phosphorus (P) and nitrogen (N) as well as trace elements, which leads to potential eutrophication and toxicity in nearby watersheds (Sims 1995; Moore et al. 1998; Rostagno and Sosebee 2001; Turner and Rabalais 2003; Sharpley et al. 2003a). Due to feed additives, besides major nutrient elements such as N and P, trace metals, especially Cu, Zn and As have been found in animal manures (Sims and Wolf 1994; Van der Watt et al. 1994; Jackson et al. 2003; Bolan et al. 2004). Copper and Zn are added to feed to prevent anemia and ensure correct eggshell deposition and feather growth, respectively, whereas As is added as roxarsone (3-nitro-4-hydroxyphenylarsonic acid) to control coccidial intestinal parasites (North and Bell 1998; Bednar et al. 2002). Copper and As are much more soluble than Zn in animal manures, however, Zn is also present in relatively high quantity (Jackson et al. 2003).

Various nutrient management plans aiming at reducing land application of animal manures have been implemented to minimize further risk of nutrient loss into water bodies (Sharpley et al. 2003b; Weld 2003). On the other hand, efforts have also been made to change the chemistry of animal manures though pre-treatment before land application (Moore and Miller 1994; Sims and Wolf 1994). Many of these pre-treatments are aimed at reducing the mobility of nutrients and immobilization of contaminants so the benefit of animal manure as nutrient sources can be maximized (Moore and Miller 1994; Schreve et al. 1995; Dao 1999; Dao et al. 2001; Wang et al. 2011). Among the pre-treatments, using industrial by-products to reduce nutrient and contaminant leaching offers an economic advantage over pure chemicals (Dao et al. 2001; Wang et al. 2011). One of the industrial by-products that has extensively been used to immobilize both inorganic and organic contaminants is bauxite residue or red mud from the aluminum refinery industry (Udeigwe et al. 2009; Wang et al. 2011).

Bauxite residue is a by-product after aluminum is extracted from bauxite ore. Currently, world-wide annual production of bauxite residues is approximately 90 million Mg and usually there is 0.6–2.0 Mg of bauxite residue generated per Mg of aluminum produced (Kumar et al. 2006). This waste material has become an environmental disposal issue for the industry (Menzies et al. 2004). Bauxite residue has two distinctive characteristics: high pH and rich contents of Fe and Al oxides and Ca carbonate. For both beneficial use and disposal, various uses of bauxite residue have been proposed ranging from large volume application as building materials (Amritphale and Patel 1987; Vincenzo et al. 2000) to specific applications as a coagulant (Orešcanin et al. 2006) or catalyst (Lamonier et al. 2005), as well as an adsorbent for anions, metals, and organics for treating wastewater effluents (Çengeloğlu et al. 2002, 2006; Li et al. 2006; Santona et al. 2006; Gupta et al. 2001; Genç-Fuhrman et al. 2004; Vaclavikova et al. 2005; Tor et al. 2006; Gupta and Ali 2006). Because of bauxite residue's alkaline nature, the material was first suggested as a liming material for acid soils (Whittaker et al. 1955). Application of the bauxite residue to acidic sandy soils increased pasture growth (Summers et al. 1996a). Bauxite residue was also investigated for possible use in nutrient retention in upland soils (Peters and Basta 1996) or in coarse soil catchments of estuaries (Summers et al. 1996b; Summers and Pech 1997). Recently, bauxite residue was used to immobilize heavy metals in abandoned mine-polluted soils (Lombi et al. 2002a, b; Brunori et al. 2005a; Friesl et al. 2006). While it may be feasible to use bauxite residue as an amendment to treat animal manure to enhance its fertilizer value by increasing nutrient retention, research data in this area are very limited. In this chapter, we will focus on effects of bauxite residue on some important processes associated with animal manure and manure-affected soils.

12.2 Properties of Bauxite Residues

Bauxite residue is produced from the Bayer process in which the bauxite ores are mixed with hot concentrated sodium hydroxide to dissolve aluminum oxide. The remaining solid residue or red mud (due to its color) from the Bayer process is separated from the process liquor in a series of thickeners and then pumped to near-by impounding sites. For low grade bauxite ores, additional processes besides the Bayer process have been used to extract aluminum. Bauxite residue has an averaged reported pH range from 10 to 13 and is mainly composed of fine particles with 90 % <75 μm (Liu et al. 2011). It has a surface area of 13–48 $m^2\ g^{-1}$ (Chvedov et al. 2001) and a cation exchange capacity (CEC) of 40–80 cmol kg^{-1} (Liu et al. 2007a, b). In addition, bauxite residue has a point of zero charge (pH_{pzc}) of 8.3, which is close to the pH_{pzc} of Fe_2O_3, Al_2O_3, and TiO_2 (Atun and Hisarli 2000). Mineral and chemical composition of bauxite residue varies with the source and refinery process. In general, bauxite residue contains many residual minerals from bauxite, such as hematite (Fe_2O_3), goethite (α-FeOOH), boehmite (γ-AlOOH), anatase/rutile (TiO_2), quartz (SiO_2), kaolinite ($Al_2SiO_5(OH)_4$), sodalite ($Na_4Al_3Si_3O1_2Cl$), calcite ($CaCO_3$), and gibbsite ($Al(OH)_3$) (Wang et al. 2008, 2011; Liu et al. 2011). Chemical analysis reveals that bauxite residue contains silica, aluminum, iron, calcium, titanium, sodium as well as an array of minor constituents, namely: K, Cr, V, Ni, Ba, Cu, Mn, Pb, and Zn (Liu et al. 2007a, b; Wang et al. 2008, 2011).

Bauxite residue is generally not considered a toxic material although it is caustic due to residual hydroxides (Wang et al. 2008). Evaluations based on the US Environmental Protection Agency (EPA) for hazardous waste characteristics: corrosivity, reactivity, ignitability and extraction procedure (EP) toxicity does not classify bauxite residue as a hazardous waste as the concentrations of all EP constituents are below the EPA regulatory levels (USEPA 1984). Nonetheless, red mud in larger quantities still presents an environmental problem due to landfilling and possible leaching, leading to the contamination of ground water under severe conditions (Wang et al. 2008). Raw or impounded bauxite residue is often treated or activated in order to increase its effectiveness as a sorbent to remove pollutants as well as minimize potential adverse environmental impacts. Common treatment or activation includes acidification (Çengeloğlu et al 2006; Huang et al. 2008), seawater washing and heating (Menzies et al. 2004; Li et al. 2006) or a combination of these processes (Liu et al. 2011). While acid neutralization removes alkali metals and other inorganic impurities as well as some organics, the procedure is generally shown to increase the surface area and pore volume, favoring adsorption (Wang et al. 2008). On the other hand, heat treatment can decompose unstable compounds and organics as well as cause particle aggregation or sintering. Seawater treatment can help neutralize red mud and reduce alkalinity without losing acid neutralization capacity. The later shifts the dominance of sodium ions to calcium and magnesium in bauxite residue (Hanahan et al. 2004; Wang et al. 2008). Combined treatment of bauxite residue using hydrogen peroxide and calcination at 500 °C increases its surface area to 108 m^2 g^{-1} (Gupta et al. 2001, 2004b; Gupta and Sharma 2002). Bauxite residue treated with seawater has been found to be environmentally compatible and the release of trapped metals is low at pHs as low as 5 based on eco-toxicological evaluations (Brunori et al. 2005b). On the other hand, organic material tends to ameliorate the adverse effect of bauxite residue on vegetation growth. Bauxite residue treated with nutrient rich sewage sludge or compost produced from a mixture of municipal green waste, pine bark and poultry manure has been found to increase the potential of direct revegetation (Wong and Ho 1994; Jones et al. 2012). Similarly, nutrient rich and strong buffering characteristics of animal manure and manure-impacted soils also likely make their interaction with bauxite residue positive and beneficial.

12.3 Impact on Properties of Manure and Manure-Affected Soils

12.3.1 Manure, Soil and Bauxite Residue Treatment

Fresh chicken litter from a poultry production farm in Louisiana and cattle manure from a cattle feedlot farm in Oklahoma were obtained. Manure-affected soils were collected from three manure-amended sites in the northern part of Louisiana and were known to have received poultry manure applications for 5–20 years with about 3–8 applications per year. An average annual application rate of approximately 9 Mg ha^{-1} has been common in this area (Gaston et al. 2003). Collected manure and soil samples were air dried, and ground to pass through a 2-mm sieve. Bauxite residues were collected from aluminum producing facilities in Arkansas, Texas, and Louisiana. The residue wastes include two types, red mud and brown mud. Red mud is a direct by-product from the Bayer process whereas brown mud is produced from red mud after additional leaching and sintering (USEPA 1984). Selected fresh bauxite residues were treated to lower the pH and/or reduce the soluble Na level using a modified seawater neutralization approach as described by McConchie et al. (2002). Selective physical and chemical properties of these animal manure and bauxite residue samples are shown in Table 12.1. For treatments, bauxite residues were mixed with manure at levels of 0, 11, 22, and 50 %, whereas they were mixed with manure-affected soils at 0, 2, and 4 % (Udeigwe et al. 2009; Wang et al. 2011). The 2 % soil treatment corresponds to approximately 40 Mg ha^{-1} of bauxite application.

12.3.2 pH and Salinity

Bauxite residue-treated manures at 11–50 % had generally much lower pH values than those of the two bauxite residues, with pHs of the brown mud-treated manures ranging from 7.2 to 8.2 and red mud-treated manures ranging from 7.0 to 7.5 in their water extracts. These pH ranges were not much different from those of original chicken litter (pH 7.3) and cattle manure (pH 7.1) as shown in Table 12.1, indicating that the high organic nature of animal manures was able to buffer the alkalinity of bauxite residues. This result suggests easy and safe handling of animal manures directly treated with bauxite residue. The observed pH range of treated manures also indicated that it was unlikely that the alkalinity of bauxite residue would cause damage to plants if applied to agricultural fields. Similar pH ranges were also found in bauxite residue-treated soils influenced by long-term application of animal manures (Table 12.2). In general, the treatments of original or seawater neutralized bauxite residues at 2–4 % increased the soil pH by 0.8–1.5 units from those of the original soils, and the impact on soil pH was slightly greater at the higher bauxite residue rate of 4 %. These results are consistent with those of others (Friesl et al. 2003; Brunori et al. 2005a).

The bauxite residue amendments did significantly ($P < 0.05$) increase the soil electrical conductivity (EC), an indicator of soluble salt concentration. As high as a twofold increase in EC was observed in the three soils treated with 4 %-seawater neutralized red mud from two different sites (NORM for site O and NARM for site A, Table 12.2). In general, the seawater neutralized red muds had more impact on EC due to their higher soluble salt contents, and in all cases, the increase in soil EC was less than 0.38 dS m^{-1}. In addition, sodium adsorption ratio (SAR), which often approximates the degree of dispersion in soil systems, was highly affected (2- to 5-fold increase) by both 2 % seawater-untreated red mud, ORM (red mud from site O) and the 4 % NORM amendments, especially the former (Table 12.2). On the other hand, Ca-dominated brown mud at both 2 and 4 % amendment rates had little effect on SAR or even decreased the soil SAR. The results suggest that the source of bauxite residue could make a difference in determining the proper choice of bauxite residue for treating manure-affected soils. The bauxite residue such as ORM may require additional treatment for sodium using Ca rich products such as

Table 12.1 Chemical and physical properties of animal manure, manure-affected soils, and bauxite residues[a]

		C	N	P	K	Ca	Mg	Na	Al	Fe	Ti	Mn	Cu	Zn	As
Material	pH	%		g kg^{-1}								mg kg^{-1}			
Manure															
CL	7.31	26	3.32	16.4	31.0	12.5	4.0	13.2	8.5	4.05	0.88	421	580	465	16
CM	7.05	21	2.07	5.18	14.9	12.0	3.9	5.03	7.6	3.78	0.79	120	17	138	3
Soil															
Briley	6.10	3.3	0.39	3.17	3.33	2.61	0.9	0.87	19	12.7	–	215	106	179	6
Darley	7.00	0.8	0.11	0.46	3.95	0.40	0.4	0.70	11	3.13	–	500	19	23	4
Smithdale	6.00	1.9	0.24	1.70	3.93	1.43	0.4	0.71	11	4.66	–	269	118	126	6
Ruston[b]	6.16	1.4	–	0.01	0.03	0.33	0.1	0.01	–	–	–	–	–	–	–
Bauxite residue															
ARM	9.17	1.2	0.03	2.39	1.52	29.5	1.1	16.6	50	316	274	5,031	115	259	44
ORM	10.5	0.7	0.04	0.40	1.23	11.6	0.4	48.1	89	225	36.0	239	66	12	11
ABM	9.95	2.9	0.03	0.82	1.12	179	1.8	7.13	41	94.7	15.2	1,610	16	47	6

Adapted from Udeigwe et al. (2009), Wang et al. (2011) and unpublished data

[a]*CL* chicken litter, *CM* cattle manure, *ARM*, red mud from site A, *ORM* red mud from site O, *ABM* brown mud from site A. All values are averages of two replicates with standard errors <5 %. pH for manure and bauxite reside based on 1:3 and 1:1 solid to solution ratio, respectively. Total C and N were determined by dry combustion and the rest of the elements were determined by digestion using HF-H_2SO_4-$HClO_4$ followed by ICP-AES analysis

[b]For Ruston soil, P was determined by Bray 2 and bases by 1 M NH_4OAc extraction

Table 12.2 Effects of various amendments on selected parameters of three manure-affected soils. Data represent the mean values with the standard error for each given in parentheses[a]

Soil treatment	pH	EC	SAR	Turbidity	BOD_5
		$dS\ m^{-1}$		NTU	$mg\ L^{-1}$
Briley	6.2 (0.03)d	0.21 (0.00)e	2.04 (0.00)c	2,127 (21)e	19.9 (1.95)a
Briley + 2 %-ORM	7.3 (0.00)b	0.35 (0.01)d	5.76 (0.27)a	4,263 (36)b	17.6 (1.20)ab
Briley + 2 %-ARM	7.1 (0.02)c	0.35 (0.00)d	1.59 (0.09)c	2,540 (4)d	13.8 (0.23)bcd
Briley + 2 %-NORM	7.1 (0.01)c	0.44 (0.02)c	3.75 (0.13)b	3,254 (38)c	14.9 (1.61)bc
Briley + 2 %-NARM	7.1 (0.02)c	0.40 (0.00)c	1.62 (0.29)c	2,190 (36)e	16.0 (1.01)abc
Briley + 4 %-NORM	7.3 (0.02)b	0.59 (0.01)a	4.15 (0.10)b	4,935 (60)a	14.4 (0.38)bc
Briley + 4 %-NARM	7.3 (0.02)b	0.49 (0.00)b	1.68 (0.01)c	2,717 (233)d	9.3 (0.86)d
Briley + 2 %-ABM	7.3 (0.02)b	0.35 (0.03)d	0.97 (0.35)d	1,937 (28)e	17.5 (2.66)ab
Briley + 4 %-ABM	7.5 (0.00)a	0.40 (0.00)c	0.98 (0.03)d	1,904 (104)e	12.7 (2.10)cd
Darley	6.5 (0.03)f	0.21 (0.00)f	1.26 (0.10)d	744 (18)e	14.2 (0.07)a
Darley + 2 %-ORM	7.6 (0.02)b	0.35 (0.00)cd	6.96 (0.35)a	4,265 (2)a	13.9 (0.82)a
Darley + 2 %-ARM	7.3 (0.01)e	0.35 (0.00)cd	1.72 (0.05)c	1,851 (15)c	9.4 (0.53)b
Darley + 2 %-NORM	7.4 (0.01)d	0.38 (0.00)c	3.60 (0.12)b	3,218 (122)b	11.1 (0.26)ab
Darley + 2 %-NARM	7.3 (0.01)d	0.38 (0.00)c	1.27 (0.06)d	1,320 (15)d	11.5 (0.22)ab
Darley + 4 %-NORM	7.5 (0.00)c	0.55 (0.02)a	3.55 (0.09)b	1,071 (114)d	10.0 (0.45)b
Darley + 4 %-NARM	7.5 (0.00)c	0.45 (0.03)b	1.50 (0.42)cd	1,001 (29)de	9.2 (2.25)b
Darley + 2 %-ABM	7.6 (0.02)b	0.32 (0.01)de	1.31 (0.04)cd	1,085 (224)d	12.7 (2.25)ab
Darley + 4 %-ABM	7.9 (0.03)a	0.30 (0.01)e	1.42 (0.03)cd	1,301 (113)d	9.0 (1.01)b
Smithdale	6.0 (0.03)f	0.21 (0.00)f	1.44 (0.04)d	708 (6)f	17.8 (1.73)a
Smithdale + 2 %-ORM	7.3 (0.03)b	0.30 (0.04)de	7.83 (0.07)a	3,792 (93)b	17.3 (0.04)a
Smithdale + 2 %-ARM	7.0 (0.02)c	0.26 (0.01)e	1.40 (0.17)d	2,256 (66)d	14.2 (0.75)ab
Smithdale + 2 %-NORM	7.1 (0.04)d	0.36 (0.01)c	3.79 (0.12)c	3,021 (15)c	14.7 (0.90)ab

(continued)

Table 12.2 (continued)

Soil treatment	pH	EC $dS\ m^{-1}$	SAR	Turbidity NTU	BOD_5 $mg\ L^{-1}$
Smithdale + 2 %-NARM	7.1 (0.02)d	0.32 (0.00)d	1.44 (0.03)d	2,094 (12)d	12.6 (1.35)bc
Smithdale + 4 %-NORM	7.2 (0.02)c	0.54 (0.00)a	4.15 (0.03)b	4,938 (138)a	7.9 (1.16)de
Smithdale + 4 %-NARM	7.2 (0.00)c	0.45 (0.00)b	1.43 (0.07)d	1,182 (21)e	10.4 (2.03)dc
Smithdale + 2 %-ABM	7.2 (0.00)c	0.31 (0.01)d	0.52 (0.12)e	1,029 (9)e	14.3 (0.86)ab
Smithdale + 4 %-ABM	7.5 (0.00)a	0.33 (0.01)cd	0.75 (0.10)e	1,184 (77)e	4.9 (0.11)e

Adapted from Udeigwe et al. (2009)

[a]*ARM* site A red mud, *NARM* neutralized site A red mud, *ORM* site O red mud, *NORM* neutralized site O red mud, *ABM* site A brown mud, *EC* electrical conductivity, *SAR* sodium adsorption ratio, BOD_5 five-day biochemical oxygen demand. Mean values in a column (within a soil) with the same lowercase letter are not statistically different at $\alpha = 0.05$

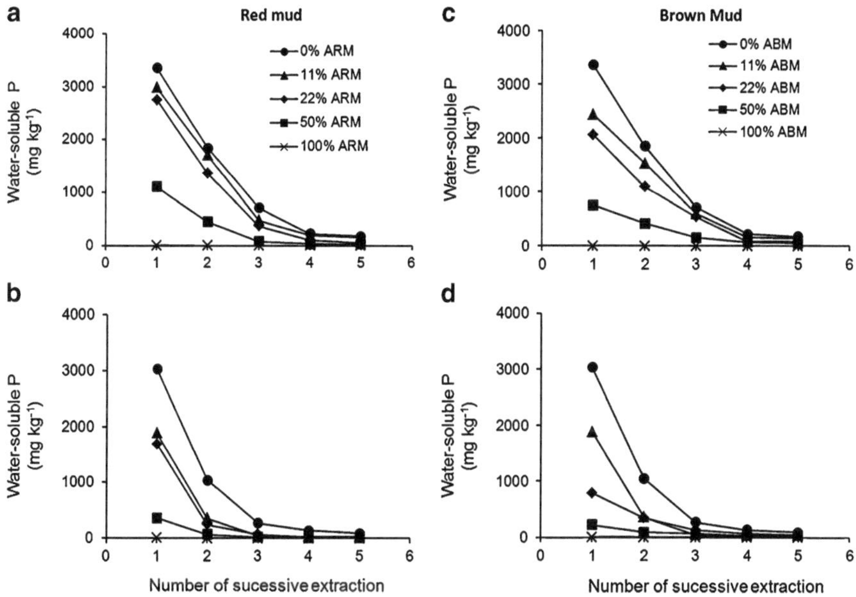

Fig. 12.1 Effect of bauxite residue on successive extractions of water-soluble P from animal manures: (**a**) Chicken litter with red mud; (**b**) Cattle manure with red mud; (**c**) Chicken litter with brown mud; and (**d**) Cattle manure with brown mud. All experimental data points are averages of two replicates with standard errors <5 % (Adapted from Wang et al. 2011)

gypsum and its direct use as amendment for manure-affected soils, even though the observed SAR impacts are still within tolerable ranges of soils for agricultural production. Among all the bauxite residue amendments evaluated, the 2 and 4 %-ABM (brown mud from site A) would tend to have the least impact on SAR and turbidity of runoff water from these manured soils (Table 12.2).

12.3.3 Impact on Phosphorus Chemistry

12.3.3.1 Leaching Characteristics of Phosphorus from Manure and Manure-Affected Soils

Bauxite residue treatment reduces water extractable P from manure, an indicator of P loss potential from leaching and runoff. As shown in Fig. 12.1, both red mud and brown mud-treated chicken litter and cattle manure showed lower P in successive water extractions than untreated manure. The higher the percentage of bauxite residue treatment, the lower the water-soluble P extracted from the manures. At the same treatment rates, greater water-soluble P reduction was observed in the cattle manure compared to chicken litter (Fig. 12.1b, d vs. Fig. 12.1a, c), which was likely due to the large difference in total P content between the two manures.

Total P in the chicken litter was three times of that in cattle manure (Table 12.1). At same 22 % treatment rate, brown mud reduced water-extractable P by 40 and 70 % relative to untreated chicken litter and cattle manure as opposed to 27 and 55 % by red mud, respectively, during five successive extractions (Fig. 12.1c, d vs. Fig. 12.1a, b). The latter indicates that brown mud is more effective than red mud in immobilizing water-soluble P of animal manures. The higher reduction in soluble P by brown mud is likely the result of higher sorption capacity of brown mud than red mud (Wang et al. 2011).

The slightly higher pH of brown mud-treated manures likely reflects the strong role of Ca or $CaCO_3$ in retaining soluble P, as P is known to be strongly adsorbed to $CaCO_3$ or precipitated as secondary Ca-P compounds at neutral to alkaline pH (Lindsay 1979; Amer et al. 1985). On the other hand, P sorption by Fe oxides was also shown to play an important role in P retention in alkaline and calcareous soil conditions (Ryan et al. 1985; Harrell and Wang 2006). Iron oxide-dominated red mud at treatment rates of 11, 22, and 50 % corresponded to total Fe/P molar ratios of 1.4, 3.0, and 9.5 in the chicken litter treatment and 4.3, 8.8, and 23.5 in the cattle manure treatments, which were greater than the minimum Fe/P molar ratio of 1 required for optimal P removal from aqueous solution by Fe/Al oxides-dominated mechanisms (Ulmgren 1975; Hsu 1976). For Ca-dominated brown mud, the same treatment rates yielded a total Ca/P ratio of 0.8, 1.0, and 2.0 in the chicken litter treatments and 2.3, 3.0, and 5.8 in the cattle manure treatments, respectively. Because precipitation of P as Ca-P minerals generally occurs with Ca/P molar ratios >1.3 (Koopmans et al. 2007), this suggests that the highest brown mud treatment rate of 50 % with this chicken litter and all three treatment rates of cattle manure had the Ca/P molar ratios that likely facilitated the precipitations as Ca-P minerals. At 50 % mixing, brown mud and red mud reduced water-soluble P by 78 and 73 % in chicken litter and by 91 and 90 % in cattle manure, respectively (Fig. 12.1). The greater increase in Al/Fe to P ratio or Ca to P ratio as a result of red or brown mud amendment likely explains the stronger treatment effect of bauxite residues on the cattle manure than on the poultry litter. Red mud and brown mud were found to adsorb 75 and 99 % of added initial P at 100 mg L^{-1} (adsorption capacities of 700 and 910 mg P kg^{-1}) even at a solution pH of 8.98 and 9.58, respectively.

Bauxite residue amendments also significantly ($P < 0.05$) reduced the water-soluble P by 58–95 % in three manure-affected soils (Fig. 12.2). It is interesting to note that red mud, ARM, showed higher reduction in soil soluble P than brown mud, ABM, a result that was different than that observed in bauxite residue-treated manures as discussed previously. This was likely due to generally slightly lower pH values observed in these brown mud-treated manure-impacted soils than those brown mud-treated manures as well as a difference due to the integrated effects of soils and manures. A slight decrease in pH significantly reduces affinity of Ca and $CaCO_3$ reacting with phosphates. In addition, these soils contain much less Ca and Mg than manures. The fact that the Fe/Al-dominated ARM reduced soluble P more than the Ca-dominated ABM in the manure-affected soils at pH 7.0–7.3 suggests that Fe/Al-oxides likely play a significant role in retaining P under neutral

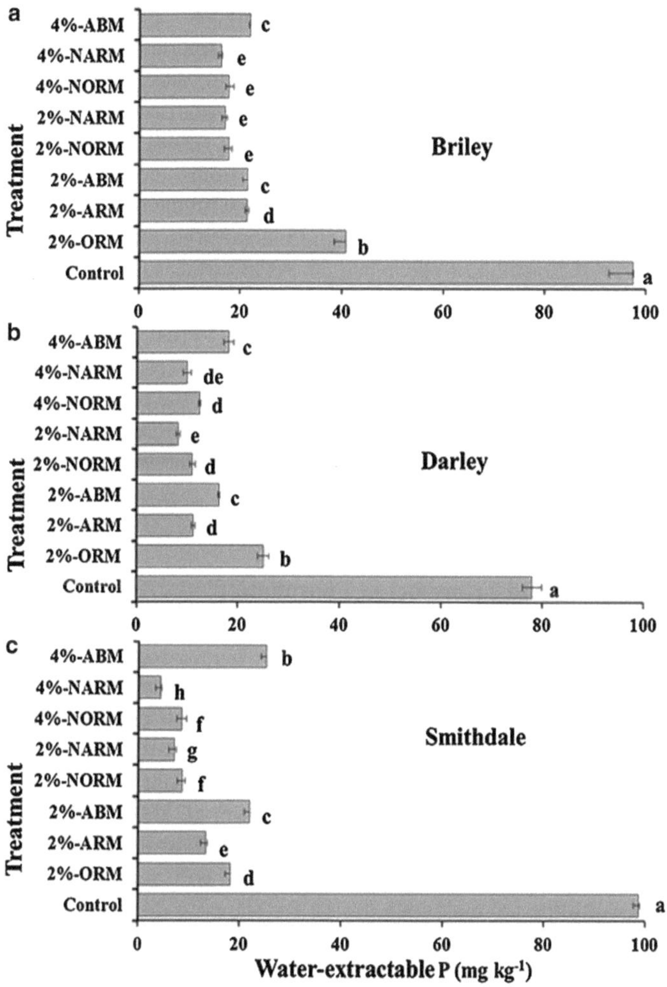

Fig. 12.2 Effects of various bauxite residue amendments and application rates on water-Extractable P of three manure-impacted soils: (**a**) Briley, (**b**) Darley, and (**c**) Smithdale. Within a soil, *bars* with the same *lowercase letter* are not statistically different (Fisher's LSD, $\alpha = 0.05$) (Adapted from Udeigwe et al. 2009)

or alkaline conditions. A similar result has been reported for calcareous soils (Harrell and Wang 2006). On the other hand, the neutralized red muds (NARM and NORM) at 2 % application rate yielded the largest reduction (≥82 %) of water-soluble P, which was 12 % more than the raw red muds (ARM and ORM). In addition, increasing the application rate from 2 to 4 % had no additional effect on soluble P reduction. This suggests that 2 % (40 Mg ha^{-1}) field application is adequate when neutralized red mud or brown mud residues are used for reducing P runoff from P-rich soils. This finding is significant as a high rate of field application of bauxite residues could cause adverse impacts on soil quality (Lombi et al. 2002a; Friesl et al. 2003) and increase operational costs.

Previous research has shown that, compared to other low-cost adsorbents, bauxite residue is effective at a wide pH range and at high phosphate concentrations (Wang et al. 2008; Liu et al. 2011). In general, about 80–90 % added P at an initial concentration of 30–100 mg/L can be sorbed. All ortho-, inorganic condensed, and organic phosphates are adsorbed by bauxite residue with the optimum efficiencies of the three phosphate types under slightly acidic conditions (Altundogan et al. 2002; Liu et al. 2011). Adsorption of phosphate on oxide surfaces of red mud is through a ligand exchange mechanism which involves exchange of an aqueous ligand for a surface hydroxyl group. This process leads to the formation of an inner sphere complex that involves coulombic interaction (Pradhan et al. 1998; Altundogan et al. 2002; Li et al. 2006). Clearly, as demonstrated by the results in Figs. 12.1 and 12.2, animal manure treated with bauxite residue will increase its P retention value whereas manure-affected soils treated with bauxite residue will reduce the potential of P loss through leaching and surface runoff.

12.3.3.2 Sorption and Transport of Phosphate in Red Mud Amended Soil

Besides loss of P to surface waters in runoff, lateral subsurface flow to seep areas may be an important route for P loading depending on depth to the restrictive subsurface horizon and its hydraulic conductivity (Walthall and Nolfe 1998; Buda et al. 2012). Red mud (Duchesne and Doye 2005; Coruh and Ergun 2010) and neutralized red mud (Munro et al. 2004; Lapointe et al. 2006; Zijlstra et al. 2010) have been examined as possible reactive permeable barriers to treat acidic mine drainage and sorb dissolved metals. Red mud mixed with soil down to a restrictive soil horizon would be an analogous design for trapping dissolved P where lateral subsurface flow substantially contributes to P loads in surface water.

We examined sorption of P from KH_2PO_4 in untreated red mud mixed with Ruston soil at different ratios and examined the applicability of batch sorption data for describing phosphate transport in these media (characterization data given in Table 12.1). Sorption was determined using a solid to solution ratio of 1:5, a range of input concentrations from 0 to 200 mg L^{-1}, and 24 h shaking. Change in sorbed phosphate concentration (ΔS, mg kg^{-1}) was calculated by change in solution concentration, and modeled by an adaptation of the Langmuir equation, $\Delta S + S_0 = kS_{Max}C/(1 + kC)$, where S_0 is initially sorbed phosphate (mg kg^{-1}), C is solution concentration (mg L^{-1}), and the parameters k (L mg^{-1}) and S_{Max} (mg kg^{-1}) are, respectively, an affinity parameter and apparent sorption maximum.

Both the Langmuir and Freundlich, $S = K_FC^N$ (where K_F and N are fitted parameters), models have been used to describe phosphate sorption onto red mud where neither a clear sorption maximum was approached nor sorption evidently continued to increase with increasing solution concentration (Li et al. 2006; Liu et al. 2007a, b). However, phosphate sorption onto Ruston soil approached a

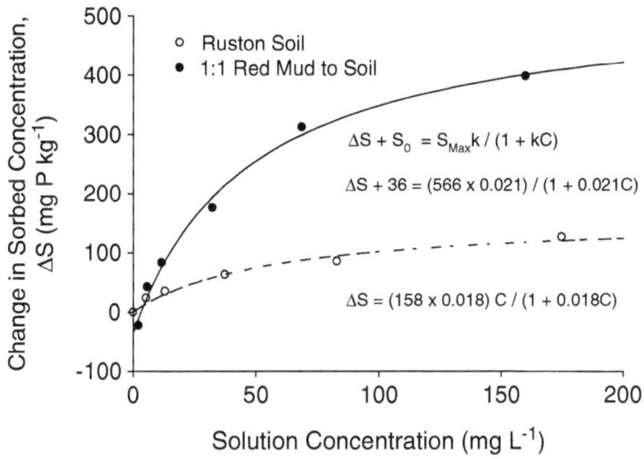

Fig. 12.3 Sorption of phosphate onto Ruston soil and a 1:1 mixture of red mud and Ruston soil. Best-fit Langmuir models are shown

maximum, so that the Langmuir would be the appropriate model (Fig. 12.3). Although there was no clear approach to maximum sorption in the 1:1 mixture, for consistency the best-fit Langmuir model is shown (Fig. 12.3). While replacing half the mass of soil with red mud increased sorption about threefold compared to the soil alone, maximum sorption was much less than might be expected based on the mass of red mud present. For example, Li et al. (2006) and Liu et al. (2007a, b) calculated sorption maxima of 100 g kg^{-1}, and Penn et al. (2011) obtained a maximum of 25 g kg^{-1}. In sorption studies that showed no maximum (i.e., where the Freundlich model was applicable) similarly large values for phosphate sorption were obtained for the highest solution concentrations (Wang et al. 2011).

Variable results for phosphate sorption likely reflect differences in chemical composition and physical properties among red mud sources. However, differences in experimental methodologies further complicate comparison among results (Cucarella and Renman 2009). The latter include wide discrepancies in ratios of red mud to solution used in batch sorption studies, and Cucarella and Renman (2009) noted a general trend of increasing P sorption capacity with decreasing mass to solution ratio. Our results are consistent with this trend, since a 1:5 ratio was used, whereas Penn et al. (2011) used a 1:15 ratio, and Li et al. (2006) and Liu et al. (2007a, b) used a 1:200 ratio. Since phosphate sorption increases with decreasing pH (Huang et al. 2008), the effect of solid to solution ratio may reflect differences in pH due to dilution of the sorbent. Particulate dispersion may also increase with dilution, increasing the surface area for sorption (van Beinum et al. 2005). Studies on the kinetics of phosphate sorption on red mud would seem to support the latter. Whereas Li et al. (2006), Liu et al. (2007a, b) and Huang et al. (2008) found no change in sorption after $\leq$6 h contact when using solid to solution ratios from 1:125 to 1:2,500, López et al. (1998) found increasing sorption up to 48 h with a 1:25 ratio. Time-dependent changes in pH may also be involved (Snars and Gilkes 2009). Despite lower sorption, a 1:5 solid to solution ratio more closely matches field conditions than wider ratios, giving data more applicable to the field (Cucarella and Renman 2009). Moreover, the low sorption

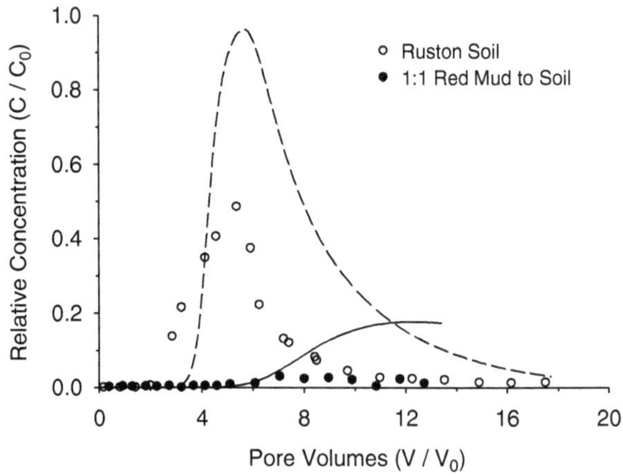

Fig. 12.4 Elution of phosphate from columns of Ruston soil and a 1 to 1 mixture of soil and red mud. Elution curves predicted assuming Langmuir equilibrium sorption/desorption are shown as smooth curves

maxima were not necessarily outliers. The Freundlich model of Wang et al. (2011) gives sorbed concentrations (about 1 g kg^{-1}) similar to our data for the solution concentrations we used.

Use of sorption data to describe the mobility of phosphate through columns of Ruston soil or 1:1 soil and red mud was next examined. After saturation with prepared native soil solution, a pulse of phosphate (100 mg L^{-1}) and Br (100 mg L^{-1}) was applied, then eluted with soil solution at steady-state flow. The convective-dispersive equation (CDE; e.g., López et al. 1998), with flux boundary conditions and assumed uniform initial concentrations was used to describe Br tracer and phosphate elution. The dispersion coefficient was determined by fitting the CDE to the Br elution data using a least-squares method (van Genuchten 1981). The CDE was expressed in implicit finite difference form, and elution of phosphate was predicted assuming that Langmuir equilibrium sorption/desorption was applicable.

Figure 12.4 compares elution of phosphate pulses from columns of the Ruston soil and the 1:1 mixture. Elution of phosphate from the mixed medium was delayed and reached a much lower maximum concentration compared to phosphate elution from the Ruston soil alone. Whereas nearly 40 % of the phosphate added to the Ruston soil was eluted, only about 4 % was eluted from the mixture. Within an overall mass balance error of about 2 %, addition of red mud to the Ruston soil increased phosphate retention from 60 to 96 %. Thus, red mud effectively trapped phosphate. However, simulations assuming equilibrium sorption (Langmuir parameters given in Fig. 12.3) did not describe phosphate mobility in either the Ruston soil or the mixture (Fig. 12.4). Measured phosphate elution was earlier or much less concentrated than predicted. Since the Br data did not indicate physical non-equilibrium (van Genuchten and Wierenga 1976) in either medium, discrepancies between experimental and modeled data suggest that sorption/desorption was likely time-dependent. While numerous studies on P transport in soil (e.g., early work by Mansell et al. (1977)) have shown that slowly reversible sorption affects P transport, it did not seem unreasonable to describe phosphate transport in the mixed medium assuming chemical equilibrium because sorption should be dominated by the red mud, for which the kinetics are either fast (Li et al. 2006; Liu et al. 2007a, b; Huang et al. 2008) or rapid sorption/desorption comprises the major fraction of total sorption (López et al. 1998). Furthermore, López et al. (1998) were able to approximately describe phosphate transport through a small column of red mud assuming equilibrium sorption. On the other hand, López et al. (1998) obtained approximately steady effluent concentration of phosphate that was only about 70 % of the input concentration, while predicted concentration continued to increase. Thus, their results may reflect slow sorption kinetics at a large solid to solution ratio (column), similar to our data.

12.3.4 *Impact on Nitrogen Retention*

As bauxite residue has generally high pH values, there is a concern over the loss of N from animal manure through NH_3 volatilization. The pH of the manure solution is often the main variable that regulates the equilibrium between NH_4^+ ions and NH_3 gas. Aerobic and alkaline conditions are noted to enhance NH_3 volatilization of manure N and could lead to as much as 9–44 % of N volatilized as NH_3 (Kirchmann and Witter 1989). Therefore, the effect of bauxite residues on NH_3 volatilization was tested by mixing with chicken litter at 10 and 30 % moisture contents, common conditions of chicken litter. In general, it was found that bauxite residue-treated chicken litter actually had slightly lower NH_3 release at both 10 and 30 % moisture contents than untreated chicken litter but the difference was not statistically significant ($P > 0.05$) (Wang et al. 2011). This result was likely due to the limited increase in pH as a result of bauxite residue treatment (Table 12.2). The latter could be attributed to the strong buffering ability of organic matter in poultry litter as discussed earlier. In addition, the large surface areas of bauxite residues could also play a role in retaining NH_3 from being evolved. In addition, bauxite residue-treated chicken litter showed lower runoff loss of NH_4^+ than alum-treated and untreated chicken litter when it was surface-applied to a pasture field (Fig. 12.5). This result is interesting as alum has extensively been used for reducing NH_3 volatilization in poultry houses and reducing P loss from surface runoff of animal manure-affected soil (Moore et al. 1995). This result could suggest that alum may have a weak role in retaining NH_4^+ when alum-treated manure is applied to the field. There are currently no data available on the role of bauxite residue on NO_3-N in manure. Nonetheless, earlier studies showed that sand treated with red mud was able to remove 30–74 % of the NO_3^- from primary and secondary sewage effluents depending on the quantity of bauxite residue treatments (Ho et al. 1992). Adsorption of NO_3^- by red mud followed Langmuir and Freundlich isotherms (Çengeloğlu et al. 2006).

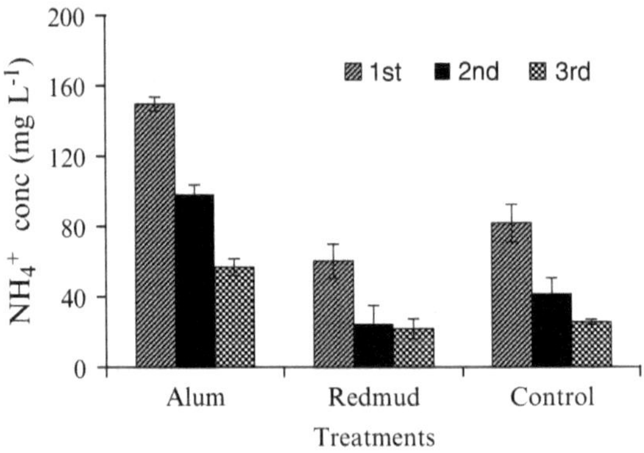

Fig. 12.5 Ammonium (NH_4^+) concentration in three consecutive runoff (first, second and third) water from pasture with surface applied chicken litter that was treated with alum and red mud (*ARM*) amendments at a rate of 12 %

12.3.5 Impact on Dissolved Organic Carbon and BOD

Runoff loss of dissolved organic carbon (DOC), in addition to N and P, is a major concern with respect to increased biological oxygen demand (BOD) in affected water bodies, an effect which has not been adequately investigated. Bauxite residue amendments in general also reduce the DOC lost from manure. As shown in Fig. 12.6, both red mud and brown mud significantly ($P < 0.05$) reduced the water-extractable (dissolved) organic C by 10–40 % in three manure-affected soils. Both Al and Fe are known to bind readily with organic matter through surface complexation. Strong correlations ($R^2 \geq 0.72$) were found between soil extractable Al/Fe and OM (Ross and Bartlett 1996; Darke and Walbridge 2000). These Al/Fe-OM complexes would likely lead to the reduction of the DOC concentration in these treated soils. The seawater neutralization significantly enhanced soluble C retention by ORM (red mud from the site O) which contained higher soluble sodium, but not by ARM, which had relatively little sodium (Udeigwe et al. 2009). In addition, the 4 % application rate offered little improvement over the 2 % application rate of the respective amendment, suggesting that the 2 % application rate of the neutralized red mud was likely adequate for reducing the release of DOC from these manure-amended soils. The effects of the bauxite residue amendments on DOC were also consistent with their effects on five-day biochemical oxygen demand (BOD_5) conducted on the extract of the 1:10 solid (soil and amendment)-water mixture (Table 12.2). Five-day BOD has often been used as a measure of the organic biodegradable materials present in a water system (Sawyer et al. 2002). Previous research has shown different degrees of relationships between BOD_5 and organic carbon in surface water and wastewater (Constable and Edward 1979). Nonetheless, this BOD_5 analysis revealed that the bauxite residue amendments decreased the resulting BOD_5, particularly with the neutralized red mud and 4 %-ABM amendments (Table 12.2).

Previous studies have also shown that bauxite residues, especially acid treated ones, can adsorb organic dye such as basic dye, methylene blue, as well as phenol

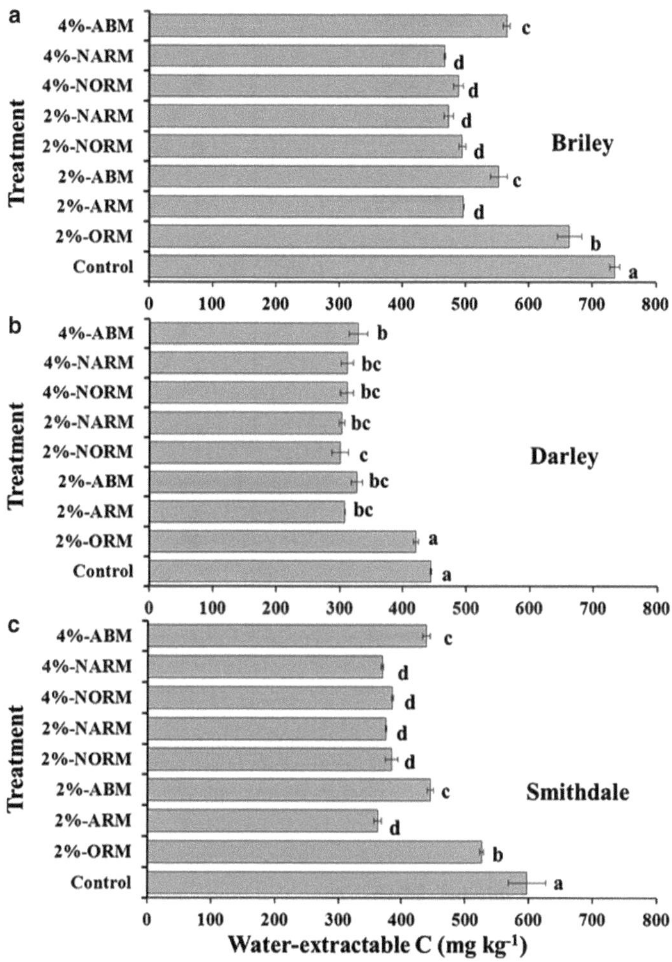

Fig. 12.6 Effects of various bauxite residue amendments and application rates on water-Extractable C of three manure-impacted soils: (**a**) Briley, (**b**) Darley, and (**c**) Smithdale. Within a soil, *bars* with the same *lowercase letter* are not statistically different (Fisher's LSD, $\alpha = 0.05$) (Adapted from Udeigwe et al. 2009)

and chlorophenol compounds such as 2-chlorophenol, 4-chlorophenol, and 2,4-dichlorophenol from wastewater. The results showed that 2,4-dichlorophenol and 4-chlorophenol were sorbed by red mud up to 94–97 %, while the removal of 2-chlorophenol and phenol was up to 50–81 % (Gupta et al. 2004a; Tor et al. 2006; Liu et al. 2011). Although not as effective as the impact on soluble P reduction, the results shown in Fig. 12.6 indicate that these Fe, Al, and Ca-dominated bauxite residues could reduce the leaching and runoff losses of soluble organic C.

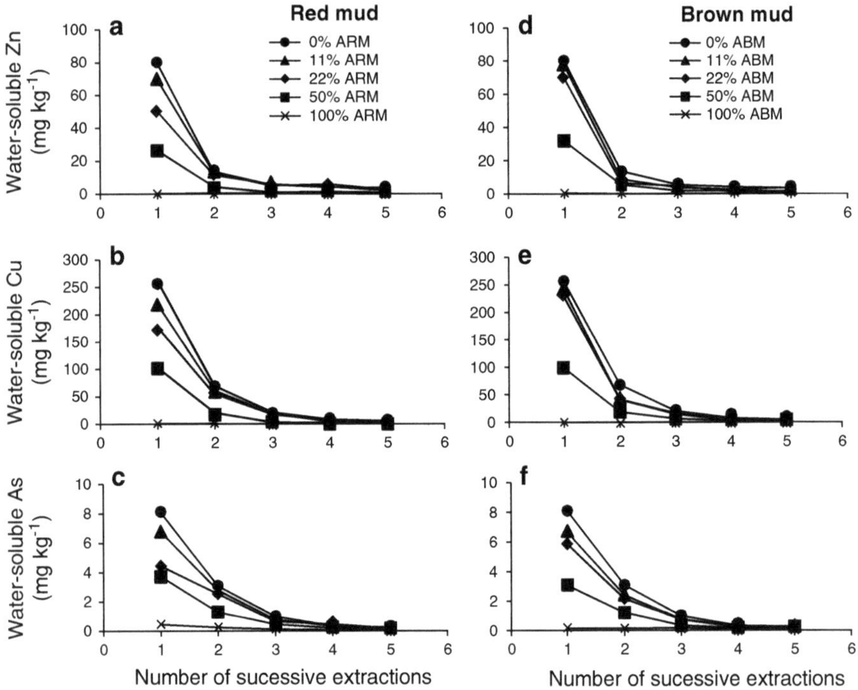

Fig. 12.7 Water extractable Zn, Cu and As from chicken litter treated with red mud (**a–c**) and brown mud (**d–f**). The y-axis is expressed as per kilogram of dry matter of chicken litter or cattle manure. All experimental data points are averages of two replicates with standard errors < 5 % (Adapted from Wang et al. 2011)

12.3.6 Impact on Metal Mobility

Treating with bauxite residue also enhanced the reduction of the water soluble fraction of Zn, Cu and As in chicken litter (Fig. 12.7). Red mud showed slightly higher reductions of extractable Zn and Cu than brown mud, especially at treatment rates of 11 and 22 % (Fig. 12.7a, b vs. Fig. 12.7d, e). This was likely due to the stronger adsorption of these two metal ions by the relatively large presence of Fe/Al oxide surfaces in red mud at pH >7.0. High pH could also promote more precipitation as metal hydroxides. There was no significant difference in water-soluble As reduction between the red mud and brown mud across three different levels of mixings (Fig. 12.7c vs. 12.7f). In all cases, the highest reduction in water-soluble metals in chicken litter was achieved at 50 % bauxite residue treatment with reductions of 60–70 % for Zn, 64–65 % for Cu, and 55–60 % for As. Similar reductions in these water-soluble fractions of these metals were also observed with bauxite residue-treated cattle manure (Wang et al. 2011). Among these three major metals often found in animal wastes, Zn is often less soluble whereas Cu and As can be very soluble (Jackson et al. 2003). The results of successive extractions shown in

Fig. 12.7 also indicate that 82 % As and 61 % Cu in chicken litter were soluble while approximately 23 % of Zn in both chicken litter and cattle manure was soluble. Nonetheless, bauxite residue treatment shows clear benefits of reducing the leaching risk of potentially toxic metals associated with animal manure.

Bauxite amendment of manure-affected soils also exhibits immobilization of Zn and Cu (Udeigwe et al. 2009). As shown in Fig. 12.8, the amendments exhibited significant ($P < 0.05$) reduction on both of the extractable metals in all three soils. Overall, the amendments reduced the water-extractable Cu by 2–98 %, and Zn by 10–90 %. At 2 % application rate, ARM (red mud from site A) was the most effective in reducing water-extractable Cu and Zn compared to ORM and ABM. In all cases, the site O red mud, ORM, was the least effective in reducing soluble metal release, especially Cu, likely due to its high background Cu concentration (Udeigwe et al. 2009). The seawater neutralized red muds, particularly the NORM, enhanced the reduction of water-extractable Cu release (Fig. 12.8), but had little improvement over the fresh ones in water-extractable Zn reduction, likely suggesting different mechanisms for retaining these two metals. As discussed earlier, the neutralization process increased the divalent cations, Mg and Ca, which could in turn enhance the binding between clay (or metal oxide) and organic matter (OM). Since Cu in manured soils is likely complexed by organic matter (Del Castilho et al. 1993; Han et al. 2000), the enhanced clay-OM binding by Mg and Ca could then increase soluble Cu retention, reducing its solubility and potential losses. On the other hand, Zn in the manured soils is likely in the exchangeable and precipitated forms (Han et al. 2000; Stietiya and Wang 2011) and would be less likely subjected to the effect of increased clay-OM binding caused by the neutralization of the red muds. In addition, increasing the application rate from 2 to 4 % increased the reduction of water-extractable Cu but had little or no impact on the reduction of water-extractable Zn for all the bauxite residue amendments. This result was consistent with generally stronger affinity of Fe/Al oxides for Cu than for Zn (Stietiya and Wang 2011).

Previous studies have also used bauxite residues to remediate metal-contaminated soils at mining sites and found that bauxite residues significantly reduce the metal concentrations in the soil pore water and shift metals from the exchangeable or "labile" form to the Fe-oxide fraction (Lombi et al. 2002a, b, 2004; Friesl et al. 2003; Garau et al. 2007). The chemisorption and decreasing acid extractability of metals were the main mechanisms for the shift (Lombi et al. 2002b). Application of 2 % red mud (w/w) in two contaminated soils decreased the uptake of Cd, Zn, Cu and Ni by oilseed rape, peas, wheat and lettuce (Lombi et al. 2002a). Field extractable fractions of Cd, Pb, and Zn were reduced by up to 96 %, 99 %, and 99 %, respectively, in bauxite residue-amended soils (Friesl et al. 2006).

12.3.7 *Impact on Fecal Coliforms*

While animal manures have indisputable nutritional values, they often contaminate waterways with fecal coliform bacteria in addition to nutrient runoff if mismanaged

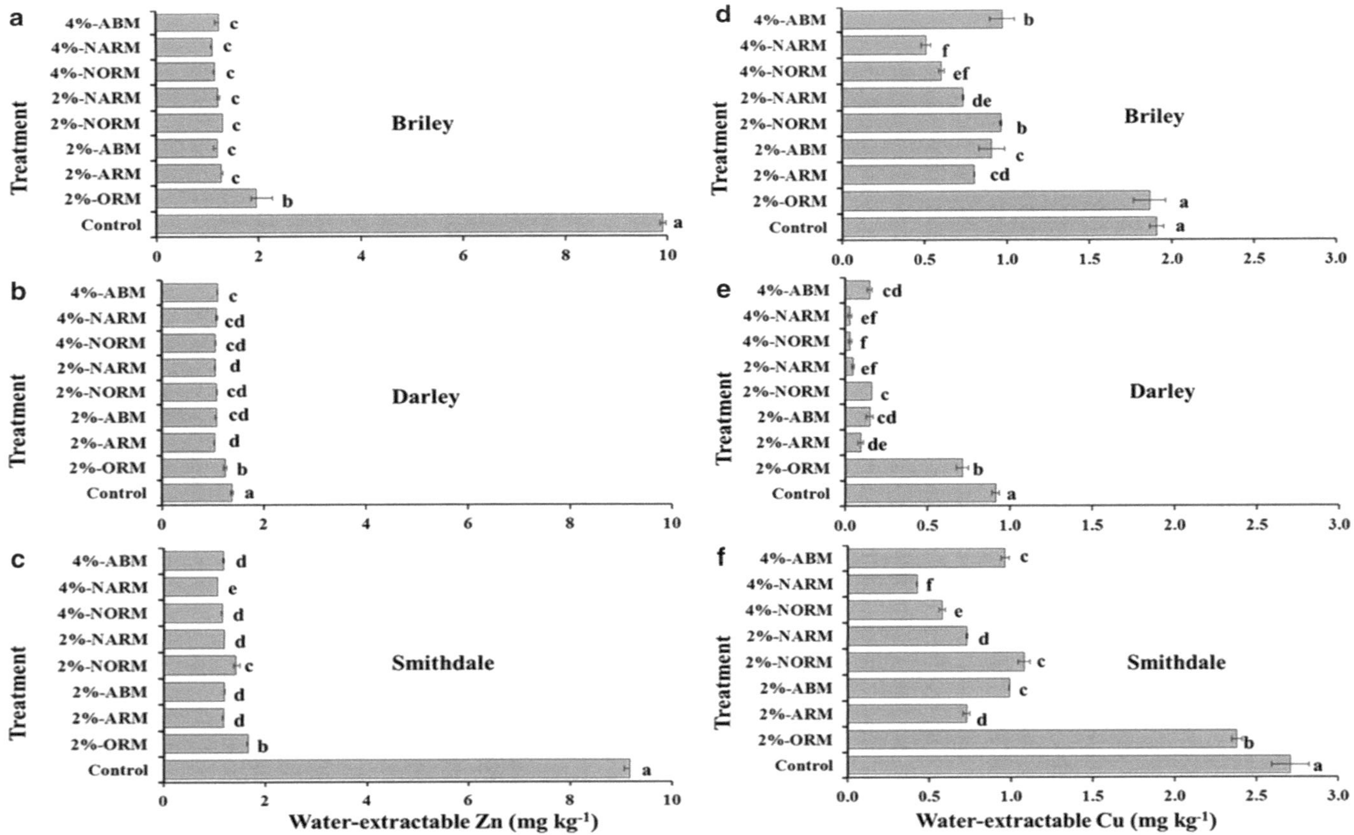

Fig. 12.8 Effect of seawater treated and untreated bauxite residue on water extractable Zn and Cu from manure-impacted soils: Briley (**a**,**d**), Darley (**b**,**e**), and Smithdale (**c**,**f**), respectively (Adapted from Udeigwe et al. 2009)

(Edwards et al. 2000; Gerba and Smith 2005; Soupier et al. 2006). To further explore the benefit of bauxite residue in improving characteristics of animal wastes for environmentally-friendly application, chicken litter was used as a model material and tested at a modest bauxite residue mixing rate of 22 % (Wang et al. 2011). Bauxite residue treatment had significant effects on fecal coliform activity in poultry litter. At 22 % mixing rate, bauxite residue reduced fecal coliform bacteria counts by more than one order of magnitude (Wang et al. 2011). Brown mud had slightly greater effect on fecal coliform bacteria die-off than red mud, although there was no statistical difference ($p > 0.05$). High pH is known to affect fecal coliform die-off, especially at a pH close to 9 in water systems (Pearson et al. 1987). The pH impact on fecal coliform population could be partly attributed to the production of toxic forms of oxygen such as hydroxyl ions at elevated pH which could damage the cytoplasmic membrane and hinder the growth of the organisms (Curtis et al. 1992). In a column study, red mud treated with 5 % gypsum and 30 % sand also showed excellent removal of *Escherichia coli*, *Salmonella adelaide* and poliovirus-1 in a secondary effluent from a wastewater treatment plant. Filtration, die-off and adsorption were thought to play roles in organism removal (Ho et al. 1991). Traditionally, composting has been commonly used to reduce fecal coliform in animal wastes. However, this process could concentrate P and therefore increase the absolute size of the dissolved P pool (Dao et al. 2001). Our results showed that direct treatment of animal manures with bauxite residues can reduce fecal coliform pollution (Wang et al. 2011). Other studies on soil application of bauxite residue showed that bauxite residue generally promoted bacterial abundance and soil enzyme activity (Lombi et al. 2002a; Garau et al. 2007). On the other hand, bauxite residue could cause a shift in the culturable bacterial population from Gram positive to Gram negative forms in soils (Garau et al. 2007).

12.3.8 Growing Plants on Soil Amended with Bauxite Residue

If red mud is added to soil to increase retention of P or other potential contaminant, this benefit must not be offset by any negative effect on plant growth. There has been extensive research on establishing vegetation on red mud (e.g., reviews by Wehr et al. (2006), Jones and Haynes (2011)) and considerable research on effects of red mud-amended soil on plants. Among the latter, Summers et al. (1996a) found that amending an acidic soil with red mud at rates to 40 Mg ha^{-1} increased yields of perennial ryegrass (*Lolium perenne*) and subterranean clover (*Trifolium subterraneum*) by increasing soil pH. While red mud amendments are beneficial to acidic pastures as liming agents up to 100 Mg ha^{-1} (Ward and Summers 1993), increased yields on acidic, sandy soils may be possible with additions up to 500 Mg ha^{-1} due to increased water-holding capacity. In more fertile soils, red mud may be used as a sink for excessive soil P (Summers et al. 1993; Udeigwe et al. 2009) without

Table 12.3 Effects of increasing rate of red mud addition to Ruston soil on the growth of transplanted switchgrass seedlings, and growth medium pH and salinity

Red mud		Salinity	Mean stem length[a]
%	pH	mg L^{-1}	cm
0	6.15	90	29 a
5	7.86	428	31 a
10	8.37	576	30 a
20	8.84	829	20 b

[a]Numbers in the column followed by the same letter are not significantly different ($\alpha = 0.05$)

adversely affecting yields (Summers et al. 1996b; Snars et al. 2004), even of horticultural crops (gypsum-treated red mud, Robertson et al. 1997).

Whether red mud is used as a topsoil amendment or a deeper band, the logical placement would be upslope from where runoff and subsurface flow discharge into surface water. This riparian area filters sediment and traps contaminants, depending on the type and stand of vegetation. Switchgrass (*Panicum virgatum*) is among several grasses that have been shown to effectively reduce contaminant loads in surface runoff (Lee et al. 1999, 2003; Kovar and Claassen 2009), and we investigated the effect of red mud concentration in Ruston soil on the growth of switchgrass seedlings. Since a permeable barrier of red mud would not cover a large area, transplanting seedlings is a reasonable option and may be preferable compared to seeding due to the negative effect of soil salinity on switchgrass seedling emergence and growth (Dkhili and Anderson 1990). Pots were filled with Ruston soil, or 19:1, 9:1 or 4:1 ratios of soil to red mud that had been amended with N, P, and K based on soil test values and recommendations for bermudagrass (*Cynodon dactylon*) hay on upland sandy loam soil. Gravimetric water content was maintained at 0.75 field capacity and Alamo switchgrass seedlings were grown for 1 month, after which soil pH and salinity were determined and stem lengths measured for growth. Results indicated that red mud addition to 10 % had no effect on the growth of transplanted switchgrass seedlings despite about 500 % greater salinity and basic pH (Table 12.3). While growth of switchgrass in the 20 % red mud medium was suppressed, all transplants survived. Since water flow through a permeable barrier would reduce salinity over time, our results tested the most severe conditions and showed that establishing transplanted switchgrass on a red mud barrier may be a viable option.

12.4 Conclusion

Reducing nutrient leaching enhances fertilizer value of animal manure while minimizing its adverse impact following long-term application to agricultural fields. This chapter demonstrated that bauxite residue (red mud and brown mud), a by-product from the aluminum refinery industry, could be used to improve the nutrient characteristics of animal manure by suppressing the leaching potential of water-soluble P and trace metals such as Zn, Cu and As. Bauxite residue also showed better retention of NH_4-N than alum-treated or untreated chicken litter following surface application to pasture fields. In addition, bauxite residue treatment of manure-affected soils significantly reduced DOC and BOD besides significantly reducing P and heavy metals in potential runoff losses. Sorption and transport studies of phosphate in soil amended with bauxite residue indicated that it could also be used for possibly making reactive permeable barriers. On the other hand, the high organic nature of animal manures appeared to be able to buffer the alkalinity contributed from the bauxite residues, preventing substantial increases in pH of the treated manures and manure-affected soils. Bauxite residue did not increase NH_3 release despite the fact that it elevated the pH of manure and manure-affected soils. Additionally, amending manures with bauxite residues decreased populations of fecal coliform in manure. Bauxite residue treatment could significantly increase the SAR of manure-affected soils if red mud containing high Na was used directly as an amendment without additional treatment. While increasing the animal manure/ bauxite mixing rates generally resulted in decreased pollutant leaching, the degree of the effect of these amendments on reducing pollutant leaching varied slightly within the bauxite residues and animal manures owing to the differences in elemental composition. Seawater neutralized bauxite residues, in general, exhibited greater retention of P and trace metals than un-neutralized ones. Bauxite residue addition up to 10 % (w/w) had no effect on the growth of transplanted switchgrass seedlings. Overall, bauxite residues could be used as a potential amendment for reducing P and other contaminant leaching in animal manures and manure-affected soils. Such a combined practice would improve beneficial uses of wastes from both animal production and the aluminum refinery industry.

12.5 Future Work

While research to date indicates that permeable barriers of bauxite residue mixed with sandy soil have a large capacity to retain input P, limits to this capacity that presumably depend on slow sorption/desorption kinetics are unknown. Experimental data are needed on the performance of barriers as affected by bauxite residue concentration, additives or pretreatments, and composition of the soil solution (P concentration, pH, etc.) so that these systems can be modeled and tentative design parameters determined. The use of bauxite residue as a dilute soil amendment to reduce loading of P and other contaminants into runoff or as a barrier to subsurface transport must be shown to have no adverse effects on plant growth or quality. Field plot studies on the long-term effects of bauxite residue amendments on performance of different pasture, riparian or biomass species in soils under long-term influence of manure application are planned. On the other hand, studies on proper incorporation of animal manure into sites of bauxite residue impoundments for revegetation of these sites are also be needed. The information from the relevant research will improve management for the sustainability of both animal production and the aluminum refinery industries.

References

Altundogan HS, Altundogan S, Tümen F, Bildik M (2002) Arsenic adsorption from aqueous solutions by activated red mud. Waste Manag 22(3):357–363

Amer F, Mahmoud AA, Sabet V (1985) Zeta-potential and surface-area of calcium-carbonate as related to phosphate sorption. Soil Sci Soc Am J 49:1137–1142

Amritphale SS, Patel M (1987) Utilisation of red mud, fly ash for manufacturing bricks with pyrophyllite. Silic Ind 2(3–4):31–35

Atun G, Hisarli G (2000) A study of surface properties of red mud by potentiometric method. J Colloids Interf Sci 228:40–45

Bednar AJ, Garbarino JR, Ranville JF, Wilderman TR (2002) Analysis and occurrence of organoarsenicals used in agriculture. In: Proceedings of the 23rd annual American Chemical Society Orland, FL

Bolan NS, Adriano DC, Mahimairaja S (2004) Distribution and bioavailability of trace elements in livestock and poultry manure by-products. Crit Rev Environ Sci Technol 34:291–338

Brunori C, Cremisini C, D'Annibale L, Massanisso P, Pinto V, Torricelli L (2005a) A kinetic study of trace element leachability from abandoned mine-polluted soil treated with SS-MSW compost and red mud: comparison with results from sequential extraction. Anal Bioanal Chem 381:1347–1354

Brunori C, Cremisini C, Massanisso P, Pinto V, Torricelli L (2005b) Reuse of a treated red mud bauxite waste: studies on environmental compatibility. J Hazard Mater B 117:55–63

Buda AR, Kleinman PJ, Bryant RB, Drohan PJ, Folmar GJ, Dewalle DR (2012) Phosphorus transport by surface and subsurface flow pathways in an upland agricultural watershed [abstract]. ASA-CSSA-SSSA annual meeting abstracts. Poster no. 335

Çengeloğlu Y, Kir E, Ersöz M (2002) Removal of fluoride from aqueous solution by using red mud. Sep Purif Technol 28:81–86

Çengeloğlu Y, Tor A, Ersöz M, Arslan G (2006) Removal of nitrate from aqueous solution by using red mud. Sep Purif Technol 51:374–378

Chvedov D, Ostap S, Le T (2001) Surface properties of red mud particles from potentiometric titration. Colloids Surf A 182:131–141

Constable TW, Edward RM (1979) BOD/TOC correlations and their application to water quality evaluation. Water Air Soil Pollut 1:363–375

Coruh S, Ergun ON (2010) Use of fly ash, phosphogypsum and red mud as liner material for the disposal of hazardous zinc leach residue water. J Hazard Mater 173:468–473

Cucarella V, Renman G (2009) Phosphorus sorption capacity of filter materials used for on-site wastewater treatment determined in batch experiments – a comparative study. J Environ Qual 38:381–392

Curtis TP, Mara DD, Silva SA (1992) Influence of pH, oxygen, and humic substances on ability of sunlight to damage fecal coliforms in waste stabilization pond water. Appl Environ Microbiol 58:1335–1343

Dao TH (1999) Co-amendments to modify P extractability and N:P ratio in feedlot manure and composted manure. J Environ Qual 28:1114–1121

Dao TH, Sikora LJ, Hasasaki A, Chaney RL (2001) Manure phosphorus extractability as affected by aluminum- and iron by-products and aerobic composting. J Environ Qual 30:1693–1698

Darke AK, Walbridge MR (2000) Al and Fe biogeochemistry in a floodplain forest: implications for P retention. Biogeochemistry 51:1–32

Del Castilho P, Chardon WJ, Salomons S (1993) Influence of cattle-manure slurry application on the solubility of cadmium, copper, and zinc in a manured acidic, loamy-sand soil. J Environ Qual 22:689–697

Dkhili M, Anderson B (1990) Salt effects on seedling growth of switchgrass and big bluestem. In: Smith DD, Jacobs CA (eds) Proceeding of the twelfth North American prairie conference: recapturing a vanishing heritage, pp 13–16. University of Northern Iowa, Cedar Fall, IA

Duchesne J, Doye I (2005) Effectiveness of covers and liners made of red mud bauxite and/or cement dust for limiting acid mine drainage. J Environ Eng 131:1230–1235

Edwards DR, Larson BT, Lim TT (2000) Runoff nutrient and fecal coliform content from cattle manure application to fescue plots. J Am Water Res Assoc 36:711–721
Friesl W, Lombi E, Horak O, Wenzel WW (2003) Immobilization of heavy metals in soils using inorganic amendments in a greenhouse study. J Plant Nutr Soil Sci 166:191–196
Friesl W, Friedl J, Platzer K, Horak O, Gerzabek MH (2006) Remediation of contaminated agricultural soils near a former Pb/Zn smelter in Austria: batch, pot and field experiments. Environ Pollut 144(1):40–50
Garau G, Castaldi P, Santona L, Deiana P, Melis P (2007) Influence of red mud, zeolite and lime on heavy metal immobilization, culturable heterotrophic microbial populations and enzyme activities in a contaminated soil. Geoderma 142:47–57
Gaston L, Drapcho C, Tapadar C, Kovar J (2003) Phosphorus runoff relationships for Louisiana Coastal Plain soils amended with poultry litter. J Environ Qual 32:1422–1429
Genç-Fuhrman H, Tjell JC, McConchie D (2004) Increasing the arsenate adsorption capacity of neutralized red mud (Bauxsol). J Colloids Interf Sci 271:313–320
Gerba CP, Smith JE (2005) Sources of pathogenic microorganisms and their fate during land application of wastes. J Environ Qual 34:42–48
Gupta VK, Ali I (2006) Removal of 2,4-dinitrophenol from wastewater by adsorption technology: a batch and column study. Int J Environ Pollut 27:104–120
Gupta VK, Sharma S (2002) Removal of cadmium and zinc from aqueous solutions using red mud. Environ Sci Technol 36:3612–3617
Gupta VK, Gupta M, Sharma S (2001) Process development for the removal of lead and chromium from aqueous solutions using red mud – an aluminum industry waste. Water Res 35:1125–1134
Gupta VK, Ali I, Saini VK (2004a) Removal of chlorophenols from wastewater using red mud: an aluminum industry waste. Environ Sci Technol 38:4012–4018
Gupta VK, Suhas Ali I, Saini VK (2004b) Removal of rhodamine B, fast green, and methylene blue from wastewater using red mud, an aluminum industry waste. Ind Eng Chem Res 43:1740–1747
Han FX, Kingery WL, Selim HM, Gerard PD (2000) Accumulation of heavy metals in a long-term poultry waste-amended soil. Soil Sci 165:260–268
Hanahan C, McConnie D, Pohl J, Creelman R, Clark M, Stocksiek C (2004) Chemistry of seawater neutralization of bauxite refinery residues (red mud). Environ Eng Sci 21:125–138
Harrell DL, Wang JJ (2006) Fractionation and sorption of inorganic phosphorus in Louisiana calcareous soils. Soil Sci 171:39–51
Ho GE, Gibbs RA, Mathew K (1991) Bacteria and virus removal from secondary effluent in sand and red mud columns. Water Sci Technol 23:261–270
Ho GE, Mathew K, Gibbs RA (1992) Nitrogen and phosphorus removal from sewage effluent in amended sand columns. Water Res 26:295–300
Hsu PH (1976) Comparison of iron (III) and aluminum in precipitation of phosphate from solution. Water Res 10:903–907
Huang W, Wang S, Zhu Z, Li L, Yao X, Rudolph V, Haghseresht F (2008) Phosphate removal from wastewater using red mud. J Hazard Mater 158:35–42
Jackson BP, Bertsch PM, Cabrera ML, Camberato JJ, Seaman JC, Wood CW (2003) Trace element speciation in poultry litter. J Environ Qual 32:535–540
Jones BEH, Haynes RJ (2011) Bauxite processing residue: a critical review of its formation, properties, storage, and revegetation. Crit Rev Environ Sci Technol 41:271–315
Jones BEH, Haynes RJ, Phillips IR (2012) Addition of an organic amendment and/or residue mud to bauxite residue sand in order to improve its properties as a growth medium. J Environ Manage 95:29–38
Kirchmann H, Witter E (1989) Ammonia volatilization during aerobic and anaerobic manure decomposition. Plant Soil 115:35–41
Koopmans GF, Chardon WJ, McDowell RW (2007) Phosphorus movement and speciation in a sandy soil profile after long-term animal manure application. J Environ Qual 36:305–315

Kovar JL, Claassen N (2009) Growth and phosphorus uptake of three riparian grass species. Agron J 101:1060–1067
Kumar S, Kumar R, Bandopadhyay A (2006) Innovative methodologies for the utilization of wastes from metallurgical and allied industries. Resour Conserv Recycl 48:301–314
Lamonier JF, Wyralski F, Leclercq G, Aboukais A (2005) Recycling of waste, red mud, as a catalyser for the elimination of volatile organic compounds. Can J Chem Eng 83:737–741
Lapointe F, Fytas K, McConchie D (2006) Efficiency of Bauxsol^{TM} in permeable reactive barriers to treat acid rock drainage. Mine Water Environ 25:37–44
Lee K-H, Isenhart TM, Schultz RC, Mickelson SK (1999) Nutrient and sediment removal by switchgrass and cool-season grass filter strips in central Iowa, USA. Agrofor Syst 44:121–132
Lee K-H, Isenhart TM, Schultz RC (2003) Sediment and nutrient removal in an established multi-species riparian buffer. J Soil Water Conserv 58:1–8
Li Y, Liu C, Luan Z, Peng X, Zhu C, Chen Z, Zhang Z, Fan J, Jia Z (2006) Phosphate removal from aqueous solutions using raw and activated red mud and fly ash. J Hazard Mater B 137:374–383
Lindsay WL (1979) Chemical equilibria in soils. Wiley, New York
Liu C, Li Y, Luan Z, Chen Z, Zhang Z, Jia Z (2007a) Adsorption removal of phosphate from aqueous solution by activated mud. J Environ Sci 19:1166–1170
Liu Y, Lin C, Wu Y (2007b) Characterization of red mud derived from a combined Bayer process and bauxite calcinations method. J Hazard Mater 146:255–261
Liu Y, Naidu R, Ming H (2011) Red mud as an amendment for pollutants in solid and liquid phases. Geoderma 163:1–12
Lombi E, Zhao FJ, Wieshammer G, Zhang GY, McGrath SP (2002a) In situ fixation of metals in soils using bauxite residue: biological effects. Environ Pollut 118:445–452
Lombi E, Zhao FJ, Zhang GY, Sun B, Fitz W, Zhang H, McGrath SP (2002b) In situ fixation of metals in soils using bauxite residue: chemical assessment. Environ Pollut 118:435–443
Lombi E, Hamon RE, Wieshammer G, McLaughlin MJ, McGrath SP (2004) Assessment of the use of industrial by-products to remediate a copper- and arsenic-contaminated soil. J Environ Qual 33:902–910
López E, Soto B, Arias M, Núñez A, Rubinos D, Barral MT (1998) Adsorbent properties of red mud and its use for wastewater treatment. Water Res 32:1314–1322
Mansell RS, Selim HM, Kanchanasut P, Davidson JM, Fiskell JGA (1977) Experimental and simulated transport of phosphorus through sandy soils. Water Resour Res 13:189–194
McConchie D, Clark MW, Davies-McConchie F (2002) New strategies for the management of bauxite refinery residues (red mud). In: Proceedings of the 6th international alumina quality workshop, Brisbane, Australia, 8–13 September 2002, pp 327–332
Menzies NW, Fulton IM, Morrell WJ (2004) Seawater neutralization of alkaline bauxite residue and implication for revegetation. J Environ Qual 33:1877–1884
Moore PA Jr, Miller DM (1994) Decreasing phosphorus solubility in poultry litter with aluminum, calcium, and iron amendments. J Environ Qual 23:325–330
Moore PA Jr, Daniel TC, Edwards DR, Miller DM (1995) Effect of chemical amendments to reduce ammonia volatilization from poultry houses. J Environ Qual 24:293–300
Moore PA Jr, Daniel TC, Gilmour JT, Shreve BR, Edwards DR, Wood BH (1998) Decreasing metal runoff from poultry litter with aluminum sulfate. J Environ Qual 27:92–99
Munro LD, Clark MW, McConchie D (2004) A Bauxsol^{TM}-based permeable reactive barrier for the treatment of acid rock drainage. Mine Water Environ 23:183–194
North M, Bell D (1998) Commercial chicken production manual, 4th edn. Van Nostrand Reinhold, New York
Oreščanin V, Nad K, Mikelic L, Mikulic N, Lulic S (2006) Utilization of bauxite slag for the purification of industrial wastewaters. Process Safe Environ 84:265–269
Pearson HW, Mara DD, Bartone CR (1987) Guidelines for the minimum evaluation of the performance of full-scale wastes stabilization pond systems. Water Res 21(9):1067–1075
Penn CJ, Bryant RB, Callahan MP, McGrath JM (2011) Use of industrial by-products to sorb and retain phosphorus. Commun Soil Sci Plant Anal 42:633–644

Peters JM, Basta NT (1996) Reduction of excessive bioavailable phosphorus in soils by using municipal and industrial wastes. J Environ Qual 25:1236–1241

Pradhan J, Das J, Das S, Thakur RS (1998) Adsorption of phosphate from aqueous solution using activated red mud. J Colloid Interf Sci 204:169–172

Robertson WJ, Jeffery RC, McPharlin IR (1997) Residues from bauxite-mining (red mud) increase phosphorus retention of a Joel sand without reducing yield of carrots. Commun Soil Sci Plant Anal 28:1059–1079

Ross DS, Bartlett RJ (1996) Field-extracted spodosol solutions and soils: aluminum, organic carbon, and pH interrelationships. Soil Sci Soc Am J 60:589–595

Rostagno CM, Sosebee RE (2001) Biosolids application in the Chihuahuan desert: effects on runoff water quality. J Environ Qual 30:160–170

Ryan J, Curtin D, Cheema MA (1985) Significance of iron oxides and calcium carbonate particle size in phosphate sorption by calcareous soils. Soil Sci Soc Am J 49:74–76

Santona L, Castaldi P, Melis P (2006) Evaluation of the interaction mechanisms between red muds and heavy metals. J Hazard Mater 136:324–329

Sawyer CN, McCarty PL, Parkin GF (2002) Chemistry for environmental engineering and science. McGraw-Hill Inc., New York

Schreve BR, Moore PA Jr, Daniel TC, Edwards DR, Miller DM (1995) Reductions of P in runoff from field-applied poultry litter using chemical amendments. J Environ Qual 24:106–111

Sharpley AN, Daniel TC, Sims JT, Lemunyon J, Stevens R, Parry R (2003a) Agricultural phosphorus and eutrophication, 2nd edn. USDA, University Park, ARS-149, 44pp

Sharpley AN, Weld JL, Beegle D, Kleinman PJA, Gburek WJ, Moore PA Jr, Mullins G (2003b) Development of phosphorus indices for nutrient management planning strategies in the United States. J Soil Water Conserv 58:137–152

Sims JT (1995) Characteristics of animal wastes and waste amended soils: an overview of the agricultural and environmental issues. In: Steele K (ed) Animal waste and the land–water interface. CRC Press, Boca Raton, pp 1–14

Sims JT, Wolf DC (1994) Poultry waste management: agricultural and environmental issues. Adv Agron 52:1–83

Snars K, Gilkes RJ (2009) Evaluation of bauxite residues (red muds) of different origins for environmental applications. Appl Clay Sci 46:13–20

Snars K, Hughes JC, Gilkes RJ (2004) The effects of the addition of bauxite red mud to soil on P uptake by plants. Aust J Agric Res 55:25–31

Soupier ML, Mostaghimi S, Yagow ER, Hagedorn C, Vaughan DH (2006) Transport of fecal bacteria from poultry litter and cattle manures applied to pastureland. Water Air Soil Pollut 169:25–136

Stietiya MH, Wang JJ (2011) Effect of organic matter oxidation on the fractionation of copper, zinc, lead and arsenic in sewage sludge and amended soils. J Environ Qual 40:1162–1171

Summers RN, Pech JD (1997) Nutrient and metal content of water, sediment and soils amended with bauxite residue in the catchment of the Peel Inlet and Harvey Estuary, Western Australia. Agric Ecosys Environ 64:219–232

Summers RN, Guise NR, Smirk DD (1993) Bauxite residue (red mud) increases phosphorus retention in sandy soil catchments in Western Australia. Fertil Res 34:85–94

Summers R, Guise NR, Smirk DD, Summers KJ (1996a) Bauxite residue (red mud) improves pasture growth on sandy soils in Western Australia. Aust J Soil Res 34:569–581

Summers RN, Smirk DD, Karafilis D (1996b) Phosphorus retention and leachates from sandy soil amended with bauxite residue (red mud). Aust J Soil Res 34:555–567

Tor A, Çengeloğlu Y, Aydin ME, Ersöz M (2006) Removal of phenol from aqueous phase by using neutralized red mud. J Colloids Interf Sci 300:498–503

Turner RE, Rabalais NN (2003) Linking landscape and water quality in the Mississippi River basin for 200 years. BioScience 53:563–572

Udeigwe TK, Wang JJ, Zhang H (2009) Effectiveness of bauxite residues in immobilizing contaminants in manure-amended soils. Soil Sci 174:676–686

Ulmgren L (1975) Swedish experiences in chemical treatment of wastewater. J Water Pollut Cont Fed 47:696–703

USEPA (1984) Overview of solid waste generation, management, and chemical characteristics in bauxite refining and primary aluminum industry. Prepared by Radon Corporation for EPA. Office of Solid Waste, EPA, Washington, DC

Vaclavikova M, Misaelides P, Gallios G, Jakabsky S, Hredzak S (2005) Removal of cadmium, zinc, copper and lead by red mud, an iron oxides containing hydrometallurgical waste. Stud Surf Sci Catal 155:517–525

van Beinum W, Beulke S, Brown CD (2005) Pesticides sorption and diffusion in natural clay loam aggregates. J Agric Food Chem 53:9146–9154

Van der Watt HVH, Sumner ME, Cabrera ML (1994) Bioavailability of copper, manganese, and zinc in poultry litter. J Environ Qual 23:43–49

van Genuchten MT (1981) Non-equilibrium transport parameters from miscible displacement experiments, Research report, 119. U.S. Salinity Laboratory, Riverside, CA

van Genuchten MT, Wierenga PJ (1976) Mass transfer studies in sorbing media. I. Analytical solutions. Soil Sci Soc Am J 40:473–480

Vincenzo MS, Renzo C, Stefano M, Giovanni JC (2000) Bauxite red mud in the ceramic industry. Part 1: thermal behavior. Eur Ceram Soc 20:235–244

Walthall PM, Nolfe JD (1998) Poultry litter amendments and the mobility and immobility of phosphorus in soil landscapes of northern Louisiana. Louisiana Agric 41:28–31

Wang S, Ang HM, Tade MO (2008) Novel applications of red mud as coagulant, adsorbent and catalyst for environmentally benign process. Chemosphere 72:1621–1635

Wang JJ, Zhang H, Schroder JL, Udeigwe TK, Zhang Z, Dodla SK, Stietiya MH (2011) Reducing potential leaching of phosphorus, heavy metals, and fecal coliform from animal wastes using bauxite residues. Water Air Soil Pollut 214:241–252

Ward SC, Summers RN (1993) Modifying sandy soils with the fine residue from bauxite refining to retain phosphorus and increase plant yield. Fertil Res 36:151–156

Wehr JB, Fulton I, Menzies NW (2006) Revegetation strategies for bauxite refinery residue: a case study of Alcan Grove in Northern Territory, Australia. Environ Manage 37:297–306

Weld JL (2003) Summary of phosphorus indices. http://www.ars.usda.gov/sp2UserFiles/Place/19020500/PhosphorousImages/Summary_Pindices.pdf. Accessed on February 25, 2014

Whittaker CW, Armiger WH, Chichilo PP, Hoffman WM (1955) "Brown mud" from the aluminum industry as a soil liming material. Soil Sci Soc Am J 19:288–292

Wong JWC, Ho G (1994) Sewage sludge as organic ameliorant for revegetation of fine bauxite refining residue. Resour Conserv Recycl 11:297–309

Zijlstra JJP, Dessi R, Peretti R, Zucca A (2010) Treatment of percolate from metal sulfide mine tailings with a permeable reactive barrier of transformed red mud. Water Environ Res 82:319–327

Chapter 13
Investigation of Compound-Specific Organic-Inorganic Phosphorus Transformation Using Stable Isotope Ratios in Phosphate

Deb P. Jaisi, Ruth E. Blake, Yuhong Liang, and Sae Jung Chang

Abstract Phosphorus (P) is one of the most important nutrients for all life forms irrespective of physiological uptake mechanism or metabolic pathway. Soils and sediments contain a variety of organic P compounds that may undergo active transformation into inorganic P and vice versa. Phosphate oxygen isotope ratios have increasingly been applied to better understand the physico-chemical and biological pathways of P cycling and its fate in agricultural and nonagricultural soils. For example, the fractionation factors during hydrolysis of structurally-similar organic P compounds studied thus far are often distinct. Therefore compound-specific isotopic composition could be a powerful tool for differentiating transformation of different P compounds, tracking their origin and fate, and ultimately to develop an integrated and quantitative understanding of P cycling in both extant and fossil ecosystems. This chapter reviews biochemical reaction mechanisms and the current state of knowledge on compound-specific isotopic effects during hydrolysis of organic P compounds. While this research is still in its infancy, a new paradigm has emerged and it is hoped that the future expansion of this research will allow development of a holistic approach to integrate transformation of organic and inorganic P over time and space in different ecosystems.

D.P. Jaisi (✉)
Department of Plant and Soil Sciences, University of Delaware, Newark, DE 19716, USA
e-mail: Jaisi@udel.edu

R.E. Blake • S.J. Chang
Department of Geology and Geophysics, Yale University, New Haven, CT 06520, USA

Y. Liang
Department of Microbiology, Mount Sinai School of Medicine, New York, NY 10029, USA

Z. He and H. Zhang (eds.), *Applied Manure and Nutrient Chemistry for Sustainable Agriculture and Environment*, DOI 10.1007/978-94-017-8807-6_13,

13.1 Introduction: Issues on P Resources and Cycling in the Environment

Phosphorus (P) is an essential nutrient and a key element involved in mediating the connection between living and non-living parts of the biosphere. Despite its occurrence at ~0.09 wt% in Earth's crust, P is a key ingredient of fertilizers and there is no other substitute for P in sustaining life and food production. Because world's P reserves and resources are declining, global issue on short and long-range sustainable use of P has become a more pressing and greater concern than before.

From the time of the agricultural revolution, the movement of P in the environment has been largely unidirectional – from mines to fertilizers for crop production, and then to processing foods and fodders for human and animal consumption (Suh and Yee 2011). In this unidirectional transport, driven primarily by human efforts, a significant fraction of extracted P is lost back to the environment. Although long-term recovery of P applied to soils may approach as high as 90 % because P fixed in the soil may change slowly into available forms (Syers et al. 2008), excess P could be released out of agricultural soils into lakes, rivers, and the ocean and deteriorate water quality. For example, eutrophication and water quality issues in the Chesapeake Bay and Gulf of Mexico in the US have caught media attention for decades and still continued to be highlighted in news and reports such as from the National Academy of Engineering (Dzombak 2011), USEPA (Ohio EPA 2010), professional magazines (Erickson 2011) as well as in a recently formed US research coordination network on sustainable P (RCN-SEES 2012). Although the double-sided threats of P (declining reserves and polluting water resources) are well-known, feasible solutions to these complex problems are seemingly intractable unless new approaches to better understand nutrient cycling and nutrient-plant interactions as well as more efficient use of P are identified.

Because of the low stoichiometric need for P as compared to other major nutrients (ca. 106C: 16N: 1P; Redfield 1958), small amounts of P addition can cause severe impacts on water quality in the receiving catchments and groundwater aquifers, and promote eutrophication (Wetzel 1983). However, understanding the sources, cycling, and transformation of P is not straightforward, compared to many other nutrients, particularly because of: (1) very low concentration of dissolved phosphate compared to the total P pool in soil and sediments and redistribution of phosphate from one phase/pool to another (Frossard et al. 1995; Bünemann and Condron 2007; Jaisi et al. 2011); (2) active but variable transformation of organic and inorganic P compounds in macro- and micro-soil environments, (3) co-occurring biotic and abiotic reactions to dissolve, precipitate, and cycle phosphate at time scales varying from a few seconds to several years (Bünemann and Condron 2007; Jaisi et al. 2011), and (4) organisms' variable strategies to uptake and store P depending both on the concentrations and compositions of organic and inorganic P compounds in the environment. These complexities not only hamper the better understanding of sources, cycling, and transformation of different P compounds, but also restrict formulation of an effective guideline for nutrient management plans that could eventually lead to the desired outcome of decreased P loss. These deeply complex issues and existing limitations to track P sources downstream have opened up debates on accountability issues on the nutrient loads released to the different open waters and ecosystems in the US and around the world. This accountability debate, unfortunately, will remain open ended unless individual sources are tracked to their origins or to specific biogeochemical processes. Therefore, fuller understanding of P sources and cycling is urgent not only for an academic interest but for greater environmental and societal importance.

In this chapter, we discuss hydrolysis pathways of organophosphorus compounds and update current progress in the compound-specific isotopic fractionation approach used to understand these pathways. We include hydrolysis in the presence of specific phosphoenzymes, ultraviolet radiation (UVR), and by intact cells, and explore how this approach could be integrated to better understand P cycling in soils and other ecosystems. While the compound-specific isotopic study of the hydrolysis of organophosphorus compounds is still in its infancy, it is hoped that the future expansion of this research along with inorganic P will collectively help develop a holistic approach to integrate transformation of organic and inorganic P over time and space in different ecosystems. Such integration may be able to identify key reactions and processes in the environments and can be exploited to address major issues on P sources and environmental concerns related to P contamination.

13.2 Organic Phosphorus and Its Cycling in the Environment

Organic P (Po) constitutes a variety of organic compounds containing P-primarily as phosphate ester or condensed P (Fig. 13.1). Some examples of organic P include nucleic acids, phospholipids, phosphoamide, phosphoproteins, phosphonic acids, and many others. In general, the composition of Po depends on nature of the environment, soil/sediment type, composition of biota, and degree of P demand, among others (e.g., Ammerman and Azam 1985; Smucker and Kim 1991; Dyhrman and Ruttenberg 2006). Diversity of Po in the environment is not well characterized and the role of Po in the global phosphorus cycle remains poorly understood. This is partly because of the lack of widely available techniques for measuring Po concentration and speciation (McKelvie 2005).

Research on the potential role of organic phosphorus compounds in the environment started in the 1970s, when attention was focused on eutrophication (Jong 2006). Follow up studies found that dissolved organic P was at least partially bioavailable to bacterioplankton and can support bacterial growth and metabolic activities, especially when the concentration of inorganic P (Pi), the preferred form of P for microbial uptake, is insufficient to meet P demand (Cotner and Wetzel 1992; Benitez-Nelson 2000; Nicholson et al. 2006; Vanderzee et al. 2007). Many Po compounds are not readily utilizable owing to their large size and complex

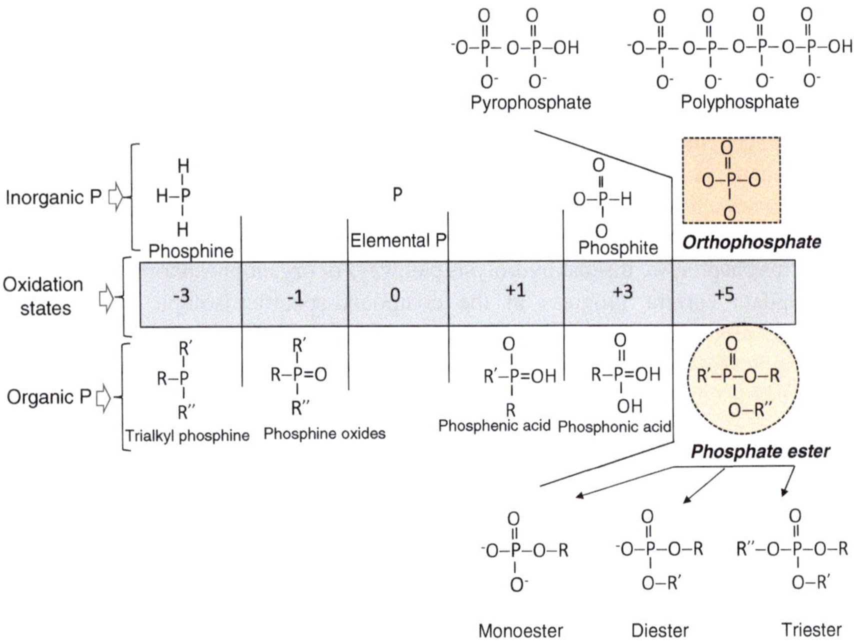

Fig. 13.1 Organic and inorganic P compounds in different oxidation states. Esters and orthophosphates are highlighted. R, R′, and R″ refer to different esters attached to the phosphate group. The variable composition of R provides the origin of different of organic P compounds in the environment

chemical structures, and have to be broken down extracellularly into a usable form before uptake (e.g., Ehrlich 1996). With the recent development of new methods to identify and quantify Po in the environment, Po has been found to be less refractory than previously assumed (e.g., Cotner and Wetzel 1992; Nausch and Nausch 2007; van Moorleghem et al. 2013).

A wide variety of phosphoenzymes present in the environment can hydrolyze Po compounds to release Pi. These enzymes can be located both outside and inside of living cells. Major phosphoenzymes that are active in terrestrial and marine ecosystems include 5′-nucleotidase, peptidase, alkaline phosphatase, aminopeptidase, phosphodiesterase, chitinase etc. (Fig. 13.2). These enzymes can be non-specific and cleave P–O–C bond of different Po compounds but some of them are highly specific to certain Po structures (e.g. nucleotides) or bond types and may act only under specific reaction conditions (e.g., pH). Some other enzymes such as C-P lyase cleaves C-P bond of a broad spectrum of phosphonates. Application of phosphate oxygen stable isotope ratios ($\delta^{18}O_p$) in Po compounds to identify sources or reaction pathways/mechanisms is not straightforward, compared to inorganic phosphate compounds, because almost all Po compounds contain extra, non-PO_4 oxygen atom in their structural sites. Therefore, phosphate ($-PO_4$) in Po has to be selectively removed from the Po structure before isotopic analysis. The breaking of a P–O–C or C–P bond at the catalytic site of an enzyme or by abiotic means involves

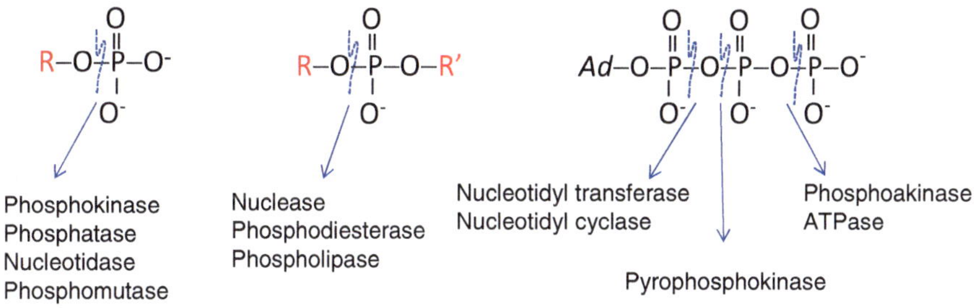

Fig. 13.2 Classes of enzymes that catalyze reactions at P in monoester, diester, and anhydride compounds. The dashed arrows show specific cleaving positions. Please note that diester hydrolysis occurs in two steps- first endonuclease cleaves the internal P diester linkage and produces oligonucleotides and second, an exonuclease further cleaves P diester bonds at the terminal nucleotide position and forms free nucleotide (monoester). The nature of the reaction catalyzed by each enzyme depends on the structure of the organophosphorus compound and enzyme specificity. 'Ad' refers to adenosine (from Knowles 1980) (Reprinted with permission from the *Annual Review of Biochemistry*, 49, 1980 by Annual Reviews)

incorporation of exogenous/outside oxygen atom (e.g., from water or other oxygen containing sources such as air) and fractionation, if any of oxygen isotopes between incorporated and residual (exogenous) oxygen atoms. Therefore, both the number of incorporated oxygen atoms and associated fractionation factors have to be taken into account to track hydrolysis of Po compounds.

Blake et al. (1997, 1998a, b) attempted the first controlled laboratory studies on the hydrolysis of Po with an aim to understand isotopic fractionation of oxygen between Pi metabolized by microbial enzymes both in-vitro and by purified cell-free enzymes (e.g., APase, inorganic pyrophosphatase) and water. From these studies, it has been found that the intracellular enzyme inorganic pyrophosphatase (PPase) is the only enzyme known, so far, to catalyze complete and temperature-dependent equilibrium exchange of O between Pi and water.

Because PPase is one of the most relevant enzymes in isotope biogeochemistry, it is relevant to discuss its distribution in different domains of life. PPase enzymes are broadly classified into four non-homologous families. Among them, three are soluble ones (Families I, II, and III) and the forth one is membrane-bound (Family IV) (Harway and Keister 1981; Kasho and Baykov 1989; Baykov et al. 1991, 1994; Baykov and Shestakov 1992; Sivula et al. 1999; Smirnova et al. 1995; Zyryanov et al. 2004). Family I PPases are most widespread in all kingdoms of life, but the family II is found to be present only in bacteria and archaea. Family IV (membrane PPase) is present in all plants and a quarter of bacteria and archaea (personal communication with Alexander Baykov). Except for soluble Family III (which is not tested for this reaction), all other PPase families catalyze reversible oxygen exchange in phosphate (Heikinheimo et al. 1996).

It is important to note that it is the PPase enzyme itself that resets PO_4 $\delta^{18}O_p$ values regardless of its location in cell membrane or in cytoplasm. Soluble PPases are rather stable enzymes and may be present in extracellular environments. Therefore when released out of cell such as due to cell lysis, death, or predation, it is intuitive to assume that they may still catalyze isotope exchange reaction. Further research is needed to understand the role of extracellular PPase in pyrophosphate synthesis and breakdown.

In contrast, alkaline phosphatase (APase) is a key enzyme involved in extracellular hydrolysis of phosphomonoesters, and is characterized by large kinetic fractionations, incomplete Pi-water O isotope exchange, and disequilibrium isotope effects as well as inheritance of original O from the Po substrate (Blake et al. 2005; Liang and Blake 2006b).

13.2.1 Enzymatic and Non-enzymatic Hydrolysis of Organophosphorus Compounds

Reactions involving P in phosphate esters are fundamental to many basic metabolic pathways in living organisms. Dubbed as 'Yin and Yang' of cellular signaling, kinase and phosphatase play central roles in the formation and hydrolysis of phosphate monoesters (Hengge 2002). The general nature of the hydrolysis reaction, particularly the type of reaction (such as a concerted reaction with transition states or a stepwise reaction with discrete intermediate phases) and variation among different esters are fairly well known. However, the differences in hydrolysis mechanisms and the nature of intermediate phases or transition states for enzyme-catalyzed (involving enzymes) vs. non-enzymatic (abiotic) reactions are still unclear. This is due largely to the variable experimental methods chosen to study hydrolysis reactions, limitations of the methods chosen, and inconsistencies among methods and terminologies used (e.g., Mildvan 1979; Knowles 1980; Hengge 2002; Cassano et al. 2004; Lassila et al. 2011). These factors have caused complications in synthesizing literature data to employ the wealth of knowledge already gained in biochemistry/enzymology to isotope geochemistry. For example, potential differences in isotopic fractionation accompanying hydrolysis of an organophosphorus compound in enzyme-catalyzed vs. non-enzymatic reactions were not studied until very recently (Liang and Blake 2006a). Differentiating enzymatic and non-enzymatic reaction mechanisms from a structural perspective provides a basis for developing a framework for identifying compound- specific isotopic fractionations accompanying hydrolysis of specific Po compounds. We review hydrolysis mechanisms that differ for or are specific to monoester, diester, and triester compounds and the current state of understanding developed in biochemistry and enzymology. It is our hope that this information can be integrated to interpret isotopic exchange and fractionations accompanying organophosphorus compound degradation and to also justify the need for the compound specific- isotope research that has recently been started.

Enzyme-catalyzed hydrolysis reactions are tremendously fast. For example, phosphatase catalyzed monoester hydrolysis reactions are $\geq$10-fold faster than uncatalyzed hydrolysis (Mildvan 1997; Wolfenden et al. 1998) and can be 27-fold faster (e.g., in methyl phosphate hydrolysis catalyzed by alkaline phosphatase,

Zalatan et al. 2008). Such an extreme increase in reaction rate results from better positioning of the nucleophile, increasing nucleophilicity, and overcoming electrostatic repulsion by enzymes (e.g., Lassila et al. 2011). These properties are aided by specific folding of enzymes and presence of specific metal ion (such as Mg, Fe, or Zn) as co-factors in the active binding sites (Knowles 1980; Admiraal and Herschlag 1999). These properties result in a tighter transition state of organophosphorus hydrolysis in enzyme catalyzed reactions than in the non-enzymatic reactions (Kirby and Jencks 1965; Knowles 1980; Matte et al. 1998).

Abiotic hydrolysis is exceptionally slow (e.g., Lad et al. 2003). However, the hydrolysis reaction rate increases by several orders of magnitude as the pH is decreased, implying that protonated phosphate monoesters (monoanionic form, i.e. $R{-}HPO_4^{1-}$) react much faster than non-protonated monoesters (phosphate monoester dianions, i.e. $R{-}PO_4^{2-}$) (Kumamoto and Westheimer 1955; Bunton et al. 1958). This is related to the overcoming of electrostatic repulsion of the O's surrounding P when these O's become protonated. Analogously, at low pH the O exchange between Pi and water is much faster because of the greater presence of protonated species like H_3PO_4 and H_2PO_4 than that at high pH (O'Neil et al. 2003). These authors also noted 5–8 ‰ difference in oxygen isotopic fractionation between protonated species and water at low pH and non-protonated species and water (like PO_4^{3-}) at high pH. Interesting, a recent study found 5–7.5 ‰ difference in $\delta^{18}O_p$ values between alkaline and acid phosphatase catalyzed degradation of two monoesters (adenosine 5′ monophosphate and glycerol phosphate) (von Sperber et al. 2014). This difference was attributed to the difference in the sources of oxygen atom incorporated (OH vs H_2O which have isotope fractionation of 40 ±3 ‰ at 25 °C, Green and Taube 1963) into the newly formed inorganic P. It is likely that the similar difference in $\delta^{18}O_p$ values between acidic and alkaline pHs both in enzymatic and non-enzymatic isotopic exchanges (O'Neil et al. 2003; von Sperber et al. 2014) is because of same sources of oxygen atoms incorporated in both cases. These results are consistent with higher exchange rates observed in the analogous sulfate–water system (Hoering and Kennedy 1957; Chiba et al. 1981).

13.2.2 Hydrolysis of Monesters, Diesters, and Triesters

Phosphoester hydrolysis can either follow a stepwise reaction [either addition-elimination ($A_N + D_N$) or elimination-addition ($D_N + A_N$) type] (see Fig. 3 in Lassila et al. 2011 for explanation) in which nucleophilic attack entails the formation of metaphosphate or phosphorane intermediates (Breslow 1993; Gerratana et al. 2000), or a concerted reaction [A_ND_N type] that involves simultaneous breaking of the leaving group (e.g., a ester, -OR) and the formation of a new bond with a nucleophile such as activated water molecule or $-OH^-$ (Kirby and Younas 1970; Hengge et al. 1995). Monoester hydrolysis has been found to follow the concerted reaction where the hydrolysis proceeds through a single transition state with simultaneous breaking and formation of bonds (Lassila et al. 2011). However, the ranges of transition states of a concerted reaction could be

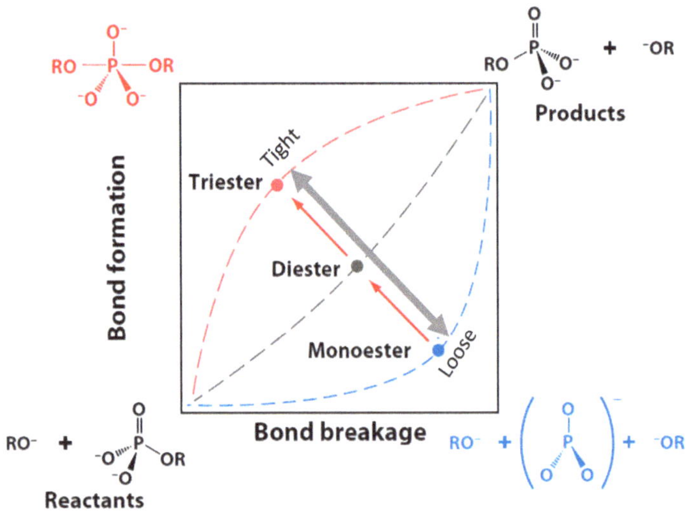

Fig. 13.3 Relative dominance of bond formation and bond breakage during hydrolysis of monoester, diester, and triester compounds, and corresponding transition states (from Lassila et al. 2011). The relative position of different esters in the two dimensional co-ordinates depends on the nature of the nucleophile and the leaving group. In general, monoesters show little variability in transition state and triesters show considerable variability (Reprinted with permission from the *Annual Review of Biochemistry*, Volume 2011 by Annual Reviews)

differentiated into loose, synchronous, and tight sub-states based on the relative dominance of bond breakage and bond formation (Fig. 13.3). These transition states depend on the degree to which the bond strength varies between the central P atom and nucleophile ion or leaving group. The tight transition state corresponds to the position in which the bond strength order is increased with a nucleophile and the leaving group, and the opposite holds true for the loose transition state (Weinhold and Landis 2005). The amount of charge developed in the transition state determines the rate of reaction (Fig. 13.3).

A monoester hydrolysis reaction follows a loose transition state because of the dominance of bond breakage and proceeds near the bottom right corner with a metaphosphate-like transition state but does not form metaphosphate intermediate (Bourne and Williams 1984; Kirby and Jencks 1965; Herschlag and Jencks 1989; Lassila et al. 2011). Linear free energy calculation, kinetic isotope analysis (both primary and secondary isotope effects) and stereochemical studies have demonstrated that the hydrolysis proceeds through the loose transition state pathway (e.g., Khan and Kirby 1970). In general, the transition states tend to be tighter from monoester to diester to triester with predominance of bond formation (Fig. 13.3). Whether this difference in transition states during hydrolysis is solely responsible for difference in isotope fractionation is yet to be measured.

Kinetic isotope effects (KIE) have long been studied in biochemistry to understand the nature of bonding and the type of bond cleavage. Please note that isotope effects are defined slightly differently in biochemistry than in isotope geochemistry.

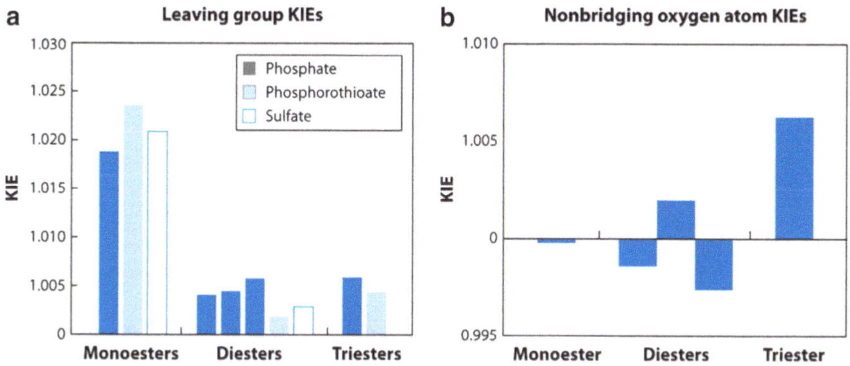

Fig. 13.4 Kinetic isotope effects (*KIE*s) for non-enzymatic hydrolysis reaction with p-nitrophenyl leaving groups. The isotope effects can be measured on leaving group (**a**) or non-bridged oxygen of the phosphate group (**b**) (from Lassila et al. 2011) (Reprinted with permission from the *Annual Review of Biochemistry*, Volume 2011 by Annual Reviews)

More often degrading anions are labeled with heavy isotopes and the rate of degradation is compared with its non-labeled counterpart. Isotope effects, that originate from the difference in vibrational energy between the heavy and light isotopes, are not present for specific bonds that are completely broken. If the bond is partially retained, however, the difference in isotope effect is still present. For example, because of a higher energy barrier for cleavage of a heavy atom in a bond, the reaction is slower and produces a normal kinetic isotope effect ($k_{light}/k_{heavy} > 1$), but this effect could be inverse ($k_{light}/k_{heavy} < 1$) if the bonding strength increases at the substituted position such as due to hydrogen bonding (Melander and Saunders 1980). Therefore, it is the usual practice in biochemistry to systematically alter the structure of a molecule and/or isotopically label a specific ion located in a specific position of a molecule and measure isotope effects during its degradation. Information gained from the difference in reaction equilibria and rates along with the isotope effects at various degrees of isotope labeling in specific structural site/s is used to identify the mechanism of hydrolysis at specific bond sites. In general, the kinetic isotope effects of the leaving group oxygens have been found to be much more pronounced for monoesters than for diesters and triesters (Fig. 13.4), but the effect is opposite for non-bridged oxygens (i.e. three other oxygen atoms in the phosphate molecule untouched in the reaction) with triesters showing the highest kinetic isotope effect.

13.2.3 *Structure-Based Mechanism of Diester Hydrolysis*

Because hydrolysis of a phosphoester is specific to its structure, composition, and enzyme specificity, a general model of hydrolysis cannot be applied to all diesters. This property can be exploited, however, to distinguish different sources of esters and origins of Pi released into the environment. For example, DNA and RNA are

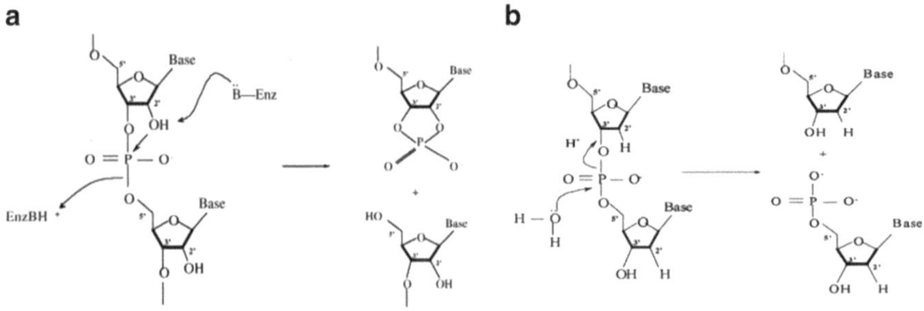

Fig. 13.5 Different hydrolysis reaction mechanisms between RNA and DNA: (**a**) RNAase catalyzed RNA hydrolysis forms cyclic 2′,3′-P-diesters. Reversibility of P diester and 3′-phosphate (formed from P diester) allows possibility of all four O's exchanging with water O; and (**b**) For DNA, absence of –OH limits the formation of cyclic diester (from Liang and Blake 2009) (Reprinted with permission of Elsevier from *Geochimica et Cosmochima Acta* 2009, 73, 3782–3794)

two of the most common diesters. As shown in Fig. 13.5, the main distinction between RNA and DNA is the presence of an –OH group at the 2′-position in RNA. Because –OH is the nucleophile for Po hydrolysis, the presence of –OH in the RNA structure allows formation of a cyclic- P-diester intermediate during hydrolysis by ribonuclease (RNase) (Walsh 1979; Mishra 2002; Liang and Blake 2009). Reversible non-enzymatic formation of a 2′,3′-phosphodiester intermediate allows extensive O isotope exchange between PO_4 and H_2O before the monoesters formed are cleaved by a monoesterase (Fig. 13.5; Liang and Blake 2009). This, however, does not occur in DNA hydrolysis because of the absence of an –OH functional group in its structure. Further research is needed with other diesters to identify any structure-specific reaction mechanism during the first step of hydrolysis so that measured isotope effects can be better explained and interpreted.

13.3 Phosphate Oxygen Isotope Ratios as a Tracer for Understanding P Sources and Cycling

Although P has 23 isotopes varying in mass from ^{24}P to ^{46}P, only one P (i.e., ^{31}P) is stable. Among radioisotopes, ^{32}P and ^{33}P have half-lives of 14.26 and 25.34 days, respectively, while all others have half-lives of less than 2.5 min. Being the only stable isotope, ^{31}P cannot be used to understand P systematics in the environment in a similar fashion to other nutrient elements – N, C, and S, which have multiple stable isotopes for analysis. However, under Earth's surface conditions, P occurs primarily as Pi and permits the use of stable isotope ratios of oxygen in Pi as a potential stable isotope tracer of P in the environment (Blake et al. 1997, 2001, 2005;

Paytan et al. 2002; Colman et al. 2005; McLaughlin et al. 2006). Oxygen isotope ratios in Pi are commonly expressed as $\delta^{18}O_p$ values defined as:

$$\delta^{18}O_p = \left(\frac{R_{sample}}{R_{standard}} - 1\right)1,000 \quad (13.1)$$

where R denotes the ratio of the less abundant to more abundant isotope ($^{18}O/^{16}O$), and R_{sample} and $R_{standard}$ refer to the ratio measured in sample and standard, respectively. The isotopic abundance is measured against a reference standard and is reported in permil (‰) relative to the VSMOW (Vienna standard mean ocean water) standard.

Stable isotope ratios of nutritional elements such as nitrogen ($\delta^{15}N$) and carbon ($\delta^{13}C$) and also oxygen ($\delta^{18}O$) in oxyanions of N and C have been classically used as a means to track nutrient sources and cycling (e.g., Böhlke and Denver 1995; Kendall 1998; Wang et al. 1998; Oelmann et al. 2007). Major breakthroughs using N and C isotopes in the field of soil sciences include the development of the nutrient retention hypothesis (Vitousek and Reiners 1975), N saturation hypothesis (Aber et al. 1989), ability to calculate gross rate of soil N mineralization, and microbial immobilization (e.g., Davidson et al. 1992; Hart et al. 1994), C3 and C4 plant respiration pathways, and proposition of the 'regulatory gate theory' (Kemmitt et al. 2008) which challenges classical theory on autochthonous and zymogenous soil microbial populations (Winogradsky 1924). While these breakthroughs are not free from controversy, they have introduced a new realm of exploration of biogeochemical processes and have led to the development of new paradigms on nutrient cycling and processes in soils and many other ecosystems. The ability of stable isotope compositions to integrate information on transfer and transformation of nutrients over time in different soil compartments and at biotic/abiotic interfaces, and more importantly, the potential for original isotopic compositions to be preserved enabling the tracing of nutrient sources, has led to their application as a major tool for disentangling mechanisms of nutrient cycling (Nadelhoffer and Fry 1994).

$\delta^{18}O_p$ compositions of P oxyanions can analogously be used as what Carreira et al. (2010) have elegantly listed the application of stable isotopes as a: (1) 'tracer' to disentangle specific biogeochemical reactions and interactions among different P forms and phases, (2) 'marker' to distinguish successive generations of P in both extant and fossil ecosystems, and (3) 'integrator' for input/output budgets and export from an ecosystem. Yet, major scientific breakthroughs for the phosphate isotopic system are only beginning to emerge. Realistic expectations for such breakthroughs are tied to the specific properties of $\delta^{18}O_p$ values, specifically, including that: (1) in abiotic processes such as sorption, desorption, transport, and mineral transformation, phosphate does not undergo any significant changes in its O isotopic composition (Tudge 1960; Lecuyer et al. 1999; O'Neil et al. 2003; Jaisi et al. 2010, 2011; Jaisi 2013); (2) during biological processes, rapid O-isotope exchange between dissolved phosphate and water may result in complete isotope exchange and produce temperature-dependent equilibrium O-isotope fractionation (Longinelli and Nuti 1973a; Blake et al. 1997, 2005). Even if P undergoes biological and abiotic cycling in an ecosystem, a distinct trend of $\delta^{18}O_p$ evolution may help identify the extent of internal cycling (Jaisi et al. 2011).

Oxygen isotopic compositions of inorganic P have been widely used to understand various biogeochemical processes such as source tracking (Markel et al. 1994; McLaughlin et al. 2006; Young et al. 2009; Elsbury et al. 2009; Gross et al. 2013); bioavailability of specific P pools and preservation of ancient intra-basinal and extrabasinal P sources including biotic and abiotic pathways of P cycling (Jaisi and Blake 2010; Blake et al. 2001, 2010; Jaisi et al. 2011); sorption, desorption, and mineral transformation (Liang and Blake 2007; Jaisi et al. 2010); and transport and remobilization (Jaisi 2013). So far, several aspects of O-isotope systematics in natural systems have been characterized (e.g., Blake et al. 1997, 2005; Paytan et al. 2002; Colman et al. 2005) including isotope signatures associated with specific P phases in soil and sediments (Zohar et al. 2010; Goldhammer et al. 2011; Tamburini et al. 2012). Enrichment and natural abundance studies have been used to identify microbial P cycling in soils (Larsen et al. 1989; Johansen et al. 1991; Melby et al. 2011; Angert et al. 2012; Tamburini et al. 2012). While inorganic P dynamics are better known and have received reasonable attention due to increasing applications of $\delta^{18}O_P$ values, research on organic-inorganic transformations and isotope effects during Po degradation have been very limited despite the fact that this research is equally important and essential to better interpret isotope data obtained from soil and other ecosystems.

13.4 Compound-Specific Organic-Inorganic Transformation: Isotope Effects

The theoretical basis for applying phosphate oxygen isotopic compositions as a tool to investigate biologically-mediated reactions came after determination of isotopic effects associated with specific enzyme(s) responsible for degradation/hydrolysis of organophosphate compounds by Blake et al. (1997). While ^{18}O-labeling technique has long been used by biochemists to elucidate oxidative phosphorylation reactions, presence of isotope exchange was first noted by Cohn (1953). Blake et al. (1998b, 2005) later found that inorganic pyrophosphatase (PPase) catalyzed the wholesale exchange of O, exchange of all 4 oxygen atoms in Pi with oxygen in ambient water, to achieve O isotopic equilibrium between Pi and water. See details of different PPase families and their roles in the environment in Sect. 13.2. Prior to these studies, enzyme-catalyzed exchange reactions were implicitly assumed to be responsible for the observed empirical relationship between mineral Pi in bones, teeth, shells, fish scales and the water ingested by organisms—the same as ambient water in the case of fish and other aquatic species, and drinking water or body water in the case of land mammals (e.g., Longinelli and Nuti 1973b; Kolodny et al. 1983). These studies marked the onset of new explorations of specific enzymes and specific phosphatic compounds to determine mechanisms of hydrolysis and associated isotope effects (Liang and Blake 2006a, b, 2009; Sandy et al. 2013; von Sperber et al. 2014).

13.4.1 Monoester Hydrolysis

Among the numerous phosphoenzymes present in natural environments that are capable of hydrolyzing Po, alkaline phosphatase (APase) is the most common and most widely studied enzyme (Perry 1972; Taft et al. 1977; Ammerman and Azam 1985; Hoppe and Ullrich 1999). It is a non-specific phosphate-scavenging enzyme that hydrolyzes a variety of organic phosphomonoester compounds as well as condensed inorganic phosphates such as pyrophosphate and even phosphite (Yang and Metcalf 2004). Unlike pyrophosphatase (PPase), however, the APase reaction is unidirectional and thus, does not promote wholesale Pi-water exchange. Furthermore, it imparts a large kinetic isotope fractionation (-30 ± 8 ‰) during conversion of Po to Pi. It is expected that the loose transition state of monoesters during degradation contributes to large kinetic isotope effects (Figs. 13.3 and 13.4). It is still unclear, however, to what extent the loose transition state can be linked to mass-dependent isotope effects.

The mechanism by which a phosphohydrolase cleaves Po is shown schematically in Fig. 13.6. The O isotope signature of enzymatic Pi regeneration from Po can be expressed as (Liang and Blake 2006b):

$$\text{For APase, } 5' \text{ nucleotidase}: \quad \delta^{18}O_P = 0.75\ \delta^{18}O_{Po} + 0.25\left(\delta^{18}O_w + F_1\right) \quad (13.2)$$

where $\delta^{18}O_P$, $\delta^{18}O_{Po}$, and $\delta^{18}O_w$ are the O-isotope composition of Pi regenerated from nucleotidase, PO_4 moiety groups bound to nucleic acids, and ambient water O-isotope composition, respectively. F_1 is the fractionation factor associated with replacement of the HO-R (alcohol) leaving group with nucleophilic O from ambient water (corresponding to steps 2 in Fig. 13.6). For the monoesters studied by Liang and Blake (2006b), one of the four oxygen atoms in Pi was found to come from water and three oxygen atoms were inherited from phosphate monoesters.

One of the key differences among different enzymes is that the fractionation between incorporated and bulk water was found to be different depending on the enzyme (e.g. APase vs. 5′ nucleotidase involved) and substrate despite the fact that monoester hydrolysis always involved incorporation of one O atom from water. For example, APase from two different sources (prokaryotic *E. coli* and eukaryotic calf) produced similar intercepts (trends overlap each other) for the same substrate both in the case of glucose-1-phosphate and β-glycerol phosphate (Fig. 13.7a). This demonstrates that for the same substrate, fractionation factor is the same irrespective of enzyme sources. However, there was ~8 ‰ difference between the intercepts of these two substrates (Fig. 13.7a). The difference of initial Po $\delta^{18}O_P$ values between these two substrates is 9 ‰ (Liang and Blake 2006b). Accounting 75 % inherited O in release Pi, the difference becomes 8.7 ‰. In other words, the different intercepts for glucose-1-phosphate and β-glycerol phosphate reflects dominantly the difference in initial isotopic composition of these 2 monoesters. In contrast, the same substrate hydrolyzed by two different enzymes, for example, when 5′-AMP is hydrolyzed by 5′-nucleotidase vs. APase (Fig. 13.7b), for

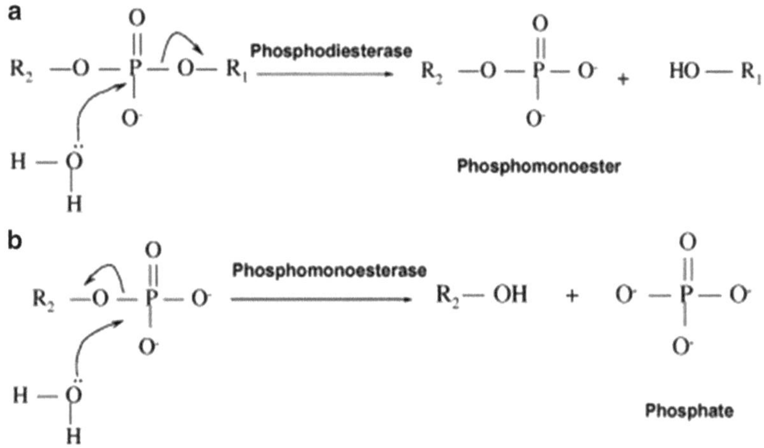

Fig. 13.6 Hydrolysis of monoester and diester model compounds. (**a**) Two-step bond cleavage reaction in diesters includes the formation of a P-monoester first and (**b**) subsequent cleavage of the P-monoester bond and release of free Pi. Oxygen incorporated in Pi can also come from an –OH on RNA structure in the case of RNase, other than H_2O (not shown in the *drawing*) (from Liang and Blake 2009) (Reprinted with permission of Elsevier from *Geochimica et Cosmochima Acta* 2009, 73, 3782–3794)

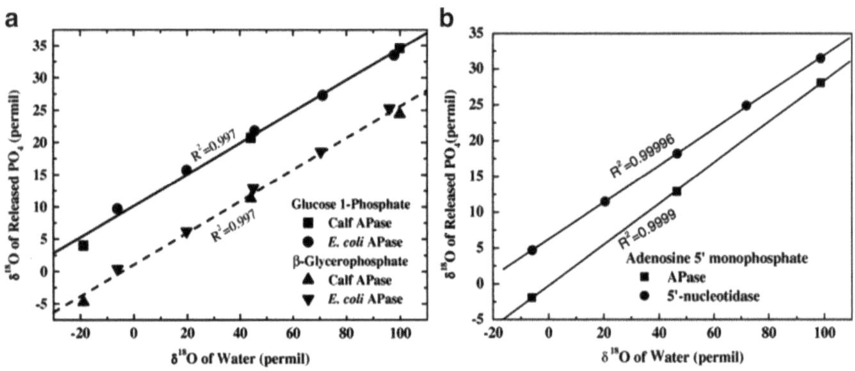

Fig. 13.7 Monoester hydrolysis involves incorporation of one oxygen atom from water and 3 oxygen atoms inherited from organic P. The y-intercept is impacted from different fractionation factors and original isotopic composition of organic P. In (**a**) the y-intercept indicates majorly the impact of inherited O on the overall isotopic composition of PO_4 but in (**b**), difference in fractionation is dominant (from Liang and Blake 2006b) (Reprinted with permission of Elsevier from *Geochimica et Cosmochima Acta* 2006, 70, 3957–3969)

5′-nucleotidase catalyzed hydrolysis, fractionation accompanying incorporation of ambient water oxygen was −10 (±1) ‰, but for APase this fractionation was ~−30(±8) ‰ (Fig. 13.7b). The fractionation factors of different substrates and enzymes measured by Liang and Blake (2006b) are listed in Table 13.1.

Table 13.1 Oxygen isotope fractionation factors (between incorporated water to Pi and ambient water) during degradation of monoesters (From Liang and Blake 2006b). The fractionation factors were independently determined by using initial Po $\delta^{18}O_P$ values by UV degradation methods and $\delta^{18}O_P$ values of PO_4 produced by each enzyme hydrolysis reaction for each of the different compounds (Reprinted with permission of Elsevier from *Geochimica et Cosmochima Acta* 2006, 70, 3957–3969)

Enzymes	Substrates	Fractionation factors
Calf APase	β-Glycerophosphate	-33 ± 2
	Adenosine 5′-monophosphate	-30 ± 8
E. coli APase	β-Glycerophosphate	-27 ± 1
Shrimp APase	β-Glycerophosphate	-31
5′-Nucleotidase	Adenosine 5′-monophosphate	-10 ± 1

It is, however, important to emphasize that the reaction mechanism and isotopic fractionation mentioned above holds true for the enzymes and substrate studied by Liang and Blake (2006b), and may not be generalized for hydrolysis of other monoesters catalyzed by other phosphohydrolases in any specific or non-specific reactions. This highlights the importance of studying other enzymes and substrates relevant to specific ecosystems and the urgency of this understanding to determine the variation in fractionation factors for different enzymes and substrates.

As illustrated above in Sect. 13.2.2, the rate of enzyme reaction depends on the charge distribution within the active site of the enzymes relative to that of the PO_4 group and metal cofactors, and primarily on the charge stabilization of the leaving group (alcohol in this case). The rates of diffusion of substrates/products and especially rates of diffusion of reactants – in this case water molecule involved in hydrolysis – is responsible for typically large kinetic fractionations in enzymatic reactions. While studies on individual O groups (e.g., leaving/bridging group and non-bridging oxygen group, see Fig. 13.4) have not been conducted so far, it is expected that research of this kind will help further characterize differences in each of these reactions and resolve variations in isotope effects. This would eventually lead to the distinction among enzymes and substrates and provide a basis for distinguishing compound- and enzyme-specific isotopic fractionations of Po hydrolysis.

13.4.2 *Diester Hydrolysis*

For diesters, two C-O-P bonds must be cleaved to release Pi (Fig. 13.6), which can be predicted based on structure alone (Fig. 13.5) to involve the incorporation of ambient water O in 2 of 4 sites in released Pi (Liang and Blake 2009) as:

$$\text{For PDase, RNase}: \ \delta^{18}O_P = 0.5\,\delta^{18}O_{Po} + 0.25\left(\delta^{18}O_w + F_1\right) + 0.25\left(\delta^{18}O_w + F_2\right) \tag{13.3}$$

Table 13.2 Oxygen isotope fractionation factors (between incorporated water to Pi and ambient water) during two step degradation of diesters (From Liang and Blake 2009, Reprinted with permission of Elsevier from *Geochimica et Cosmochima Acta* 2009, 73, 3782–3794)

		Fractionation factors		
Substrate	Monoesterase	F1	F2	F
DNA	Nucleotidase	−20	−10	−15
	Apase	−20	−30	−25
RNA	Nucleotidase	+20	−10	+5
	Apase	+20	−30	−5

where all other symbols are the same as in Eq. 13.2, except F_2 which is the fractionation factor associated with breaking the P-diester bond (corresponding to steps 1 in Fig. 13.6). Results presented by Liang and Blake (2006b, 2009) suggest that the O isotope effect of APase and 5′ nucleotidase monoesterase catalyzed hydrolysis depends on compound structure, but not on the chemical composition, whereas hydrolysis of phosphodiesters such as DNA and RNA was compound-specific. This finding prompted these authors to develop structure-reaction-based models to predict isotope effects during cell-free enzyme reactions/hydrolysis as well as whole cell hydrolysis of Po. This is due to the structure of RNA, specifically due to the presence of –OH at the 2′-position in ribose (Fig. 13.5). Therefore, during the first C-O-P bond cleavage in RNA catalyzed by RNase, a cyclic P-diester intermediate is formed by an intra-molecular attack of the 2′ –OH on P without incorporation/involvement of O from water (see also Figure 3 in Liang and Blake 2009). The transformation between cyclic P-diester and 2′ and 3′ P-monoesters does not require enzymatic catalysis and is reversible; therefore, all four oxygen atoms in PO_4 moieties could possibly exchange with water O during this process. It should be noted that this reversibility and incorporation of more than one O from water into the PO_4 moiety during the first step of RNA hydrolysis catalyzed by RNase was observed *only* in laboratory experiments performed with a single, cell-free, enzyme without monoesterase present to immediately catalyze the second C-O-P hydrolysis of 2′ and 3′ nucleotides. Thus, RNase can promote extensive Pi-water exchange similar to PPase by creating the cyclic diester intermediate, but RNase is not likely to be present in nature and in whole cells alone without 5′-nucleotidase or APase present to immediately catalyze the second hydrolysis step. Also, the nature of the PO_4 moiety-water exchange (equilibrium/non-equilibrium, temperature dependent) that occurs in the cyclic diester has not been measured specifically. In contrast to this (RNase) enzyme-specific isotopic fractionation, during DNA and RNA hydrolysis by the non-specific phosphodiesterase, phosphodiesterase 1 (PDase), compound-specific O isotopic fractionations are observed. The first C-O-P bond is cleaved by hydrolysis and ^{16}O from ambient water is preferentially incorporated into the individual nucleotides released from DNA; whereas ^{18}O is preferentially incorporated into nucleotides released from RNA. The second C-O-P cleavage step for both RNase and PDase catalyzed reactions is catalyzed by a phosphomonoesterase such as a non-specific enzyme like APase or a specific enzyme like 5′-nucleotidase with attendant isotope effects described above. The fractionation factors of DNA and RNA degradation with two different monoesterases are shown in Table 13.2.

Fig. 13.8 Chemical structure of two phosphonates: glyphosate and phosphonoacetic acid. C-P bond is hydrolyzed by water and reactive oxygen produced from UV light

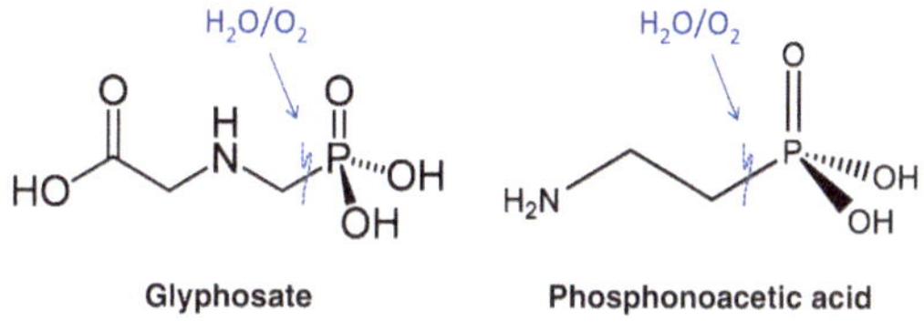

13.4.3 *Hydrolysis of Phosphonates*

Phosphonates are a group of single chain organophosphorus compounds which contain a direct C–P bond (Fig. 13.8). Phosphonates comprise about 5 % of soil organic P (Cade-Menun et al. 2002) and a similar fraction of marine organic P (e.g., Kolowith et al. 2001; Dyhrman et al. 2009). Because of their high sorbent capacity, they have been used as chelating agents in household detergents and cleaning products, and to prevent scaling in industrial processes such as wastewater treatment as well as carriers of selected radionuclide ions in cancer therapy. Glyphosate and phosphonoacetic acids are anthropogenic phosphonates, with glyphosate being the most widely used non-selective systemic herbicide (Cox 2004). Because of the nature of the direct C-P bond, these organophosphorus compounds persist for extended times in the environment.

Hydrolysis of two phosphonates, glyphosate and phosphonoacetic acid, has been studied recently under conditions of intensive UV radiation (1,200 W) (Sandy et al. 2013). These authors found a relatively higher rate of UV hydrolysis of phosphonoacetic acid compared to glyphosate which they interpreted as being due to its relatively simple structure consisting of an acetate group attached to a phosphate moiety through a single C-P bond (Fig. 13.8). Under UV radiation, the reaction was likely aided by singlet oxygens formed by the breakdown of atmospheric oxygen (Braun et al. 1986).

The slopes of plots of oxygen isotope compositions of released phosphate vs. ^{18}O-labeled water used to carry out experiments were positive suggesting the incorporation of oxygen from ambient water (Fig. 13.9). While the slopes varied considerably among different organophosphorus compounds, with almost negligible (0.016) for G-1-P (glucose 1-phosphate) to small (e.g., 0.18 for glyphosate) slopes, a clear distinction was observed among monoesters, diesters, and phosphonates studied by these authors, with phosphonates incorporating relatively higher proportions of ambient water oxygen during UV-promoted hydrolysis. Interestingly, in all cases the number of oxygen atoms incorporated was less than 1 (i.e. slope < 0.25). This means that hydrolysis of the C–P bond must also involve the incorporation of O from an external source other than ambient water such as partial contribution of singlet oxygen generated from atmospheric O_2 during UV excitation (Braun et al. 1986). By assuming that both water and atmospheric O are incorporated directly into regenerated Pi without any fractionation, this method

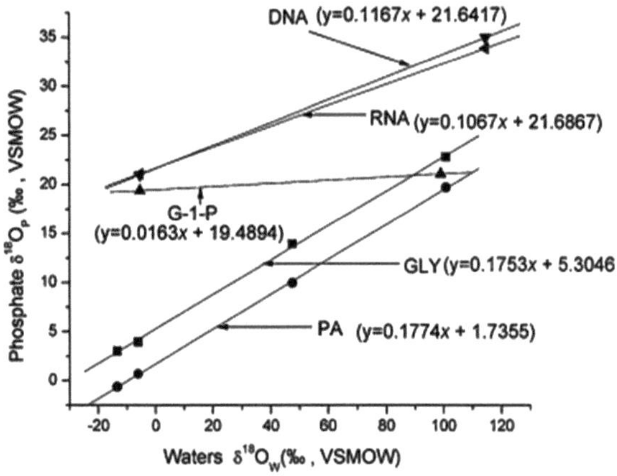

Fig. 13.9 Slope of phosphates vs water $\delta^{18}O$ during UV photo-oxidative hydrolysis of different organic compounds. The intercepts are indicative of fractionation of water used in hydrolysis. Studied organophosphorus compounds include glyphosate (*Gly*), phosphonoacetic acid (*PA*), DNA and RNA (from Sandy et al. 2013) (Reprinted with permission of Elsevier from *Journal of Hazardous Materials* 2013, 260, 947–954)

provided a means of calculating the original isotopic composition of PO_4 moiety groups bound to phosphonates ($\delta^{18}O_{Ph}$) as:

$$\delta^{18}O_{Ph} = \frac{1}{0.75}\left[\delta^{18}O^*_P - \left\{X\delta^{18}O^*_w + (0.25 - X)\delta^{18}O^*_{O_2}\right\}\right] \tag{13.4}$$

where $\delta^{18}O^*{}_P$, and $\delta^{18}O^*{}_w$, and $\delta^{18}O^*{}_{O2}$ are O-isotope composition of measured/released Pi (regenerated from phosphonates), experimental water, and atmospheric oxygen, respectively. Similarly, X and (0.25 − X) refer to the fraction of oxygen incorporated from water and atmospheric O_2, respectively. It is straightforward to calculate the fractional contribution of oxygen from atmospheric O_2 directly from the slope as (0.25 − X) (e.g. 7 % for the slope of 0.18 in Fig. 13.9).

13.4.4 Distinction Between Enzyme, UV, and Microbial Hydrolysis of Organic P

Microbial metabolism is the major pathway of hydrolysis of organophosphorus compounds in natural environments. The role of free enzymes could be substantial in the environment but depends on various factors, primarily the release of these enzymes and their stability/perseverance in the environment. These enzymes, however, are often not freely exposed but rather bound to particles or cell surfaces, or shielded by mineral surfaces/structures to which they attach. The solar UV irradiation could also play an important role in organic P degradation particularly when these compounds are associated with metal oxides that

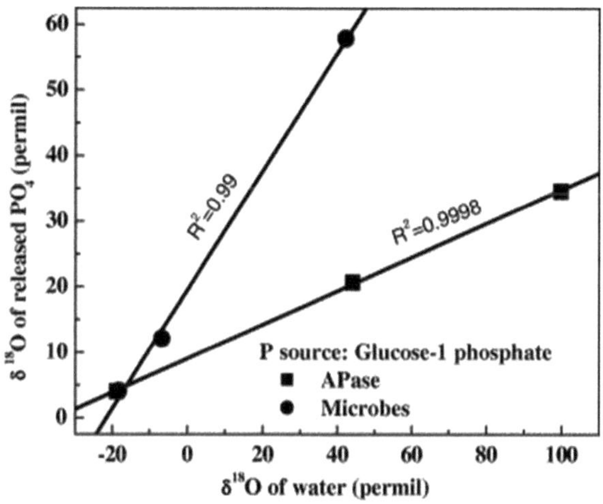

Fig. 13.10 Oxygen isotope effects during hydrolysis of Po measured in laboratory systems with single cell-free enzymes vs. whole microbial cells under conditions of forced extracellular hydrolysis of Po with subsequent metabolism of released Pi (Blake et al. 1998a, 2005; Liang and Blake 2006b). The increase in slope from ~0.25 in the APase catalyzed reaction to 0.90 in whole-cell experiments with forced extracellular hydrolysis of Po followed by uptake and cycling of Pi through the cells indicates that cellular cycling of Pi regenerated from Po causes Pi to evolve toward wholesale exchange of all four oxygen atoms in Pi with ambient water and equilibrium isotopic compositions (Reprinted with permission of Elsevier from *Geochimica et Cosmochima Acta* 2006, 70, 3957–3969)

catalyze photo-oxidation of these compounds (e.g. Mopper and Kieber 2002). One of the important distinctions between cell-free enzyme-catalyzed reactions *vs.* metabolism by intact microbial cells is that the latter involves a series of phosphoenzymes acting together including mono- and di-esterases (including RNase) as well as cytoplasmic pyrophosphatase (PPase), which can overprint the large kinetic fractionations produced by diesterase and monoesterase enzymes (Fig. 13.10). Therefore the intracellular production of Pi from Po with large isotope fractionation resulting in very low $\delta^{18}O_P$ values may evolve to achieve equilibrium with water inside the cells by the action of PPase. It is likely that a suite of $\delta^{18}O_P$ values could be produced depending on the timing of Po hydrolysis. The isotopic evolution from fresh regeneration to complete recycling could be observed in some environments.

13.5 Perspectives on Compound-Specific Isotope Research in Agricultural Environments

Agricultural expansion over the next 50 years is expected to be accompanied by a 2.4–2.7-fold increase in nitrogen and phosphorus driven eutrophication of terrestrial and coastal environments (Tilman et al. 2001). Such ecosystem damage is caused primarily by nutrients released from fertilizer and manure applied to agricultural fields (Carpenter et al. 1998). Temporally and spatially variable inputs of point and nonpoint source nutrient pollutants and their mixing and recycling in an ecosystem pose a greater challenge for efficient ecosystem management (Ator et al. 2011) because of the unknown sources of these nutrients and the underlying process of their delivery. When P from a distinct source is released to the environment and different sources mix with each other, the origin and history of P are largely lost. Given that the different sources of P applied or present in agricultural soils (e.g., inorganic fertilizers, manures, municipal wastes, and rock weathering) often have distinct isotopic compositions (e.g., Gruau et al. 2005; Young et al. 2009), stable isotopes can be most useful tool to trace these sources and track biological cycling of specific P sources. Therefore, stable oxygen isotope ratios of inorganic P, and compound-specific oxygen isotopic compositions as well as fractionation factors are expected to be a most valuable analytical tool for understanding P dynamics in the environment. While compound-specific isotope studies of organic P are still in their infancy, a new paradigm has emerged and it is hoped that the future expansion of this research will provide a 'missing link' to develop a holistic approach that integrates transfer and transformation of organic and inorganic P compounds and eventually leads to sustainable agriculture and more healthy ecosystems.

Acknowledgments Authors would like to thank editors for the invitation to contribute this chapter. This work was supported by research grants from the US Department of Agriculture (NIFA awards 2012-67019-19320 and 2013-67019-21373) and ACS-PRF (53469-DNI2) to DPJ, and National Science Foundation (OCE 0928247) to REB.

References

Aber JD, Nadelhoffer KJ, Steudler P, Melillo JM (1989) Nitrogen saturation in northern forest ecosystems. BioScience 39:378–386

Admiraal SJ, Herschlag D (1999) Catalysis of phosphoryl transfer from ATP by amine nucleophiles. J Am Chem Soc 121:5837–5845

Ammerman JW, Azam F (1985) Bacterial 5′-nucleotidase in aquatic ecosystems: a novel mechanism of phosphorus regeneration. Science 227:1338–1340

Angert A, Weiner T, Mazeh S, Sternberg M (2012) Soil phosphate stable oxygen isotopes across rainfall and bedrock gradients. Environ Sci Technol 46:2156–2162

Ator SW, Brakebill JW, Blomquist JD (2011) Sources, fate, and transport of nitrogen and phosphorus in the Chesapeake Bay watershed: an empirical model. U.S. Geological Survey Scientific Investigations Report 2011–5167, 27pp

Baykov AA, Shestakov AS (1992) Two pathways of pyrophosphate hydrolysis and synthesis by yeast inorganic pyrophosphatase. Eur J Biochem 206:463–470

Baykov AA, Shestakov AS, Kasho VN, Vener AV, Ivanov AH (1990) Kinetics and thermodynamics of catalysis by the inorganic pyrophosphatase of Escherichia coli in both directions. Eur J Biochem 194:879–887

Baykov AA, Kasho VN, Bakuleva NP, Rea PA (1994) Oxygen exchange reactions catalyzed by vacuolar H^+-translocating payrophosphatase. FEBS Lett 350:323–327

Benitez-Nelson CR (2000) The biogeochemical cycling of phosphorus in marine systems. Earth Sci Rev 51:109–135

Blake RE, O'Neil JR, Garcia GA (1997) Oxygen isotope systematics of microbially mediated reactions of phosphate: I. Degradation of organophosphorus compounds. Geochim Cosmochim Acta 61:4411–4422

Blake RE, O'Neil JR, Garcia GA (1998a) Enzyme-catalyzed oxygen isotope exchange between inorganic phosphate and water: reaction rates and temperature dependence at 5.7–30 °C. Miner Mag 62:163–164

Blake RE, O'Neil JR, Garcia GA (1998b) Effects of microbial activity on the $\delta^{18}O_p$ of dissolved inorganic phosphate and textural features of synthetic apatites. Am Mineral 83:1516–1531

Blake RE, Alt JC, Martini AM (2001) Oxygen isotope ratios of PO4: an inorganic indicator of enzymatic activity and P metabolism and a new biomarker in the search for life. Proc Nat Acad Sci USA 98:2148–2153

Blake RE, O'Neil JR, Surkov AV (2005) Biogeochemical cycling of phosphorus: insights from oxygen isotope effects of phosphoenzymes. Am J Sci 305:596–620

Blake RE, Chang SJ, Lepland A (2010) Phosphate oxygen isotopic evidence for a temperate and biologically active Archaean ocean. Nature 464:1029–1032

Böhlke JK, Denver JM (1995) Combined use of groundwater dating, chemical, and isotopic analyses to resolve the history and fate of nitrate contamination in two agricultural watersheds, Atlantic coastal plain, Maryland. Water Resour Res 31:2319–2339

Bourne N, Williams A (1984) Evidence for a single transition state in the transfer of the phosphoryl group ($-PO_3{}^{2-}$) transfer to pyridines from isoquinoline-N-phosphonate. J Am Chem Soc 105:3357–3358

Braun AM, Frimmel FH, Haigne J (1986) Singlet oxygen analysis in irradiated surface waters. Int J Environ Anal Chem 27:137–149

Breslow R (1993) Kinetics and mechanism in RNA cleavage. Proc Natl Acad Sci U S A 90:1208–1211

Bünemann EK, Condron LM (2007) Phosphorus and sulfur cycling in terrestrial ecosystems. In: Marschner P, Rengel Z (eds) Nutrient cycling in terrestrial ecosystems, Soil biology. Springer, Heidelberg

Bunton CA, Llewellyn DR, Oldham KG, Vernon CA (1958) The reactions of organic phosphates: the hydrolysis of methyl dihydrogen phosphate. J Chem Soc 3574–3587

Cade-Menun BJ, Liu CW, Nunlist R, McColl JG (2002) Soil and litter phosphorus-31 nuclear magnetic resonance spectroscopy: extractants, metals, and phosphorus relaxation times. J Environ Qual 31:457–465

Carpenter SR, Caraco NF, Correl DL, Howarth RW, Sharpley AN, Smith VH (1998) Nonpoint pollution of surface waters with phosphorus and nitrogen. Ecol Appl 8:559–568

Carreira JA, Vinegla B, Blanes MC, Garcia-Ruiz R (2010) Stable isotopes and changing paradigms on soil nitrogen and carbon biogeochemistry. Ecosistemas 29:14–23

Cassano AG, Anderson VE, Harris ME (2004) Understanding the transition states of phosphodiester bond cleavage: insights from heavy atom isotope effects. Biopolymers 73:110–129

Chiba H, Kusakabe M, Hirano SI, Matsuo S, Somiya S (1981) Oxygen isotope fractionation factors between anhydrite and water from 100 °C to 550 °C. Earth Planet Sci Lett 53:55–62

Cohn M (1953) A study of oxidative phosphorylation with ^{18}O-labeled inorganic phosphate: J Biol Chem 201:735–750

Colman AS, Blake RE, Karl DM, Fogel ML, Turekian KK (2005) Marine phosphate oxygen isotopes and organic matter remineralization in the oceans. Proc Natl Acad Sci U S A 102:13023–13028

Cotner JB, Wetzel RG (1992) Uptake of dissolved inorganic and organic phosphorus compounds by phytoplankton and bacterioplankton. Limnol Oceanogr 37:232–243
Cox C (2004) Herbicide factsheet-glyphosate. J Pest Reform 24:10–15
Davidson EA, Hart SC, Firestone MK (1992) Internal cycling of nitrate in soils of a mature coniferous forest. Ecology 73:1148–1156
Dyhrman ST, Ruttenberg KC (2006) Presence and regulation of alkaline phosphatase activity in eukaryotic phytoplankton from the coastal ocean: implications for dissolved organic phosphorus re-mineralization. Limnol Oceanogr 51:1381–1390
Dyhrman ST, Benitez-Nelson CR, Orchard ED, Haley ST, Pellechia PJ (2009) A microbial source of phosphonates in oligotrophic marine systems. Nat Geosci 2:696–699
Dzombak DA (2011) Nutrient control in large-scale US watersheds: the Chesapeake Bay and Northern Gulf of Mexico. Bridge Nat Acad Eng 41:13–22
Ehrlich HL (1996) Geomicrobiology, 3rd edn. Deckker, New York
Elsbury KE, Paytan A, Ostrom NE, Kendall C, Young MB, McLaughlin K, Rollog ME, Watson S (2009) Using oxygen isotopes of phosphate to trace phosphorus sources and cycling in Lake Erie. Environ Sci Technol 43:3108–3114
Erickson BE (2011) Cleaning the Chesapeake Bay. Chem Eng News 89:10–14
Frossard E, Brossard M, Hedley MJ, Metherell A (1995) Reactions controlling the cycling of P in soils. In: Tiessen H (ed) Phosphorus cycling in terrestrial and aquatic ecosystems: a global perspective. SCOPE/Wiley, New York
Gerratana B, Sowa GA, Cleland WW (2000) Characterization of the transition state structures and mechanisms for the isomerization and cleavage reactions of 3′m-nitrobenzyl phosphate. J Am Chem Soc 122:12615–12621
Goldhammer T, Brunner B, Bernasconi SM, Ferdelman T, Zabel M (2011) Phosphate oxygen isotopes: insights into sedimentary phosphorus cycling from the Benguela upwelling system. Geochim Cosmochim Acta 75:3741–3756
Green M, Taube H (1963) Isotopic fractionation in the OH–H_2O exchange reaction. J Phys Chem 67:1565–1566
Gross A, Nishri A, Angert A (2013) Use of phosphate oxygen isotopes for identifying atmospheric-P sources: a case study at Lake Kinneret. Environ Sci Technol 47:2721–2727
Gruau G, Legeas M, Riou C, Gallacier E, Martineau F, Hénina O (2005) The oxygen isotopic composition of dissolved anthropogenic phosphates: a new tool for eutrophication research? Water Res 39:232–238
Hart SC, Stark JM, Davidson EA, Firestone MK (1994) Nitrogen mineralization, immobilization, and nitrification. In: Weaver RW, Angle JS, Bottomley PJ, Bezdicek DF, Smith MS, Tabatabai MA, Wollun AG (eds) Method of soil analysis, Part 2. Microbiological and biochemical properties, Soil Science Society of America book series. Soil Science Society of America, Wisconsin, pp 985–1018
Harway GW, Keister DL (1981) Energy-linked reactions in photosynthetic bacteria: Pi ⇋ HOH oxygen exchange catalyzed by the membrane-bound inorganic pyrophosphatase of Rhodospirillum rubrum. Arch Biochem Biophys 208:426−430
Heikinheimo P, Lehtonen J, Baykov A, Lathi R, Cooperman BS, Goldman A (1996) The structural basis for pyrophosphatase catalysis. Structure 4:1491−1508
Hengge AC (2002) Isotope effects in the study of phosphoryl and sulfuryl transfer reactions. Acc Chem Res 35:105–112
Hengge AC, Tobin AE, Cleland WW (1995) Studies of transition-state structures in phosphoryl transfer reactions of phosphodiesters of p-nitrophenol. J Am Chem Soc 214:5919–5926
Herschlag D, Jencks WP (1989) Evidence that metaphosphate monoanion is not an intermediate in solvolysis reactions in aqueous solution. J Am Chem Soc 111:7579–7586
Hoering TC, Kennedy JW (1957) The exchange of oxygen between sulfuric acid and water. J Am Chem Soc 79:56–60
Hoppe H-G, Ullrich S (1999) Profiles of ectoenzymes in the Indian Ocean: phenomena of phosphatase activity in the mesopelagic zone. Aquat Microb Ecol 19:139–148

Jaisi DP (2013) Isotope effects during phosphate transport in packed-bed sediment columns. J Contam Hydrol 154:10–19

Jaisi DP, Blake RE (2010) Tracing sources and cycling of phosphorus in Peru Margin sediments using oxygen isotopes in authigenic and detrital phosphates. Geochim Cosmochim Acta 74:3199–3212

Jaisi DP, Blake RE, Kukkadapu RK (2010) Fractionation of oxygen isotopes in phosphate during its interactions with iron oxides. Geochim Cosmochim Acta 74:1309–1319

Jaisi DP, Kukkadapu RK, Stout LM, Varga T, Blake RE (2011) Biotic and abiotic pathways of phosphorus cycling in minerals and sediments: insights from oxygen isotope ratios in phosphate. Environ Sci Technol 45:6254–6261

Johansen HS, Middleboe V, Larsen S (1991) Delabeling of ^{18}O enriched phosphate added to soil as a function of biological activities in the soils. In: Proceedings of the stable isotopes in plant nutrition, soil fertility and environmental studies. IAEA, Vienna, pp 553–559

Jong F (2006) Marine eutrophication in perspective. Springer, New York

Kasho VN, Baykov AA (1989) Two pathways for phosphate-water oxygen by yeast inorganic pyrophosphatase. Biochem Biophys Res Commun 161:475–480

Kemmitt SJ, Lanyon CV, Waite IS, Wen Q, Addiscott TM, Bird NA, O'Donnell AG, Brookes PC (2008) Mineralization of native soil organic matter is not regulated by the size, activity or composition of the soil microbial biomass – a new perspective. Soil Biol Biochem 40:61–73

Kendall C (1998) Tracing nitrogen sources and cycles in catchments. In: Kendall C, McDonnell JJ (eds) Isotope tracers in catchment hydrology. Elsevier Science, Amsterdam, pp 519–576

Khan SA, Kirby AJ (1970) Reactivity of phosphate esters. Multiple structure-reactivity correlations for reactions of triesters with nucleophiles. J Chem Soc B 1172–1182

Kirby AJ, Jencks WP (1965) Reactivity of nucleophilic reagents toward p-nitrophenyl phosphate dianion. J Am Chem Soc 87:3209–3216

Kirby AJ, Younas M (1970) The reactivity of phosphate esters. Diester hydrolysis. J Chem Soc 3:510–513

Knowles JR (1980) Enzyme-catalyzed phosphoryl transfer reactions. Annu Rev Biochem 49:877–919

Kolodny Y, Luz B, Navon O (1983) Oxygen isotope variations in phosphate of biogenic apatites: 1. Fish bone apatite & rechecking the rules of the game. Earth Planet Sci Lett 64:398–404

Kolowith LC, Ingall ED, Benner R (2001) Composition and cycling of marine organic phosphorus. Limnol Oceanogr 46:309–320

Kumamoto J, Westheimer FH (1955) The hydrolysis of mono- and dibenzyl phosphates. J Am Chem Soc 77:2515–2518

Lad C, Williams NH, Wolfendan R (2003) The rate of hydrolysis of phosphomonoester dianions and the exceptional catalytic proficiencies of protein and inositol phosphatase. Proc Natl Acad Sci U S A 100:5607–5610

Larsen S, Middelboe V, Johansen HS (1989) The fate of ^{18}O labeled phosphate in soil plant-systems. Plant Soil 117:143–145

Lassila JK, Zalatan JG, Herschlag D (2011) Biological phosphoryl-transfer reactions: understanding mechanism and catalysis. Ann Rev Biochem 80:669–702

Lecuyer C, Greansjean P, Sheppard MF (1999) Oxygen isotope exchange between dissolved phosphate and water at temperatures <135 °C: inorganic versus biological fractionations. Geochim Cosmochim Acta 63:855–862

Liang Y, Blake RE (2006a) Oxygen isotope composition of phosphate in organic compounds: isotope effects of extraction methods. Org Geochem 37:1263–1277

Liang Y, Blake RE (2006b) Oxygen isotope signature of Pi regeneration from organic compounds by phosphomonoesterases and photooxidation. Geochim Cosmochim Acta 70:3957–3969

Liang Y, Blake RE (2007) Oxygen isotope fractionation between apatite and dissolved aqueous phosphate: 20–45 °C. Chem Geol 238:121–133

Liang Y, Blake RE (2009) Compound and enzyme-specific phosphodiester hydrolysis mechanisms revealed by delta O-18 of dissolved inorganic phosphate: implications for marine P cycling. Geochim Cosmochim Acta 73:3782–3794

Longinelli A, Nuti S (1973a) Revised phosphate–water isotopic temperature scale. Earth Planet Sci Lett 19:373–376

Longinelli A, Nuti S (1973b) Oxygen isotope measurements of phosphate from fish teeth and bones. Earth Planet Sci Lett 20:337–340

Markel D, Kolodny Y, Luz B, Nishri A (1994) Phosphorus cycling and phosphorus sources in Lake Kineret: tracing by oxygen isotopes in phosphate. Israel J Earth Sci 43:165–178

Matte A, Tari LW, Delbaere LJ (1998) How do kinases transfer phosphoryl groups? Structure 6:413–419

McKelvie ID (2005) Separation, preconcentration and speciation of organic phosphorus in environmental samples. In: Turner BL, Frossard E, Baldwin DS (eds) Organic phosphorus in the environment. CABI Publishing, London, pp 1–20

McLaughlin K, Kendall C, Silva SR, Young M, Paytan A (2006) Phosphate oxygen isotope ratios as a tracer for sources and cycling of phosphate in North San Francisco Bay, California. J Geophys Res Biogeosci 111:3003

Melander L, Saunders WH (1980) Reaction rates of isotopic molecules. Wiley, New York

Melby ES, Soldat DJ, Barak P (2011) Synthesis and detection of oxygen-18 labeled phosphate. PLoS One 6:e18420

Mildvan AS (1979) The role of metals in enzyme-catalyzed substitutions at each phosphorus atom of ATP. Adv Enzymol 49:103–126

Mildvan AS (1997) Mechanisms of signaling and related enzymes. Proteins 24:401–416

Mishra NC (2002) Nucleases: molecular biology and application. Wiley, New York

Mopper K, Kieber DJ (2002) Photochemistry and the cycling of carbon, sulfur, nitrogen, and phosphorus. In: Hansell DA, Carlson CA (eds) Biogeochemistry of marine dissolved organic matter. Academic Press, Amsterdam, pp 455–508

Nadelhoffer KJ, Fry B (1994) Nitrogen isotope studies in forest ecosystems. In: Lajtha K, Michener R (eds) Stable isotopes in ecology and environmental science. Blackwell Scientific Publications, Boston, pp 23–44

Nausch M, Nausch G (2007) Bioavailable dissolved organic phosphorus and phosphorus use by heterotrophic bacteria. Aquat Biol 1:151–160

Nicholson D, Dyhrman S, Chavez F, Paytan A (2006) Alkaline phosphatase activity in the phytoplankton communities of Monterey Bay and San Francisco Bay. Limnol Oceanogr 51:874–883

O'Neil JR, Vennemann TW, McKenzie WF (2003) Effects of speciation on equilibrium fractionations and rates of oxygen isotope exchange between $(PO_4)_{(aq)}$ and H_2O. Geochim Cosmochim Acta 67:3135–3144

Oelmann Y, Kreutziger Y, Bol R, Wilcke W (2007) Nitrate leaching in soil: tracing the NO_3^- sources with the help of stable N and O isotopes. Soil Biol Biochem 39:3024–3033

Ohio EPA (2010) Ohio Lake Erie phosphorus task force final report. Ohio Environmental Protection Agency, Columbus, 109pp

Paytan A, Luz B, Kolodny Y, Neori A (2002) Biologically mediated oxygen isotope exchange between water and phosphorus. Global Biogeochem Cycles 16:131–137

Perry MJ (1972) Alkaline phosphatase activity in Subtropical Central North Pacific waters using a sensitive fluorometric method. Mar Biol 15:113–119

RCN-SEES (2012) Coordinating phosphorus research to create a sustainable food system. http://sustainablep.asu.edu/prcn. Accessed on 8/11/2013

Redfield AC (1958) The biological control of chemical factors in the environment. Am Sci 46:1–221

Sandy EH, Blake RE, Chang SJ, Jun Y, Yub C (2013) Oxygen isotope signature of UV degradation of glyphosate and phosphonoacetate: tracing sources and cycling of phosphonates. J Hazard Mater 260:947–954

Sivula T, Salminena A, Parfenyev AN, Pohjanjoki P, Goldman A, Cooperman BS, Baykov A, Lathi R (1999) Evolutionary aspects of inorganic pyrophosphatase. FEBS Lett 454:75–80

Smirnova IN, Kasho VN, Volk SE, Ivanov AH, Baykov AA (1995) Rates of elementary steps catalyzed by rat liver cytosolic and mitochondrial inorganic pyrophosphatases in both directions. Arch Biochem Biophys 318:340–348

Smucker RA, Kim CK (1991) Chitinase activity in estuarine waters. In: Chrost RJ (ed) Microbial enzymes in aquatic environments. Springer, New York, pp 249–269

Suh S, Yee S (2011) Phosphorus use-efficiency of agriculture and food system in the US. Chemosphere 84:806–813

Syers JK, Johnson AE, Curtin D (2008) Efficiency of soil and fertilizer phosphorus use: reconciling changing concepts of soil phosphorus behavior with agronomic information. FAO Fertilizer and Plant Nutrition Bulletin, 123 pp

Taft JL, Loftus ME, Taylor WR (1977) Phosphate uptake from phosphomonoesters by phytoplankton in the Chesapeake Bay. Limnol Oceanogr 22:1012–1021

Tamburini F, Pfahler V, Bunemann EK, Guelland K, Bernasconi SM, Frossard E (2012) Oxygen Isotopes unravel the role of microorganisms in phosphate cycling in soils. Environ Sci Technol 46:5956–5962

Tilman D, Fargione J, Wolff B, D'Antonio C, Dobson A, Howath R, Schinder D, Schlesinger WH, Simberloff D, Swackhamer D (2001) Forecasting agriculturally driven global environmental change. Science 292:281–284

Tudge AP (1960) A method of analysis of oxygen isotopes in orthophosphate and its use in measurement of paleotemperatures. Geochim Cosmochim Acta 18:81–93

van Moorleghem C, De Schutter N, Smolders E, Merckx R (2013) Bioavailability of organic phosphorus to *Pseudokirchneriella subcapitata* as affected by phosphorus starvation: an isotope dilution study. Water Res 47:3047–3056

Vanderzee C, Roevros N, Chou L (2007) Phosphorus speciation, transformation and retention in the Scheldt estuary (Belgium/The Netherlands) from the freshwater tidal limits to the North Sea. Mar Chem 106:76–91

Vitousek PM, Reiners WA (1975) Ecosystem succession and nutrient retention: a hypothesis. Bioscience 25:370–381

von Sperber C, Kries H, Tamburini F, Bernasconi SM, Frossard E (2014) The effect of phosphomonoesterases on the oxygen isotope composition of phosphate. Geochim Cosmochim Acta 125:519–527

Walsh C (1979) Enzymatic reaction mechanisms. W H Freeman and Co, San Francisco

Wang Y, Huntington TG, Osher LJ, Wassenaar LI, Trumbore SE, Amundson RG, Harden JW, McKnight DM, Schiff SL, Aiken GR, Lyons WB, Aravena RO, Baron JS (1998) Carbon cycling in terrestrial environments. In: Kendall C, McDonnell JJ (eds) Isotope tracers in catchment hydrology. Elsevier Science, Amsterdam, pp 577–610

Weinhold F, Landis CR (2005) Valency and bonding: a natural bond orbital donor-acceptor perspective. Cambridge University Press, Cambridge

Wetzel RG (1983) Limnology. WB Sounders and Company, Philadelphia

Winogradsky S (1924) Sur la microflora autochtone de la terre arable. Comptes rendus hebdomadaires des seances de l'Academie des Sciences 178:1236–1239

Wolfenden R, Ridgway C, Young G (1998) Spontaneous hydrolysis of ionized phosphate monoesters and diesters and the proficiencies of phosphatases and phosphodiesterases as catalysts. J Am Chem Soc 120:833–834

Yang K, Metcalf WW (2004) A new activity for an old enzyme: *Escherichia coli* bacterial alkaline phosphatase is a phosphite-dependent hydrogenase. Proc Natl Acad Sci U S A 101:7919–7924

Young MB, McLaughlin K, Kendall C, Stringfellow W, Rollog M, Elsbury K, Donald E, Paytan A (2009) Characterizing the oxygen isotopic composition of phosphate sources to aquatic ecosystems. Environ Sci Technol 43:5190–5196

Zalatan JG, Fenn TD, Herschlag D (2008) Comparative enzymology in the alkaline phosphatase superfamily to determine the catalytic role of an active-site metal ion. J Mol Biol 384:1174–1189
Zohar I, Shaviv A, Klass T, Roberts K, Paytan A (2010) Method for the analysis of oxygen isotopic composition of soil phosphate fractions. Environ Sci Technol 44:7583–7588
Zyryanov AB, Vener AV, Salminen A, Goldman A, Lahti R, Baykov AA (2004) Rates of elementary catalytic steps for different metal forms of the family II pyrophosphatase from Streptococcus gordonii. Biochemistry 43:1065–1074

Chapter 14
Chemical Characteristics of Custom Compost for Highbush Blueberry

Dan M. Sullivan, David R. Bryla, and Ryan C. Costello

Abstract Recent development of markets for blueberry (*Vaccinium corymbosum* L.) produced under Organic certification has stimulated interest in production of composts specifically tailored to its edaphic requirements. Blueberry is a calcifuge (acid-loving) plant that responds favorably to mulching and incorporation of organic matter into soil. Many composts are high in pH and soluble nutrients, and may not be suited to blueberry. This chapter describes recent improvements in developing chemical criteria for composts that indicate potential suitability for blueberry. The experimental work, which was conducted in western Oregon, USA, confirmed that acidic pH (<5.5) is the most important characteristic needed in a custom compost for blueberry. Composts with pH < 6 are rare, so a testing protocol to quantify the pH buffering capacity of compost (CBC) and the quantity of acidity needed to reduce compost pH to 5.0 was developed. Median compost buffering capacity ($n = 36$) was 0.20 mol H+/kg/pH unit, and median elemental S (S^o) addition required for acidification to pH 5 was 8 g S^o/kg, assuming full reaction of S^o to H^+. In compost acidified with S^o to a final pH of 5–6, EC was increased 1.6 fold, accompanied by increased solubility for K and Na (1.3–1.4×); P, Ca, and Mg (3.2–3.6×); and SO_4-S (5.2×). Blueberry plants accumulated K supplied by compost, accompanied by reduced plant uptake of Mg, reduced growth, and possible Mg deficiency. Compost acidification to below pH 6 improved blueberry plant growth and Mg uptake. We conclude that compost can be used to increase soil organic matter for blueberry, but that compost N must be limited to low analysis values (total N < 20 g/kg) in order to avoid problems with high pH, EC, and excess K. Because all compost feedstocks that met USDA-Organic certification requirements needed acidification to reach the desired pH level for blueberry

D.M. Sullivan (✉) • R.C. Costello
Department of Crop and Soil Science, Oregon State University, 3017 Agriculture and Life Science Bldg., Corvallis, OR 97331, USA
e-mail: Dan.Sullivan@oregonstate.edu

D.R. Bryla
Horticultural Crops Research Unit, U.S. Department of Agriculture, Agricultural Research Service, 3420 NW Orchard Avenue Corvallis, Corvallis, OR 97330, USA

Z. He and H. Zhang (eds.), *Applied Manure and Nutrient Chemistry for Sustainable Agriculture and Environment*, DOI 10.1007/978-94-017-8807-6_14,

(<5.5), future research should focus on economical, safe, and reliable methods for compost acidification that are acceptable under Organic certification rules.

14.1 Introduction

Consumption of blueberries has increased rapidly in recent years, primarily due to promotion of their health benefits and to greater availability of the fruit year round. Currently, there are more than 77,000 ha of cultivated blueberry (*Vaccinium* sp.) grown worldwide, and production is expected to increase by another 46 % over the next 5 years (Brazelton 2011). Eighty-seven percent of the crop is produced in North America and South America, but plantings are expanding into less traditional growing regions, including Mexico, Brazil, Peru, southern Europe, northern and southern Africa, Asia, and Australia. By 2015, total global production is predicted to reach 635,000 metric tonnes of blueberries per year.

Blueberry is a long-lived perennial crop (30+ years) categorized as a calcifuge, well-adapted to acidic soil conditions. The plants acquire primarily the NH_4 form of N over NO_3-N and tolerate relatively low levels of P, K, Ca, and Mg in the soil and high concentrations of plant-available metals such as Mn and Al (Korcak 1988). The plants are also known to respond positively to organic matter when grown in mineral soils, including sawdust, pine bark, and peat moss, although the exact mechanism responsible for the benefit remains unclear (Clark 1991; Clark and Moore 1991; Lareau 1989; Moore 1979; White 2006). In most cases, the benefit is attributed to better drainage and soil aeration when the organic matter is incorporated in the soil prior to planting, and to improved weed control and reduced fluctuations in soil temperatures when it is applied as a surface mulch. Organic matter may also affect pH and availability of soil nutrients to blueberry (Larco et al. 2013b). Traditionally, most blueberry plantings were grown with either fir or pine obtained as either bark or sawdust from local lumber industries. However, with increasing costs of bark and sawdust, the use of weed mat (perforated landscape fabric) is becoming a common alternative. When blueberry beds are covered with weed mat, it is usually not feasible to add organic matter, except at planting or when the weed mat is replaced. Because weed mat is typically replaced every 3–5+ years, a high rate of compost application would be needed at the time of replacement.

Finished compost has many characteristics similar to that of soil organic matter (C:N of 12–20; high CEC; high stability). As compost decomposes, it slowly releases mineral N, with 3–10 % of the total compost N converted to mineral N for several years after application (Sikora and Szmidt 2001). Previous field research has demonstrated that a high rate of application of compost can build soil N mineralization rates for 7+ years after application (Sullivan et al. 2003). However, composts with high N (or low C:N ratios) are also usually high in soluble nutrients and electrical conductivity (EC). Many are also high in pH. For example, composts with C:N ratios <12 such as those produced from poultry litter or grass clippings are very high in soluble nutrients and pH and therefore probably unsuitable for high rate application to blueberry. Thus, choosing appropriate feedstocks to develop a custom compost for blueberry is a compromise between supplying N and other nutrients and avoiding high pH and high EC (Table 14.1). Composts with C:N ratios

Table 14.1 Linkage between N supplied by compost and its potential for use in field-scale highbush blueberry production systems[a]

Feedstocks	Typical N analysis		Feasible application rate	Expected change in soil organic matter	Factor most limiting application rate	Other first year limiting factors	Long-term issues	
	C:N	N (g/kg)					High pH	Excess K
Manure or vegetative plants only	<12	20+	Low	0	Excess N	Weeds, high EC, rapid decomposition	+	+
Typical composts (mixed feedstocks)	12–25	10–20	High	++	High pH and high EC	Weeds, excess N	++	++
Raw conifer wood chips/sawdust[b]	200+	<2	High	0	Deficient N	Possible P and S deficiency	0	0

0, +, ++, = zero to relatively large change
[a]Based on author experience
[b]Not composted

of 12–25 can be produced with a wide variety of feedstocks. Such composts may be suitable for high-rate incorporation into blueberry planting beds or may be applied as mulch under the weed mat. For a high rate of compost application, the ideal compost will contain manageable levels of soluble salts and will not increase soil pH above that tolerated by blueberry (i.e., pH $>$ 5.5). Composts with C:N $>$ 25 are usually undesirable as a soil amendment due to increased N fertilizer requirements; however, composted woody mulch with large-sized particles and a C:N of 25–40 usually has little effect on N fertilizer requirements.

Recent experimental work in western Oregon (Costello 2011a, b; Costello and Sullivan 2013) has improved current understanding of the linkage between compost feedstocks and compost chemical characteristics and the growth and nutrient uptake of blueberry. Compost acidification has also been investigated as a tool for making compost more suitable for blueberry.

14.2 Overview of Experimental Methods

Below is a brief summary of the key methods used in the research of Costello (2011a, b) and Costello and Sullivan (2013).

14.2.1 Compost Sources

Composts were collected from farms or municipal composting facilities in western Oregon and Washington with the goal of representing a wide range of feedstocks, including: dairy manure solids recovered from mechanical separation of flushed dairy manure + bedding (Dairy), horse manure + bedding (Horse), chipped woody prunings mixed with grass and other plant materials from urban landscape maintenance (Yard), deciduous tree leaves from urban street sweepings (Leaf), peppermint (*Mentha piperita*) leaves and stems recovered after steam distillation of mint oil (Mint), or mixed feedstocks (Other). Bark:biosolids compost was produced from a mix (v/v) of 60 % aged conifer bark, 20 % fine softwood sawdust, and 20 % USEPA Class A dewatered (25 % DM) municipal biosolids by the City of Tacoma, WA. Composts were held in moist condition at 10–30 °C prior to chemical analyses and preparation of soil + compost mixes for plant growth trials.

14.2.2 Chemical Analyses of Compost and Plant Tissue

Compost total nutrient analyses were performed by Brookside Laboratories (New Bremen, OH) using methods described in *Test Methods for Composts and Composting* (TMECC; Thompson et al. 2001). Compost total P, Ca, Mg, K and Na were determined by digesting samples in nitric/perchloric acid, with determination by an inductively coupled plasma (ICP) spectrophotometer (TMECC 4.03,

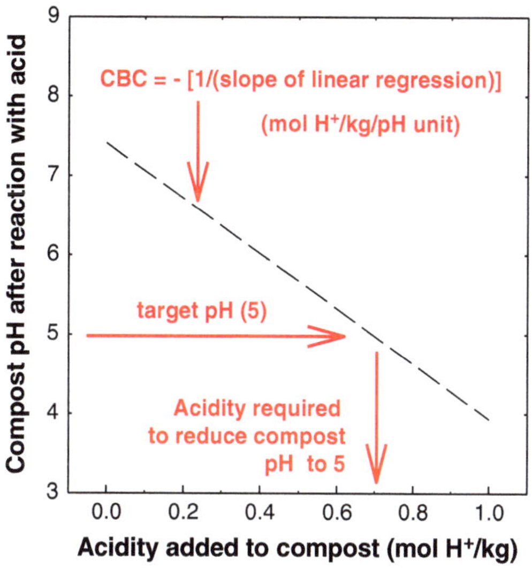

Fig. 14.1 Illustration of acid titration of compost to determine compost pH buffering capacity (CBC) and the quantity of acidity needed to reduce compost pH to a target value (Adapted from Costello and Sullivan 2013)

4.04, 4.05); total C and N were determined by combustion analysis (TMECC 4.01-A; 4.02-D). Combustion analysis was also used to determine N in plant tissue. Other nutrient concentrations in plant tissue were determined by digesting in nitric/perchloric acid, with determination by an ICP spectrophotometer (Gavlak et al. 1994).

Compost pH, EC, and soluble nutrients were determined using an adaptation of the saturated media extract (SME) method (Warncke 1986; Brookside Laboratory soil-less media package S003). Using the SME method, a 200-g "as-is" compost sample (sample is not dried or ground) is moistened to saturation (compost begins to flow when tilted and has minimal free water on the surface). The SME extract is obtained by filtration under vacuum, and soluble nutrients are determined by an ICP spectrophotometer and expressed on an extract solution concentration basis (mg/L).

14.2.3 Determination of Compost Buffering Capacity

To determine compost buffering capacity (CBC), 5 g of compost was reacted with 50 mL of diluted H_2SO_4 at addition rates of 0.1–1.0 mol H^+ per kg for 72 h (Costello and Sullivan 2013). The CBC was calculated as the negative reciprocal of the slope of the linear regression between compost pH (y-axis) and acid addition (x-axis) (Fig. 14.1). The rate of elemental sulfur (S^o) required to reach target compost pH (g S^o per kg compost) was then calculated as:

$$(\text{Initial compost pH} - \text{Target final compost pH}) \times \text{CBC} \times \text{Seq}$$

where:

CBC = mol H^+ per kg compost per pH unit
Seq = equivalent wt of S^o (16 g/mol H^+)

14.2.4 *Plant Growth Trials*

In preparation for plant growth trials, cured composts were acidified by adding wettable S^o dust that was allowed to react with moist compost for 70 days at 20–30 °C. Prior to S^o addition, compost organic matter was stable, as indicated by net CO_2 evolved in a 21-day incubation in soil at 22 °C (<3 % of compost-C respired). All composts had more NO_3-N than NH_4-N, indicating that compost had cured at mesophilic temperatures conducive to nitrification.

Two plant growth trials were conducted in pots to evaluate compost suitability. Organic amendment (compost or sawdust) was added to pots at a high rate (3 parts compost or sawdust to 7 parts soil; v/v). Trials were performed in winter in a greenhouse (4-L pots; Jan-May) and in summer in an outdoor can yard (14-L pots placed on a gravel bed; May-Sept). A silt loam soil (pH 5.3; 40 g/kg OM) was used in our growth trials (instead of nursery media) because our goal was to evaluate compost suitability for field application.

During each plant growth trial, N was supplied by fertigation every 14 days (Greenhouse Trial) or every 7 days (Can Yard Trial). Fertilizer N was supplied via mineral NH_4-N salts in the Greenhouse Trial, and by fish emulsion in the Can Yard Trial. More N was supplied to plants via fertigation in the Can Yard Trial (5 g/plant) than in the Greenhouse Trial (<0.1 g/plant).

Whole plants were harvested at the end of each plant growth trial. Shoots were separated into old stem (present in transplant) and new growth (leaves + stem produced after transplanting). In the Greenhouse Trial, root growth was measured from the original root ball using a cm scale (1 = no root growth; 7 = maximum root growth extended across entire pot diameter). In the Can Yard Trial, soil was removed with a high-pressure water spray, and then root dry weight was determined.

14.3 Compost Nutrient Analyses and Their Relationship to Blueberry Growth

Composts that were highest in N (low C:N) such as Dairy and Mint compost also had characteristics considered undesirable for blueberry, including high pH and EC as well as high K (Table 14.2). Bark:biosolids, Leaf, and Yard composts, on the other hand, had lower pH and EC, and a C:N ratio near 20, and were closest to that of organic materials typically used for blueberry such as peat and sawdust.

Blueberry plant growth was superior with the Bark:biosolids, Leaf, and Yard composts (Table 14.3). Mint and Dairy, in contrast, hosted very limited root growth in both the Greenhouse Trial and the Can Yard Trial. Under low fertilizer N supply (Greenhouse Trial), plant top growth was superior with all composts except Mint than with the grower-standard amendment of sawdust. However, under greater fertilizer N supply (Can Yard Trial), plant growth was better with sawdust than

Table 14.2 Nutrient analysis of composts compared to other organic amendments suited to highbush blueberry

Organic source (*n*)	pH[a]	EC[a] (dS/m)	C (g/kg)	N (g/kg)	C:N ratio	P (g/kg)	Ca (g/kg)	Mg (g/kg)	K (g/kg)	Na (g/kg)
Compost										
Dairy (9)	7.6	6.1	310	22	14	5	30	6	14	3
Horse (5)	7.8	3.3	300	12	24	3	11	4	16	3
Yard (5)	7.0	4.0	280	13	21	3	17	4	8	1
Leaf (2)	7.4	2.2	200	10	19	2	24	3	3	0
Mint (2)	7.7	11.6	380	47	8	7	23	9	30	1
Other (5)	6.8	6.0	200	16	12	5	14	5	10	5
Bark:biosolids (3)	6.5	2.5	280	11	26	7	12	4	2	1
Other organic amendment										
Sphagnum peat moss (1)	4.8	0.7	480	13	37	1	10	2	1	0.1
Sawdust (1)[b]	5.2	0.4	490	1	490	0.3	0.4	0.3	<0.1	<0.1

[a]Compost pH and EC were determined using the SME method
[b]Suitable for highbush blueberry with supplemental N fertilizer

Table 14.3 Shoot and root growth in compost-amended soil pot trials with highbush blueberry

	% of bark:biosolids compost treatment[a]			
	Greenhouse trial (low N)		Can yard trial (adequate N)	
Compost	New shoots and leaves	Roots	New shoot and leaves	Roots
Bark:biosolids	100	100	100	100
Dairy	62	39	84	55
Horse	96	96	52	51
Leaf	88	120	94	75
Mint	45	45	78	56
Yard	88	93	78	82
Sawdust	35	157	87	90

[a]Values <80–85 % relative yield reflect plant growth inferior to Bark:biosolids-based substrate

with several composts. Plants grew well with Bark:biosolids compost in both the low and adequate N trials, demonstrating its suitability across a variety of management regimes.

Compost also increased N uptake under low N supply conditions relative to sawdust and soil only (Table 14.4). Blueberry plants likewise responded to compost application by increasing K uptake and reducing Mg uptake, especially for composts with the highest K concentrations (Mint, Dairy, Horse; Table 14.2). Consequently, compost had a strong effect on cation ratios in the shoots. The K:Mg ratio was more sensitive to K supply than the K:Ca ratio. These results demonstrate that K can accumulate to high levels in blueberry amended with compost (luxury consumption), and may lead to problems with plant Mg deficiency. Bark:biosolids

Table 14.4 Effect of compost on total aboveground plant biomass and N and cation uptake by highbush blueberry plants grown in a greenhouse with a low N fertilizer rate

		Nutrient uptake (mg/plant)					
Compost	Total biomass[a] (g DW/plant)	N	K	Mg	Ca	K:Mg ratio	K:Ca ratio
Horse	6.7	91	95	7.1	20	13	4.7
Mint	3.3	65	117	3.1	9	38	12.6
Dairy	4.6	64	95	3.4	13	28	7.5
Leaf	6.3	59	46	6.4	20	7	2.3
Yard debris	6.3	59	57	5.9	18	10	3.2
Bark:biosolids	6.4	57	37	7.4	21	5	1.7
Sawdust	3.0	16	17	4.1	13	4	1.3
Soil only	5.2	42	25	5.9	17	4	1.5

[a]Biomass (aboveground) at harvest: old stem + new stem + leaves (Costello 2011a)

compost was a notable exception, having low compost K analysis (Table 14.2) and little effect on plant K or Mg uptake (Table 14.4).

Competition between K^+ and Mg^{2+} for plant uptake, resulting in elevated plant K uptake, and reduced Mg uptake has been reported in most plant species (Dibb and Thompson 1985), including tree fruit and small fruit crops (Cummings 1985). Highbush blueberry, a calcifuge plant, has lower K requirements than most fruit crops. Forge et al. (2013) reported that highbush blueberry leaf K was increased from 5 g/kg (with sawdust mulch) to 8 g/kg with 7–10-cm yard debris compost mulch, accompanied by a decrease in leaf Mg from 1.4 to 1.1 g/kg. Leaf K of 4–7 g/kg and Mg of 1.3–2.5 g/kg are considered adequate for blueberry in the Pacific Northwest USA/Canada (Hart et al. 2006).

14.4 Determination of Compost Buffering Capacity and Amount of Elemental Sulfur Required to Acidify the Compost

The response of compost pH to dilute acid addition was reasonably fit with a linear regression function (Fig. 14.2). Linear regression was highly significant ($P < 0.001$) for all composts tested, and each r^2 value exceeded 0.87. Compost buffering capacity, as determined by dilute acid titration, ranged from 0.17 to 0.28 mol H^+ per kg per pH unit across feedstock categories (Table 14.5) and from 0.10 to 0.39 among individual compost samples ($n = 36$; Fig. 14.3). Mint and dairy composts had notably higher CBC than other feedstocks. Variation in CBC among composts made from the same feedstock was considerable. For example, for dairy composts, the 25th and 75th percentile CBC values were 0.23–0.33 mol H^+ per kg (Fig. 14.3). The quantity of S^o needed to acidify compost to pH 5, assuming complete oxidation of S^o to H^+, ranged from 2 to 17 g/kg (Fig. 14.3).

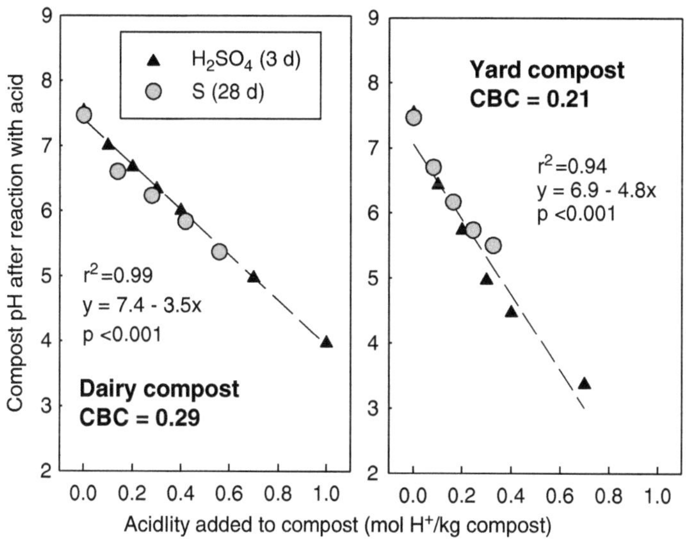

Fig. 14.2 Compost pH buffering capacity (CBC), as determined by titrating compost with dilute sulfuric acid for 3 days at 22 °C. Similar pH values were obtained by incubating compost with wettable elemental sulfur dust for 28 days at 22 °C

Table 14.5 Acidity to reduce compost pH to 5.0 (as determined by titration) vs. total cation analysis of compost

	Titration	Titration		Titration
	CBC[a]	H^+ to pH 5	Total K + Mg + Ca	H^+/Cation(+)
Compost (*n*)	(mol H^+/kg/pH unit)	(mol H^+/kg)	[mol (+)/kg]	Ratio
Dairy (9)	0.28	0.73	2.51	0.29
Horse (5)	0.17	0.48	1.35	0.36
Yard (5)	0.18	0.41	1.42	0.29
Leaf (2)	0.20	0.50	1.55	0.32
Mint (2)	0.28	0.72	2.81	0.26
Other (5)	0.18	0.33	1.40	0.24
Bark:biosolids (3)	0.21	0.29	0.92	0.32

[a]CBC = Compost pH buffer capacity to pH 5.0 as determined by dilute acid titration (Costello and Sullivan 2013)

The relationship between the 3-day dilute acid measurement of CBC and calculated S^o required for acidification was verified via laboratory incubations of compost + wettable S^o dust (Costello and Sullivan 2013). For the eight composts evaluated, S^o addition to acidify compost by 1 pH unit (as predicted by CBC) resulted in an actual pH decline after 28 days of 1.1 units (SE = 0.2), and predicted S^o addition to acidify it by 2 pH units resulted in an actual decline of 1.7 units (SE = 0.2).

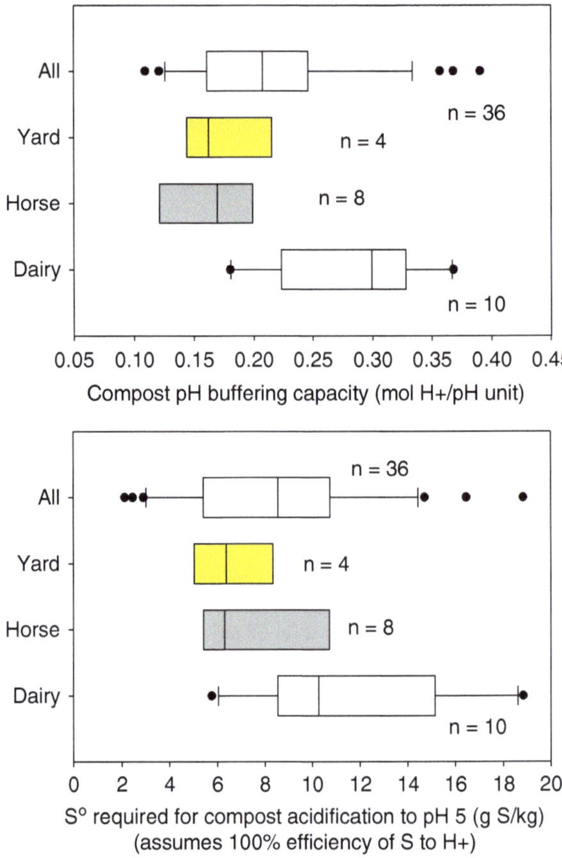

Fig. 14.3 Compost pH buffering capacity (CBC) as determined by dilute acid titration of compost samples (*top*) and elemental sulfur (S^o) required for compost acidification to pH 5.0, assuming complete oxidation of S^o to H^+ (*bottom*). *Center line in box* = median value. Outside of box are 25th and 75th percentile values. Whiskers at 5th and 95th percentile. Outliers = *dots*

Compost buffering capacity was correlated with the total cation analysis of compost (K, Mg, Ca, Na), expressed as mol (+)/kg ($P > 0.001$). The molar ratio of CBC to cations [mol H^+/mol cation(+)] ranged from 0.24 to 0.36 across feedstock categories, with an average value of 0.30 (Table 14.5).

Linear regression analysis yielded a significant ($r^2 = 0.51$; $P > 0.001$) equation to predict S^o requirement based on the total cation analysis of compost (Fig. 14.4). For acidification to pH 5, approximately 5 g of S^o per kg was required per mol of total cations. Correlations between total cation content of organic matter and its liming effect have been reported previously (Wong et al. 1998; Naramabuye and Haynes 2006), but these studies reported data for only a few composts.

Compost buffering capacity, when expressed on an OM basis ranged from 0.5 to 0.7 mol H+/kg OM/pH unit for eight composts from western Oregon (Costello and Sullivan 2013), a similar or greater pH buffer capacity than has been reported for peat (Jeong 2010), for soil organic matter in forest soils (James and Riha 1986; Magdoff et al. 1987), and for soil organic matter in agricultural soils (Helling et al. 1964; Curtin and Rostad 1997).

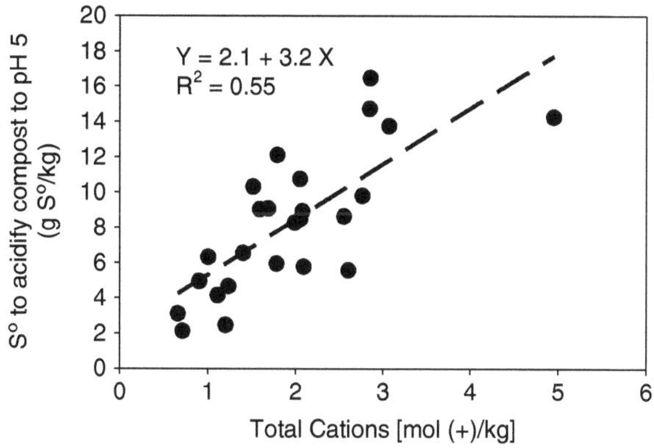

Fig. 14.4 Relationship between compost total cation analysis (K, Ca, Mg, Na) and amount of elemental sulfur (S°) required for compost acidification, as determined by dilute acid titration. The S° requirement estimate (y-axis) assumes complete oxidation of S° to H^+

Table 14.6 Elemental sulfur (S°) reaction with compost increases acidity, EC, and soluble nutrients

	Compost		Change
Measurement[a]	No S°	With S°	(With S°/no S°)
pH	7.5	5.8	
EC (dS/m)	5.6	9.1	1.6
P (mg/L)	25	80	3.2
Ca (mg/L)	130	470	3.6
Mg (mg/L)	90	290	3.3
K (mg/L)	980	1,380	1.4
Na (mg/L)	260	330	1.3
SO_4-S (mg/L)	410	2,170	5.2
Mn (mg/L)	0.2	5.5	25
Zn (mg/L)	0.3	0.4	1.3

[a]Average of SME extract for 24 compost samples from Dairy (8), Horse (5), Yard debris (3) and Other (8)

14.5 Effect of Compost Acidification on Blueberry Growth and Nutrient Uptake

Reaction with S° increased EC and nutrient solubility across all composts (Table 14.6). The effect of acidification on solubility was larger for divalent cations (Ca^{2+} and Mg^{2+}) than for monovalent cations (K^+ and Na^+). Soluble Mg increased threefold in compost after acidification, while K increased only 1.4-fold. Phosphorus solubility was similarly increased by more than threefold, indicating dissolution of mineral P, and Mn solubility was increased by more than tenfold. Compost feedstocks with high initial soluble nutrient concentrations (before acidification) showed the largest increases in EC and soluble nutrients after acidification. Because compost acidification increased soluble Ca and Mg more than K, the acidified compost had a lower soluble K:Mg and K:Ca ratios than the non-acidified compost.

Table 14.7 Effect of compost acidification on compost analysis and blueberry plant growth in a 78-day pot trial[a]

Measurement	Compost acidification		Change[b]
	No S°	With S°	(S°/no S°)
Compost analysis[c]			
pH	7.2	5.3	–
EC (dS/m)	1.9	3.5	1.8
Plant growth			
Leaves (g DW/plant)	2.9	3.4	1.17
Shoot (g DW/plant)	5.2	5.9	1.13
Root rating (cm of growth)	3.6	4.6	1.28

[a]Compost analysis and plant biomass are averaged across 10 composts (Costello 2011a)
[b]Compost analysis and plant growth responses were significant at $P < 0.001$
[c]Compost pH and EC were determined by the 1:10 (compost:water; w/w) method

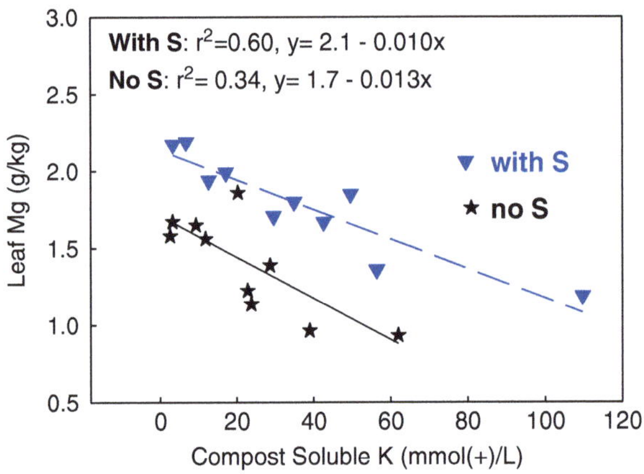

Fig. 14.5 Compost acidification with elemental sulfur (S°) increases Mg concentration in young blueberry leaves. Compost was acidified prior to soil incorporation. Young leaves were collected from plants at 50 days after transplanting into compost-amended soil (Costello 2011a). Soluble K was determined by SME analysis. Leaf Mg of 1.3–2.5 g/kg is considered adequate for recent fully expanded leaves (Hart et al. 2006)

Blueberry leaves, shoots and roots had greater growth in soil amended with acidified compost, even though compost EC was increased by acidification (Table 14.7). The improvement in plant growth (about 20 %) with compost acidification was observed across a wide range of compost feedstocks. One mechanism of benefit to blueberry from acidified compost appears to be improved plant uptake of Mg. Leaf Mg was higher in plants growing in soil amended with acidified compost (Fig. 14.5). Plants grown in soil amended with compost containing high concentrations of soluble K had low leaf Mg concentrations (<1.5 g/kg), indicating possible plant Mg deficiency (Hart et al. 2006).

14.6 Summary and Future Research

14.6.1 Choosing Feedstocks to Produce Custom Compost for Blueberry

Pot trials presented here have confirmed the generally held idea that organic matter sources for blueberry should be low in soluble salts and have pH < 5.5. Only a small subset of available compost feedstocks will be suitable for partial replacement of the current blueberry industry standard organic substrates, conifer sawdust (for field application) and peat moss (for containers). Manures and plant materials that contained total N > 20 g/kg were not suitable for blueberry because of associated characteristics of high pH, EC, and K (Table 14.8). The manure composts evaluated in the present studies (i.e., horse manure with woody bedding and dairy manure solids separated by gravity from a liquid handling system) were lower in EC than whole manures (no bedding). A high-N crop residue (i.e., mint) was also poorly suited to blueberry.

Composts derived from municipal yard debris, leaves from street sweepings, or dewatered municipal biosolids composted with conifer bark need to be further evaluated as potential materials to replace sawdust in field-scale production systems. Yard debris composts are likely the best candidate for use by growers because they have relatively consistent chemical analyses, can be acidified to pH 5 with a moderate rate of S^o, and composters have the ability to control particle size by screening. Cation analysis of yard debris compost is reflective of local soil and vegetation. Yard debris composts from more arid regions may not be suitable for blueberry compost because of higher cation analyses, higher CBC, and high EC.

Bark:biosolids and Leaf composts used in this study had near-ideal chemical properties for blueberry. They were the only composts that supplied low levels of K (<5 g/kg) and low levels of other soluble salts (EC < 3 dS/m). Most, but not all dewatered municipal biosolids will be low in K because K follows the liquid stream at wastewater treatment plants. Cation analyses of biosolids (and CBC) will likely be higher in regions with "hard" water. The major limitation to use of biosolids compost is public perception. Biosolids compost is specifically prohibited in rules governing Organic certification (USDA 2000), although USEPA considers biosolids composted to meet Class A standards to be suitable for use with any food crop (USEPA 2002; Sullivan et al. 2007). Compost analyses and plant growth response was also favorable for leaves collected from street sweepings, but availability of this material is more limited than yard debris. Leaves collected from some urban areas may also have unacceptably high levels of trace elements due to soil contamination (e.g., Pb and Cd; Heckman and Kluchinski 1996) and therefore should be tested.

Table 14.8 Relative suitability of composts for highbush blueberry for high rate incorporation into soil

Major feedstock	Description	Compliant with USDA-Organic?[a]	Evaluated in pot trials?	Major limitation	Unsuitable[b]	Trial use[c]
Manure	Poultry litter (manure + woody bedding)	Yes	No	pH, EC	●	
	Dairy solids from mechanical separator	Yes	Yes	pH, EC	●	
	Horse manure + bedding	Yes	Yes	pH, EC	●	
Plant material	Plant residue from peppermint distillation	Yes	Yes	pH, EC	●	
	Street sweepings (deciduous tree leaves)	Yes	Yes	pH, rapid decomposition		●
	Yard debris	Yes	Yes	pH, K		●
Biosolids	Lime-stabilized municipal biosolids	No	No	pH, EC	●	
	Conifer bark + dewatered biosolids with Fe or Al addition	No	Yes	Consumer perception		●

[a]USDA National Organic Program (USDA 2000)
[b]Unsuitable = reduced growth of highbush blueberry in pot trials when incorporated at 30 % (v/v) into soil
[c]Trial use = maintained or improved growth of highbush blueberry in pot trials when incorporated at 30 % (v/v) into soil

14.6.2 Recommended Compost Analysis Methods

pH: Compost pH values are relatively consistent across different methods. Similar pH values can be obtained with either SME or 1:5 or 1:10 methods.

pH buffering capacity: The pH buffering capacity of compost can be accurately determined using equilibration with dilute acid for 3 days (Costello and Sullivan 2013). This method may be too time-consuming for implementation by commercial laboratories, but it is simple enough for use by nurseries and other large plant production facilities that have a scale and a well-calibrated pH meter. Because the relationship between acid addition and compost pH is approximately linear, it may be feasible to use a single acid addition rate to estimate pH buffering capacity, rather than a regression line based on multiple acid addition rates. The dilute acid test may also prove useful to formulators of potting mixes, as an indicator of the lime substitution value of specific composts. The median value for the liming value of compost (pH > 5) was equivalent to 25 g of $CaCO_3$ per kg and ranged from 6 to 50 g/kg.

Electrical conductivity: Compost EC values are strongly method dependent. The EC determined by SME is two- to three-times higher than that determined by other methods using a fixed compost (dry) to water ratio (1:5 or 1:10 methods; VanderGheynst et al. 2004). Composts that are unsuitable for blueberry have EC values >4 dS/m with the SME method, >2 dS/m with the 1:5 method, and >1–1.5 dS/m with the 1:10 method.

Cations: The most common method offered by commercial laboratories, total nutrient analysis, is suitable for identifying composts with excessive K and composts with high buffering capacity (via sum of cations). Calcium carbonate equivalency tests can be used to identify composts that contain high quantities of carbonate minerals, which are likewise unsuitable for blueberry. We do not recommend compost CEC testing as an indicator of pH buffering capacity because the pH buffering capacity of compost resides in both mineral and organic matter fractions, and CEC is usually determined at pH 7.

Respiration rate and inorganic N: An assessment of compost respiration rate and NH_4-N + NO_3-N determination will provide information about the likely effect of N dynamics on compost pH (Sullivan and Miller 2001). Composts that are unstable and have high concentrations of NH_4-N show a decline in pH with further curing, as NH_4 is converted to NO_3-N via nitrification, or NH_3 is lost to the atmosphere.

14.6.3 Interpretation of Recommended Compost Analyses for Blueberry

Our research demonstrates the value of pH, EC, K, and pH buffering capacity tests as screening criteria to identify potential composts for high application on blueberry beds. We consider composts with total K > 10 g/kg, total cations > 1.5 mol (+)/kg, compost acidification requirement to pH 5 (as determined by pH buffering capacity test) > 0.5 mol H + per kg, and EC > 4 dS/m (by SME) to be a risk for short-term plant damage when applied at high rates (>5 cm depth of compost applied to the beds). Note that these limits are based on experience with composts available in western Oregon and should be regarded as preliminary. Composts that meet all of the criteria considered ideal for blueberry are rare. Therefore, composts that meet "most" of the analytical criteria should be considered for trial use. For example, yard debris composts in western Oregon have analysis values that are approximately equal to the given limits. Results from a recent study on organic blueberry indicated that a 4-cm deep layer of yard debris compost covered with 5 cm of sawdust mulch produced more shoot growth and higher early yields than 9 cm of sawdust alone but reduced net cumulative returns relative to weed mat (landscape fabric) or sawdust alone by as much as $7,800/ha, primarily due to higher weed management costs (Julian et al. 2012; Larco et al. 2013a). The yard debris compost with sawdust also increased soil pH and organic matter content and resulted in higher levels of soil NO_3-N, P, K, Ca, B, Cu, and Zn than sawdust alone and increased leaf K and B (Larco et al. 2013a). Forge et al. (2013) observed, however, that a 7.5–10 cm layer of yard debris compost without sawdust reduced plant vigor in blueberry compared to sawdust and attributed the reduction to increased soil pH and EC levels, Mg deficiency in the plants, and to less colonization by mycorrhizal fungi in the roots. Thus, the use of yard debris compost in blueberry can be beneficial, provided weeds are managed economically and the compost application rate is not excessive.

14.6.4 Recommendations for Future Research

The data presented here illustrate the challenge of developing compost test criteria applicable to production of a specialty crop like highbush blueberry. To achieve confidence in compost analyses as predictors of compost suitability, growth trials designed to evaluate the relationship between compost analysis and crop performance are essential. Developing custom uses for compost will require close collaboration among horticulturalists, agronomists, and soil scientists.

Composts that contain high concentrations of N are often considered desirable by growers but also generally have a high EC and pH buffering capacity and contain high levels of K (undesirable characteristics). Future research should evaluate the feasibility of enriching low N composts by using NH_3 captured from animal manure storage or animal manure composting. Woody debris is commonly used as biofilter material to trap NH_3 and odors at large composting facilities (Williams and Miller 1993; Rynk and Richard 2001). It may be possible to trap N in biofilter material, then utilize the spent biofilter material as N-enriched woody compost feedstock.

Future research and technology development to improve compost suitability for blueberry should also address safe and effective methods for on-farm compost acidification. Compost pH is critical to the use of compost in blueberry and may be customized by using S^o for acidification. In studies reported here, we used wettable S^o dust to acidify compost at a laboratory scale. Sulfur dust is potentially hazardous to farm workers, however, and it can be explosive when handled improperly. Another potential problem with S^o addition to large compost piles is emission of odiferous reduced S compounds such as organic sulfides, mercaptans, and hydrogen sulfide (Rosenfeld et al. 2004; Buyuksonmez et al. 2007). Thus, alternative methods for compost acidification need to be developed for on-farm use. Perhaps the most feasible method to acidify compost with S^o is to add it to compost at the time of field application. Regardless of the timing of S^o addition, commercial formulations (pellets, prills, etc.) need to be evaluated for efficacy in rapidly acidifying compost or in compost-amended soil. In a pilot study, we found that S-bentonite pellets did not acidify compost after 10 weeks in a 40 °C compost windrow, while S^o dust nearly reacted completely with compost after 2–4 weeks. Binding compounds used in some pelleted S^o formulations also disqualify these products for use under Organic certification.

Accumulation of K in soil appears to be a major drawback to use of yard debris compost in blueberry. Our results showed that K from the compost can accumulate in blueberry plants and possibly cause Mg deficiency. Once applied to soil, K resists leaching by adsorption to CEC, so it is likely to accumulate in soil over time when compost is routinely used. Research is needed on how to reduce K in compost. Because most of the K is present in water soluble form, it may be possible to remove K by leaching the compost at a composting facility. This would require technology, expense, and an acceptable method for utilizing the compost leachate.

References

Brazelton C (2011) World blueberry acreage & production. U.S. Highbush Blueberry Council, Folsum

Buyuksonmez, F, Yucel A, Rynk R, Cotton M (2007) A literature review: generation, measurement, and mitigation of odors from composting facilities. CA Integrated Waste Management Board Publication 442-07-001, Sacramento, CA. www.ciwmb.ca.gov/Publications

Clark J (1991) Rabbiteye and southern highbush blueberry response to sawdust mulch. Ark Farm Res 40:3

Clark JR, Moore JN (1991) Southern highbush blueberry response to mulch. HortTechnology 1:52–54

Costello RC (2011a) Compost acidification increases growth and nutrient uptake of highbush blueberry under a low N fertilizer regime. In: Costello RC (ed) Suitability of diverse composts as soil amendments for highbush blueberry (*Vaccinium corymbosum* L.). M.S. thesis, Oregon State University, Corvallis, OR, pp 15–47. http://hdl.handle.net/1957/26590

Costello RC (2011b) Evaluating diverse composts for highbush blueberry under an organic fertilization regime. In: Costello RC (ed) Suitability of diverse composts as soil amendments for highbush blueberry (*Vaccinium corymbosum* L.). M.S. thesis, Oregon State University, Corvallis, OR, pp 48–79. http://hdl.handle.net/1957/26590

Costello RC, Sullivan DM (2013) Determining the pH buffering capacity of compost via titration with dilute sulfuric acid. Waste Biomass Valorization (in press). doi:10.1007/s12649-013-9279-y
Cummings GA (1985) Potassium nutrition of deciduous and small fruits. In: Munson R (ed) Potassium in agriculture. Soil Science Society of America, Madison, pp 1087–1104
Curtin D, Rostad HP (1997) Cation exchange and buffer potential of Saskatchewan soils estimated from texture, organic matter and pH. Can J Soil Sci 77:621–626
Dibb DW, Thompson WR Jr (1985) Interaction of potassium with other nutrients. In: Munson R (ed) Potassium in agriculture. Soil Science Society of America, Madison, pp 515–533
Forge T, Temple W, Bomke A (2013) Using compost as mulch for highbush blueberry. Acta Hortic 1001:369–374
Gavlak RG, Horneck DA, Miller RO (1994) Plant, soil and water reference methods for the Western region, Western Region Extension Publication 125 (WREP-125). University of Alaska-Fairbanks, Fairbanks, Alaska
Hart J, Strik B, White L, Yang W (2006) Nutrient management for blueberries in Oregon. Oregon State University Extension Service Publication EM 8918, Corvallis, OR
Heckman JR, Kluchinski D (1996) Chemical composition of municipal leaf waste and hand collected urban leaf litter. J Environ Qual 25:355–362
Helling CS, Chesters G, Corey RB (1964) Contribution of organic matter and clay to soil cation-exchange capacity as affected by the pH of the saturating solution. Soil Sci Soc Am J 28:517–520
James BR, Riha SJ (1986) pH buffering in forest soil organic horizons: relevance to acid precipitation. J Environ Qual 15:229–234
Jeong KY (2010) Establishment and stabilization of pH in container root substrate. M.S. thesis, North Carolina State University, Raleigh, NC, pp 21–41
Julian JW, Strik BC, Larco HO, Bryla DR, Sullivan DM (2012) Costs of establishing organic northern highbush blueberry: impacts of planting method, fertilization, and mulch type. HortScience 47:866–873
Korcak RF (1988) Nutrition of blueberry and other calcifuges. Hortic Rev 10:183–227
Larco HO, Strik BC, Bryla DR, Sullivan DM (2013a) Mulch and fertilizer management practices for organic production of highbush blueberry. I. Plant growth and allocation of biomass during establishment. HortScience 48:1250–1261
Larco HO, Strik BC, Bryla DR, Sullivan DM (2013b) Mulch and fertilizer management practices for organic production of highbush blueberry. II. Impact on plant and soil nutrients during establishment. HortSci 48:1484–1495
Lareau M (1989) Growth and productivity of highbush blueberries as affected by soil amendments, nitrogen fertilization and irrigation. Acta Hortic 241:126–131
Magdoff FR, Bartlett RJ, Ross DS (1987) Acidification and pH buffering of forest soils. Soil Sci Soc Am J 51:1384–1386
Moore JN (1979) Highbush blueberry culture in the upper South. In: Proceedings of the fourth North American Blueberry Research Workers conference, University of Arkansas, Fayetteville, Arkansas, 16–18 October 1979, pp 84–86
Naramabuye FX, Haynes RJ (2006) Effect of organic amendments on soil pH and Al solubility and use of laboratory indices to predict their liming effect. Soil Sci 171:754–763
Rosenfeld P, Grey M, Sellew P (2004) Measurement of biosolids compost odor emissions from a windrow, static pile, and biofilter. Water Environ Res 76:310–315
Rynk R, Richard TL (2001) Commercial compost production systems. In: Stofella PJ, Kahn BA (eds) Compost utilization in horticultural cropping systems. CRC Press, Boca Raton, pp 51–93
Sikora LJ, Szmidt RAK (2001) Nitrogen sources, mineralization rates, and nitrogen nutrition benefits to plants from composts. In: Stofella PJ, Kahn BA (eds) Compost utilization in horticultural cropping systems. CRC Press, Boca Raton, pp 287–305
Sullivan DM, Miller RO (2001) Compost quality attributes, measurements and variability. In: Stofella PJ, Kahn BA (eds) Compost utilization in horticultural cropping systems. CRC Press, Boca Raton, pp 95–120

Sullivan DM, Bary AI, Nartea TJ, Myrhe EA, Cogger CG, Fransen SC (2003) Nitrogen availability seven years after a high-rate food waste compost application. Compost Sci Util 11:265–275
Sullivan, DM, CG Cogger, AI Bary (2007) Biosolids quality. In: Fertilizing with biosolids. Pacific Northwest Extension Publication 508-E. Oregon State University, Corvallis, OR, pp 12–13
Thompson WH, Leege P, Millner P, Watson M (eds) (2001) Test methods for the examination of composting and compost (TMECC). U.S. Composting Council, Rokonkoma
United States Environmental Protection Agency (USEPA) (2002) Biosolids technology fact sheet: use of composting for biosolids management. US EPA/832-F–02-024. Municipal Technology Branch, Office of Water, Washington, DC
USDA Agricultural Marketing Service (2000) National Organic Program. CFR Part 205. Federal Register 65:246. http://www.ams.usda.gov/AMSv1.0/nop
VanderGheynst JS, Pettygrove S, Dooley TM, Arnold KA (2004) Estimating electrical conductivity of compost extracts at different extraction ratios. Compost Sci Util 12:202–207
Warncke DD (1986) Analyzing greenhouse growth media by the saturation extract method. HortScience 21:223–225
White LD (2006) The effect of pre-plant incorporation with sawdust, sawdust mulch, and nitrogen fertilizer rate on soil properties and nitrogen uptake and growth of 'Elliott' highbush blueberry. M.S. thesis, Oregon State University, Corvallis, OR
Williams TO, Miller FC (1993) Composting facility odor control using biofilters. In: Hoitink HAJ, Keener H (eds) Science and engineering of composting: design, environmental, microbiological and utilization aspects. Renaissance Publications, Worthington, pp 262–281
Wong MTF, Nortcliff S, Swift RS (1998) Method for determining the acid ameliorating capacity of plant residue compost, urban waste compost, farmyard manure, and peat applied to tropical soils. Commun Soil Sci Plant Anal 29:2927–2937

Chapter 15
Distribution and Biodegradability of Water Soluble Organic Carbon and Nitrogen in Subarctic Alaskan Soils Under Three Different Land Uses

Mingchu Zhang, Aiqin Zhao, and Zhongqi He

Abstract Water-extractable organic matter (WEOM) contains labile organic carbon (C) and nitrogen (N) and is sensitive to soil management. However, knowledge about quantitative changes of water soluble organic C (WSOC) and N (WSON) impacted by land use conversion is still limited. In this chapter, the level and degradability of WSOC and WSON in soils sampled from subarctic Alaska under three different land use managements (i.e. forest, agriculture; and grassland converted from agricultural use and under a Conservation Reserve Program-CRP) were examined. There was no statistical difference of WSOC and WSON among three land uses. Furthermore, WSOC was dominated by large (>0.45 μm) and small (<1 kDa) size molecules in CRP and forest soils, while small molecules predominated in agricultural soils. All WEOM fractions displayed three similar fluorophore components (two humic acid-like, fulvic-like and a protein-like), indicating that they were major components of WSOC and WSON, and the impact of land uses was mainly on the quantity, rather than on the composition, of WEOM. Laboratory incubation revealed that the biodegradability of cold or hot WEOC was significantly correlated ($p < 0.01$) with the protein-like component, indicating that a protein-like fluorophore represented a labile fraction in cold and hot WEOM pools. High correlation was found between cold and hot WSON and other chemically extracted potential mineralizable N indexes, showing that hot WSON was a

M. Zhang (✉)
School of Natural Resources and Agricultural Sciences,
University of Alaska Fairbanks, Fairbanks, AK 99775, USA
e-mail: mzhang3@alaska.edu

A. Zhao
Editorial Department of the Transactions of the Chinese Society of Agricultural Engineering,
No. 41 Maizidian Street, Chaoyang District, Beijing 100125, China

Z. He
Southern Regional Research Center, USDA-ARS, New Orleans, LA 70124, USA

Z. He and H. Zhang (eds.), *Applied Manure and Nutrient Chemistry for Sustainable Agriculture and Environment*, DOI 10.1007/978-94-017-8807-6_15,

part of acid or alkaline hydrolysable N. Information derived from this chapter increased understanding of subarctic soil WEOM properties and its biodegradability.

15.1 Introduction

Water-extractable organic matter (WEOM) is generally regarded as the most labile and mobile form of soil organic matter fractions (Zsolnay 1996; Zhang et al. 2011b). WEOM is a heterogeneous mixture of complex geomolecules and biomolecules (Hunt et al. 2007a, b; Ohno et al. 2010). Among them, dissolved or water soluble organic C (DOC or WSOC) and N (DON or WSON) are the important components (Rousk and Jones 2010; Zhang et al. 2011b; Ghani et al. 2013). The quantity and biological nature of WEOM are affected by extraction temperatures (Nkhili et al. 2012) so that WEOM is frequently extracted by water at ambient (20 °C) temperature (cold WEOM), and at a temperature of 80 °C (hot WEOM). The cold and hot water extractions can be performed separately (Landgraf et al. 2006; Ohno et al. 2009; Bu et al. 2010) or sequentially (Gregorich et al. 2003).

Various methods are developed for evaluating WEOM. Spectroscopic characteristics of WEOM are usually analyzed by techniques such as UV/visible (Zhang et al. 2011a), fluorescence (Ohno et al. 2009; Ohno and He 2011), FTIR (Bu et al. 2010; He et al. 2011a) and NMR spectroscopy (Tatzber et al. 2008; He et al. 2009a). Chemical composition and degradability are typically characterized by element (especially C and N content and C/N ratio) analysis (He et al. 2011a; Zhang et al. 2011b), laboratory incubation (Hunt et al. 2007a), sequential filtration (Smolander and Kitunen 2002; Kiikkilä et al. 2006), and chromatographic techniques (Hunt et al. 2007a; He et al. 2010; Tatzber et al. 2011; Nkhili et al. 2012). Generally speaking, there are plenty of research studies on soil WSOC but not enough research on soil WSON (McDowell 2003). Partly for this reason, the information on C:N ratios of WEOM are often contradictory. Some show that the C:N ratios in hot WEOM are narrower than C:N-ratios of cold WEOM and whole soil organic matter (Kaiser and Zech 2000; Curtin et al. 2006), but others report a wider C:N ratio in hot WEOM than in cold WEOM (Landgraf et al. 2006; Chantigny et al. 2010). Whereas Curtin et al. (2006) demonstrate that hot water extracted organic N can serve as an index for estimating soil potential mineralizable N, we still don't know the properties of these hot water extracted organic N.

Climate change and demand of bioenergy have stimulated new aspirations on agricultural production in the subarctic regions. The Delta Junction area (64°49′ N, 147°52′ W) is a major small grain production region in the subarctic Alaska, USA (Zhang et al. 2012). The arable land was developed in 1978 by clearing native forest. Later, part of the arable land was converted to grassland through the Conservation Reserve Program (CRP) administrated by United States Department of Agriculture (USDA). Based on USDA regulations, the CRP land must be maintained in a condition so that the land can be easily returned to crop production.

Table 15.1 Land uses and related management practices in subarctic Alaskan soils

Land use	Management practices	Plant species
Forest	No	Black spruce (*Picea mariana*), poplars (*Populus balsamifera*, *Populus tremuloides*), and various under story plant species, mostly sphagnum moss
Agriculture	N fertilization, tillage	Continuous barley (*Hordeum valgare* L.) planting every year since the clearance of the land in 1979
CRP[a]	Mow every 5 years	A mixture of shrubs (Bebb's willow (*Alix bebbiana*), paper birch (*Betula apyrifera*), quacking aspen (*Populus tremuloides*), forbs (common yarrow (*Achillea millifolium*)); grass of Parnassus (*Parnassia palustris*), winter green (*Pyrola* spp.), and grasses (bluejoint reedgrass (*Calamagrostis Canadensis*); wheatgrass (*Elymus trachycaulus*) and sedge (*Cares* spp.)); a history of 18 years at time of soil sampling

[a]Grassland land under a Conservation Reserve Program administrated by United states Department of Agriculture

The details of land uses and related management practices were summarized (Table 15.1). The studies in high latitude often focused on impact of land use on the stored total soil carbon pool (Grünzweig et al. 2003; Grünzweig et al. 2004) or on effect of vegetation on WSOM in tundra forest soils (Neff and Hooper 2002). Little information is available on WEOM for different land uses in the high latitude area. Therefore, increased understanding of impact of different land uses on characteristics and biodegradability of soil WEOM will provide information for future land and water quality management in the subarctic region. Based on published and unpublished data (Zhao 2011; Zhao et al. 2012, 2013), this chapter systematically presents and discusses the quantity, distribution, and features of soil WEOM, especially its critical components, organic C and N, as affected by the three land uses in the subarctic Alaskan region. Knowledge and insight derived from this work may also be helpful in enhancing the sustainability of the agricultural land management in other climatic regions.

15.2 Wet Chemistry of Water Extracted Organic Matter

15.2.1 Soil Basic Properties, Organic C and N Extracted by Cold and Hot Water

Fourteen soil samples were taken from the major agricultural areas (Fairbanks, Delta Junction, and Palmer) of subarctic Alaska, USA, representing different land uses (Table 15.2). These soils were loam to sandy loam in texture, with a pH ranging from 5.0 to 6.9, total soil C content from 2.3 to 8.3 %, soil total N content from 0.2 to 0.5 %, and soil C:N ratios from 12 to 24. The cold and hot WSOC concentrations ranged from 274.1 to 755.5 mg kg^{-1} and 608.9 to 1,700.6 mg kg^{-1},

Table 15.2 Soil properties and water extractable organic C (WEOC) and N (WEON) extracted by cold (22 °C) and hot (80 °C) water from different land use soils in Alaska (Zhao et al. 2013)

			Soil	TOC[b]	TON[c]	Soil	Cold WEOC	Hot WEOC	Cold WEON	Hot WEON	Cold water	Hot water
Soils	Location	Soil texture	pH[a]	%	%	C:N	mg kg^{-1}	mg kg^{-1}	mg kg^{-1}	mg kg^{-1}	C:N	C:N
CRP												
1	Delta Junction	Sandy loam	5.4	4.5	0.2	22.5	423.1	1,304.3	24.1	76.6	17.5	17.0
2	Delta Junction	Sandy loam	5.4	4.1	0.2	20.5	547.4	1,516.2	27.6	80.2	19.8	18.9
3	Delta Junction	Loam	5.0	5.0	0.3	16.7	326.4	974.7	24.3	57.4	13.2	17.0
Mean			5.3	4.5	0.2	19.9	432.3 b	1,265.1 a	25.3 b	71.4 a	16.9	17.6
Forest												
4	Delta Junction	Sandy loam	5.2	4.7	0.2	23.5	351.6	1,387.5	15.1	62.7	23.2	22.1
5	Delta Junction	Loam	5.4	4.8	0.2	24.0	511.4	1,383.3	22.5	59.2	22.7	23.4
6	Delta Junction	Loam	5.4	4.8	0.2	24.0	318.9	796.1	14.9	35.9	21.4	22.2
Mean			5.3	4.8	0.2	23.8	394.0 b	1,189.0 a	17.5 b	52.6 a	22.4	22.5
Agriculture												
7	Palmer	Sandy loam	5.4	5.7	0.3	19.0	755.5	1,700.6	46.6	104.4	16.2	16.3
8	Delta Junction	Loam	5.4	3.8	0.2	19.0	554.4	1,653.6	30.5	92.7	18.2	17.8
9	Fairbanks	Loam	5.3	8.3	0.5	16.6	411.4	1,282.2	25.8	70.7	15.9	18.2
10	Fairbanks	Loam	5.3	3.1	0.2	15.7	465.7	958.8	25.7	61.0	18.2	15.7
11	Fairbanks	Loam	5.1	3.3	0.2	19.4	274.1	707.9	10.3	39.3	26.5	18.0
12	Delta Junction	Loam	6.9	2.3	0.2	12.0	440.5	608.9	24.8	45.1	17.8	13.5
13	Delta Junction	Loam	6.2	2.9	0.2	15.4	394.3	688.4	26.9	54.5	14.7	12.6
14	Delta Junction	Sandy loam	6.7	3.6	0.2	18.1	360.2	729.9	22.1	48.8	16.3	15.0
Mean			5.6	4.3	0.2	19.2	457.0 b	1,041.1 a	26.6 b	64.6 a	18.0	15.9
Sign.[d]			NS[e]	NS	NS		NS	NS	NS	NS		

Different letters in columns indicate significance at $p < 0.05$ between cold and hot WEOM for each property

[a]DI water (solution:soil = 1:1 (v:w))

[b]*TOC* total organic carbon

[c]*TON* total nitrogen

[d]Significance test among land use for each property except for C:N ratio

[e]*NS* no significant difference ($p < 0.05$) among land uses

respectively. The cold and hot WSOC averaged 1.0 % and 2.6 % of soil total organic C, respectively. The cold and hot WSON concentrations ranged from 10.3 to 46.6 mg kg^{-1} and 35.9 to 104.4 mg kg^{-1}, respectively. The cold and hot WEON represented 1.2 % and 3.2 % of soil total organic N, respectively. The hot WEOC and WEON concentrations were significantly ($p < 0.05$) higher than cold WSOC and WSON for all soils. On average, the concentration of hot WSOC was about four times higher than that of cold WSOC in all soils, and the concentration of hot WSON was about six times higher than that of cold WEON in all soils. The C:N ratios were 13.2–26.5 for cold WEOM to 12.6–23.4 for hot WEOM, respectively. Increases of WSOC and WSON in hot water extractions than cold water extractions has been reported before (Curtin et al. 2006; Ros et al. 2009; Chantigny et al. 2010), and this increase in soil may be attributed to released soluble organic C and N by depolymerization of SOM, decreased viscosity of soil humus, physical state change of soil organic matter, and killed microbes (Chantigny et al. 2010). The similarity of C:N ratios in hot WEOM and whole soils was also found by Chantigny et al. (2010). This similarity is attributed to the preferential dissociation of soil organic C at high temperature (Curtin et al. 2006; Chantigny et al. 2010).

15.2.2 Molecular Size Distribution of WSOC and WSON of Water Extractable Organic Matter

Three molecular size fractions of cold WEOM were separated: small molecules: <1 kDa, medium molecules: 1 kDa–0.45 μm, and large molecules: <2.5 μm but >0.45 μm (Table 15.3). The WSOC content within each size fraction was rather similar among soils of three land uses for both field moist and air-dried samples. In contrast, the WSON content was statistically different ($p < 0.05$) in medium (1 kDa–0.45 μm) and small (<1 kDa) size fractions for the field moist soil samples, and no difference for all size fractions from the air-dried soil samples. For the field moist soils, most WSOC and WSON were distributed in large and small sizes, whereas, for air-dried soil samples, most WSOC was distributed in small size fraction, and WSON was less changed among size fractions. Calculated from their contents, the C:N ratio of WSOM from the field moist soil samples was widened when molecule sizes decreased in CRP and forest soils, whereas, the C: N ratio of WSOM in agricultural soil changed little. In contrast, in the air-dried soil samples, the C:N ratio increased prominently in the small size fraction. Air-drying process did not decrease large size fraction of WSOC for all land uses, but did significantly increase this pool in medium and small size fractions of WSOC for all land uses. Also, air-drying process decreased the WSON-pool in large size fraction, but increase in the medium size fractions for all land uses. Only in the agricultural land use, small size fraction of WSON was decreased by air drying. This study demonstrated that size distribution of WSOC and WSON was affected by land uses. Forest soils and CRP contained a relative large proportion of large molecules in

Table 15.3 Size distribution of water soluble organic C (WSOC), and N (WSON) of cold (22 °C) WEOM from the field moist (F) and air-dried (A) soil samples with CRP, forest and agricultural land uses

	WSOC (mg kg^{-1} of dry soil)						WSON (mg kg^{-1} of dry soil)					
	L		M		S		L		M		S	
Land use	F	A	F	A	F	A	F	A	F	A	F	A
CRP	68.94	24.58	10.06	55.09	107.36	374.15	8.36	2.26	0.52	5.49	3.02	5.13
Forest	88.71	17.56	8.26	59.69	146.95	311.02	7.59	1.80	0.35	3.25	1.94	1.27
Agricultural	36.41	25.15	7.20	113.16	118.90	305.87	4.26	1.05	1.12	6.30	9.94	3.11
Prob. (F test)	0.35	0.16	0.56	0.56	0.22	0.49	0.42	0.04	0.02	0.13	0.002	0.03
LSD (0.05)	NS	NS	NS	NS	NS	NS	NS	0.92	0.53	NS	4.31	2.85

Data are adapted from Zhao et al. (2012)

L Large size fractions (>0.45 μm), *M* medium size fractions (1 kDa–0.45 μm), *S* small size fractions (<1 kDa)

Values are means of nine replicates. Statistical analysis was applied to WSOC and WSON among different land uses within each size fraction. *NS* not significant at $P = 0.05$

WSOM while agricultural soils consisted of primarily small molecules. There are two ways that forest floor organic matter entering the beneath mineral soils: through leaching or mixing by large soil faunas (i.e. bioturbation). Kaiser et al. (2002) reported a significant amount of high molecular weight in pine and beech forest floor in autumn. Since there is lack of bioturbation caused by large soil fauna in the forest soil in subarctic area in this study (Zhang et al. 2012), the organic matter in forest mineral soil might be mostly from the downward movement of dissolved organic matter in the forest floor, and this leached dissolved organic matter might contain large molecules (aromatic and aliphatic). Agricultural soils contained preferably small molecules due to frequent disruption of tillage, enhanced decomposition, and N fertilization (Delprat et al. 1997).

15.3 Fluorescence Excitation-Emission Matrix (EEM) Spectroscopy

15.3.1 Fluorophore Component Features of WEOM

Fluorescence excitation and emission matrix spectrum is a method often used to characterize soluble organic matter. The method uses full excitation and emission wavelengths to identify fluorophores in a solution. The output of the EEM is a three dimensional contour curve with excitation and emission wavelength respectively as x and y coordinator and fluorescence intensity as z coordinator. The EEM spectra for all soil samples were remarkably similar. Parallel factor analysis (PARAFAC) revealed that three fluorophore components were present in these cold and hot WEOMs (Fig. 15.1). The spectral features of the three components are similar to those in WEOM of Cecil sandy loam soils from Cotton-corn fields in Georgia, USA (Zhang et al. 2011b). Specifically, the component 1 in cold WEOM and the component 2 in hot WEOM had similar broad peak (Em 474–480 nm/Ex 240 and 348 nm), which was typical terrestrial humic substance-like fluorophore (He et al. 2006b; Fellman et al. 2008). The component 2 in cold WEOM and the component 1 in hot WEOM were similar with a shorter emission wavelength (Em 400–440 nm) than humic acid-like, which is considered as fulvic acid-like component (He et al. 2006b; Santin et al. 2009). It is also attributed to the marine or fresh water humic acid-like component (Yamashita and Tanoue 2004; Stedmon and Markager 2005). This pattern also was observed in the water extracts of soil, plant biomass and animal manure (Ohno and Bro 2006). The spectral contours of the component 3 showed different protein contours between cold and hot WEOM. Cold WEOM showed a strong emission signal at wavelengths ranging from 335 to 340 nm at the comparably low excitation wavelength 240 nm which is assigned to tryptophan-like components (Chen et al. 2003; Yamashita et al. 2008).

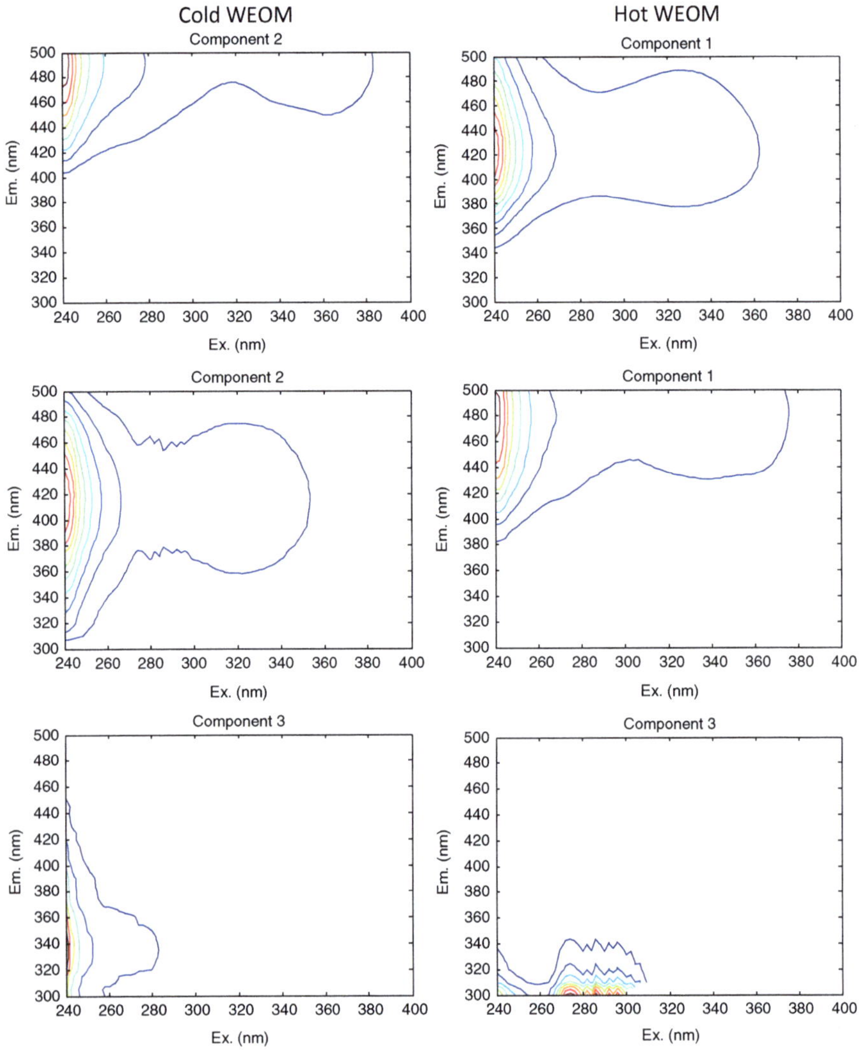

Fig. 15.1 The PARAFAC fluorophore components for cold (22 °C) (*left*) and hot (80 °C) (*right*) WEOM from sample soils. Ex. means excitation wavelength; Em. means emission wavelength (Zhao et al. 2013)

Hot WEOM included unknown fluorophores with a shorter emission wavelength (300 nm) for the maximum at the lower excitation wavelength. According to operationally defined fluorophore locations by Chen et al. (2003), this component is most likely composed of tyrosine-like components. When the cold WEOM was further fractionated to the three size fractions, the EEM-PARAFAC analysis revealed similar three-fluorophore-component features (Zhao et al. 2012).

15.3.2 Distribution of Fluorophore Components in Different WEOM Fractions

An average of 94 % and 89 % of the fluorescence intensity was contributed by humic acid-like and fulvic acid-like components in cold and hot WEOM, respectively. However, the distribution of the two components in cold and hot WEOM was markedly different, with the fulvic acid-like fluorophore being higher than humic acid-like fluorophore for the cold WEOM but lower than humic acid-like fluorophore in hot WEOM. Increasing extraction temperature resulted in a significant ($p < 0.05$) decrease in the contribution of fulvic acid-like fluorophore for forest and agricultural soils, and a significant increase in the humic-like fluorophore for all soils and in the protein-like fluorophore for CRP and agriculture soils. As hot water extracted more humic acid-like and less fulvic acid-like than cold WEOM, this reflected the presence of significant quantities of water soluble humic acid as extraction temperature increased. Curtin et al. (2006) also found that a large proportion of organic N extracted by hot water is resistant to acid hydrolysis, which was assigned to the fraction of humic acids.

Statistical difference ($p = 0.05$) was found in the relative loadings of the terrestrial- and marine humic acid- like components in the <1 kDa fraction of WEOM from different land uses. The relative loading of the terrestrial humic acid-like component in the field moist soil samples was similar for agricultural and CRP soils among the three size fractions, but this component increased as the size went smaller (Fig. 15.2). For the air-dried soil samples, this component increased in the forest and the agriculture but not for CRP as size fraction decreased. As for the relative loading of marine/fresh water humic acid-like component, it increased as size fraction decreased for forest and the agricultural soils but not for CRP of both field moist and air-dry soil samples. The relative loadings of the protein (tyrosine)-like component in all land uses were reduced as molecule sizes decreased especially for the air dried soil. Statistical differences were observed between the CRP, forest and agricultural land uses among different sizes for WEOM from air-dried soil samples, but such difference was only found in the fresh forest soil samples. Air-dried process significantly decreased the tyrosine-like component in WSOM of all three sizes.

15.4 Degradation of Organic C and N of WEOM in Laboratory Incubation

15.4.1 Loss of Organic C and N of WEOM

The incubation for cold and hot WEOM was carried out following a method by Kalbitz et al. (2003). In brief, the extracted solution (diluted to <20 mg C L^{-1}) was

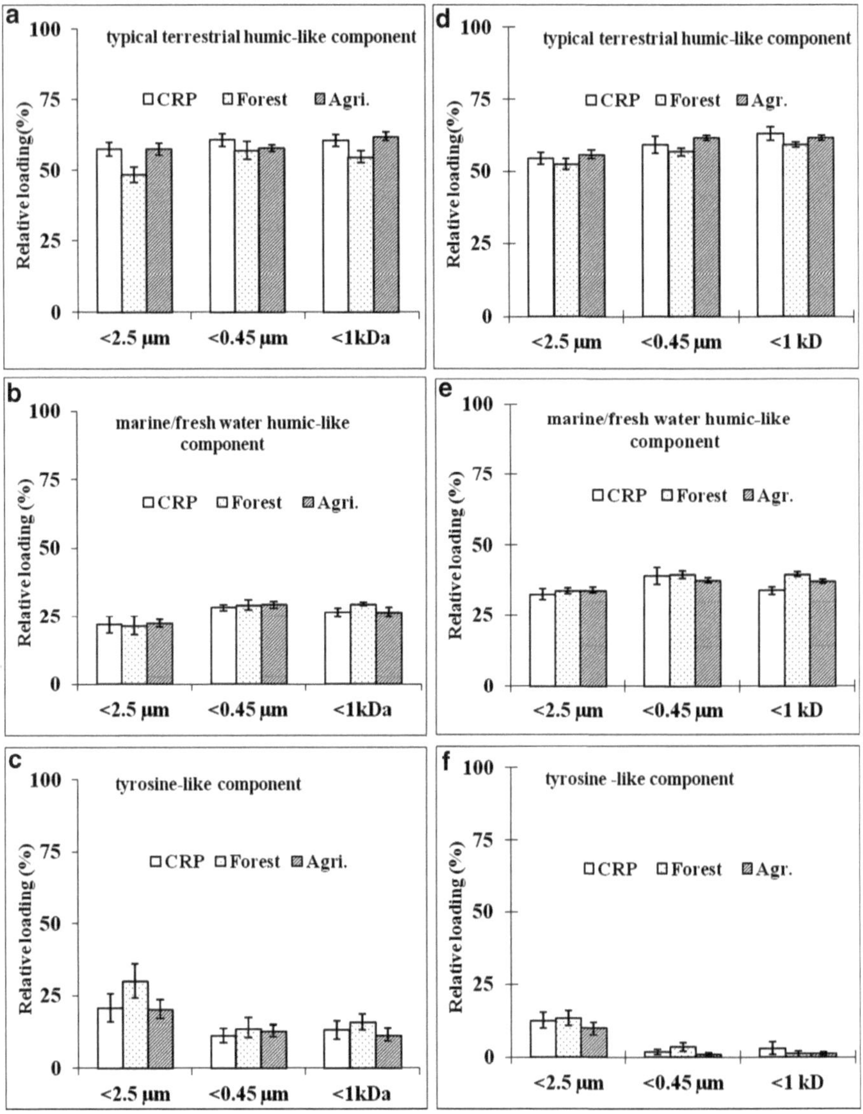

Fig. 15.2 Relative fractional loadings of PARAFAC components of WEOM under different land uses and different size fractions from field moist soil samples (**a**–**c**) and air-dried soil samples (**d**–**f**), (**a**, **d**): typical terrestrial humic acid-like component; (**b**, **e**): marine/fresh water humic acid-like component; (**c**, **f**): tyrosine-like component. *Vertical bars* denote standard errors for nine replicates (Zhao et al. 2012)

incubated in a flask in dark at 15 °C. Prior to the incubation, a mixed inoculum extracted (2 mM $CaCl_2$, soil:solution ratio, 1:2) from the 14 experimental soils was added in the solution (1 % v/v). The concentration of organic carbon before and after the 21-day incubation was determined and the differences were the C loss

Table 15.4 The loss of cold (22 °C) and hot (80 °C) water Extractable C (WEOC) and N (WEON) after a 21-day incubation at 15 °C for Alaska soils

	%WEOC loss		%WEON loss	
Soils	Cold	Hot	Cold	Hot
Mean for CRP (n = 3)	14.72 b	24.19 a	54.61	56.24
Mean for forest (n = 3)	9.54	12.10	13.68 b	52.26 a
Mean for agriculture (n = 8)	11.66 b	24.87 a	61.95	63.11
Significance of land uses	NS[a]	NS	$p = 0.007$	NS

Data are adapted from Zhao et al. (2012)
Different letters indicate significance at $p < 0.05$ between loss of cold and hot water extractable organic matter for each land use
[a]*NS* no significant difference ($p < 0.05$) among land uses for tested parameters

during the incubation. McDowell et al. (2006) suggested that a 7-day incubation at 20 °C can lead to a relatively labile DOC-pool and a 42-day incubation at the same temperature can determine labile and relatively refractory components. In this study, the soil solution was incubated at a relatively low temperature of 15 °C which is an average growing season temperature in the study region. The mineralization rate at 15 °C could be lower than that at 20 °C hence a longer (21-day) incubation was chosen to insure that most of the labile WEOC and WEON and little of refractory component of WEOC and WEON was determined.

The losses of cold and hot WEOC during 21 day incubation (at 15 °C) ranged from 9.5 to 14.7 % and from 12.1 to 24.9 %, respectively (Table 15.4). These changes were quite different to the range of 5.3–31.8 % of mineral soils reported by Kalbitz et al. (2003) in a 90 day 20 °C incubation. The losses of cold and hot WEON during 21 day incubation ranged from 13.7 to 62.0 % and from 52.3 to 63.1 %, respectively. These outcomes were wider than the range of 12–20 % of hot WEOC loss reported by Bu et al. (2010). The greater losses of WEON suggested that the fraction of WEON was more biodegradable than the fraction of WEOC. Previous studies also reported a higher biodegradation of WEON than WEOC in the humus layer under coniferous forest (Norway spruce and Scots pine) (Kiikkilä et al. 2006), and in maize-cropped agricultural soils (Gregorich et al. 2003).

Compared between cold and hot water extraction, hot water extraction increased biodegradable organic C in agriculture and CRP but not forest soils (Table 15.4). Quite contrary, hot water extraction increased biodegradable organic N in forest but not agriculture and CRP soils. Comparing among the three land uses, only difference ($p < 0.05$) was found with cold WEON where forest soil had the lowest fraction of biodegradable WEON (Table 15.4). It appeared that the increase in biodegradability of hot WEOC did not couple with an increase in hot WEON in CRP and agricultural soils, and the increasing biodegradable hot WEON did not couple with the biodegradability of the hot WEOC in forest soil. This decoupling suggested that hot WEOC may have totally different turnover rate than hot WEON.

Table 15.5 Means, standard deviations and ranges of potential mineralizable N indices in Alaska soils

Index	Mean	SD[a]	Minimum	Maximum	C.V.%[b]
$N_{24weeks}$[c], mg N kg^{-1}	50.5	26.0	12.3	86.1	56
$N_{24weeks}$/total organic N, %	2.4	1.1	0.6	4.3	51
N_a[d], mg N kg^{-1}	16.5	9.0	1.9	30.2	55
N_s[e], mg N kg^{-1}	57.2	30.8	18.9	120.5	54
(N_a + N_s)/total organic N, %	3.7	1.4	0.9	5.6	52
Organic N_hot water, mg N kg^{-1}	63.4	20.4	35.9	104.1	31
Organic C_hot water, mg C kg^{-1}	1,120.8	395.6	608.9	1,700.6	34
Organic N_cold water, mg N kg^{-1}	24.3	8.8	10.3	46.6	35
Organic C_cold water, mg C kg^{-1}	438.2	125.4	274.1	755.5	28
NaOH hydrolyzable N, mg N kg^{-1}	133.6	26.5	91.8	183.3	37
HCl hydrolyzable N, mg N kg^{-1}	73.9	31.0	36.0	156.2	43
$NaHCO_3$_260, UV absorbance	0.65	0.12	0.46	0.96	18
$NaHCO_3$_205, UV absorbance	1.55	0.55	0.93	2.66	36

[a]Standard deviation
[b]Coefficient of variance
[c]The mineralized N during an incubation time of 24 weeks
[d]The labile N_a pool estimated from modified double exponential model
[e]The labile N_s pool estimated from modified double exponential model

15.4.2 Soil Mineralizable N Parameters

As Curtin et al. (2006) demonstrated that hot WEON can serve as a potential soil mineralizable N index, another 24-week incubation was carried out based on the method of Wang et al. (2004). This was performed for estimation of total soil potential mineralizable N in the 14 investigated subarctic Alaskan soils. During the 24-week incubation, total mineralizable N in the 14 soils ($N_{24weeks}$) ranged from 12 to 86 mg kg^{-1} with a CV of 56 % (Table 15.5). The $N_{24weeks}$ accounted for between 0.6 and 4.3 % (average of 2.4 %) of total soil organic N. The active mineralizable N pool (N_a) size estimated from a fixed double exponential model (Zhao 2011) ranged from 1.9 to 30.2 mg N kg^{-1} with an average of 16.5 mg N kg^{-1}. The slow mineralizable N pool (N_s) ranged from 18.9 to 120.5 mg N kg^{-1} with an average of 57.2 mg N kg^{-1}. The proportion of the active N pool (N_a) in the total mineralizable N-pool (N_a + N_s) averaged only 22 %, indicating slow N pool was the majority of potential mineralizable N. During the 24-week incubation, an average of 69 % of total mineralizable N (N_a + N_s) was released. However, the total mineralizable N pool size (N_a + N_s) had an average proportion of 3.7 % of the total soil organic N which was relatively small (Table 15.5).

The proportion of the total mineralizable N pool (N_a + N_s) in total soil organic N (3.7 %) was lower than reported by Sharifi et al. (2007, 2008) (6–7.5 %) and Benedetti and Sebastiani (1996) (10 %). This may attribute to at least three reasons. First, the incubation in this study was at 15 °C, which was lower than other studies (25 °C for Sharifi et al. (2007); and 30 °C for Benedetti and Sebastiani (1996)).

The mineralization rate increases exponentially with temperature-increase from 5 to 30 °C (Stanford and Smith 1972). Second, leaching procedure before incubation can lead to overestimation of potentially mineralizable N pool sizes because the leachate N may include mineral N that could have been immobilized (Wang et al. 2003). Third, C:N ratios of soils here (12–24) were higher than that reported by Benedetti and Sebastiani (7–15, 1996) and Sharifi et al. (7–13, 2007), which may result in less net N mineralization in soils of this study.

The 1 M NaOH hydrolysable N content ranged from 92 to 183 N kg^{-1} with a mean of 134 mg N kg^{-1} (Table 15.5). The NaOH hydrolysable N accounted for 6.7 % of total soil organic N. This is in line with the study of Wang et al. (2001), who obtained values ranging from 30 to 400 mg N kg^{-1}. The NaOH hydrolysable N extraction is similar to the procedure used for Illinois soil nitrogen test (ISNT), except for an extraction of Extractable NH_4^+-N (Sharifi et al. 2007). The ISNT can recover amino sugar N derived primarily from bacterial and fungal cell walls and some α-amino acid (Khan et al. 2001; Greenfield 2002). The NaOH hydrolysable N extracted the highest proportion of total organic N in soils among the chemical methods used (Table 15.5). The 1 M HCl hydrolysable N content, another potential mineralizable N index, ranged from 36 to 156 with a mean of 74 mg N kg^{-1} (Table 15.5). The HCl hydrolysable N accounted for 3.7 % of the total soil organic N, similar to the reported 25–150 mg N kg^{-1} in 19 arable soils by Wang et al. (2001).

The UV absorbance at 260 nm for 0.01 M $NaHCO_3$ hydrolysable organic matter ranged from 0.46 to 0.96 with a mean of 0.65 and the UV-absorbance at 205 nm was from 0.93 to 2.66 with a mean of 1.55. The two UV absorbances were used in the past and they are positively related with the amount of N mineralized in situ (Hong et al. 1990; Serna and Pomares 1992).

15.4.3 Correlations Between Water Soluble C and N with the Other Chemically Determined Mineralizable N

The hot water extracted N was related to alkaline hydrolysable N (NaOH extraction) ($r = 0.72, p < 0.01$), and acid hydrolysable N (HCl extractable) ($r = 0.86, p < 0.01$), and the absorbance at 260 nm for $NaHCO_3$ extraction ($r = 0.61$, $p < 0.05$) (Table 15.6). The cold water extracted organic N, on the other hand was related to these potential mineralizable N indexes, but additionally, the cold water extracted N was also related to the absorbance at 205 nm for $NaHCO_3$ extraction. In addition, the water extracted C and N was related to each other for both cold and hot water extraction. The water extractable organic N was not related to the total soil organic N.

Cold water extractable organic N often consists of labile hydrophilic compounds (e.g. free amino acids, amino sugars, small carbonxylic acids, proteins, sugars) due to the selective sorption of recalcitrant aromatic and hydrophobic compounds by

Table 15.6 Pearson correlation coefficient between water extractable organic C and N and the other indices of N mineralization

Index	HWEON[a]	HWEOC[b]	CWEON[c]	CWEOC[d]
HWEON[a], mg N kg^{-1}	1			
HWEOC[b], mg N kg^{-1}	0.89**	1		
CWEON[c], mg N kg^{-1}	0.80**	0.53*	1	
CWEOC[d], mg N kg^{-1}	0.84**	0.72**	0.89**	1
NaOH hydrolysable N, mg N kg^{-1}	0.72**	0.66**	0.63*	0.50
HCl hydrolysable N, mg N kg^{-1}	0.86**	0.84**	0.80**	0.93**
$NaHCO_3$_260, UV absorbance	0.61*	0.47	0.83**	0.81**
$NaHCO_3$_205, UV absorbance	0.27	0.01	0.59*	0.41
TOC[e], g kg^{-1}	0.15	0.50	0.21	0.37
TON[f], g kg^{-1}	0.09	0.22	0.29	0.26

[a]Organic N in hot water
[b]Organic C in hot water
[c]Organic N in cold water
[d]Organic C in cold water
[e]Total C
[f]Total organic N
*Significant correlation at $p < 0.05$; **Significant correlation at $p < 0.001$

the solid fraction of organic matter from the soil solution (Qualls and Haines 1991; Kalbitz et al. 2003). The hot-water extractable organic matter contains proportions deriving from dead microbes, soluble soil carbohydrates and other simple compounds which may account for the labile fraction of soil organic matter (Ghani et al. 2013). The HCl hydrolysable organic N and NaOH hydrolysable organic N are assumed to be indicators of biologically active soil organic matter (Xu et al. 1997; Wang et al. 2001). The amount of N from these extractions were only a small proportions (<6.7 %) of total soil organic N. The amount of N extracted in total soil organic N was in an order of alkaline hydrolysable N (6.7 %) > acid hydrolysable N (3.7 %) > hot water Extractable N (3.2 %) > cold water Extractable N (1.2 %). The addition of cold water and hot water extraction was 4.4 %, higher than acid hydrolysable N but lower than alkaline hydrolysable N. Due to its small pool size, water extractable N was not related to the total organic N in soil. However, water extractable organic N (22, 80 °C), HCl hydrolysable organic N and NaOH hydrolysable organic N were similar in their extracted nitrogen-quantities and were correlated to each other, indicating these extraction solution might attack similar N pools.

Earlier studies reported poor correlations between water extractable organic C and N at 22 °C (Campbell et al. 2000; Curtin et al. 2006). The poor correlation between water extractable organic C and N is possibly due to different variables driving dissolved organic C and N production in soil such as N fertilization and ecosystem age (McDowell 2003). For example, N fertilization may increase dissolved organic N but not dissolved organic C. In this study, samples were taken before N fertilization. Apparently, a good correlation of water extractable organic C and N (22 °C) reflected little influence of factors such as N fertilization or soil age on dissolved organic C and N production. A high correlation between water extractable organic C and N (80 °C) has been reported by Curtin et al. (2006). A high correlation between water extractable organic C and N (80 °C) was also found in this study.

Both UV_{260} and UV_{205} of $NaHCO_3$ extracted soil organic matter serve as potential mineralizable N indicator. The UV_{260} of $NaHCO_3$ indicates the amount of water extractable organic matter but the UV_{205} of $NaHCO_3$ reflects the amount of inorganic NO_3 and organic carbon in water extraction (Hong et al. 1990; Serna and Pomares 1992). As such, the correlation of water extractable organic C extracted at 22 °C (i.e. cold water extraction) was higher with the UV_{260} of $NaHCO_3$ than with the UV_{205} of $NaHCO_3$. In general, the UV absorbance at longer wavelengths determines the aromaticity of organic matter. Compared with water extractable organic matter at 22 °C, water extractable organic C at 80 °C showed a poor correlation with UV values at 260 and 205 nm of $NaHCO_3$, which may reflect a low aromaticity of hot water extractable soil organic matter.

15.4.4 Relationships Between Chemically Extractable Organic N with Model Estimated Mineralizable N

The amount of N released in 24-week incubation was 50.5 mg N kg $soil^{-1}$ (Table 15.5). This was only 69 % of the potential mineralizable N in soil as estimated by an unbiased double potential model (Zhao 2011),

$$N_t = N_a\left(1 - e^{-0.693t}\right) + N_s\left(1 - e^{-0.051t}\right)$$

where two pools are used to describe the active mineralizable N, N_a (mg kg^{-1}) and the less readily mineralizable N, N_s (mg kg^{-1}), and the total potential mineralizable N for a time period is N_t.

The active N pool N_a was linearly related ($p < 0.05$) with hot water extractable organic N, NaOH hydrolysable organic N, and cold water extractable organic N (Table 15.7), indicating a portion of hot and cold water extractable N, and NaOH hydrolysable organic N was in the active N pool.

The total potential mineralizable N, $N_a + N_s$, was linearly related ($p < 0.05$) with hot water organic C and N, cold water organic C and N, NaOH hydrolysable N, and HCl hydrolysable N (Table 15.7). There was no significant ($p > 0.05$) relationship between $N_a + N_s$ and other chemically determined organic N including UV absorbance at 205 and 260 nm of soil organic matter extracted with $NaHCO_3$, and total soil organic C and N.

In summary, Quality and quantity of soil hot (80 °C) and cold (22 °C) WEOM were related to the three land uses. For all land uses, there were three fluorescence components in both hot and cold WEOM, but their biodegradability was different. Hot and cold WEON were also related to other potential mineralizable indices,

Table 15.7 Linear regression of tested indices of soil N mineralization to total mineralizable N using model estimation based on 24-week incubation

	N_a[a]		$N_a + N_s$[b]	
	mg N kg^{-1} soil		mg N kg^{-1} soil	
Index	R^2	*p* value	R^2	*p* value
Organic N_hot water, mg N kg^{-1} soil	0.42	0.01	0.70	0.0002
NaOH hydrolyzable N, mg N kg^{-1} soil	0.40	0.01	0.62	0.0008
Organic C_hot water, mg N kg^{-1} soil	0.17	0.14	0.46	0.007
Organic N_cold water, mg N kg^{-1} soil	0.36	0.02	0.43	0.01
Organic C_cold water, mg N kg^{-1} soil	0.14	0.18	0.32	0.03
HCl hydrolyzable N, mg N kg^{-1} soil	0.14	0.18	0.32	0.04
$NaHCO_3$_260, UV absorbance	0.14	0.18	0.13	0.19
$NaHCO_3$_205, UV absorbance	0.28	0.05	0.12	0.22
Total organic C, g kg^{-1}	0.20	0.11	0.18	0.79
Total organic N, g kg^{-1}	0.33	0.03	0.21	0.73

[a]Labile organic N pool size estimated through the double exponential model with the fixed rate constants

[b]Total mineralizable N ($N_a + N_s$) estimated from the double exponential model with the fixed rate constants

especially acid (1 M HCl) and alkaline (1 M NaOH) hydrolizable N. The hot plus cold WEON was a small proportion (4.4 %), higher than acid hydrolizable N (3.7 %) but lower than alkaline hydrolysable N (6.7 %) to the total soil organic N. Both hot and cold WEON was related to mineralizable N obtained from laboratory incubation, therefore, either can be used as a potential soil mineralizable N index.

15.5 Future Research

This research mainly focused on the total level, size distribution and bio-degradability of WSOC and WSON in WEOMs in the subarctic Alaskan soils and the impacts by land uses. A part from soil WEOM, particulate soil organic matter (POM) is another indicator sensitive to soil management and land uses (Cambardella and Elliott 1992). The approaches for defining this SOM-fraction are different (e.g. physical dispersion vs. chemical dispersion, Liao et al. (2006); Billings (2006)). However, researchers found that the chemical dispersion (sodium hexametaphosphate) separated particulate soil organic matter is sensitive to soil management practice. Whereas research on particulate soil organic matter focus on management practices such as no till or land use changes on active carbon pool so that terrestrial carbon cycles can be quantified and modeled; the water extractable soil organic matter, on the other hand, is also found to be a sensitive indicator for soil management such as land use changes and tillage practices (Zhang et al. 2011b). Nevertheless, studies on water extracted soil organic matter emphasize more on its availability to microorganisms or plants (Zhao et al. 2013), and its relationship to surface water quality such as spring runoff from forest or agriculture areas (Royer et al. 2007; Wilson and Xenopoulos 2009). However, the link between these two soil organic matter fractions is not clear. Will POM positively or negatively relate to WEOM? What is the proportion of this WEOM in this POM fraction? Another missing link is the protected or occluded soil organic matter in relation to WEOM. Hence a further question is how much of this WEOM attributes to the protected or occluded soil organic matter. Further knowledge in these areas will help land management practice to achieve the goal of both high soil fertility and high water quality.

Whereas characterization of WSOC and WSON would contribute to better understanding of WEOM properties and their biodegradability in soil and other environments, one should be aware that another nutrient P may associate or bind to some organic matter fractions (Mahieu et al. 2000; He et al. 2006b, 2009b, 2011a, b). These organic matter-bound P could be released biotically or abiotically by enzymatic hydrolysis and UV irradiation (He et al. 2009b, b). In other words, these organic matter-associated P forms are bioavailable. It would be of interest to investigate if and how P is present in these WEOMs from the subarctic Alaskan soils, and how the composition and bio availability would change by different land uses. In addition, the chemical interaction between soil organic matter and P is important for reactions of P at the soil solid phase, in soil solution, and in natural waters (Hens and Merckx 2001; He et al. 2006a; Gerke 2010). Decomposed WEOM is reported to inhibit P sorption to a greater extent than did WEOM from fresh materials (Hunt et al. 2007b). Thus, investigating the soil WEOM-metal-P interaction in Alaskan field conditions and laboratory decomposition incubation will certainly provide valuable additional insight into the management and conservation strategies for the subarctic, perhaps also other soils under different climate.

References

Benedetti A, Sebastiani G (1996) Determination of potentially mineralizable nitrogen in agricultural soil. Biol Fertil Soils 21:114–120

Billings SA (2006) Soil organic matter dynamics and land use change at a grassland/forest ecotone. Soil Biol Biochem 38:2934–2943

Bu X, Wang L, Ma W, Yu X, McDowell WH, Ruan H (2010) Spectroscopic characterization of hot-water extractable organic matter from soils under four different vegetation types along an elevation gradient in the Wuyi Mountains. Geoderma 159:139–146

Cambardella CA, Elliott ET (1992) Particulate soil organic-matter changes across a grassland cultivation sequence. Soil Sci Soc Am J 56:777–783

Campbell JL, Hornbeck JW, McDowell WH, Buso DC, Shanley JB, Likens GE (2000) Dissolved organic nitrogen budgets for upland, forested ecosystems in New England. Biogeochemistry 49:123–142

Chantigny MH, Curtin D, Beare MH, Greenfield LG (2010) Influence of temperature on water-extractable organic matter and ammonium production in mineral soils. Soil Biol Biochem 74:517–524

Chen W, Westerhoff P, Leenheer JA, Booksh K (2003) Fluorescence excitation-emission matrix regional integration to quantify spectra for dissolved organic matter. Environ Sci Technol 37:5701–5710

Curtin D, Wright CE, Beare MH, McCallum FM (2006) Hot water extractable nitrogen as an indicator of soil nitrogen availability. Soil Sci Soc Am J 70:1512–1521

Delprat L, Chassin P, Linères M, Jambert C (1997) Characterization of dissolved organic carbon in cleared forest soils converted to maize cultivation. Eur J Agron 7:201–210

Fellman JB, D'Amore DV, Hood E, Boone RD (2008) Fluorescence characteristics and biodegradability of dissolved organic matter in forest and wetland soils from coastal temperate watersheds in southeast Alaska. Biogeochemistry 88:169–184

Gerke G (2010) Humic (organic matter)-Al(Fe)-phosphate complexes: an underestimated phosphate form in soils and source of plant-available phosphate. Soil Sci 175:417–425

Ghani A, Sarathchandra U, Ledgard S, Dexter M, Lindsey S (2013) Microbial decomposition of leached or extracted dissolved organic carbon and nitrogen from pasture soils. Biol Fertil Soils 49:747–755

Greenfield L (2002) The origin and nature of organic nitrogen in soil as assessed by acidic and alkaline hydrolysis. Eur J Soil Sci 52:575–583

Gregorich EC, Beare MH, Stocklas U, St-Georges P (2003) Biodegradability of soluble organic matter in maize cropped soils. Geoderma 113:237–252

Grünzweig JM, Sparrow SD, Chapin ES III (2003) Impact of forest conversion to agriculture on carbon and nitrogen mineralization in subarctic Alaska. Biogeochemistry 64:271–296

Grünzweig JM, Sparrow SD, Yakir D, Chapin FS III (2004) Impact of agricultural land use change on carbon storage in boreal Alaska. Global Change Biol 10:452–542472

He Z, Dao TH, Honeycutt CW (2006a) Insoluble Fe-associated inorganic and organic phosphates in animal manure and soil. Soil Sci 171:117–126

He Z, Ohno T, Cade-Menun BJ, Erich MS, Honeycutt CW (2006b) Spectral and chemical characterization of phosphates associated with humic substances. Soil Sci Soc Am J 70:1741–1751

He Z, Mao J, Honeycutt CW, Ohno T, Hunt JF, Cade-Menun BJ (2009a) Characterization of plant-derived water extractable organic matter by multiple spectroscopic techniques. Biol Fertil Soils 45:609–616

He Z, Olk DC, Honeycutt CW, Fortuna A (2009b) Enzymatically- and ultraviolet-labile phosphorus in humic acid fractions from rice soils. Soil Sci 174:81–87

He Z, Ohno T, Olk DC, Wu F (2010) Capillary electrophoresis profiles and fluorophore components of humic acids in Nebraska corn and Philippine rice soils. Geoderma 156:143–151

He Z, Honeycutt CW, Zhang H (2011a) Elemental and Fourier transform infrared spectroscopic analysis of water and pyrophosphate extracted soil organic matter. Soil Sci 176:183–189

He Z, Olk DC, Cade-Menun BJ (2011b) Forms and lability of phosphorus in humic acid fractions of Hord silt loam soil. Soil Sci Soc Am J 75:1712–1722

Hens M, Merckx R (2001) Functional characterization of colloidal phosphorus species in the soil solution of sandy soils. Environ Sci Technol 35:493–500

Hong S, Fox R, Piekielek W (1990) Field evaluation of several chemical indexes of soil nitrogen availability. Plant Soil 123:83–88

Hunt JF, Ohno T, He Z, Honeycutt CW, Dail DB (2007a) Influence of decomposition on chemical properties of plant- and manure-derived dissolved organic matter and sorption to goethite. J Environ Qual 36:135–143

Hunt JF, Ohno T, He Z, Honeycutt CW, Dail DB (2007b) Inhibition of phosphorus sorption to goethite, gibbsite, and kaolin by fresh and decomposed organic matter. Biol Fertil Soils 44:277–288

Kaiser M, Zech W (2000) Sorption of dissolved organic nitrogen by acid subsoil horizons and individual mineral phases. Eur J Soil Sci 51:403–411

Kaiser K, Guggenberger G, Haumaier L, Zech W (2002) The composition of dissolved organic matter in forest soil solutions: changes induced by seasons and passage through the mineral soil. Org Geochem 33:307–318

Kalbitz K et al (2003) Changes on properties of soil-derived dissolved organic matter induced by biodegradation. Soil Biol Biochem 35:1129–1142

Khan S, Mulvaney R, Hoeft R (2001) A simple soil test for detecting sites that are nonresponsive to nitrogen fertilization. Soil Sci Soc Am J 65:1751–1760

Kiikkilä O, Kitunen V, Smolander A (2006) Dissolved soil organic matter from surface organic horizons under birch and conifers: degradation in relation to chemical characteristics. Soil Biol Biochem 38:113–127

Landgraf D, Leinweber P, Makeschin F (2006) Cold and hot water extractable organic matter as indicators of litter decomposition in forest soils. J Plant Nutr Soil Sci 169:76–82

Liao JD, Boutton TW, Jastrow JD (2006) Storage and dynamics of carbon and nitrogen in soil physical fractions following woody plant invasion of grassland. Soil Biol Biochem 38:3184–3196

Mahieu N, Olk DC, Randall EW (2000) Analysis of phosphorus in two humic acid fractions of intensively cropped lowland rice soils by ^{31}P-NMR. Eur J Soil Sci 51:391–402

McDowell WH (2003) Dissolved organic matter in soils-future directions and unanswered questions. Geoderma 113:179–186

McDowell WH, Zsolnay A, Aitkenhead-Peterson JA, Gregorich EG, Jones DL, Jödemann D, Kalbitz K, Marschner B, Schwesig D (2006) A comparison of methods to determine the biodegradable dissolved organic carbon from different terrestrial sources. Soil Biol Biochem 38:1933–1942

Neff JC, Hooper DU (2002) Vegetation and climate controls on potential CO_2, DOC and DON production in northern latitude soils. Global Change Biol 8:872–884

Nkhili E, Guyot G, Vassal N, Richard C (2012) Extractability of water-soluble soil organic matter as monitored by spectroscopic and chromatographic analyses. Environ Sci Pollut Res 19:1–8

Ohno T, Bro R (2006) Dissolved organic matter characterization using multiway spectral decomposition of fluorescence landscapes. Soil Sci Soc Am J 70:2028–2037

Ohno T, He Z (2011) Fluorescence spectroscopic analysis of organic matter fractions: the current status and a tutorial case study. In: He Z (ed) Environmental chemistry of animal manure. Nova Science Publishers, New York, pp 83–103

Ohno T, He Z, Tazisong IA, Senwo ZN (2009) Influence of tillage, cropping, and nitrogen source on the chemical characteristics of humic acid, fulvic acid, and water-soluble soil organic matter fractions of a long-term cropping system study. Soil Sci 174:652–660

Ohno T, He Z, Sleighter RL, Honeycutt CW, Hatcher PG (2010) Ultrahigh resolution mass spectrometry and indicator species analysis to identify marker components of soil- and plant biomass-derived organic matter fractions. Environ Sci Technol 44:8594–8600

Qualls RG, Haines BL (1991) Geochemistry of dissolved organic nutrients in water percolating through a forest ecosystem. Soil Sci Soc Am J 55:1112–1123

Ros GH, Hoffland E, van Kessel C, Temminghoff EJM (2009) Extractable and dissolved organic nitrogen – a quantitative assessment. Soil Biol Biochem 41:1029–1039

Rousk J, Jones DL (2010) Loss of low molecular weight dissolved organic carbon (DOC) and nitrogen (DON) in H_2O and 0.5 MK_2SO_4 soil extracts. Soil Biol Biochem 42:2331–2335

Royer I, Angers DA, Chantigny MH, Simard RR, Cluis D (2007) Dissolved organic carbon in runoff and tile-drain water under corn and forage fertilized with hog manure. J Environ Qual 36:855–863

Santin C, Yamashita Y, Otero XL, Alvarez MA, Jaffe R (2009) Characterizing humic substances from estuarine soils and sediments by excitation-emission matrix spectroscopy and parallel factor analysis. Biogeochemistry 96:131–147

Serna M, Pomares F (1992) Evaluation of chemical indices of soil organic nitrogen availability in calcareous soils. Soil Sci Soc Am J 56:1486–1491

Sharifi M, Zebarth BJ, Burton DL, Grant CA, Cooper JM (2007) Evaluation of some indices of potentially mineralizable nitrogen in soil. Soil Sci Soc Am J 71:1233–1239

Sharifi M, Burton DL, Grant CA, Bittman S, Drury C, McConkey BG, Ziadi N (2008) Response of potentially mineralizable soil nitrogen and indices of nitrogen availability to tillage system. Soil Sci Soc Am J 72:1124–1131

Smolander A, Kitunen V (2002) Soil microbial activities and characteristics of dissolved organic C and N in relation to tree species. Soil Biol Biochem 34:651–660
Stanford G, Smith SJ (1972) Nitrogen mineralization potentials of soils. Soil Sci 36:465–472
Stedmon CA, Markager S (2005) Resolving the variability in dissolved organic matter fluorescence in a temperate estuary and its catchment using PARAFAC analysis. Limnol Oceanogr 50:686–697
Tatzber M, Stemmer M, Spiegel H, Katzlberger C, Haberhauer G, Gerzabek MH (2008) Impact of different tillage practices on molecular characteristics of humic acids in a long-term field experiment – an application of three different spectroscopic methods. Sci Total Environ 406:256–268
Tatzber M, Mutsch F, Mentler A, Leitgeb E, Englisch M, Gerzabek MH (2011) Capillary electrophoresis characterisation of humic acids: application to diverse forest soil samples. Environ Chem 8:589–601
Wang WJ, Smith CJ, Chalk PM, Chen D (2001) Evaluating chemical and physical indices of nitrogen mineralization capacity with an unequivocal reference. Soil Sci Soc Am J 65:368–376
Wang WJ, Smith CJ, Chen D (2003) Towards a standardized procedure for determining the potentially mineralizable nitrogen of soil. Biol Fertil Soils 37:362–374
Wang WJ, Smith CJ, Chen D (2004) Predicting soil nitrogen mineralization dynamics with a fixed double exponential model. Soil Sci Soc Am J 68:1256–1265
Wilson HF, Xenopoulos MA (2009) Effect of agricultural land use on the composition of fluvial dissolved organic matter. Nat Geosci 2:37–41
Xu J, Cheng H, Koskinen W, Molina J (1997) Characterization of potentially bioreactive soil organic carbon and nitrogen by acid hydrolysis. Nutr Cycl Agroecosys 49:267–271
Yamashita Y, Tanoue E (2004) Chemical characteristics of amino acid-containing dissolved organic matter in seawater. Org Geochem 35:679–692
Yamashita Y, Jaffé R, Maie N, Tanoue E (2008) Assessing the dynamics of dissolved organic matter (DOM) in coastal environments by excitation emission matrix fluorescence and parallel factor analysis (EEM-PARAFAC). Limnol Oceanogr 53:1900–1908
Zhang M, He Z, Zhao A (2011a) Ultraviolet-visible absorptive features of water extractable and humic fractions of animal manure and relevant compost. In: He Z (ed) Environmental chemistry of animal manure. Nova Science Publishers, New York, pp 61–81
Zhang M, He Z, Zhao A, Zhang H, Endale DM, Schomberg HH (2011b) Water-extractable soil organic carbon and nitrogen affected by tillage and manure application. Soil Sci 176:307–312
Zhang M, Sparrow SD, Veldhuizen RV, Masiak DT (2012) Soil quality under different land uses in a subarctic environment in Alaska. J Land Use Sci 7:109–121
Zhao A (2011) Assessment and prediction of potentially mineralizable organic nitrogen for subarctic Alaska soils. Ph.D. thesis, University of Alaska Fairbanks, Fairbanks, AK, USA
Zhao A, Zhang M, He Z (2012) Characteristics of soil water-soluble organic C and N under different land uses in Alaska. Soil Sci 177:683–694
Zhao A, Zhang M, He Z (2013) Spectroscopic characteristics and biodegradability of cold and hot water extractable soil organic matter under different land uses in subarctic Alaska. Commun Soil Sci Plant Anal 44:3030–3048
Zsolnay A (1996) Dissolved humus in soil waters. In: Piccolo A (ed) Humic substances in terrestrial ecosystems. Elsevier, Amsterdam, pp 171–223

Chapter 16
Remote Sensing of Nutrient Concentrations of Soils and Crops in Biosolid Amended Soils

B.B. Maruthi Sridhar, Fengxiang X. Han, and Robert K. Vincent

Abstract The accumulation of heavy metals in the biosolid amended soils and the risk of their uptake into different plant parts is a topic of great concern. This chapter summarizes the accumulation of several heavy metals and nutrients in soils and in plants grown on biosolid applied soils and the use of remote sensing to monitor the metal uptake and plant stress. Field and greenhouse studies were conducted with soybeans grown on soils applied with biosolids at varying rates. The data indicated that the chemical concentrations in soils and in soybean shoots and roots increased significantly with increase in applied biosolid concentrations. Copper (Cu) and Molybdenum (Mo) accumulated significantly in the shoots and roots of the metal-treated plants. The spectral and Landsat TM data indicated that the Normalized Difference Vegetative Index (NDVI) can be used to distinguish the metal stressed plants. The NDVI showed significant negative correlation with increase in soil Cu concentrations followed by other elements. This data indicated that the remote sensing can be used to monitor soybean stress patterns and to indirectly assess soil chemical characteristics. The data from remote sensing analysis of the biosolid applied soils indicated that the satellite imagery of bare-soil fields can be used to quantify and map the spatial variation of total phosphorous and copper concentration in surface soils.

B.B.M. Sridhar (✉)
Department of Environmental and Interdisciplinary Sciences,
Texas Southern University, Houston, TX 77004, USA
e-mail: bhaskarm@tsu.edu

F.X. Han
Department of Chemistry and Biochemistry,
Jackson State University, Jackson, MS 39217, USA

R.K. Vincent
Department of Geology, Bowling Green State University, Bowling Green, OH 43403, USA

Z. He and H. Zhang (eds.), *Applied Manure and Nutrient Chemistry for Sustainable Agriculture and Environment*, DOI 10.1007/978-94-017-8807-6_16,

16.1 Introduction

The agricultural use of biosolids (sewage sludge) for crop production is widely favored as it is an inexpensive source of several soil nutrients. Biosolids are known to improve soil physical and chemical properties (Epstein et al. 1975; Wei et al. 1985), and supply the nutrients required for crop growth (Sommers 1977; Singh and Agrawal 2008). However, biosolids also contain some toxic heavy metals (Logan et al. 1997). The effect of these heavy metals on the plant uptake and growth varies with the type of plant species. Logan et al. (1997) observed a plateau-type response in accumulation of Cd, Cu and Zn concentrations in corn and a linear response in lettuce plants. Consequently, Logan et al. (1997) considered that the metal concentration in plants may follow either a linear or plateau-type relationship, depending on the plant species. Accordingly, Dowdy et al. (1994) found no increase in heavy metal concentrations in corn after 19 years of sludge application and Chang et al. (1987) showed that the Cd concentration in leaves of radish and swiss chard increased yearly with each sludge application. Heckman et al. (1987) reported that the Zn, Cd, Cu and Ni concentrations in soybeans increased linearly with the increase in rate of sludge application. Similar results were also reported by Reddy et al. (1989) and Ham and Dowdy (1978) in soybean. McBride et al. (2000) showed higher concentration of Mo in peas, soybeans and canola than in corn plants. Soybeans grown on biosolids amended soils also showed high Mo accumulation in the grain (O'Connor et al. 2001).

Similar to biosolid land disposal, agricultural land application offers the best solution for the management of enormous amounts of Poultry Litter (PL) generated on U.S. poultry farms each year. Although PL is one of the best organic fertilizer sources available, excessive land application of PL can cause environmental problems (Bolan et al. 2010; Moore et al. 1995). Land application of PL is known to improve soil physical and chemical properties, such as plant nutrient availability, soil reaction (pH), organic matter content, cation exchange capacity, water holding capacity, structural stability and soil tilth (Bolan et al. 2010; Williams et al. 1999; Moore et al. 1995). Poultry waste contains all essential nutrients and it has been well documented that it provides a valuable source of plant nutrients (Kelley et al. 1996; Williams et al. 1999; Chan et al. 2008; Harmel et al. 2009). Apart from high concentrations of N, P and K and micro nutrients, PL also contains heavy metals such as As, pesticide residues, pharmaceuticals such as coccidiostats, endocrine disruptors and microorganisms (Bolan et al. 2010; Moore et al. 1995).

Excessive accumulation of heavy metals such as Zn, Cd, Cu and Ni cause changes in physiological status and alters the growth rate of plants (Berti and Jacobs 1996). The increased metal accumulation causes changes in pigment concentration, water content, dry weight and growth of plants. All these changes in plant characteristics result in distinct light absorption and reflectance characteristics of the plants which can thus be used as an indicator of soil contamination. A number of field and green house studies have demonstrated that metal stress in plants cause changes in the spectral reflectance of the vegetation (Collins et al. 1983; Labovitz et al. 1983; Milton et al. 1989; Kooistra et al. 2004; Su et al. 2007; Sridhar et al. 2007a, b). In general, the plant spectral response varies with soil and plant type under study, level of soil contamination, and the sensitivity of the plant to different metal concentrations.

16.1.1 Remote Sensing of Vegetation

Remote sensing is the process of inferring surface parameters from distant measurements of upwelling electromagnetic radiation of the land surface (Schmugge et al. 2002). Remote sensing has been used as an alternative method for determining and mapping the physical and chemical characteristics of the soil and plants at a spatial scale.

Plant reflectance is governed by leaf surface properties and internal structure, as well as by the concentration and distribution of foliar pigments and biochemical components (Penuelas and Filella 1998; Raun et al. 1998). Thus analysis of remotely sensed reflected light can be used to assess the changes in plant characteristics such as biomass and the physiological status of plants (Raun et al. 1998). Spectral reflectance of leaves is relatively low in the visible region (400–700 nm) where light absorption by leaf pigments (primarily due to chlorophyll) is the determining factor. The absorption maxima occur in the blue and red at 470 and 680 nm respectively, while the familiar green reflectance peak occurs at 550 nm. In the near and middle infrared regions, these pigments are transparent and internal leaf structure and biochemical composition controls the reflectance. The reflectance spectrum of principal biological interest occurs in the near infrared (NIR) region between 700 and 1,300 nm, where reflectance is high and absorption is minimal (with two minor water absorption bands at 975 and 1,175 nm); beyond 1,300 nm, major water absorption bands (at 1,450 and 1,950 nm) become significant.

All plants respond to increase in supply of plant nutrients in basically the same way, through the increase in growth rate and increase in pigment concentration. Several studies have indicated that the increase in plant biomass and chlorophyll concentration resulted in an increase in NIR reflectance affecting the red edge of spectral curve compared to normal plants. NDVI (Rouse et al. 1974), a widely used vegetative index in remote sensing, can be used to monitor plant changes as a result of biosolid application.

16.1.2 Remote Sensing of Soil

Remote sensing has been used as an alternative method for determining and mapping the physical and chemical characteristics of the soil at a spatial scale. High resolution aerial imagery were used to map the organic carbon (Chen et al. 2000), clay content (Sullivan et al. 2005), organic matter (OM) and Bray-1 phosphorus (P) concentration in soils (Varvel et al. 1999), and satellite imagery has been used to map total soil phosphorus, copper and sulfur concentrations (Sridhar et al. 2009a, b) in bare soils.

Nanni and Dematte (2006) used Landsat TM imagery to estimate the various physical and chemical properties of bare soils. Soil samples were collected from a total of 184 ha study area with soils belonging to five different soil orders and significant correlations were obtained with all the soil parameters using Landsat single band reflectance values (Nanni and Dematte 2006). However, no soil maps were presented in that study. Nanni and Dematte (2006) suggested that the success of their methodology for estimating the soil attributes depends on factors such as atmospheric image correction, geometric corrections, transforming the data to real reflectance values.

Varvel et al. (1999) collected soil spectral data using multispectral digital cameras on a field containing two different soil types to map organic matter and Bray-1 P. They used digital cameras to record brightness values for each pixel in the blue (450–520 nm), green (520–600 nm) and NIR (775–900 nm) bands. Soil OM, Bray-1 P were correlated with the bare soil aerial image. The brightness values, which can vary from date to date due to sun angle, sensor gain, atmospheric conditions, were used in that study. In spite of these potential problems, Varvel et al. (1999) noted that the differences in brightness values could still be used for mapping OM and soil test P. Significant correlations between brightness values of blue, green and NIR bands with OM and Bray-1 P of soils and surface maps for OM and Bray-1 P were produced (Varvel et al. 1999) and a correlation coefficient of 0.57 was reported between soil OM and Bray-1 P.

Applications of airborne imaging systems have shown that soil properties such as clay mineral content, mineral composition, salt content, pH, OM levels and cation exchange capacity (CEC) can be extracted directly from observations of bare soil surfaces or through physiological conditions in the overlying vegetation cover (Bogrekci and Lee 2007, 2005; Chang et al. 2001; Dalal and Henry 1986; Henderson et al. 1992; Lobell and Asner 2002; Dematte et al. 2003; Morra et al. 1991; Post et al. 2000; Reeves et al. 2002).

There are several concerns about the potential buildup of heavy metal concentrations in soils and thereby affecting the crop growth from repeated organic amendment applications (Berti and Jacobs 1996). Without extensive and time consuming field work, it is often difficult to obtain an overview of the spatial distribution of soil chemical concentrations and its effect on crop growth.

As the agricultural fields are covered with vegetation for most of the year, remote sensing of crop-covered biosolid applied fields helps in making indirect assessment of soil concentrations through mapping the spatial and temporal heterogeneity of crop growth. Remote sensing also improves the possibility of monitoring several fields simultaneously. Based on published data (Sridhar et al. 2009a, 2011), this chapter summarizes the impact of the accumulation of metals and nutrients in soils and in soybeans grown on biosolid amended soils and the relationship between soybean vegetation reflectance and different levels of soil contamination.

16.2 Soybean Growth on Biosolid Amended Soils

A greenhouse study was conducted in parallel with a field study, to assess the phytotoxic effects of increase in soil chemical concentrations on soybean growth and to study the changes in plant spectral reflectance. The soil used for the pot experiment was a Miracle-Gro® potting mix (Miracle Grow Lawn Products Inc., Marysville, OH) and the biosolid used for making the sludge treatments was Millorganite® (Milwaukee Metropoliton Sewerage District, Milwaukee, WI). The soybean seeds were of a commercial variety obtained locally. A total of 30 potted plants consisting of five sludge treatments and a control or untreated group (P0) were arranged in a completely randomized design, with 5 replicates in each treatment. Sludge treatments were made by adding the biosolids at the rate of 2, 5, 10, 25 and 50 t/acre on dry weight basis to the potting-soil and these groups were subsequently referred as P2, P5, P10, P25 and P50 respectively. The soils were air dried, homogenized and about 2.0 kg of the soil was weighed and transferred into plastic pots (10″ × 7″ in size). The seedlings were thinned to two plants per pot at the 2-3-leaf stage. Two plants per pot were allowed to grow upto 57 days after sowing (DAS).

The field study was carried out in a 65 acre production farm in NW Ohio which had been applied with biosolids for the past 20 years. The soybeans were grown in the field from May through October of 2007, following the land application of biosolids. The selected parts of the field were applied with biosolids at the rate of 5 and 10 t/acre and these are subsequently referred as F5 and F10 respectively. A part of the field that never received biosolids was selected as untreated or control (F0). The control field had similar soil and crop types and identical management history like that of the other treatments except that no biosolids were land applied (Sridhar et al. 2011).

Surface soil samples (0–15 cm) and aboveground shoot samples were collected from 18 locations across the field on September 15, 2007 and the field sampling was conducted on the day of Landsat TM overpass. All the sampling locations were marked using an eTrex Venture® (Garmin Limited, KS, USA) global positioning system (GPS) receiver.

16.2.1 Nutrient and Metal Accumulation in Soybeans

At the end of the greenhouse study (after 57 DAS), the plants were cut about 2 cm above the soil level, shoots (stems and leaves) and roots were harvested and washed with de-ionized water and the samples were dried at 80 ºC in an oven for 24 h. The samples were then ground, passed through a 2 mm sieve and a representative subsample was analyzed for chemical analysis. The soil samples collected were also dried and passed through a 2 mm sieve. From the field study, a total of 36 soil and plant samples were collected from 18 georeferenced locations on the day of Landsat TM overpass. These samples were processed in the same steps as described for the greenhouse study (Sridhar et al. 2011).

Table 16.1 Chemical concentration in soils, shoots and roots (in mg kg^{-1} dry weight) of soybean treated with biosolids at the end of the pot experiment

Treatment	Ba	Cd	Cu	Mo	P	S	Zn
Soil samples							
P0	17 d[a]	0.4 d	5 d	1.1 d	786 d	972 c	22 d
P2	14 d	0.3 d	5 d	1.1 d	861 d	904 c	19 d
P5	22 c	0.6d	8 d	1.5 c	1,087 c	1,181 b	30 c
P10	26 c	0.8 c	12 c	1.8 c	1,445 c	1,332 b	37 c
P25	38 b	1.5 b	24 b	2.3 b	2,113 b	1,759 a	58 b
P50	53 a	2.4 a	42 a	3.1 a	3,354 a	1,955 a	83 a
Shoot samples							
P0	4 c[a]	0.2 b	3 d	2 b	7,651 a	3,901 c	28 b
P2	5 c	0.3 b	4 c	3 b	9,093 a	3,289 c	27 b
P5	6 c	0.5 b	5 c	4 a	6,183 a	3,750 c	26 b
P10	9 b	0.5 b	8 b	4 a	6,235 a	3,651 c	27 b
P25	10 b	0.8 a	10 b	5 a	7,375 a	4,406 b	41 a
P50	13 a	0.9 a	17 a	5 a	8,831 a	5,507 a	45 a
Root samples							
P0	4 a[a]	0.2 a	3 b	0.7 b	13,306 a	2,006 a	50 a
P2	4 a	0.3 a	3 b	0.6 b	13,903 a	2,320 a	50 a
P5	3 a	0.2 a	3 b	1.1 b	13,308 a	2,197 a	46 a
P10	3 a	0.2 a	4 b	1.2 b	13,454 a	2,481 a	48 a
P25	3 a	0.2 a	6 a	1.2 b	10,080 a	2,236 a	55 a
P50	2 a	0.3 a	7 a	3.1 a	11,155 a	2,709 a	53 a

The P0, P2, P5, P10, P25 and P50 treatments were applied with biosolids at the rates of 0, 2, 5, 10, 25 and 50 t/acre, respectively on dry weight basis

The given are mean values (n = 5) of five replicates (Sridhar et al. 2011)

[a]Means followed by a different letter are significantly different at the 0.05 probability level, grouped into classes a, b and c

The soybean plants in the P0, P2, P5, P10, P25 and P50-treated groups grew steadily and chlorosis was not visually observed during the pot study. Accumulation of Ba, Cd, Cu, Mo, P, S and Zn in soils, roots and shoots are summarized in Table 16.1. The metal and nutrient accumulation in soils increased significantly ($p < 0.05$) with increase in applied sludge (Millorganite) concentration in all the sludge-treated groups (Table 16.1). Accumulation of Ba, Cd, Cu, Mo, S and Zn in roots increased with increase in their soil chemical concentrations (Table 16.1). Accumulation of Cu and Mo in shoots increased with increase in the applied sludge concentrations (Table 16.1). Metal concentrations remained high in soils, followed by roots and shoots while the nutrient (P and S) concentrations remained high in shoots and roots followed by soils.

In the field study, the chemical concentration of the surface soils and shoot samples of untreated (F0), F5 and F10-treated fields are shown in Table 16.2. Among all the chemicals that were analyzed, the accumulation of Ba, Cu, P, S and Zn were significantly ($p < 0.05$) higher in the surface soils of F5 and F10-treated fields compared to F0 (Table 16.2). In soybean shoots, accumulation of Cu and S increased with increase in sludge concentration.

Table 16.2 Chemical concentration in soils and shoots (in mg kg^{-1} dry weight) of soybean treated with biosolids at the end of the field experiment

Treatment	Ba	Cd	Cu	Mo	P	S	Zn
Soil samples							
F0	103 c[a]	3 a	24 b	3 a	547 b	325 c	81 b
F5	128 b	3 a	40 a	3 a	3,073 a	511 b	118 a
F10	156 a	3 a	47 a	4 a	3,477 a	627 a	125 a
Shoot samples							
F0	3 a[a]	0.02 a	6 b	1 a	7,646 a	1,168 b	44 a
F5	2 a	0.03 a	8 a	1 a	7,948 a	1,768 a	49 a
F10	2 a	0.03 a	9 a	2 a	8,197 a	1,601 a	52 a

The F0, F5 and F10 treatments were applied with biosolids at the rates of 0, 5 and 10 t/acre, respectively on dry weight basis
The given are mean values (n = 6) of six replicates
[a]Means followed by a different letter are significantly different at the 0.05 probability level, grouped into classes a, b and c (Sridhar et al. 2011)

The analytical results showed that the accumulation of soil and plant chemical concentrations increased with the increase in biosolid application rates in both the greenhouse (Table 16.1) and field studies (Table 16.2). This confirms the results of increased accumulation of Cu, Zn (Reddy and Dunn 1986, Reddy et al. 1989; Heckman et al. 1987), Mo (McBride et al. 2000; O'Connor et al. 2001) in soil and tissues of soybean grown on biosolid amended soils. According to Heckman et al. (1987) the soybean tissue concentrations of Zn, Cd, Cu and Ni exhibited significant linear increase with sludge application rate from 0 to 112 Mg/hac. The high chemical concentration in soils of biosolid amended fields (F5 and F10) was due to the historical land application of biosolids for over 20 years in the study area. Several studies reported that heavy metal and nutrient concentrations in the surface soils increased with long-term application of biosolids (Mantovi et al. 2005; Udom et al. 2004; Nyamangara and Mzezewa 1999). Metal accumulation in roots increased with the increase in sludge concentration. The accumulation of Cu and Mo are significantly higher in the shoots of P50-treated plants in the pot study and in the F5 and F10-treated plants in the field study. This indicates that the translocation of Cu and Mo into soybean shoot increases with increase in their soil concentration. The significant positive correlation of shoot concentrations of Cu, Zn and Mo with soil concentrations also confirms their increased translocation rate compared to other metals. This agrees with the results of Reddy et al. (1989) that the Cu concentration in soybean seeds increased from 4 to 11 ppm with increase in rate of biosolid application from 0 to 120 Mg/hac. The Millorganite is widely used as a garden fertilizer along with potting mix and soils for plant growth.

16.2.2 Spectral Reflectance of Soybean

A Fieldspec Pro spectroradiometer (ASD Inc., Boulder, CO, USA) with a spectral range of 350–2,500 nm was used to obtain the reflectance spectra of each individual

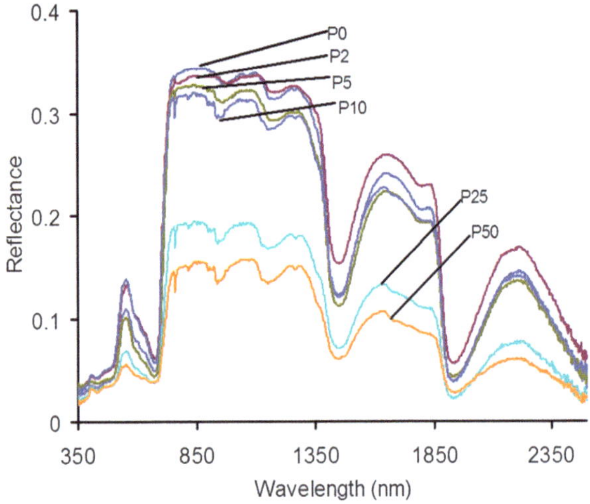

Fig. 16.1 Averaged (n = 5) canopy reflectance spectra of control (P0), P2-, P5-, P10-, P25- and P50-treated groups on the last day (57th day) of pot experiment. All the spectra were obtained inside the green house (Sridhar et al. 2011)

potted plant in the green house, with a quartz-tungsten-halogen (QTH) lamp as a light source. Diffused light from the 100 W Lowell Pro-Light was used to illuminate the plant canopy at 45° angles when spectra were collected in the laboratory (Sridhar et al. 2011).

The spectral measurements of each treatment group (n = 5) of the pot experiment were averaged to overcome individual variations. Changes in spectral reflectance along with metal accumulation in soils, plant roots and shoots were used to characterize the metal stress in plants. The averaged NDVI of each treatment group was calculated for all the days of spectral collection. The spectra obtained in the green house study were resampled over the same spectral band intervals as used by the Landsat TM sensor to simulate the satellite spectral bands. Similarly, NDVI of each treatment group of the field study was calculated from all the Landsat TM images obtained during the period of study. The significant differences between the treatments were evaluated through analysis of variance (ANOVA). Pearson's correlation coefficients have been calculated to identify statistically significant relationships between spectral indices and concentration of metal content in soil and plant tissues. Duncan's multiple range test was performed using SAS statistical analysis software (SAS Institute Inc., Cary, NC).

The averaged (n = 5) reflectance spectra of canopies of all the treated groups from the last day (57th day) of green house study are given in Fig. 16.1. The reflectance spectra of the P25 and P50-treated groups showed a decrease in reflectance in the 800–1,300 nm, 1,470–1,850 nm, and 2,000–2,500 nm regions (Fig. 16.1). The reflectance spectra from the P5- and P10-treated groups showed less decrease in the 800–1,300 nm and 1,470–1,850 nm regions compared to untreated (P0) plants (Fig. 16.1). The NDVI results obtained during the sludge treatment period are given in Fig. 16.2. The P10, P25 and P50-treated plants show significant ($p < 0.05$) decrease in NDVI compared to untreated plants (P0) by last day (57th) of the study (Fig. 16.2). The P2 and P5-treated plants show no significant

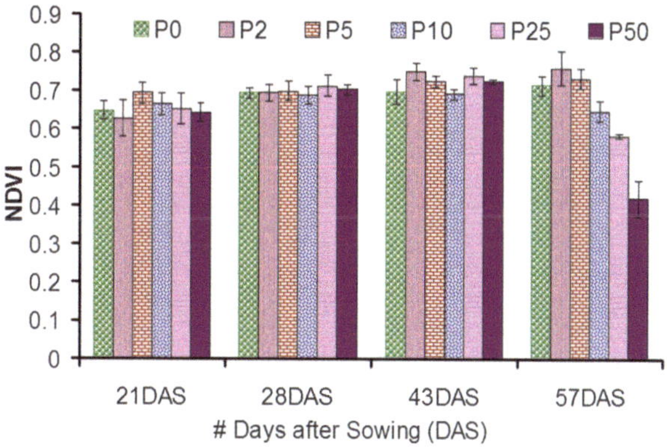

Fig. 16.2 Changes in NDVI of control (P0), P2-, P5-, P10-, P25- and P50-treated plant groups that were grown on biosolid amended soils at 21, 28, 43 and 57 days after sowing (*DAS*). Bars are ± standard error of five replicates (Sridhar et al. 2011)

difference in NDVI from untreated plants during the entire plant growing period (Fig. 16.2).

Pearson's correlation coefficients were calculated in order to identify the significant correlations between the metal concentration in soil and shoots of the plants. Among all the chemicals that were analyzed, Cu showed significant positive correlation ($r = 0.83^{**}$; $p < 0.01$) followed by Zn ($r = 0.64^{**}$; $p < 0.01$) and Mo (0.62^{**}; $p < 0.01$). This indicates that the Cu, Zn and Mo concentrations in plant tissue increased linearly with increase in soil concentration of the respective metals. Among correlations calculated between NDVI and soil metal concentrations, NDVI showed significant negative correlation with soil Cu concentration ($r = -0.64^{**}$; $p < 0.01$).

16.2.3 Satellite Data Collection and Analysis

To determine the changes in seasonal variation in soybean growth, all the cloud free Landsat TM images available from May through October of 2007 were downloaded and processed for analysis. The study area was located within the Landsat TM overpass region of Path 20, Row 31. The georeferenced and terrain corrected Landsat TM images were downloaded from the USGS EROS Data Center. The Landsat TM images were processed using the ERMapper image processing software, a commercial product of Earth Resources Mapping, Inc. (now part of ERDAS).

Based on the locations of the sampling points, the dark object subtracted (DOS) pixel values corresponding to the Landsat TM bands 1–5 and 7 were derived from the corresponding imagery. The spectral range of these Landsat TM bands are as follows: Band 1: 450–520 nm; Band 2: 520–600 nm; Band 3: 630–690 nm; Band 4: 760–900 nm; Band 5: 1,550–1,750 nm; and Band 7: 2,080–2,350 nm. The dark object of each spectral band is defined as one value less than the minimum digital number found in all the pixels of the image (Vincent et al. 2004). The detailed

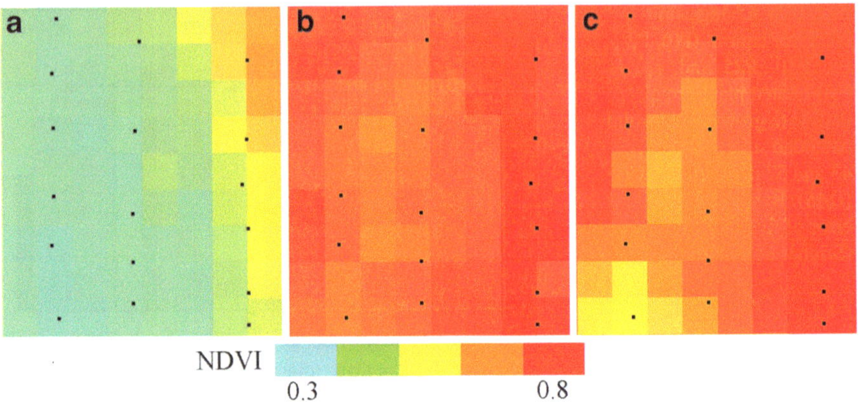

Fig. 16.3 The NDVI images of the field study site that were derived from the Landsat TM data acquired on July 13 (**a**); July 28 (**b**) and Sept 15 of 2007 (**c**), which were of 44, 60 and 108 DAS respectively. The sampling transects corresponds to control (F0), F10 and F5-treatments from right to left in each image (**a–c**) respectively (Sridhar et al. 2011)

procedure for DOS and its effects on removal of atmospheric haze has been presented elsewhere in Vincent (1997), Vincent et al. (2004) and Sridhar et al. (2009a, b). From the DOS-corrected digital number (DN) values of the six 30-m resolution Landsat single bands, the normalized difference vegetative index (NDVI) was calculated. The NDVI was calculated using the formula NDVI = ((Band4 − Band3)/(Band4 + Band3)) (Rouse et al. 1974).

The DOS NDVI spectral ratio images of the fields obtained during the soybean growth period are given in Fig. 16.3. The Landsat TM NDVI images of the plants grown in biosolid amended soils show a consistent and systematic difference from the plants grown in untreated soils starting from 44 DAS. The plants grown in F5 and F10-treated fields show significant ($p < 0.05$) decrease in NDVI compared to untreated plants (F0) starting from 44 DAS (Fig. 16.4).

The significant ($p < 0.05$) decrease in the NDVI values of P10, P25 and P50-treated plants compared to the control (P0) plants of the pot study and F5 and F10-treated plants compared to the control (F0) of the field study indicates that the NDVI decreases with increase in metal accumulation in soybean tissues and thus with increase in soil chemical concentrations. Although no visible phytotoxic symptoms were observed, the plants at higher treatments clearly showed stunted growth compared to the untreated plants in both greenhouse and field studies. Also the canopy-level spectra of the P25 and P50-treated plants were clearly lower in the entire spectral range (from 350 to 2,500 nm) compared to the control (P0) plants (Fig. 16.1). This data indicates how heavy metals can pose an additive toxic effect when accumulated at higher concentrations.

Among correlations calculated between NDVI and soil metal concentrations of the field, NDVI showed significant negative correlation with soil Cu concentration (Fig. 16.5) during all the stages of crop growth at 44, 60 and 108 DAS, respectively. Regression analyses were conducted between the soil Cu concentrations and the

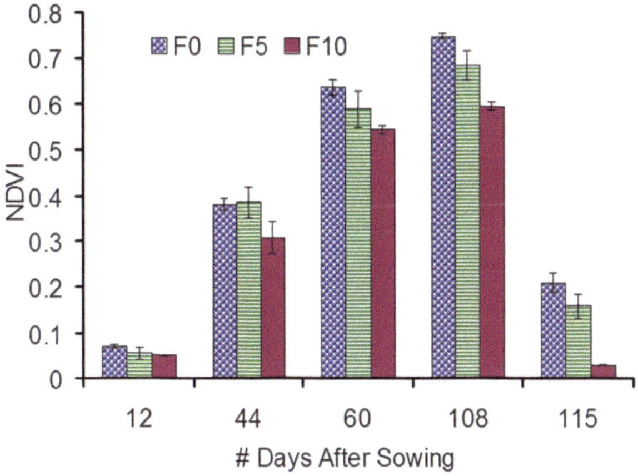

Fig. 16.4 Changes in NDVI of control (F0), F5-, and F10-treated plant groups that were grown on biosolid amended plots at 44, 60 and 108 days after sowing (DAS). Bars are ± standard error of five replicates (Sridhar et al. 2011)

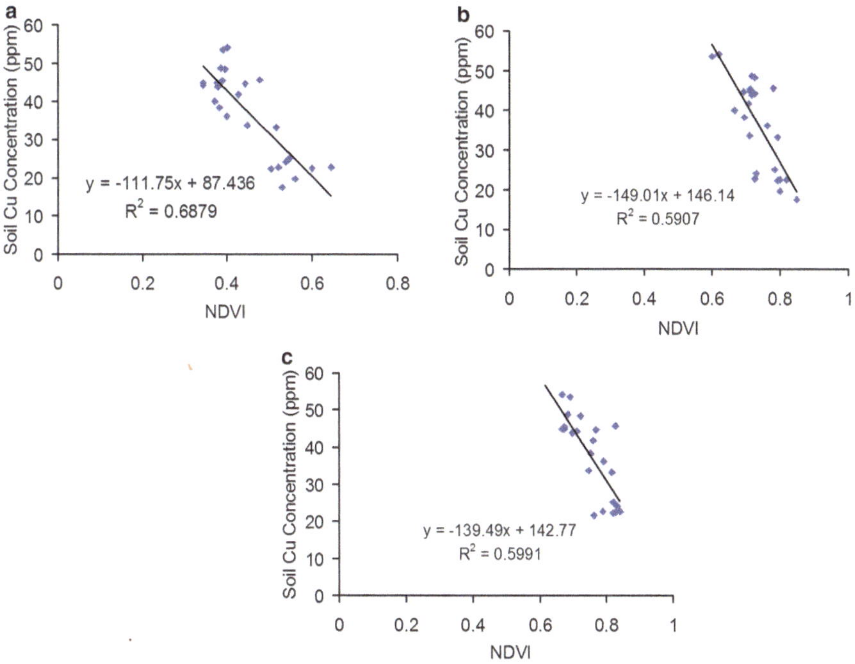

Fig. 16.5 The relationship between the NDVI that were derived from the Landsat TM data acquired on July 13 (**a**); July 28 (**b**) and Sept 15 of 2007 (**c**), which were of 44, 60 and 108 DAS respectively and the soil Cu concentration

NDVI of the crop plants during each of the crop growth stages at 44, 60 and 108 DAS respectively (Fig. 16.5). Application of the regression equations to the respective Landsat images resulted in mapping the soil Cu concentrations based on the reflectance characteristics of the standing soybean crop (Fig. 16.6). The spatial

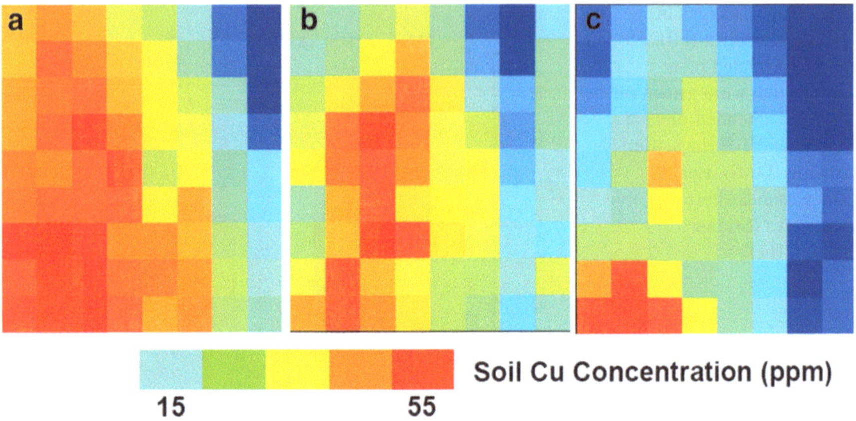

Fig. 16.6 The images of the field study site showing the soil Cu concentrations that were derived from the Landsat TM data acquired on July 13 (**a**); July 28 (**b**) and Sept 15 of 2007 (**c**), which were of 44, 60 and 108 DAS respectively

distribution of the soil Cu concentration in the fields which was indirectly calculated based on the plant spectral and NDVI characteristics (Fig. 16.6) clearly represents higher soil Cu concentrations in the F5 and F10-treated fields.

In summary, the NDVI of plants will decrease with the increase in the shoot and root metal concentrations and these phenomena were evident from the data reported in both the green house and field studies. The P2 and P5-treated potted plants showed a slight increase in NDVI compared to the other treatments, this is because the biosolids at lower rate supplied the soil nutrients at the optimum rate required for plant growth. Besides finding a relationship between patterns of NDVI and soil metal concentrations, the Landsat TM imagery can also be used to determine the seasonal variation in plant stress and its effect on the different vegetative and reproductive stages of the crop. Thus remote sensing can be used to monitor metal stress in plants and also for indirect assessment of soil contamination.

16.3 Mapping the Chemical Concentration of Soils

Application of biosolids to agricultural land has become a prominent and acceptable method of waste utilization in recent years. Biosolids are known to improve soil physical characteristics (Epstein et al. 1975; Wei et al. 1985), increase soil organic matter content and cation exchange capacity, and supply the nutrients required for crop growth (Sommers 1977; Singh and Agrawal 2008). However, the potential for excess application of biosolids, resulting in a build up of nitrogen, phosphorus (Mantovi et al. 2005), zinc, copper, lead (Mantovi et al. 2005; Udom et al. 2004; Nyamangara and Mzezewa 1999) and cadmium (Bergkvist et al. 2003) in the surface soils of agricultural fields continues to be an area of concern. Accumulation of phosphorus (P) to high concentrations is a major environmental

Table 16.3 Concentration of selected elements and moisture content in the surface soils applied with 34 t/acre^{-1} (F34) and 11 t acre^{-1} (F11) of class B biosolids

Treatment	Ba (mg/kg)	Cd (mg/kg)	Cu (mg/kg)	S (mg/kg)	P (mg/kg)	Moisture (%)
F34	161	5.7	55	405	2,550	9.7
	(± 28.5)	(± 0.5)	(± 5)	(± 57)	(± 625)	(± 7.3)
F11	98	3.6	37	265	988	5.5
	(± 19.1)	(± 1.9)	(± 6.6)	(± 68)	(± 303)	(± 2.5)

The given values are means ± standard deviation of 35 replicates (Sridhar et al. 2009a)

concern, as it affects the water quality of lakes and rivers in the event of runoff (Shober and Sims 2003). Hence, there is an increasing need to continuously monitor the extent of soil contamination in biosolid and animal manure-applied fields. Even though conventional methods of soil sampling and testing are being used for this purpose, they are often expensive, time-consuming and impractical for mapping soil contamination over large areas. Based on the published data (Sridhar et al. 2009a), this section of the chapter summarizes the impact of elemental concentrations on soils amended with biosolids and the use of Landsat data to map these elemental concentrations of the soils when they are not covered by vegetation (bare).

16.3.1 Soil Chemical Characteristics

Two adjacent agricultural fields, F34 and F11 that received a cumulative amount of 34 t acre^{-1} (76 Mg hac^{-1}) and 11 t acre^{-1} (25 Mg hac^{-1}) of Class B biosolids on a dry weight basis during the period of 1985–2002 were selected. Soil samples were collected at surface, 30, and 50 cm depths from each of the 70 sampling locations across the two fields. These fields were selected because they are representative of large areas of northwest Ohio where land application of biosolids has become an important agricultural practice. The soil samples were collected on May 19 of 2005, one day prior to Landsat over pass, and the sampling locations were marked using a Trimble GeoExplorer (Trimble Navigation Limited, CA, USA) global positioning system (GPS) receiver. The collected soil samples were dried and passed through a 2 mm sieve (Sridhar et al. 2009a).

Soil samples (approximately 0.5 g) were digested with concentrated HNO_3, according to USEPA method SW846-3051A (USEPA 1998) using a Mars Xpress microwave digestion unit (CEM, Matthews, NC, USA). Analysis of variance (ANOVA) was used to compare the accumulation of each element in the F34 and F11 fields using SAS version 9.1 statistical software (SAS Institute Inc., Cary, NC, USA). An alpha level of 0.05 was used to determine the significance.

The chemical concentration of the surface soils and the moisture content of the surface soils in both the F34 and F11 treated fields are shown in Table 16.3. Among all the elements that were analyzed, the accumulation of Ba, Cd, Cu, S and P in the surface soils of F34 were significantly ($p < 0.05$) higher than those of F11 (Table 16.3). However, the moisture content of the surface soils in both the fields was similar (Table 16.3). The soils were of the prevalent latty silty clay type with surface soils having 40–55 % of clay and 3–5 % of organic matter (Soil Survey Staff 2007).

The analytical results showed that the accumulation of P in surface soil samples of F34 was about 2.6 times of F11 soils. This confirms the report of Chang et al. (1983) that 5 years continuous biosolids application in two California soils at 0, 22.5, 45 and 90 t/hectare increased the total P concentration of surface soil (0–15 cm) from 515–540 mg/kg to 1,092–1,312, 1,657–2,163 and 2,617–3,470 mg/kg, respectively. Similarly Maguire et al. (2000) reported that the concentration of total soil P in surface soils (0–20 cm) of biosolid amended soils was 738 mg/kg, or nearly double the concentrations in unamended soils, where the total soil P was 403 mg/kg. High concentrations of Cd and Cu in the surface soils of F34 compared to F11 were consistent with the reports of Nyamangara and Mzezewa (1999), that the long-term application of biosolid increased the accumulation of Cd and Cu in the surface soils.

16.3.2 Satellite Model Development

The Landsat image frames of May 20 and June 5, 2005, covering the study area were downloaded soon after soil sampling. They were then processed with the ER Mapper image processing software. The study area was located within the Landsat overpass region of Path 20, Row 31. The locations of all the 70 soil sampling points collected one day prior to LANDSAT 5 overpass were marked with a GPS unit. The study site was dry, without any vegetation, implying that image spectral reflectances represent the spectral reflectance of a bare soil (Sridhar et al. 2009a). The procedure for developing the GIS database of the Class B biosolid permitted fields in Wood and Lucas counties of northwest Ohio was reported in detail by McNulty (2005).

Based on the locations of the 70 soil samples, the dark object subtracted (DOS) pixel values corresponding to the LANDSAT TM bands 1–5 and 7 were derived from the original May 20, 2005 image. The spectral range of these LANDSAT TM bands were as follows: Band 1: 450–520 nm; Band 2: 520–600 nm; Band 3: 630–690 nm; Band 4: 760–900 nm; Band 5: 1,550–1,750 nm; and Band 7: 2,080–2,350 nm. The dark object of each spectral band is defined as one value less than the minimum digital number found in all the pixels of the image (Vincent et al. 2004). The detailed procedure for DOS and its effects on removal of atmospheric haze was given in Vincent (1997) and Vincent et al. (2004). From the DOS-corrected digital number (DN) values of the six LANDSAT single bands, 15 non-reciprocal spectral ratios were calculated. These spectral ratios are: $R_{2,1}$; $R_{3,1}$; $R_{3,2}$; $R_{4,1}$; $R_{4,2}$; $R_{4,3}$; $R_{5,1}$; $R_{5,2}$; $R_{5,3}$; $R_{5,4}$; $R_{7,1}$; $R_{7,2}$; $R_{7,3}$; $R_{7,4}$; $R_{7,5}$ where R represents the ratio and the numbers represent the LANDSAT TM band numbers

(Vincent 1997). The spectral ratios were calculated using the MINITAB statistical software (MINITAB Inc., State College, PA, USA).

The relationship between the chemical concentrations of the surface soil samples and the DOS DN values corresponding to the six single bands and the 15 - non-reciprocal spectral ratios were developed by regression analysis. Using the MINITAB regression analysis component the best subsets regression was employed, and only the top two models with highest R^2 adjusted values were chosen to report for each number of variables. The best subsets procedure was used for sequentially entering independent variables one at a time to improve the regression equation's predictive ability. The reported models from the best subset regression output were tested for autocorrelation with a Durbin-Watson (DW) statistical test (Durbin and Watson 1951). This tests for autocorrelation in the input parameters. Finally, the model which had the highest R^2 adjusted and that also passed the DW test was selected as the best model for given inputs. This procedure was reported in detail elsewhere by Vincent (2000) and Vincent et al. (2004). The identified best model was then applied to the same May 20, 2005, LANDSAT image, which was used in developing the model to map the elemental concentration of the surface soils. The model was also applied and validated using the June 5, 2005, LANDSAT image, which was obtained 17 days after the soil sampling. In the LANDSAT images that were applied with the best model, masks were created to limit the display to only bare soil fields.

Regression equations were established to determine the chemical concentrations of Ba, Cd, Cu, S and P, which are significantly ($p < 0.05$) higher in the surface soils of F34 compared to F11, using the DOS-corrected six TM bands and the 15 non-reciprocal spectral ratios. The best spectral ratio input model for P and Cu that passed the DW test of significance along with their R^2 adjusted and standard error values were P (mg/kg) $= 4{,}156 - 1{,}690\ (R_{51}) + 2{,}257\ (R_{73})$, R^2 adjusted value (67.9 %), standard error of 531.2 mg/kg while for Cu (mg/kg) $= 75 - 17.9\ (R_{51}) + 21.9\ (R_{73})$, R^2 adjusted value (59 %), standard error of 6.9 mg/kg. Phosphorus had the highest R^2 adjusted value (67.9 %) among the chemical attributes that passed the DW test followed by Cu. Hence, only the P and Cu results were shown in this chapter. The P and Cu values obtained from chemical analysis of the 70 surface soil sampling locations versus the predicted values of P and Cu for the same locations obtained by applying the P and Cu spectral ratio models were given in Fig. 16.7a, b. The P and Cu spectral ratio model were also applied to the June 5, 2005 LANDSAT image frame and the predicted P and Cu values were plotted against the P and Cu values obtained by the soil analysis (Fig. 16.7c, d). The model performed well in predicting the P and Cu concentrations of surface soil when applied to either of the Landsat TM images (Fig. 16.7).

The application of the best P and Cu spectral ratio models to the Landsat TM frame of May 20, 2005, which was also used in developing the model, is shown in Fig. 16.8. The redder color in this image corresponds to higher amounts of P and Cu in surface soil. The application of the best P and Cu spectral ratio models to the Landsat TM image, reveals all the fields and other locations at the watershed scale that have higher concentrations of P and Cu (Fig. 16.8).

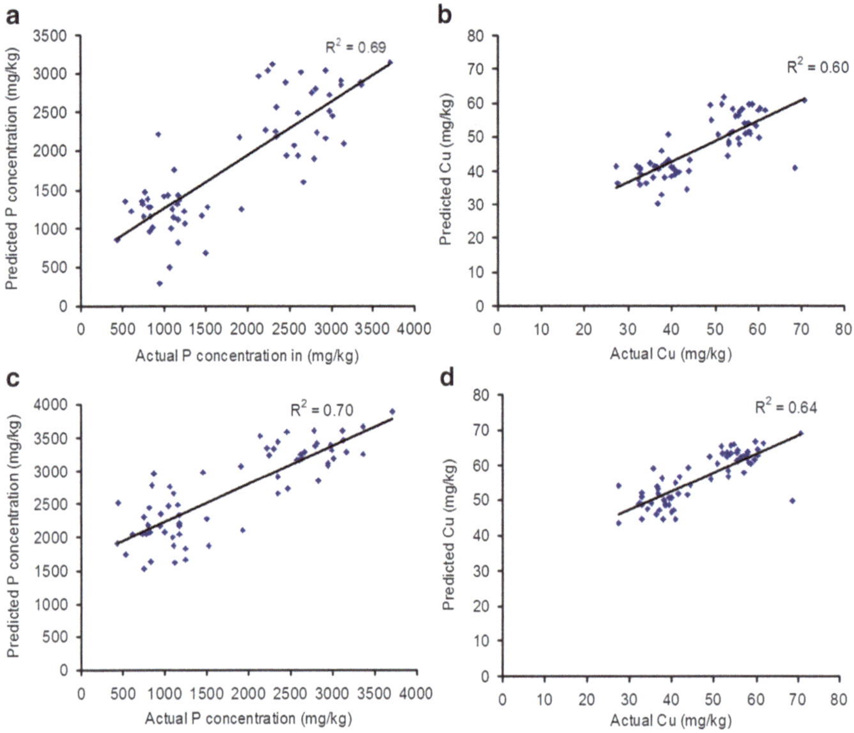

Fig. 16.7 Actual versus predicted P and Cu concentrations (in mg/kg) of surface soil samples using the dark object subtracted best P and Cu spectral ratio models being applied to the Landsat TM frame of May 20, 2005 (**a**, **b**), which was also used for developing the model. Also the P and Cu spectral ratio models being applied to a different Landsat TM frame of June 5, 2005 (**c**, **d**), which was also used for evaluating the models (Sridhar et al. 2009a)

16.3.3 Satellite Model Application

Nanni and Dematte (2006) have successfully employed Landsat TM data to estimate the sand, silt, clay, organic matter, cation exchange capacity (CEC) and sum of cations in Brazilian soils. They derived spectral reflectance values from the corrected LANDSAT image to develop multiple regression equations in order to estimate the different physical and chemical characteristics of the soils; however, no soil maps were presented in that study (Nanni and Dematte 2006).

Aerial imagery was used to map the organic carbon (Chen et al. 2000), clay content (Sullivan et al. 2005), OM and Bray-1 P (Varvel et al. 1999) and Landsat TM imagery was used to estimate the physical and chemical properties (Nanni and Dematte 2006) of surface soils in the previous studies. However, the algorithms developed in these studies were based on the reflected image intensity values of the soils, which required correction for atmospheric haze with atmospheric models before applying the algorithms to another date. The P spectral ratio models

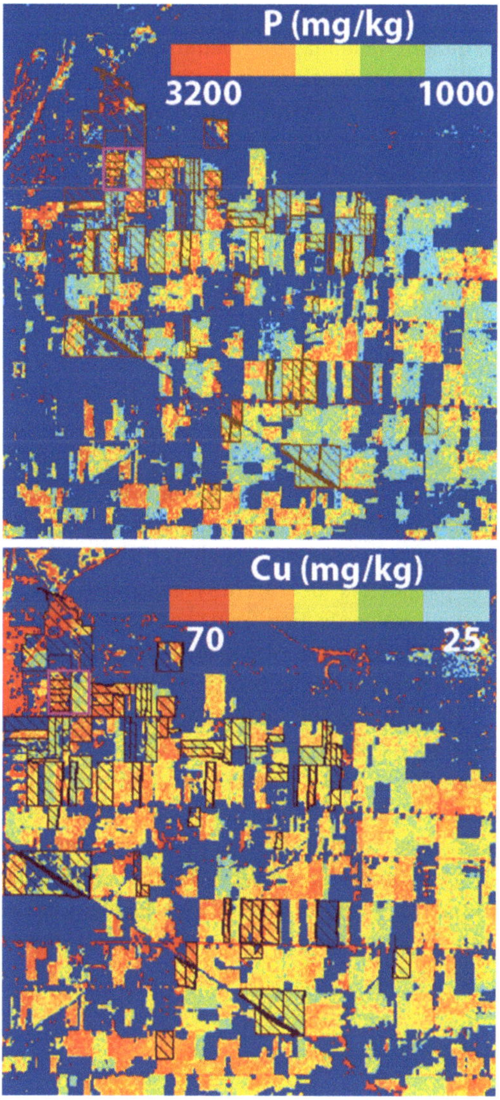

Fig. 16.8 The total P and Cu concentration (mg/kg) in surface soil samples of the bare soil fields in the eastern part of the Lucas County of northwest Ohio, which is a part of the drainage basin of Lake Erie, which is located at the northern side (*top*) of the image. The image was obtained by applying the P and Cu spectral ratio models to the Landsat TM frame of May 20, 2005 which was used for developing the model. The fields permitted for Class B biosolid application in the vicinity are outlined in the image. The fields marked with a *pink border* are the experimental fields used in this study. Not all of the permitted fields in this image would have recently received biosolids, and some non-permitted fields could have recently received other types of fertilizers (Sridhar et al. 2009a)

(Sridhar et al. 2009a, b) were based on the DOS-corrrected spectral ratios and are more robust than any model that could be derived from a combination of single spectral bands. Vincent et al. (2004) showed that the DOS spectral ratio models were more robust than single band models and can be applied with reasonable accuracy to different times of data collection, although their subject was about cyanobacteria blooms in lakes or streams, and the present context of this chapter is about P and Cu concentrations in bare soils on dry land.

In summary, the Landsat TM data can be used to estimate and map selected chemical characteristics of soils, such as total P and Cu. The summarized data enables us to conclude that remotely sensed imagery of bare soil fields can be used to quantify and map the spatial variation of total P and Cu concentration in surface soils. The technology is simple enough to be applied to the entire watershed. The P and Cu spectral ratio models were more robust and reliable than the single band input models and can be applied to bare soil fields with low soil moisture.

16.4 Conclusions

Satellite remote sensing has the potential to monitor the effect of elevated chemical concentration in soybeans grown on biosolid amended soils and also to map the chemical concentrations in bare soils. The EPA part 503 regulations (USEPA 2002) require monitoring of heavy metal additions to biosolids amended soils. However, they do not require any testing and monitoring of the crops grown on these soils. The effects of biosolid application on soil and crop quality have rarely been monitored in non-research field conditions, such as the regular production agriculture farm. Most of the biosolid research has been conducted on experimental fields where all the parameters are carefully controlled and such controls are absent in regular agriculture production farms and hence remote sensing is a means to monitor such fields. Mapping the crop growth patterns and indirectly assessing the soil chemical characteristics can help farmers in planning future biosolid application rates.

By applying the satellite spectral ratio models, nutrient and heavy metal concentrations, such as P and Cu data shown in this chapter, can be identified and mapped in surface soils as a result of biosolid application. Because P accumulation in soils can also result from the application of biosolids, animal manures, and man-made fertilizers, the application of remote sensing has significant implications in identifying the fields with high concentrations of surface soil P, thus helping implementation of P-based management practices on agricultural fields, with an aim toward reduction of P runoff into nearby surface water bodies.

16.5 Future Research

Several states and territories in the United States now have regulations to restrict the land application of biosolids, based on phosphorus concentration in soils. Most of these states have established actual numerical limits for soil test phosphorus (STP), with an aim to cease the application of biosolids once these limits are reached. However, monitoring the P concentrations and several other nutrient and heavy metal concentrations at field and landscape level is very difficult. Further research is required to extend and apply the research of satellite nutrient and metal sensing at a large scale, and to establish the predictable model for plant available P or STP. Since STP is used to guide animal manure and biosolids application instead of total soil P. The spectral ratio models also need to be tested on fields with higher moisture contents. Further, these models need to be applied to surface soils of other soil types in the different agricultural zones. With respect to the application of remote sensing to monitor the metal stress in plants and the relationship between the plant spectral reflectance and the soil chemical concentrations are specific to the type of plant species; the results summarized in this chapter are applicable to soybeans only. Another aspect that needs to be considered was that the soil samples were taken from surface (0–15 cm); however' roots of plants may penetrate deeper and considerable vertical variation in soils is known. Further research should be done with different plant types grown on biosolid applied soils to determine their metal uptake and changes in spectral features.

References

Bergkvist P, Jarvis N, Berggren D, Carlgren K (2003) Long-term effects of sewage sludge applications on soil properties, cadmium availability and distribution in arable soil. Agric Ecosyst Environ 97:167–179

Berti WR, Jacobs LW (1996) Chemistry and phytotoxicity of soil trace elements from repeated sewage sludge applications. J Environ Qual 25:1025–1032

Bogrekci I, Lee WS (2005) Spectral soil signatures and sensing phosphorus. Biosyst Eng 92:527–533

Bogrekci I, Lee WS (2007) Comparison of ultraviolet, visible and near infrared sensing for soil phosphorus. Biosyst Eng 96:293–299

Bolan NS, Szogi AA, Chuasavathi T, Seshadri B, Rothrock MJ, Panneerselvam P (2010) Use and management of poultry litter. World Poult Sci J 66:673–698

Chan KY, Van Zwieten L, Meszaros L, Downie A, Joseph S (2008) Using poultry litter biochars as soil amendments. Aust J Soil Res 46:437–444

Chang AC, Page AL, Sutherland FH, Grgurevic E (1983) Fractionation of phosphorus in sludge affected soils. J Environ Qual 12:286–290

Chang AC, Page AL, Warneke JE (1987) Long-term sludge applications on cadmium and zinc accumulation in swiss chard and radish. J Environ Qual 16:217–221

Chang CW, Laird DA, Mausbach MJ, Hurburgh CR Jr (2001) Near-infrared reflectance spectroscopy – principal component regression analysis of soil properties. Soil Sci Soc Am J 65:480–490

Chen F, Kissel DE, West LT, Adkins W (2000) Field-scale mapping of surface soil organic carbon using remotely sensed imagery. Soil Sci Soc Am J 64:746–753

Collins W, Chang SH, Raines G, Canney F, Ashley R (1983) Airborne biogeophysical mapping of hidden mineral deposits. Econ Geol 78:737–749

Dalal RC, Henry RJ (1986) Simultaneous determination of moisture, organic carbon, and total nitrogen by near infrared reflectance spectrophotometry. Soil Sci Soc Am J 50:120–123

Dematte JAM, Pereira HS, Nanni MR, Cooper M, Fiora PR (2003) Soil chemical alterations promoted by fertilizer application assessed by spectral reflectance. Soil Sci 168:730–747

Dowdy RH, Clapp CE, Linden DR, Larson WE, Halbach TR, Polta PC (1994) Twenty years of trace metal partitioning on the Rosemont sewage sludge watershed. In: Clapp CE (ed) Sewage sludge: land utilization and the environment. ASA/CSSA/SSSA, Madison, pp 149–158

Durbin J, Watson GS (1951) Testing for serial correlation in least squares regression: II. Biometrica 38:159–178

Epstein E, Taylor JM, Chaney RL (1975) Effects of sewage sludge on some soil physical properties. J Environ Qual 4:139–142

Ham GE, Dowdy RH (1978) Soybean growth and composition as influenced by soil amendments of sewage sludge and heavy metals: field studies. Agron J 70:326–330
Harmel RD, Smith DR, Haney RL, Dozier M (2009) Nitrogen and phosphorus runoff from cropland and pasture fields fertilized with poultry litter. J Soil Water Conserv 64:400–412
Heckman JR, Angle JS, Chaney RL (1987) Residual effects of sewage sludge on soybean: I. Accumulation of heavy metals. J Environ Qual 16:113–117
Henderson TL, Baumgardner MF, Franzmeieir DP, Stott DE, Coster DC (1992) High dimensional reflectance analysis of soil organic matter. Soil Sci Soc Am J 56:865–872
Kelley TR, Pancorbo OC, Merka WC, Thompson SA, Cabrera ML, Brnhat HM (1996) Elemental concentrations of stored whole and fractionated broiler litter. J Appl Poultry Res 5:276–281
Kooistra L, Salas EAL, Clevers JGPW, Wehrens R, Leuven RSEW, Nienhuis PH, Buydens LMC (2004) Exploring field vegetation reflectance as an indicator of soil contamination in river flood plains. Environ Pollut 127:281–290
Labovitz ML, Masuoka EJ, Bell R, Siegrist AW, Nelson RF (1983) The application of remote sensing to geobotanical exploration for metal sufides- results from the 1980 field season at mineral. Virginia Econ Geol 78:750–760
Lobell DB, Asner GP (2002) Moisture effects on soil reflectance. Soil Sci Soc Am J 66:722–727
Logan TJ, Lindsay BJ, Goins LE, Ryan JA (1997) Field assessment of sludge metal bioavailability to crops: sludge rate response. J Environ Qual 26:534–550
Maguire RO, Sims JT, Coale FJ (2000) Phosphorus fractionation in biosolids-amended soils: relationship to soluble and desorbable phosphorus. Soil Sci Soc Am J 64:2018–2024
Mantovi P, Baldoni G, Toderi G (2005) Reuse of liquid, dewatered, and composted sewage sludge on agricultural land: effects of long-term application on soil and crop. Water Res 39:289–296
McBride MB, Richards BK, Steenhuis T, Spiers G (2000) Molybdenum uptake by forage crops grown on sewage sludge amended soils in the field and green house. J Environ Qual 29:848–854
McNulty WS (2005) The creation of a GIS database and the determination of sludge's spectral signature in an agricultural setting. M.S. thesis, Department of Geology, Bowling Green State University, Bowling Green
Milton NM, Ager CM, Eiswerth BA, Power MS (1989) Arsenic and selenium induced changes in spectral reflectance and morphology of soybean plants. Remote Sens Environ 30:263–269
MINITAB (2007–2008) Statistical software version 15. MINITAB Inc, State College
Moore PA, Daniel TC, Sharpley AN, Wood CW (1995) Poultry manure management: environmentally sound options. J Soil Water Conserv 50:321–327
Morra MJ, Hall MH, Freeborn LL (1991) Carbon and nitrogen analysis of soil fractions using near infrared reflectance spectroscopy. Soil Sci Soc Am J 55:288–291
Nanni MR, Dematte JAM (2006) Spectral reflectance methodology in comparison to traditional soil analysis. Soil Sci Soc Am J 70:393–407
Nyamangara J, Mzezewa J (1999) The effect of long-term sewage sludge application on Zn, Cu, Ni and Pb levels in a clay loam soil under pasture grass in Zimbabwe. Agric Ecosyst Environ 73:199–204
O'Connor GAO, Granato TC, Basta NT (2001) Bioavailability of biosolids molybdenum to soybean grain. J Environ Qual 30:1653–1658
Penuelas J, Filella I (1998) Visible and near infrared reflectance techniques for diagnosing plant physiological status. Trends Plant Sci 3:151–156
Post DF, Fimbres A, Matthias AD, Sano EE, Accioly L, Batchily AK, Ferreira LG (2000) Predicting soil albedo from soil color and spectral reflectance data. Soil Sci Soc Am J 64:1027–1034
Raun WR, Johnson GV, Sembiring H, Lukina EV, Laruffa JM, Thomason WE, Phillips SB, Solie JB, Stone ML, Whitney RW (1998) Indirect measures of plant nutrients. Commun Soil Sci Plan 29:1571–1581

Reddy MR, Dunn SJ (1986) Heavy metal absorption by soybean on sewage sludge treated soil. J Agric Food Chem 34:750–753

Reddy MR, Lameck D, Rezaina ME (1989) Uptake and distribution of copper and zinc by soybean and corn from soil treated with sewage sludge. Plant Soil 113:271–274

Reeves JB III, McCarty GW, Mimmo T (2002) The potential of diffuse reflectance spectroscopy for the determination of carbon inventories in soil. Environ Pollut 116:S264–S277

Rouse JW, Haas RH, Schelle JA, Deering DW, Harlan JC (1974) Monitoring the vernal advancement and retrogradation (green wave effect) of natural vegetation. Type III Final Report, NASA Goddard Space Flight Center, Green Belt

SAS Institute (2002–2003) SAS software version 91. SAS Institute, Inc, Cary

Schmugge TJ, Kustas WP, Ritchie JC, Jackson TJ, Rango A (2002) Remote sensing in hydrology. Adv Water Resour 25:1367–1385

Shober AL, Sims JT (2003) Phosphorus restrictions for land application of biosolids: current status and future trends. J Environ Qual 32:1955–1964

Singh RP, Agrawal M (2008) Potentail benefits and risks of land application of sewage sludge. Waste Manage 28:347–358

Soil Survey Staff (2007) Natural Resources Conservation Service, United States Department of Agriculture Web Soil Survey. http://websoilsurvey.nrcs.usda.gov

Sommers LE (1977) Chemical composition of sewage sludges and analysis of their potential use as fertilizers. J Environ Qual 6:225–232

Sridhar BBM, Han FX, Diehl SV, Monts DL, Su Y (2007a) Monitoring the effects of Arsenic- and Chromium- accumulation in Chinese brake fern (*Pteris vittata*) using microscopy and near infrared spectral reflectance. Int J Remote Sens 28:1055–1067

Sridhar BBM, Han FX, Monts DL, Diehl SV, Su Y (2007b) Spectral reflectance and leaf internal structure changes of barley plants due to phytoextraction of zinc and cadmium. Int J Remote Sens 28:1041–1054

Sridhar BBM, Vincent RK, Witter JD, Spongberg AL (2009a) Mapping the total phosphorus concentration of biosolid amended surface soils using LANDSAT TM data. Sci Total Environ 407:2894–2899

Sridhar BBM, Vincent RK (2009b) Mapping and estimation of phosphorus and copper concentrations in fly ash spill area using LANDSAT TM Images. Photogramm Eng Remote Sens 75:1030–1033

Sridhar BBM, Vincent RK, Roberts SJ, Czajkowski K (2011) Remote sensing of soybean stress as an indicator of chemical concentration of biosolid amended surface soils. Int J App Earth Obs 13:676–681

Su Y, Sridhar BBM, Han FX, Diehl SV, Monts DL (2007) Effect of bioaccumulation of Cs and Sr natural nuclides and impact on foliar structure and plant spectral reflectance of Indian mustard (*Brassica juncea*). Water Air Soil Pollut 180:65–74

Sullivan DG, Shaw JN, Rickman D (2005) IKONOS imagery to estimate surface soil property variability in two Alabama physiographies. Soil Sci Soc Am J 69:1789–1798

Udom BE, Mbagwu JSC, Adesodun JK, Agbim NN (2004) Distributions of zinc, copper, cadmium and lead in a tropical ultisol after long-term disposal of sewage sludge. Environ Int 30:467–470

US Environmental Protection Agency (1998) Test methods for evaluating solid waste. Office of Solid Waste and Emergency Response, Washington, DC

US Environmental Protection Agency (2002) Standards for the use or disposal of sewage sludge. Office of Water, Washington, DC

Varvel GE, Schlemmer MR, Schepers JS (1999) Relationship between spectral data from an aerial image and soil organic matter and phosphorus levels. Precision Agric 1:291–300

Vincent RK (1997) Fundamentals of geological and environmental remote sensing. Prentice Hall, Upper Saddle River

Vincent RK (2000) Forecasts of monthly averaged daily temperature highs in Bowling Green, Ohio from monthly sea surface temperature anomalies in Eastern Pacific ocean during the previous year. Photogramm Eng Remote Sens 66:1001–1009

Vincent RK, Qin X, McKay RML, Miner J, Czajkowski K, Savino J, Bridgeman T (2004) Phycocyanin detection from LANDSAT TM data for mapping cyanobacterial blooms in Lake Erie. Remote Sens Environ 89:381–392

Wei QF, Lowery B, Peterson AE (1985) Effect of sludge application on physical properties of a silty clay loam soil. J Environ Qual 14:178–180

Williams CM, Barker JC, Sims JT (1999) Management and utilization of poultry wastes. Rev Environ Contam Toxicol 162:105–157

Chapter 17
Cotton Production Improvement and Environmental Concerns from Poultry Litter Application in Southern and Southeastern USA Soils

Haile Tewolde and Karamat R. Sistani

Abstract Much of the poultry litter generated in the US is concentrated in six southern and southeastern states which also produce the majority of the nation's cotton. The enormous amount of litter generated in the region can be an environmental burden if not managed properly. Research has been conducted in the region to encourage the utilization of poultry litter as a row crop fertilizer away from the traditional application to pastures around chicken houses. This chapter reviews results of the research on the effectiveness of poultry litter as cotton fertilizer and environmental concerns associated with its land application. The results which have been published in at least 55 refereed journal articles have shown that poultry litter is an excellent cotton fertilizer. Litter supplies all essential plant nutrients, builds soil organic matter, does not further acidify low pH soils unlike synthetic fertilizers, and may control or suppress harmful plant parasitic nematodes. Fertilizing cotton with poultry litter has been shown to increase lint yield above that possible with synthetic fertilizers in many soils in the region. Land-applying litter as cotton fertilizer should benefit cotton and other row crop farmers by increasing farm profits and poultry producers by generating new income if its use can be effectively and fully integrated into the cropping systems of the region. However, the potential of litter-derived P, Cu, and Zn to accumulate in the soil is a concern that future research should address.

H. Tewolde (✉)
Genetics and Precision Agriculture Research Unit, USDA-ARS,
Mississippi State, MS 39762, USA
e-mail: haile.tewolde@ars.usda.gov

K.R. Sistani
Food Animal Environmental Systems Research Unit, USDA-ARS,
230 Bennett Lane, Bowling Green, KY 42104, USA

Z. He and H. Zhang (eds.), *Applied Manure and Nutrient Chemistry for Sustainable Agriculture and Environment*, DOI 10.1007/978-94-017-8807-6_17,

17.1 Introduction

Animal manures have been used as fertilizers or soil conditioners throughout the history of agriculture. However, their integration into modern agriculture has been haphazard at best partly because manures were produced in small quantities and used on a small scale. Since the transformation of animal agriculture from small-holding farms to large scale animal feeding operations, manures have been generated in large quantities in small, confined spaces. The generation of manures in quantities exceeding the local assimilation capacity means it was necessary to find safe, efficient, and sustainable ways of using the manures away from the point of generation to prevent local environmental pollution. Recent manure research is predicated on this reality.

Much of the poultry litter (a combination of manure and one of several bedding materials) generated in the US is concentrated in six southern and southeastern states. According to the United States Department of Agriculture, National Agricultural Statistics Service, six states in the southern and southeastern region (Georgia, Alabama, Arkansas, North Carolina, Mississippi, and Texas) produce approximately 65 % of the broiler chickens in the US (USDA-NASS 2012). Assuming each of the 5.6×10^9 birds generated 1.46 kg litter (Sharpe et al. 2004) with 30 g N kg^{-1} and 12.5 g P kg^{-1} litter (van der Watt et al. 1994), these six states in 2011 generated 8.2×10^9 kg litter from broiler chickens alone containing 2.46×10^8 kg N, 1.02×10^8 kg P, and other agriculturally important mineral nutrients.

Until about 10 years ago, much of this litter was land-applied on pastures within a few miles of the poultry houses (usually belonging to the same poultry producers), primarily because disposing of the litter as close to the point of generation as possible is more convenient and less expensive. Litter that exceeded the local soil assimilative capacity had to be given away free of charge to those who could pay for its transportation away from the local area. Research has been conducted in the region, more intensively in the last 10–20 years, to encourage the integration of poultry litter as a row crop fertilizer into the region's cropping systems and alleviate environmental burden on pastures and hay fields around the poultry houses which have been the traditional litter application sites. Today, as a result of this research and the rising cost of synthetic fertilizers, poultry litter is no longer freely available. It is usually sold to row crop farmers who use it extensively as a fertilizer in place of synthetic fertilizers. Often these are farmers who cultivate much of the nation's cotton and substantial amount of corn and soybean. This chapter reviews results of current research of poultry litter use for cotton production systems in the southern and southeastern US.

17.2 Poultry Litter as Fertilizer for Cotton

Much of the cotton land area in the US is located in the same region that also produces the majority of poultry. Although land under cotton production has declined by more than 50 % in the last 5 years, the cotton production system can still assimilate a substantial amount of the litter generated in the region benefitting not only the poultry industry but also the profitability of cotton production as research has shown litter to be a valuable fertilizer in many soils.

17.2.1 Litter Is As Good As or Better than Conventional Inorganic Fertilizers for Cotton

A cursory literature search revealed that about 55 referred journal articles have been published since 1990 reporting results from the research of poultry litter on cotton in the region. The results from this research show that poultry litter is an excellent fertilizer for cotton production either in combination with commercial fertilizers or as a stand-alone fertilizer (Sistani et al. 2004; Tewolde et al. 2007). In some soil types, it is even a better fertilizer than the synthetic fertilizers (Table 17.1). For example, Tewolde et al. (2008) compared the growth and lint yield of cotton fertilized with broiler litter or inorganic fertilizers under no-till or conventional tillage system in an Atwood silt loam soil in northern Mississippi. Their results showed that cotton fertilized with litter produced 10 % more lint under conventional tillage and 14 % more lint under no-till than cotton fertilized with inorganic fertilizers following local fertilization practices. The results also showed that plants grew larger when fertilized with poultry litter than with inorganic fertilizers, further confirming that litter is a superior fertilizer for cotton in this soil compared to conventional inorganic fertilizers. In a Cecil sandy loam soil in Georgia, Endale et al. (2002) reported yield increases of up to 50 % when cotton was fertilized with poultry litter under no-till compared with conventional-till system fertilized with conventional inorganic fertilizers. The cotton fertilized with poultry litter was also taller and had greater leaf area than cotton fertilized with conventional inorganic fertilizers. In another Mississippi soil considered marginally productive, Tewolde et al. (2010) fertilized cotton with six litter rates ranging between 2.2 and 13.4 Mg ha^{-1} and with six NH_4NO_3–N rates ranging between 34 and 168 kg ha^{-1} so that peak lint yield would fall within these ranges. Lint yield calculated based on fitted models peaked at 1,156 kg ha^{-1} when fertilized with 8.7 Mg ha^{-1} litter and at only 1,033 kg ha^{-1} when fertilized with 94 kg ha^{-1} NH_4NO_3–N. The 123 kg ha^{-1} difference in the two peak lint yields showed that poultry litter is a superior cotton fertilizer in this soil, resulting in 12 % more lint yield than that possible with any rate of the inorganic NH_4NO_3.

Other research results from the region confirm that cotton production in many marginally productive soils could benefit from use of poultry litter. In Alabama, Reddy et al. (2007) reported that cotton produced more lint yield when fertilized with fresh poultry litter (1,492 kg ha^{-1}) than when fertilized with urea (1,391 kg ha^{-1}) in a 5-year study on a Decatur silt loam soil. In a heavier texture soil in Mississippi, Tewolde et al. (2009b) found that cotton fertilized with poultry litter produced as much as 26 % more lint than cotton fertilized with inorganic fertilizers but it required an application of as much as 13.5 Mg ha^{-1} litter. Unlike soils in other

Table 17.1 Lint yield increase of cotton fertilized with poultry litter relative to that fertilized with synthetic fertilizers in different soil types in Mississippi and Alabama as compiled from different sources

				Fertilization				
				None	Synthetic N fertilizer	Poultry litter	Increase, litter over synthetic N	
Location	Soil type	Year	Tillage	Lint yield, kg ha^{-1}			%	Source
Pontotoc, MS	Atwood silt loam	2003–2005	NT[a]	870	991	1,130	14	Tewolde et al. (2008)
Pontotoc, MS	Atwood silt loam	2003–2005	CT	1,105	1,129	1,242	10	Tewolde et al. (2008)
Belle Mina, AL	Decatur silt loam	1997–2001	NT	845	1,421	1,597[b]	12.4	Reddy et al. (2004)
Starkville, MS	Marietta fine sandy loam	2003–2005	CT	669	975	1,053[c]	8	Tewolde et al. (2009c)
Verona, MS	Catalpa silty clay loam	2003–2005	CT	886	1,240	1,466[b]	18.2	Tewolde et al. (2009b)
Holy Springs, MS	Loring silt loam	2002–2004	CT	877	1,066	1,161	8.9	Tewolde et al. (2010)
Coffeeville, MS	Ariel silt loam	2002–2004	NT	785	1,285	1,453	13.1	Tewolde et al. (2007)
Belle Mina, AL	Decatur silt loam	1994–1998	Not known	932	1,391	1,492	7.2	Reddy et al. (2007)
Cruger, MS	Dubbs silt loam	2002–2004	CT	1,045	1,681	1,690	0.6	Tewolde et al. (2007)

[a]*NT* no-till, *CT* conventional till
[b]Applied litter exceeded the rate estimated to supply equivalent N as the synthetic fertilizer rate
[c]Litter applied by subsurface banding

locations where $\leq$9.0 Mg ha^{-1} litter was adequate to produce yield equal to standard inorganic fertilization, cotton lint yield in this clayey soil increased in direct proportion to the litter rate up to 13.5 Mg ha^{-1}, which suggests yield in heavier soils may be optimized with greater litter fertilization rate than in lighter soils. Such yield increases may not be possible with synthetic N fertilizers regardless of the N rate (Tewolde et al. 2010).

Poultry litter has also been shown to be an effective fertilizer in more productive soils in the region. In the rich and highly productive soils of the Mississippi Delta, Tewolde et al. (2007) fertilized cotton with several litter rates with or without supplemental inorganic N and reported 4.5 Mg ha^{-1} litter supplemented with 67 kg ha^{-1} inorganic N resulted in lint yield equivalent to the farm standard fertilization with inorganic fertilizers. As much as 6.7 Mg ha^{-1} litter without supplemental N was inadequate for optimum lint yield in this soil. Unlike in many less productive soils in the region, there was no clear advantage of poultry litter over inorganic fertilization in the rich Mississippi Delta soil. However, the results showed that litter supplemented with inorganic N sources can substitute for all inorganic fertilizers even in such fertile soils. Overall, current research from the region has shown that fertilizing cotton with poultry litter results in lint yield equal to or greater than fertilizing with inorganic fertilizers.

17.2.2 Cotton Appearance Not Indicator of Performance of Poultry Litter

Cotton plants fertilized with poultry litter in almost all situations appear less green and contain less tissue N concentration than cotton plants fertilized with inorganic N fertilizers. For example, Tewolde et al. (2009b) found that cotton fertilized with 13.5 Mg ha^{-1} litter in a clayey soil in northern Mississippi had less green foliage and much less bulk leaf N concentration but produced as much as 26 % more lint yield than cotton that received the locally recommended inorganic fertilization. Cotton fertilized with 123 kg N ha^{-1} as urea-ammonium-nitrate solution (UAN) in this soil consistently looked much greener but produced only 5.6 % more lint yield (not statistically significant) than cotton fertilized with only 4.5 Mg ha^{-1} poultry litter (Fig. 17.1). In a lighter Atwood silt loam soil also in northern Mississippi, cotton that received twice the amount of total litter N had consistently less N concentration in leaves, stems, and reproductive parts than cotton that received standard inorganic fertilizers (Tewolde et al. 2009a). Both visually and based on chlorophyll index measurements, cotton that was fertilized with poultry litter appeared under-fertilized much similar to cotton that received no fertilization. However, lint yield and leaf area index results showed cotton that received the litter-only fertilization was as productive as or more productive than the cotton fertilized with standard inorganic fertilizers (Tewolde et al. 2008). In another marginally productive upland soil in Mississippi, cotton fertilized with poultry litter had much less green foliage and less leaf N concentration but greater lint yield than cotton fertilized with NH_4NO_3–N

Fig. 17.1 Cotton fertilized with 123 kg N ha^{-1} in the form of the synthetic inorganic N fertilizer urea-ammonium-nitrate solution (UAN) consistently looked much greener but produced only 5.6 % more (statistically not significant) lint yield than cotton fertilized with only 4.5 Mg ha^{-1} poultry litter. This shows cotton lint yield cannot be accurately predicted by how green the cotton appears during the growing season

(Tewolde et al. 2010). These results suggest that greener foliage or greater leaf N concentration may not necessarily translate into greater growth and lint production (Fig. 17.1). It may also be an indication that nutrients other than N, P, and K were limiting to yield of the cotton fertilized with the regionally recommended synthetic inorganic fertilizers. Applying the litter at planting may also be an advantage to early-season growth relative to conventional inorganic N fertilizers which are usually applied later at the first square stage. These results show that appearance of the cotton is not a good indicator of the yield potential and should not be used as a criterion of the performance of cotton fertilized with poultry litter.

17.2.3 Why Poultry Litter Is a Better Cotton Fertilizer than Synthetic Fertilizers

Possible reasons for cotton fertilized with poultry litter to grow better and produce more lint while appearing less green and contain lower tissue N concentration than cotton fertilized with equivalent rates of synthetic fertilizers is explored below:

17.2.3.1 Poultry Litter Is a Slow Release Fertilizer

Poultry litter is like a slow-release fertilizer supplying nutrients, N in particular, in smaller amounts but on a continuous basis throughout the growing season. As much as 90 % of the total litter N is in organic forms and therefore is not readily available for plant uptake immediately after application. Once the litter is applied to the soil, the organic N fraction is converted to mineral forms by soil microbial action. The process is gradual, however, and may take years for the entire organic N in litter to be fully mineralized. Much of the mineralization research has been conducted under incubation conditions for periods shorter than 1 year. In the field, residual effects of litter and other animal manures are often reported years after application which suggests some fraction of the organic N persists in the soil for long periods.

Unlike litter N, 100 % of the N from inorganic N fertilizers is already in available forms at the time of application. Typically, commercial N fertilizers are applied to cotton either 100 % within a few days of planting or may be split-applied, 1/3–1/2 at planting and the remainder later around the first square stage. This practice exposes cotton fertilized with inorganic N fertilizers to large doses of available N early and may lead to initial luxury-consumption and deep greening of the crop. The greater mid-season tissue N concentration and greener foliage of cotton fertilized with inorganic N than cotton fertilized with poultry litter indicate that the cotton received larger doses of N from inorganic N fertilizers than from poultry litter (Tewolde et al. 2008, 2009a, b).

Several incubation studies have shown that N from poultry litter mineralizes gradually (Bitzer and Sims 1988; Hartz et al. 2000; Sistani et al. 2008). In a study in the field, Schomberg et al. (2011) applied either poultry litter or inorganic N fertilizers to a Cecil sandy loam soil which was planted with corn. They measured the inorganic N content of the soil in the top 10 cm for up to 15 weeks and found that the inorganic N content of the soil that received commercial synthetic N had greater inorganic N early in the season than the soil that received litter. The inorganic N content of the soil that received litter exceeded that of the soil fertilized with synthetic N later in the season, which shows that the readily available inorganic N of the synthetic fertilizer was depleted at a faster rate while litter released mineralized N gradually throughout the growing season. Schomberg et al. (2011) reported that aboveground biomass of corn that was fertilized with litter was greater than that of corn fertilized with commercial inorganic N, suggesting that the gradual availability of inorganic N from litter may have benefited corn growth. Typically, the mineralization of litter N shortly after application is high but stabilizes after the initial flush of inorganic N (Preusch et al. 2002; Sanchez and Mylavarapu 2011). As shown by Schomberg et al. (2011), the amount from this initial surge of inorganic N from litter is not great enough to equal that from the application of commercial synthetic N. Therefore, it may be much more ideal for crop growth and yield if fertilized with litter than with one-time application of synthetic N fertilizers whose N is 100 % available at the time of application.

Fig. 17.2 Cotton grown in a 2:1 sand:vermiculite inert media with (**a**) no fertilization showing the media is near inert, (**b**) full Hoagland's nutrient solution (Hoagland and Arnon 1950), and (**c**) poultry litter. The large difference in plant growth between (**a**) and (**c**) is the addition of 90 g of litter to (**c**). The similarity in plant growth between (**b**) and (**c**) shows that poultry litter supplied nutrients equivalent to the formulated Hoagland's solution

17.2.3.2 Poultry Litter Supplies All Essential Plant Nutrients

When analyzed in the laboratory, poultry litter is known to contain all elements essential for plant growth (Collins et al. 1999) although not always in the most ideal proportions. Jackson et al. (2003) analyzed 40 litter samples collected from three southeastern states (Georgia, Alabama, and South Carolina) and reported the concentration and water solubility of several essential plant nutrients. The samples contained not only the macro- and micro-nutrients needed for plant growth but also heavy metals that may be of concern when the litter is land-applied. The phyto-availability of the elements usually found in poultry litter was demonstrated by Tewolde et al. (2005b) with cotton. Tewolde et al. (2005b) grew cotton in an inert growth media consisting of washed sand and vermiculite and fertilized with litter, Hoagland's nutrient solution, or no fertilizers. While cotton fertilized with nutrient solution grew normally to form bolls (Tewolde et al. 2005b), cotton that received no fertilization never grew beyond the first true leaf, confirming that the media was near inert with little or no essential plant nutrients (Fig. 17.2a). Cotton fertilized with poultry litter in the same media grew normally to about the same plant size as cotton fertilized with the full nutrient solution (Fig. 17.2b, c), showing that the poultry litter supplied all essential nutrients for normal cotton growth and development. This was confirmed with measurements of nutrient concentration in the different plant parts (Tewolde et al. 2005a, b). The ability of poultry litter to supply all essential plant nutrients means that fertilizing crops in the field with litter would correct nutrient deficiency that may not be easily identified by regular soil nutrient analysis. Typically, soil analysis and corrective action are focused on the major nutrients N, P, and K and occasionally on S and Mg. It is very easy to overlook minor deficiencies of the micronutrients including Zn, B, and Cu. Applying poultry litter intended to supply N, P, and K can also easily meet unidentified deficiencies of these and other essential elements.

17.2.3.3 Poultry Litter Reduces or Maintains Soil Acidity

Many soils in the southern and southeastern US are acidic and are often limed to raise the soil pH to levels ideal for crop production. Applying the most commonly used fertilizers such as NH_4NO_3 and UAN are known to further acidify the soil. Poultry litter, unlike synthetic fertilizers, has the opposite effect – it increases soil pH or maintains the initial pH level (Adeli et al. 2010; Kingery et al. 1994; Materechera and Mkhabela 2002; Moore and Edwards 2005; Van der Watt et al. 1994). Adeli et al. (2010) reported that applying NH_4NO_3 to an already acidic silt loam upland soil with pH of 4.92 in northern Mississippi further reduced the pH to as low as 4.76. Applying poultry litter to the same soil raised the pH to as high as 5.41. Similarly, Moore and Edwards (2005) found that, after 7 years, NH_4NO_3 application reduced soil pH and poultry litter increased the pH in direct proportion to the rate of application. Tewolde et al. (2011) attributed better yield performance of cotton fertilized with poultry litter than with NH_4NO_3–N in a low pH soil, in part, to a more ideal soil pH and favorable tissue Mn concentration. It is very clear that correction of the acidic southern and southeastern soils by liming is necessary if fertilizing cotton and other crops with inorganic N fertilizers such as NH_4NO_3 in the region. Much of the body of research to date shows that poultry litter may serve as a liming material (Table 17.2) in addition to its use as a source of all plant nutrients. Thus, fertilizing with poultry litter eliminates the need to add liming materials, reduces the cost of production, and increases farm profit (Tewolde et al. 2011) and should be the preferred fertilizer material for the southern and southeastern US cropping systems where low soil pH is a concern. After conducting long-term research, Moore and Edwards (2005) also reached at the same conclusion that poultry litter may be a more sustainable fertilizer than NH_4NO_3.

Table 17.2 Poultry litter increases or maintains soil pH while inorganic N fertilizers reduce it

				Fertilization			
		Years of	Initial	None	Synthetic N fertilizers	Poultry litter	
Soil type	Tillage	application	Soil pH				Source
Loring silt loam	CT[a]	3	4.92	4.96	4.76	5.41	Adeli et al. (2010)
Captina silt loam	Pasture	7	5.20	–	3.90	5.80	Moore and Edwards (2005)
Rhodic haplustox	Incubation	16 weeks	–	4.11	–	4.28	Materechera and Mkhabela (2002)
Dubbs silt loam	CT	6	6.60	5.50	5.52	6.30	Adeli et al. (2011)
Ariel silt loam	NT[a]	6	5.70	5.58	5.35	6.72	Adeli et al. (2011)
Catalpa silty clay loam	CT	3	5.23	5.11	5.02	5.49	Adeli et al. (2009)

[a]*CT* conventional till, *NT* no-till

17.2.3.4 Poultry Litter Increases Soil Organic Matter and Improves Soil Physical Properties

Poultry litter and other manures are applied to the soil not only as sources of nutrients but also to help build soil organic matter and improve soil physical properties. Poultry litter like many manure types is known to increase soil organic matter at least on a short-term basis (Adeli et al. 2009; Kingery et al. 1994; Watts et al. 2010). Usually, the organic matter increase is proportional to the rate of application. For example, Adeli et al. (2009) found that the organic C content of a clayey soil in the top 0–15 cm in Mississippi increased from 16.0 g C kg^{-1} soil when no litter was applied, to 23.1 g C kg^{-1} soil when 9.0 Mg ha^{-1} litter was applied, and to 25.6 g C kg^{-1} soil when 13.4 Mg ha^{-1} litter was applied.

Other soil benefits associated with soil organic matter improvement due to litter and other manure applications include lower bulk density, greater water holding capacity, better water-stable soil aggregation, and improved porosity (Adeli et al. 2009; Mbagwu 1992; Tejada et al. 2006; Weil and Kroontje 1979). Tejada et al. (2006) applied poultry manure yearly for 5 years to a saline soil in Spain and found that the bulk density of the soil decreased progressively throughout the 5-year period. At the end of the fifth year, bulk density decreased by 22 % relative to the unamended soil. Weil and Kroontje (1979) applied high rates of poultry manure to a clay soil in Virginia and reported the bulk density of the soil decreased from 1.1 to 0.8 g cm^{-3}, water-stable aggregates increased from 73 to 94 %, and water holding capacity increased from 32 to 42 %. In a clayey soil in northern Mississippi,

Adeli et al. (2009) found that agronomic rates of broiler litter increased microbial biomass C, total porosity, and aggregate stability. In Nigeria, Mbagwu (1989, 1992) reported that the addition of high rates of poultry manure (50 Mg ha^{-1} in the field and 5 or 10 % of soil weight in an incubation study) increased total porosity and macroporosity, saturated hydraulic conductivity, water retention at different potentials and decreased soil bulk density of a degraded Ultisol soil.

Although the positive impact of poultry litter on soil physical properties is widely recognized, its impact on crop productivity associated with the improvement in soil physical properties is not easily measurable. Typically, the value of litter is measured by the amount of its N content, and sometimes by its P and K content, which is determined analytically in the laboratory. All other benefits are ignored because of the difficulty in quantifying them. But, usually, the value of poultry litter as a fertilizer to farmers is greater than the value determined based on its N, P, and K content alone (Tewolde et al. 2010). Tewolde et al. (2010) determined the fertilizer replacement value of litter as cotton fertilizer in northern Mississippi to be 27 % more than the value calculated based on its N, P, and K content.

17.2.3.5 Poultry Litter May Suppress Harmful Plant Parasitic Nematodes

The ability of poultry litter to suppress plant parasitic nematodes may be another reason for the better lint yield performance of cotton fertilized with litter than cotton fertilized with inorganic fertilizers. Damage to cotton yield by parasitic nematodes is a major concern to the extent that nematicides are often a key component of best management practices in the southern and southeastern US. Some research indicates poultry litter may be effective in suppressing certain plant parasitic nematodes on cotton and other crops (Koenning and Barker 2004; Morant et al. 1997; Riegel and Noe 2000; Sumner et al. 2002) and may prevent yield loss due to nematode damage. The effectiveness of litter in suppressing harmful nematodes may be associated with its stimulating effect on beneficial soil microorganisms (Riegel and Noe 2000). Applying litter to soil has been shown to stimulate soil microorganisms (Riegel and Noe 2000; Pratt and Tewolde 2009; Tewolde et al. 2009b) and it has also been shown to reduce harmful nematodes (Riegel and Noe 2000). Litter may also suppress nematodes if it releases nematicidal levels of NH_3 to the soil environment (Oka et al. 2007). Oka (2010) reviewed research on use of organic soil amendments including poultry litter for controlling nematodes and proposed the following possible modes of action: generation of nematicidal compounds such as ammonia and fatty acids, enhancement and/or introduction of microorganisms antagonistic to nematodes, improvement of plant tolerance to nematodes, and creating soil environments unsuitable for nematode activity. Poultry litter may not always suppress nematodes with one or two applications, but its consistent and repeated use may create a healthier soil environment that may lead to a gradual decline of harmful nematodes and improve cotton productivity.

17.3 Environmental Concerns of Land-Applying Poultry Litter

Land application of poultry litter to pasture and row crops is the outlet most favored by the poultry industry for the massive amount of litter generated in small localized areas. If done properly, land-application as a fertilizer is probably the most effective way of handling litter. However, concern exists regarding its proper handling much of which arises from viewing litter as a waste instead of as a resource. One important concern is the potential of litter-derived pathogens to contaminate food and drinking water supplies. While this is an ever-existing threat, the greatest concern of land-applying poultry litter is environmental–the potential to degrade soil, air, and water quality.

Poultry litter is a rich source of nearly all nutrients essential for normal plant growth and development. The nutrients, however, do not exist in a balance proportional to plant needs. The environmental threat of land-applying litter arises from this nutrient imbalance which leads to undesirable accumulation of certain nutrients in the soil. In cotton and other row crops, the imbalance of N and P is a key issue that has received some level of research attention. If litter is applied to meet crop N needs, it has been shown to result in elevated soil P levels after only a few years of application (Adeli et al. 2011; Kingery et al. 1994; Mitchell and Tu 2006; Schomberg et al. 2009). The soil P elevation occurs because crop plants including cotton and corn need much less P than N. At harvest, cotton accumulates as much as 6–8 kg N for every 1 kg P while many soils in the region supply adequate P. Poultry litter supplies only approximately 2 kg N for every 1 kg P. Therefore, applying litter to meet the N need always leads to excess P application and to elevated soil P levels (Table 17.3), particularly if the application is repeated for consecutive seasons. In addition to causing soil nutrient imbalance, elevated soil P levels can pollute surface water by way of runoff and leaching. This has been the greatest concern of fertilizing row crops repeatedly with poultry litter in certain regions such as the eastern US. A toxic outbreak of microbes in 1997 in the Chesapeake Bay was associated with P pollution as one factor and the poultry industry became a focus of attention as a source of the P (Parker 2000). The ecosystems in the southern and southeastern US have not become as sensitive to nonpoint source pollution as other ecosystems such as that of the Chesapeake Bay, but nutrient pollution of the Gulf of Mexico and other coastal areas in the region will be of concern as the use of poultry litter expands and soil levels of P and other nutrients continue to elevate in the Mississippi River watershed.

Other nutrients that accumulate in the soil and may become an environmental concern when poultry litter is used as a fertilizer include Cu and Zn. The trace elements Cu and Zn are present in poultry litter at elevated levels from the addition of copper sulfate and zinc sulfate to chicken feed to enhance health and performance of the chickens (Skrivan et al. 2005). Therefore, applying poultry litter elevates the levels of these nutrients in the soil beyond what the crop can absorb and utilize. The accumulation of Cu and Zn in the soil has been well documented

Table 17.3 The concentration of Extractable P in cropped soils is accelerated by as few as three yearly applications of poultry litter

Soil type	Tillage	Years of application	Initial	Fertilization: None	Fertilization: Inorganic fertilizers	Fertilization: Poultry litter	Source
			Extractable soil P, mg/kg				
Dubbs silt loam	CT[a]	6	56	55	111	145	Adeli et al. (2011)
Ariel silt loam	NT	6	31	22	61	97	Adeli et al. (2011)
Compass fine sandy loam	CT/CON	10	25	–	–	125	Mitchell and Tu (2006)
Atwood silt loam	NT	3	12.7	15.2	16.4	33	Adeli et al. (2008)
Catalpa silty clay loam	CT	3	24.6	20.3	21	36.3	Adeli et al. (2009)

[a]*CT* conventional till, *CON* conservation till, *NT* no-till

(Adeli et al. 2009, 2008; Mitchell and Tu 2006; Schomberg et al. 2009), but the environmental consequence of such accumulation has not been well researched. Extremely high soluble soil Cu levels (≈3,000 mg/kg) is known to cause deficiency of other plant nutrients such as Fe (McBride and Martinez 2000) but only when the levels are extreme such as in contaminated soils. Metal element accumulations that have the potential to reach toxic levels can also occur when poultry litter is applied for an extended period (Van der Watt et al. 1994).

Nitrogen, if mismanaged, can be an environmental hazard. However, the application of litter can be controlled so that the target crop can utilize the N before leaving the field. Other concerns of land-applying poultry litter are degradation of air quality due to emissions of greenhouse gases including CO_2 and NH_3. However, these concerns are beyond the scope of this chapter and will not be discussed here.

17.4 Suggested Management Strategies and Future Research

Poultry litter may be land-applied as a fertilizer safely, effectively, and sustainably with the proper use of management strategies. To date, the most widely studied and recommended strategy for continuous use of litter without the buildup of excess nutrients is applying the litter to meet crop P needs, often described as P-based fertilization (Eghball and Power 1999; Maguire et al. 2008). However, limiting litter application rate to meet just the P need leads to under-supplying N, which necessitates supplemental inorganic N application and increases the cost of fertilization. Management strategies that are economical and do not cause unsustainable soil nutrient buildup when adequate litter is applied to fertilize cotton should be developed as alternatives to P-based litter management. Such strategies may include crop and fertilizer rotation schemes that enhance the uptake and removal of excess nutrients, P in particular.

17.5 Conclusion

Poultry litter has proven to be an excellent fertilizer for cotton and other row crops in the southern and southeastern US. It supplies all essential plant nutrients, maintains or increases the pH of acidic soils, builds soil organic matter, and may control or suppress harmful plant parasitic nematodes. The lint yield improvement reported across the region if cotton is fertilized with poultry litter relative to that fertilized with synthetic fertilizers may be associated with all these benefits of litter. If effectively integrated into the cropping systems of the region, litter should benefit not only cotton and other row crop farmers but also the poultry producers in the region. The entire region should also benefit environmentally by recycling the enormous amount of waste generated in the poultry industry if effective and sustainable strategies of managing excess soil nutrients can be developed.

References

Adeli A, Shankle MW, Tewolde H, Sistani KR, Rowe DE (2008) Nutrient dynamics from broiler litter applied to no-till cotton in an upland soil. Agron J 100:564–570

Adeli A, Tewolde H, Sistani KR, Rowe DE (2009) Broiler litter fertilization and cropping system impacts on soil properties. Agron J 101:1304–1310

Adeli A, Tewolde H, Sistani KR, Rowe DE (2010) Comparison of broiler litter and commercial fertilizer at equivalent N rates on soil properties. Commun Soil Sci Plant Anal 41:2432–2447

Adeli A, Tewolde H, Rowe DE, Sistani KR (2011) Continuous and residual effects of broiler litter application to cotton on soil properties. Soil Sci 176:668–675

Bitzer CC, Sims JT (1988) Estimating the availability of nitrogen in poultry manure through laboratory and field studies. J Environ Qual 17:47–54

Collins Jr ER, Barker JC, Carr LE, Brodie HL, Martin Jr JH (1999) Poultry waste management handbook. NRAES-132. Natural Resources, Agriculture, and Engineering Service Cooperative Extension, Ithaca

Eghball B, Power JF (1999) Phosphorus- and nitrogen-based manure and compost applications: corn production and soil phosphorus. Soil Sci Soc Am J 63:895–901

Endale DM, Cabrera ML, Steiner JL, Radcliffe DE, Vencill WK, Schomberg HH, Lohr L (2002) Impact of conservation tillage and nutrient management on soil water and yield of cotton fertilized with poultry litter or ammonium nitrate in the Georgia Piedmont. Soil Till Res 66:55–68

Hartz TK, Mitchell JP, Giannini C (2000) Nitrogen and carbon mineralization dynamics of manures and composts. HortScience 35:209–212

Hoagland DR, Arnon DI (1950) The water-culture method for growing plants without soil. Circ. 347. California Agricultural Experiment Station, University of California, Berkeley

Jackson BP, Bertsch PM, Cabrera ML, Camberato JJ, Seaman JC, Wood CW (2003) Trace element speciation in poultry litter. J Environ Qual 32:535–540

Kingery WL, Wood CW, Delaney DP, Williams JC, Mullins GL (1994) Impact of long-term land application of broiler litter on environmentally related soil properties. J Environ Qual 23:139–147

Koenning SR, Barker KR (2004) Influence of poultry litter applications on nematode communities in cotton agroecosystems. J Nematol 36:524–533

Maguire RO, Mullins GL, Brosius M (2008) Evaluating long-term nitrogen- versus phosphorus-based nutrient management of poultry litter. J Environ Qual 37:1810–1816

Materechera SA, Mkhabela TS (2002) The effectiveness of lime, chicken manure and leaf litter ash in ameliorating acidity in a soil previously under black wattle (*Acacia mearnsii*) plantation. Bioresour Technol 85:9–16

Mbagwu JSC (1989) Effects of organic amendments on some physical properties of a tropical ultisol. Biol Waste 28:1–13

Mbagwu JSC (1992) Improving the productivity of a degraded ultisol in Nigeria using organic and inorganic amendments. Part 2: changes in physical properties. Bioresour Technol 42:167–175

McBride MB, Martinez CE (2000) Copper phytotoxicity in a contaminated soil: remediation tests with adsorptive materials. Environ Sci Tech 34:4386–4391

Mitchell CC, Tu S (2006) Nutrient accumulation and movement from poultry litter. Soil Sci Soc Am J 70:2146–2153

Moore PA, Edwards DR (2005) Long-term effects of poultry litter, alum-treated litter, and ammonium nitrate on aluminum availability in soils. J Environ Qual 34:2104–2111

Morant MA, Casasola JL, Brooks CB, Philip ET, Mitchell VG, Orr CR (1997) Poultry litter enhances soybean productivity in field infested with soybean cyst nematode. J Sustain Agric 11:39–51

Oka Y (2010) Mechanisms of nematode suppression by organic soil amendments-A review. Appl Soil Ecol 44:101–115

Oka Y, Shapira N, Fine P (2007) Control of root-knot nematodes in organic farming systems by organic amendments and soil solarization. Crop Prot 26:1556–1565

Parker D (2000) Controlling agricultural nonpoint water pollution: costs of implementing the Maryland Water Quality Improvement Act of 1998. Agric Econ 24:23–31

Pratt RG, Tewolde H (2009) Soil fungal population levels in cotton fields fertilized with poultry litter and their relationships to soil nutrient concentrations and plant growth parameters. Appl Soil Ecol 41:41–49

Preusch PL, Adler PR, Sikora LJ, Tworkoski TJ (2002) Nitrogen and phosphorus availability in composted and uncomposted poultry litter. J Environ Qual 31:2051–2057

Reddy CK, Nyakatawa EZ, Reeves DW (2004) Tillage and poultry litter application effects on cotton growth and yield. Agron J 96:1641–1650

Reddy KC, Malik RK, Reddy SS, Nyakatawa EZ (2007) Cotton growth and yield response to nitrogen applied through fresh and composted poultry litter. J Cot Sci 11:26–34

Riegel C, Noe JP (2000) Chicken litter soil amendment effects on soilborne microbes and *Meloidogyne incognita* on cotton. Plant Dis 84:1275–1281

Sanchez JF, Mylavarapu RS (2011) Potential nitrogen mineralization in sandy soils under long-term poultry litter management. Commun Soil Sci Plant Anal 42:424–434

Schomberg HH, Endale DM, Jenkins MB, Sharpe RR, Fisher DS, Cabrera ML, McCracken DV (2009) Soil test nutrient changes induced by poultry litter under conventional tillage and no-tillage. Soil Sci Soc Am J 73:154–163

Schomberg H, Endale D, Jenkins M, Fisher D (2011) Nutrient source and tillage influences on nitrogen availability in a Southern Piedmont corn cropping system. Biol Fertil Soils 47:823–831

Sharpe RR, Schomberg HH, Harper LA, Endale DM, Jenkins MB, Franzluebbers AJ (2004) Ammonia volatilization from surface-applied poultry litter under conservation tillage management practices. J Environ Qual 33:1183–1188

Sistani KR, Rowe DE, Johnson J, Tewolde H (2004) Supplemental nitrogen effect on broiler-litter-fertilized cotton. Agron J 96:806–811

Sistani KR, Adeli A, McGowen SL, Tewolde H, Brink GE (2008) Laboratory and field evaluation of broiler litter nitrogen mineralization. Bioresour Technol 99:2603–2611

Skrivan M, Skrivanova V, Marounek M (2005) Effects of dietary zinc, iron, and copper in layer feed on distribution of these elements in eggs, liver, excreta, soil, and herbage. Poult Sci 84:1570–1575

Sumner DR, Hall MR, Gay JD, MacDonald G, Savage SI, Bramwell RK (2002) Root diseases, weeds, and nematodes with poultry litter and conservation tillage in a sweet corn-snap bean double crop. Crop Prot 21:963–972

Tejada M, Garcia C, Gonzalez JL, Hernandez MT (2006) Use of organic amendment as a strategy for saline soil remediation: influence on the physical, chemical and biological properties of soil. Soil Biol Biochem 38:1413–1421

Tewolde H, Sistani KR, Rowe DE (2005a) Broiler litter as a micronutrient source for cotton: concentrations in plant parts. J Environ Qual 34:1697–1706

Tewolde H, Sistani KR, Rowe DE (2005b) Broiler litter as a sole nutrient source for cotton: nitrogen, phosphorus, potassium, calcium, and magnesium concentrations in plant parts. J Plant Nutr 28:605–619

Tewolde H, Sistani KR, Rowe DE, Adeli A, Johnson JR (2007) Lint yield and fiber quality of cotton fertilized with broiler litter. Agron J 99:184–194

Tewolde H, Shankle MW, Sistani KR, Adeli A, Rowe DE (2008) No-till and conventional-till cotton response to broiler litter fertilization in an upland soil: lint yield. Agron J 100:502–509

Tewolde H, Shankle MW, Adeli A, Sistani KR, Rowe DE (2009a) Macronutrient concentration in plant parts of cotton fertilized with broiler litter in a marginal upland soil. Soil Till Res 105:1–11

Tewolde H, Buehring N, Adeli A, Sistani KR, Rowe DE, Pratt RG (2009b) Cotton response to chicken litter in rotation with corn in clayey soil. Agron J 101:626–634

Tewolde H, Armstrong S, Way TR, Rowe DE, Sistani KR (2009c) Cotton response to poultry litter applied by subsurface banding relative to surface broadcasting. Soil Sci Soc Am J 73:384–389

Tewolde H, Adeli A, Sistani KR, Rowe DE, Johnson JR (2010) Equivalency of broiler litter to ammonium nitrate as a cotton fertilizer in an upland soil. Agron J 102:251–257

Tewolde H, Adeli A, Sistani KR, Rowe DE (2011) Mineral nutrition of cotton fertilized with poultry litter or ammonium nitrate. Agron J 103:1704–1711

USDA-NASS (2012) Poultry – production and value summary. USDA, Washington, DC. http://usda01.library.cornell.edu/usda/current/PoulProdVa/PoulProdVa-04-29-2013.pdf. Accessed 30 Apr 2013

Van der Watt HVH, Sumner ME, Cabrera ML (1994) Bioavailability of copper, manganese, and zinc in poultry litter. J Environ Qual 23:43–49

Watts DB, Torbert HA, Prior SA, Huluka G (2010) Long-term tillage and poultry litter impacts soil carbon and nitrogen mineralization and fertility. Soil Sci Soc Am J 74:1239–1247

Weil RR, Kroontje W (1979) Physical condition of a Davidson clay loam after five years of heavy poultry manure applications. J Environ Qual 8:387–392

About the Editors

ZHONGQI HE is a Research Chemist at the United States Department of Agriculture-Agricultural Research Service, Southern Regional Research Center, New Orleans, Louisiana. He was a recipient of the National Research Council postdoctoral fellowship hosted by the United States Air Force Research Laboratory, Tyndall Air Force Base, Florida. The author or co-author of over 150 research articles, patents, and book chapters, he has actively pursued basic and applied research in chemistry and biochemistry of agricultural products, byproducts, and plant nutrients. Previously, Dr. He organized and served as the sole or principal editor of four books. He has provided peer review services for more than 50 journals and served in an editorial board. He received a B.S. degree (1982) in applied chemistry from Chongqing University, China, M.S. degrees (1985 and 1992) in applied chemistry from South China University of Technology, Guangzhou, and in chemistry from the University of Georgia, Athens, USA, and a Ph.D. degree (1996) in biochemistry from the University of Georgia, Athens, USA.

HAILIN ZHANG is a Regents Professor and Nutrient Management Extension Specialist at the Department of Plant and Soil Sciences, Oklahoma State University (OSU). He is fellow both of American Society of Agronomy and Soil Science Society of America. His research and extension activities focus on agricultural testing and interpretations, nutrient management, and environmental quality protection. He served as the Director of Agricultural Testing and Research Laboratory at the Navajo Agriculture Product Industry in Farmington New Mexico prior to joining OSU in 1996. He has published two extension books, authored or coauthored over 200 research and extension publications. Dr. Zhang received his B.S. degree (1982) from Nanjing Agricultural University, China, graduated from Iowa State University and the University of Minnesota with an M.S. (1986) and Ph.D. (1990) degree, respectively.

Z. He and H. Zhang (eds.), *Applied Manure and Nutrient Chemistry for Sustainable Agriculture and Environment*, DOI 10.1007/978-94-017-8807-6,

Index

Z. He and H. Zhang (eds.), *Applied Manure and Nutrient Chemistry for Sustainable Agriculture and Environment*, DOI 10.1007/978-94-017-8807-6,

R

S

MIX
Papier aus verantwortungsvollen Quellen
Paper from responsible sources
FSC® C105338

If you have any concerns about our products,
you can contact us on
ProductSafety@springernature.com

In case Publisher is established outside the EU,
the EU authorized representative is:
Springer Nature Customer Service Center GmbH
Europaplatz 3, 69115 Heidelberg, Germany

Printed by Libri Plureos GmbH
in Hamburg, Germany